Introducing Human Geographies

EDITED BY

Paul Cloke

School of Geographical Sciences,
University of Bristol

Philip Crang

Department of Geography,
University College London

Mark Goodwin

Institute of Geography and Earth Sciences,
University of Wales, Aberystwyth

ARNOLD

A member of the Hodder Headline Group
LONDON
Co-published in the United States of America by
Oxford University Press Inc., New York

First published in Great Britain in 1999
This impression printed in 2003 by
Arnold, a member of the Hodder Headline Group,
338 Euston Road, London NW1 3BH

http://www.arnoldpublishers.com

Co-published in the United States of America by
Oxford University Press Inc.,
198 Madison Avenue, New York, NY 10016
Oxford is a registered trademark of Oxford University Press

British Library Cataloguing in Publication Data
A catalogue record for this book is available from the British Library

Library of Congress Cataloging-in-Publication Data
A catalog record for this book is available from the Library of
Congress

ISBN 0 340 69192 1 (hb)
ISBN 0 340 69193 X (pb)

6 7 8 9 10

Production Editor: Julie Delf
Production Controller: Sarah Kett
Cover Design: T. Griffiths

Typeset in 10.5/13 Bembo by Phoenix Photosetting, Chatham, Kent
Printed and bound in India by Replika Press Pvt. Ltd.

What do you think about this book? Or any other Arnold title?
Please send your comments to feedback.arnold@hodder.co.uk

Contents

PART III   Contexts   235

Contents

List of contributors

EDITORS

Paul Cloke
    University of Bristol, UK
Philip Crang
    University College London, UK
Mark Goodwin
    University of Wales, Aberystwyth, UK

CONTRIBUTORS

William M. Adams
    University of Cambridge, UK
Jacquie Burgess
    University College London, UK
Ruth Butler
    University of Hull, UK
Stuart Corbridge
    University of Miami, USA
Mike Crang
    University of Durham, UK
Tim Cresswell
    University of Wales, Aberystwyth, UK
Luke Desforges
    University of Wales, Aberystwyth, UK
Felix Driver
    Royal Holloway, University of London, UK
Claire Dwyer
    University College London, UK
Jon Goss
    University of Hawaii at Manoa, USA
Pyrs Gruffudd
    University of Wales, Swansea, UK
Chris Hamnett
    King's College London, UK
Ken Hillis
    University of North Carolina at Chapel Hill, USA

Nuala C. Johnson
    Queen's University Belfast, UK
Andrew Jordan
    University of East Anglia, UK
James Kneale
    University of Exeter, UK
Roger Lee
    Queen Mary and Westfield College, University of
    London, UK
Catherine Nash
    Royal Holloway, University of London, UK
Tim O'Riordan
    University of East Anglia, UK
Miles Ogborn
    Queen Mary and Westfield College, University of
    London, UK
Richard Phillips
    University of Wales, Aberystwyth, Wales
Sarah A. Radcliffe
    University of Cambridge, UK
Kevin Robins
    University of Newcastle Upon Tyne, UK
Paul Routledge
    University of Glasgow, UK
Joanne P. Sharp
    University of Glasgow, UK
Susan J. Smith
    University of Edinburgh, UK
Peter J. Taylor
    University of Loughborough, UK
Adam Tickell
    University of Southampton, UK
Michael Watts
    University of California at Berkeley, USA
Sarah Whatmore
    University of Bristol, UK

Acknowledgements

It might be expected that the creation of a book with three editors and thirty other contributors will be a shared endeavour. To state this, however, does not do justice to the collective nature of this particular project. From its very beginnings, this book has been a communal venture, and the editors would like to take this opportunity to extend their heartfelt thanks to all those who helped us see it through to completion. Two groups of people deserve special thanks – the contributors, and those involved in the publication process at Arnold.

The contributors have been marvellous in responding to our many and varied requests concerning the shape and content of their chapters. Initially, they all deserve thanks for agreeing to be involved in the project. Since then they have helped enormously to give the book its shape and integrity, and have done so within the very strict deadlines imposed by both editors and publishers. We are aware of the ever increasing pressures on academics, and we will always be grateful for the manner and speed in which this particular set of contributors responded to us when they almost certainly had far more pressing concerns and engagements elsewhere.

Those at Arnold also warrant our warmest thanks. Laura McKelvie, the former Geography and Environmental Science editor at Arnold merits special praise. She worked alongside us from the very beginning, and we suspect that she had an idea for a book of this nature long before we did. She then coaxed us through the various stages of writing and re-writing with patience, good humour and unfailing support. Co-ordinating thirty three authors within a strict timetable is a thankless task, yet she managed it through a combination of hard work and enormous skill. The book should really be dedicated to her.

At the production stage others from Arnold played key roles. Julie Delf, as Production Editorial Manager somehow managed to pull together a disparate range of materials, helped by Liz Gooster and Emma Heyworth-Dunn. Milly Neate has done a superb job in marketing and publicizing the book and Jenny Ungless proved her skill time and time again as a picture researcher in negotiating the use of key images and diagrams.

There is one more group to thank, and these are the various cohorts of first year students we have taught over the years. The idea for the book began to form when we were all teaching first year Human Geography courses at the University of Wales, Lampeter. We witnessed the students there being inspired by the preoccupations of contemporary Human Geography; but we also saw them having to develop that inspiration through materials not written or designed for them. Quite rightly, they often wondered why inspiration had to go hand in hand with such extreme mental perspiration. In subsequent years, at Bristol, UCL and Aberystwyth, students have voiced the same complaint regarding the lack of suitable first year texts which accessibly address contemporary research in the discipline. We hope that this book will help to meet their concerns, and in so doing provide future first year students with some indication of the excitement and relevance of contemporary Human Geography.

On a personal note, special thanks are due to Viv, Katharine and Anne, who shared in the project from its outset and who helped in so many ways to bring it to fruition.

Introduction

Starting a university course is always an exciting and stressful time. Even without the changes it may bring about in your life, the academic challenges faced are in themselves enough. New ways of learning are often required. New ideas and issues are confronted. Even a subject studied at school or college is likely to look and feel very different at the university level. This is certainly the case with Human Geography. Sometimes this is wonderfully energizing, but it can also be disorientating. There is so much to take in and make sense of. The aim of this book is to help new Human Geography students have the excitement without the stress, the inspiration without the feeling of being totally lost. *Introducing Human Geographies* provides you with a 'travel guide' into the academic subject of Human Geography (marked typographically through capitalization) and the worldly human geographies it investigates (marked typographically in the lower case). It maps out the 'big questions' that Human Geographers past and present are fascinated by; explores in more depth the key research topics that are being pursued in the subject today; and takes you to some of the geographical sites in which these topics come together.

The overall effect, we hope, is to represent a subject that is relevant, thought-provoking, challenging and dynamic. The Human Geography you will be introduced to here is a million miles from some of the popular images of the subject. It is not a dry compendium of facts about the world, its countries, capital cities, and so on. This book probably won't be a great deal of help in getting the Geography questions right in a game of 'Trivial Pursuit'. Sorry. Nor do we have much space for the sorts of simplistic models of spatial laws and forms that for a while came to dominate many anglophonic school curricula. The

Human Geography we want to introduce you to, and the Human Geography you will find to be predominant in the subject's intellectual arteries of research journals and books, avoids the easy and ultimately dull options of retreating into worlds of compiled fact or modelled fantasy. It engages with real life and real lives, embracing their wonderful complexity. It seeks to do more than record or model; it tries to explain, understand, question, interpret and maybe even improve these human geographies.

In this general Introduction we want to say a little about why and how we are going to introduce this sort of Human Geography. We begin by addressing that thorny question 'What is Human Geography?', the bane but also the inspiration for all of us who have to explain to friends, family and colleagues the things we are interested in as Geographers. Our response focuses in particular on how Human Geography combines substantive breadth with a coherent intellectual core. We then briefly consider the character of contemporary Human Geography, highlighting the subject's dynamism and explaining some of the key features of recent developments within it. Here, we also begin to move away from issues of content – what Human Geographers study – and onto questions of approach and styles of thought. We try to give you a feel for what is being asked of you as Human Geography students, highlighting ways of thinking about the world that you can apply across the full range of Human Geography's substantive concerns. Finally, we explain the layout of the book itself, both in terms of structure and in terms of the presentational style of chapters. Our intention has been to produce a book that gives you the cutting edge of Human Geographical thought, but in an accessible and usable form.

WHAT IS HUMAN GEOGRAPHY?

A fairly common exercise for an early Human Geography tutorial or seminar is a request to mine a week's news coverage and to come back with an example of something that seems to you to be 'human geography'. Have a go at doing this now. Think about the last week's news. Draw up a short list of two or three stories that strike you as the kinds of things Human Geography would study. Then reflect on how you decided on these, and what you thought was geographical about them. Now read on.

Literally, if one goes back to the word's Greek origins, Geography means 'to write (graphien) the earth (geo)'. It is hardly surprising, then, that both Human and Physical Geography actually have a vast range of substantive topics. In this book you will find chapters based on contents as diverse as Third World development, international finance, sustainability and environmental policy, nationalism, landscape painting and appreciation, natural history films, heritage tourism, colonialism, shopping, cyberspace, and (wait for it) oven-ready chickens, to pick out just a few highlights. You could, then, have picked any of these and more in your hypothetical seminar exercise. It is quite common to have mixed feelings about this range. On the one hand, many people choose to study Geography because of it, appreciating the wider understanding of human life it seems to offer. On the other hand, though, if this wider view is to be anything more than an intellectual pick and mix, it is also important to have a sense of what makes Human Geography cohere and what binds these apparently diverse topics together. One response is of course to look at these topics and to see some as more geographical than others. You may be doing that right now. But we need to think very carefully about how we make such judgements. Perhaps all too often we base them more on convention than on any serious and sustained thinking about the intellectual contribution of Human Geography. So, for example, because we are used to Human Geographers studying the urban planning issues of out of town developments, we think of that as Geography; whereas a discussion of, say, eighteenth-century landscape art does not have that familiarity and hence feels less like 'normal' and 'proper' Geography.

Especially as university level Geography is going to throw up a huge number of topics you may be unfamiliar with, we want to suggest that mere convention alone is an insufficient criterion through which to assess whether something is Human Geography or not. Having an understanding of the traditions of Human Geography is enormously valuable, but one of the crucial lessons we learn from that historical understanding is that what counts as Human

Figure i   Which of these photographs of work looks more like it should be in a Human Geography textbook to you? Why?
Credit: Popperfoto

Geography has always been subject both to change and to contestation (see Livingstone 1992 for an excellent, sustained analysis of this). For instance, for much of its history Human Geography, reflecting the social worlds it was being produced in, largely ignored over half the world's human beings, i.e. women. Economic geographers ignored the domestic work done by women at home; development geographers paid too little attention to the gendered nature of both development problems and practice; and so on (see Fig. i). At the level of research and teaching there are still more men than women in Human Geography, though it has to be said that some progress has been made, and certainly the contributions of women have tended to make up in quality, and hence influence, much of what they have lacked in simple quantity. Few of you, we imagine, would today say that Human Geography should be about and done by men (though for an analysis of how this is a continuing issue see Domosh, 1991). But it would be naïve to pretend that the predominance of men in the research and teaching levels of the subject, both past and present, has not moulded, for the worse, our senses of what 'conventional' Human Geography is. Topics and ways of studying them that are seen as more 'feminine' still tend to be seen as less central to the subject. Bizarrely, given that we all have a gender, this can even include the whole issue of gender itself, which is all too often seen as something that concerns only a few 'feminist geographers' rather than being an absolutely central facet of what it means to be human, and hence to have human geographies. (And to clarify, the problem here is not a product of '**Feminist Geography**' but of its sometime ghettoization as a fringe geographical activity rather than as something all Human Geographers need to engage with – if you want to follow this up, see the review of work on 'feminist geographies' by members of the Women and Geography Study Group of the RGS–IBG, 1997).

How, then, can we approach the question of 'What is Human Geography' in more open and thoughtful ways? To begin with, we have to move back beyond the more superficial questions of topic, by returning to that basic definition of Geography – 'to write the earth' – with which we began. If we think about the meanings of the 'geo' or the 'earth' in Geo-graphy we can identify two interconnected facets at the core of Human Geography's intellectual project (see also Cosgrove, 1994). First, there is earth as 'Mother Earth' (there is gender again), as 'the living planet earth', as 'soil'. What this signals is Human Geographers' preoccupation with the relations between human beings and the 'nature' we are also part of. Second, however, there is also earth as 'the whole earth', as 'the world'. These meanings lead us into Human Geographers' fascination with the peoples and places that make up our world, with the differences and similarities between them, with the ways they are connected to each other across space, and with the processes through which the world is structured into identifiable peoples, spaces and places in the first place.

In summary, branching off from the 'Geo' in Geography are two foundational concerns of Human Geography: human–nature relations, and human life's constitution through socially constructed spaces and places (or as they are often termed '**spatialities**') (*see* Fig. ii). These are the connective threads that hold together the seemingly diverse contents of Human

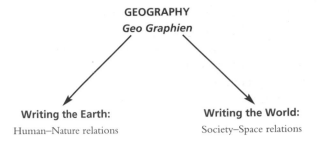

GEOGRAPHY
Geo Graphien

Writing the Earth:
Human–Nature relations

Writing the World:
Society–Space relations

Figure ii Human Geography: writing the earth and writing the world

Geography courses and texts. They are also why, returning to one of our earlier examples, an oven-ready chicken (Fig. iii) is doubly geographical. After all, and without wishing to steal the thunder from Michael Watts' fascinating analysis in Chapter 32, this oven-ready chicken is an embodiment of some particular ways for human beings to relate to nature: based on logics and practices of domestication, commodification, industrialized production, purposive modification and untroubled consumption that reach well beyond this one member of the animal kingdom. It is also an embodiment of some particular sorts of spatiality: in which different people and places are all connected together through the economic systems of the chicken (the consumers eating it, the farmers raising it, the large companies controlling its production and its distribution, the scientists genetically modify-

Figure iii Human Geography? Credit: British Chicken Information Service

ing it); but also one in which these connections are forgotten (by many) through a distancing of production and consumption (so even avid meat eaters would be unlikely to want to see video footage of broiler production and death as they tuck into their evening meal). Again, these are geographies which have far wider resonances, offering us a lens through which to think of a world that both connects and divides up people, places and activities. So, Geography is not just to be found on the library shelves or indeed in the newspaper; it is all around us, even right there in our kitchens and on our plates.

Summary

- One of Human Geography's main intellectual contributions is to understand the relations between human beings and the natural world of which we are a part.

- Another of Human Geography's main intellectual contributions is to understand how all facets of human life – the economic, the environmental, the political, the social, the cultural, the historical among others – are bound up with questions of space and place or 'spatiality'.

- These central concerns of Human Geographers are pursued across diverse and changing subject matters. We would encourage you to be open to that diversity and change rather than defining Human Geography in terms of topics and approaches that are already familiar to you.

DOING HUMAN GEOGRAPHY TODAY

Our approach in this book is therefore to introduce you to a Human Geography that has a strong intellectual coherence at its heart, but which applies this with an invigorating catholicism. More particularly, in traversing the subject's substantive diversity we have been especially concerned to give you a feel for the kinds of topics that are coming to the fore in contemporary Human Geography, offering you a bridge into the debates which are being thrashed out, often in rather more obtuse and difficult languages, in research journals and books. Human Geography has always been a dynamic field of enquiry, but the last ten years have seen particularly rapid changes. In the somewhat hyperbolic words of one British commentator, 'Geography has been re-invented' (Wadham-Smith, 1996). The most common rubric applied to these recent developments is the 'cultural turn'. This signifies the ways in which questions of 'culture' – by which is meant questions of meaning, language, discourse,

and representation – have become central not just to the long-standing sub-discipline of Cultural Geography but to Geographers' studies of environmental, economic, political, social and historical processes.

This book bears the imprint of that cultural turn, both in terms of what is in it and what is not. We deliberately wanted to convey some of the best qualities that it has brought to Human Geography: its attention to the practices and objects of everyday life; its determination to analyse critically the ways we conventionally think about the world and its geographies; its emphasis on human geographies as constructed and contested rather than innate; and its sensitivity to the relations between our geographical knowledges and forms of power. But we are uneasy about simply replicating the label of the 'cultural turn'. For a start, the phrase 'cultural turn' is actually a rather misleading way of describing recent developments. It has a tendency to suggest that other aspects of life – such as the economic, the environmental or political – have been turned away from. That is certainly not the case, as we hope this book demonstrates. In fact, we would suggest it is more accurate to view the 'cultural turn' as a recent twist on a number of longer-running trends in approaches to Human Geography. Two such trends are especially important: a questioning of the 'facticity' or factual character of human geographies; and a commitment to 'critical' thinking. Expressed like this, these are fairly abstract, so let us elaborate, using as an illustrative sketch how they might frame our understandings of the human geographies of a country such as the United States of America.

We begin with Human Geography's move beyond the factual. The 'Trivial Pursuit' model of the Human Geography of the USA would be that it involves learning key geographical facts: names and locations of states and state capitals perhaps; population distributions, maybe divided up into categories such as racial and ethnic groups, age bands (showing the retirement belts of Florida, for example) and class; locations of different industries and forms of employment, and so on. Now, knowing these kinds of facts is of some value, not least because the activities required to find them out – reading decent newspapers, finding useful sites amongst the deluge of information on the Internet, reading books – promote a more generally inquiring attitude to the world we live in. But in themselves they are a pretty pathetic sort of geographical knowledge. They convert human geographies into the thinnest of forms, not even beginning to scratch the surface of people's real lives, their values and beliefs, their daily preoccupations, their hopes and dreams, their loves and hates. For that one has to look to much more people-centred and **Humanistic Geography** approaches (for exemplary collections see Ley and Samuels, 1978; Meinig, 1979). As simple geographical facts they also tell us nothing about how these geographies have come into being and how and why they might change in the future. For that we have to think about how ways of relating to nature and of organizing activities spatially are bound up with wider American ways of life: ways of making a livelihood; ways of seeking enjoyment through leisure; ways of politically organizing, and so on (for an accessible example of American ways of relating to nature, see Wilson, 1992). In a country such as the USA this centrally involves understanding capitalist ways of life; something that has been the broader project for the last thirty years of what are termed '**political–economic** approaches' in Human Geography (for a summary review, see Peet and Thrift, 1989).

Simple geographical facts also tell us nothing about the phenomena they claim to measure and how they have come into being as categories of thought and life. For that we have to turn to various '**discursive** approaches' (for a collection of these, see Barnes and Duncan, 1992). We have to think carefully about how so-called facts – like race or ethnicity – are actually produced and contested as part of human life, rather than being aspects of it that Human Geographers can map as obvious and unproblematic. The USA, for example, actually has different understandings of race and ethnicity from those in Africa, Asia or even the UK. Even the geographical units of the USA – both its regions and its own status as a nation–state – cannot simply be assumed but are part of what needs investigating. After all, the regions of the USA are not simple facts of nature but dynamic political, economic and vernacular constructions (see Shortridge, 1991; Warf, 1988); and the USA itself has to be questioned as a 'fact' to focus on, in so far as developments within it can only be understood in terms of its relationship to the wider world political–economy (this argument is developed at length in Agnew, 1987b).

So, long-standing trends in Human Geography – towards humanistic, political–economic and discursive approaches – have emphasized kinds of geographical knowledge that go well beyond the recording and mapping of facts. One important consequence of this is that the doing of Human Geography involves far more than compiling data that provide unproblematically right answers. Instead, it requires 'critical thinking'. It is easy to misunderstand this phrase. When we say someone is being critical often we mean they are being negative, just finding fault; and at times it can seem that what academics most love to do is indeed to nit-pick, demonstrating their own intelligence by dismantling the ideas of others. But that is not what we mean here. True critical thought is as much about seeing strengths as weaknesses. What it does require is not taking things for granted, questioning the assumptions held by others and, crucially, oneself. It means – and this is a tricky balance – combining a determined scepticism with a profound openness to unfamiliar ideas and voices. It means recognizing that the 'rightness' of answers to geographical questions is something that has to be justified, and is as dependent on values, beliefs and perspective as it is on the 'facts' of the matter. Within the subject, a number of intellectual movements have emphasized this, perhaps most importantly the various 'radical geographies' of feminist, Marxist and postcolonial writers (for a sense of these you could flick through past and present volumes of the 'radical journal of Geography', *Antipode*). But critical thinking wears no particular political colours. It is more a style of thought, a way of doing Human Geography. Now at times a requirement for critical thought can be infuriating. All Geography students have probably had moments, especially around exam time, when they wished there was a simple right answer that someone could just tell them and they could repeat. However, in the end, it is the need to think critically that makes doing a Human Geography course more than time spent learning information to be forgotten soon afterwards. It is what ensures that you really can learn something that stays with you. So, to get even more evangelical in tone, we would urge you not to take the apparently easy way out, not to settle for second-hand ideas and conventional wisdoms. Be willing to give and gain more than that. Do Human Geography properly, and get critical.

Summary

- Human Geography is a dynamic subject that has changed rapidly in the last few years. This book introduces you to the approaches and issues preoccupying Human Geographers as we enter the twenty-first century.

- These contemporary approaches build on longer-standing bodies of work which have emphasized how geographical questions rarely have simple, right or wrong answers.

- Doing a Human Geography course therefore means learning to 'think critically' about the world and your own and other people's understandings of it.

STRUCTURE AND STYLE OF THE BOOK

The thirty-four chapters in this book are organized into three parts: *Foundations*, *Themes* and *Contexts*, each with its own brief editorial introduction. The six chapters in *Foundations* give you the latest thinking on some of the 'big questions' of substance and approach that have always concerned Human Geographers. Rather than deal with these through potted histories of the subject, or through abstract theoretical pieces, all the authors weave together conceptual ideas with accessible examples and illustrations. The relations discussed – between society and nature, society and space, the local and the global, structure and agency, the same and the different, and image and reality – are not likely to match with particular, substantive lectures in a first year course. But in many ways they deal with the most important questions to think about as a new Human Geography student, and introduce you to ideas that you will be able to use across a range of more substantive courses. Those substantive areas of the subject are mostly addressed in the second and largest part of this book, *Themes*. Here, eighteen chapters, themselves divided into six sections each with a brief editorial introduction, cover the major thematic 'sub-disciplines' of Human Geography: development geographies, economic geographies, environmental geographies, historical geographies, political geographies and social and cultural geographies. This part of the book provides you with thought-provoking arguments on the key issues currently being debated within these sub-disciplines, as well as giving you a feel for the kind of Human Geography undertaken within each.

Thematic sub-disciplines are one of the major ways in which teaching curricula are organized and research activity structured, to the extent that Geographers are often labelled according to these specialisms (as Economic Geographers, Political Geographers, and so on). However, there is another principle of organization within Geographical teaching and research, that of area-based specialisms. The final part of the book, *Contexts*, develops this regional spirit, having ten chapters that are organized not in terms of theme but in terms of generic kinds of places, spaces and environments. Some of these will probably be familiar geographical topics to you; the city, for example. But contemporary Human Geography increasingly has a wider range of contexts that it focuses on: attention is being paid to previously ignored places and spaces of human life, such as the body; there is a recognition that individual regions have long been integrated within continental, worldwide and **transnational** geographies; and there is a move towards thinking about not only the places we live in but the **'spaces of flows'** that connect and bring those places into being (Castells, 1989).

Stylistically, every chapter obviously has its own authorial signature, but all the contributions combine discussions of challenging ideas and issues with accessible presentation. Unfamiliar academic terminology is kept to a minimum, but where central to an argument is marked in bold type and defined in the Glossary at the back of the book. Chapters are deliberately short, but can easily be followed up using the suggested readings included at the end of each.

Further readings are perhaps an appropriate place to stop introducing *Introducing Human Geographies*. For they signal that the most important contribution this book can make is to encourage readers to move on. We hope you enjoy and are just a little bit inspired by this collection, but we know we are a stepping stone to other things. At the risk of sounding like a dating agency, think of this book as setting you up with other Human Geographies. It would be nice if you remembered us fondly, of course, but we hope there will be a time when you want to hang out with those other books and articles in the library, maybe even falling in love with some of them. Most importantly, use our introduction to their geographical ideas, themes and contexts as an excuse to invite *them* out for a breath of fresh air. They get bored in their ivory towers. Take those geographical ideas to your favourite haunts and see what they make of them. Find out how you get on, and if you don't seem to connect, agonize over whether that is their fault or yours. Have a relationship with Human Geography. In the end, we are sure you will hit it off with some bits of it, learning and changing in the process. After all, as our contributors have demonstrated so admirably, Human Geography today is vibrant, lively, fun and thought-provoking. Some even say it is intellectually sexy right now, but of course, not being shallow we love it for its inner self, not its superficial aura. Sure, at times it can be a bit challenging and difficult, but anything worth having a relationship with is. So time for us to shut up now, and let you and Human Geography get to know each other a little better. Here's to a magnificent romance.

Further reading

Obviously, our primary suggested reading is the rest of this book! But useful complementary texts, that fulfil slightly different functions, are the following:

- Agnew, J. A., Livingstone, D. and Rogers, A. (eds) (1996) *Human geography: an essential anthology*. Oxford: Blackwell.
An excellent collection of classic geographical writings, both old and new. This anthology demonstrates the longstanding but changing Human Geographical interest in questions both of culture, nature and landscape and of region, place and space.

- Daniels, S. and Lee, R. (eds) (1996) *Exploring human geography: a reader*. London: Arnold.

A collection of relatively recent geographical writings, giving a feel for contemporary Human Geography's 'critical' spirit and its determination to explain and understand, rather than just describe. As this is a reader, the contents are not specifically designed for a student audience, but all are relatively accessible and well set in context by the editors.

- Johnston, R.J., Gregory, D., Pratt, G. and Watts, M. (eds) (2000) *The dictionary of human geography*, 4th edition. Oxford: Blackwell.

This dictionary has concise but comprehensive definitions and explanations relevant to almost every aspect of Human Geography. As a reference tool it is invaluable. A book you will be able to use throughout your time studying Human Geography.

- Kneale, P. (1999) *Study skills for geographers*. London: Arnold.

A guide to the basic study skills Human Geography students need at the university level. In this chapter we have suggested Human Geographers need to be able to 'think critically'. This book helps you with the hard work of actually doing that. Well worth a look early on in your degree programme.

- Livingstone, D. (1992) *The geographical tradition*. Oxford: Blackwell.

A scholarly rendition of the history of Human Geography, a topic we pay little attention to in this collection. Livingstone concentrates on the longer-term history of the subject rather than on its recent developments. Throughout one gets fascinating insights into how the concerns of Human Geographers have run in parallel with wider social currents.

PART I

Foundations

INTRODUCTION

Sometimes the start of Human Geography textbooks, and indeed courses, can be very daunting. This is because of the perception by some of the authors of the books and courses concerned that it is necessary to throw in a load of theoretical stuff at the beginning, before getting on with the more interesting stuff. While it may indeed be preferable that certain theoretical foundations are laid before dealing with systematic issues, the net result is likely to be that the reader/course-attender can either be bored to tears, or bemused by the abstract nature of those foundations. Well, here's the bad news – we have also decided to begin this book with some theoretical dimensions. But, here's the good news – we utterly reject the false division between abstract theory and the substantive issues of everyday life. Indeed, we believe that our everyday lives are simply teeming with the kinds of issues and questions which are often pigeon-holed as theory. Much of the excitement and value in Human Geography is to address these issues and questions by thinking through aspects of our own lives and of the world(s) in which we live.

As an illustration, here is a very short account of a typical journey to work for one of us – Paul Cloke. Neither the story nor the journey is in any way special; that is the point of narrating it. It could be any part of your everyday experience, whoever you are, or wherever you live. What it does show is that different sets of Human Geography relationships crop up all over the place, and certainly not just in the abstract treatments of theory in books and lectures. So, imagine if you can the leafy suburb of Stoke Bishop in North West Bristol . . .

An alarm clock rings on a Monday morning. It is part of the routine to make sure that my son and daughter are awake and up in time for school. This often involves bringing them breakfast in bed! My son's room is decorated with photos of sports stars – footballers from English and Scottish clubs, and basketball players from the USA – and team shirts from US ice hockey and basketball. My daughter has grown out of wall-to-wall pictures, but one or two remnants from the international worlds of film and music remain. In both rooms, music sets the atmosphere, focusing variously on the anarchy of Chumbawamba, the politics of Black American rap or the trip-hop of Bristol's music scene. A little later, I walk the dog – a Border Collie of Welsh farming stock – and feed the pond fish – goldfish of Japanese breeding stock.

Later again, I head for the front door, tripping over my wife's Traidcraft boxes – one of her passions is fair trade issues, and she works hard as a voluntary representative, channelling fairly trading goods from 'developing' countries to people in our local church and beyond. Then, it's onto the bike and up the hill to the Downs, a large tract of open parkland often thought of as a lungful of fresh air in the heart of the city. By day, this is a site of many different kinds of leisure and recreation, but by night it is a landmark in the geography of women's (and some men's) fear, not to be traversed alone on foot. At the edge of the Downs is Bristol Zoo, which attracts large visitor numbers and houses 'exotic' animals.

After the Downs, there is a welcome downhill stretch from Blackboy Hill past the Police Station to Whiteladies Road. The very names of the streets resonate with Bristol's history as a port and a centre of slave-trading. Maybe, in a different way, the names also suggest a present set of social differences in the city. Local shops hug the side of the road, including a knot of travel agents displaying posters of far-flung places, and offering the passer-by a distinctive gaze on worlds of the imagination. They may not be 'real worlds' but they are real enough if that's the only experience of them you've got. Then comes 'the strip', Bristol's hot development of designer pubs and clubs, with Irishness here, and Henry Africas there, interspersed with the by-now unremarkable Indian and Chinese restaurants. Designer label beer and wine from all over the world is spilt here over designer T-shirts from all over the world.

At the bottom of the hill are two strange bed-fellows. BBC Bristol – home of local news broadcasting and 'nature' programming amongst other things – stands across the road from an Army headquarters building sometimes adorned with parked fieldguns and trucks in full battle-dress. A little further on come the banks and building societies, competing to serve the needs of local business and specifically placed to attract business from those connected to the university. Inside the nearby big department store are expensive items from all over the world, while outside someone sells the local newspaper, and another the *Big Issue*. This is often 'Chuck', a homeless man who is very fed up with the compassion fatigue of passers-by, who are quite capable of avoiding eye contact with the humanity, and the issues, of street homelessness.

Finally, it is up to the department, passing through the multinational, but somehow overwhelmingly middle-class throng of students in the precinct. Once inside my office, the first move is to fetch a cup of (fairly traded) coffee, switch on the PC and check my e-mails, hardly noticing the rows of shelves loaded with the production of particular knowledges about governments, policies, plans and politics, and how the lives of real people in real places intersect with so much in the geographical world.

There is so much else that I could (and perhaps should) have mentioned. Never mind, this brief glimpse is sufficient to suggest everyday prompts about different dimensions of human geography. The journey encounters nature in many forms, and as Sarah Whatmore suggests in Chapter 1, nature is shaped by the human imagination and filtered by the categories and conventions – for example, pets, zoos, parklands – by which we represent ways of seeing the components and landscapes of nature in our culture. Culture–nature relations, then, lie at the heart of our everyday (as well as more intermittent) experiences. So, too are relations between society and space. In travelling from home through suburb and key street-routes, my narrative is jam-packed with references to gender class and race – some of which are intended, and some suggested by your own powers of interpretation. In Chapter 2, Susan Smith shows how spatial patterns can reflect social structures, and how spatial

processes can be used as an index of social relations. However, she warns that social categories cannot be taken for granted. Such categories are constructed socially, politically, culturally, and are mediated by the organisation of space. Thus a leafy suburb might be thought of as geographically and socially different from an inner city housing estate, but equally Chuck occupies a socially constructed position of difference from those who pass by only inches from him. Moreover, we can no longer rely on two-dimensional maps of society and space. Beyond the obvious, there is complexity, ambiguity and multi-dimensional identity. Class, race and gender, for example, will cross-cut and intersect in different ways at different times and places.

Equally, two-dimensional geographies of place can no longer suffice. Phil Crang in Chapter 3 discusses the relations between the global and the local, and the sights, sounds, histories and commodities of the global crop up time and again in the local story of my journey to work. Local places get their distinctive character from their past and present connections to the rest of the world, and therefore we need a global sense of the local. Conversely, global flows of information, ideas, money, people and things are routed into local geographies. We therefore also need a local sense of the global.

'Structures' are often hidden in the story – the invisible skeleton which restricts and permits all kind of life-opportunities in our society. However, the actions of governments and the associated agencies of governance, the police, the army, financial institutions, and so on can contribute structural constraints and opportunities within which human beings make their decisions and non-decisions. Mark Goodwin's account of structure and agency in Chapter 4 suggests a very important dimension of who gets what, where, how and why in the everyday human geographies of our lives and the lives of others. These geographies can be seen both from the outside looking in, and from the inside looking out.

In Chapter 5, Paul Cloke explores the importance of 'self' and 'other' in these various lookings. Being reflexive about the self is a vital part of understanding how our knowledge of human geographies is situated. Our experience, politics, spirituality, identities and so on can add to our stories about the world, and denying their importance in search of 'objectivity' could well be dishonest. My journey to work will not be the same as yours, even if it follows much the same route. However, there is also a danger that we only see the world in terms of our selves and those who are the same as us, thus creating categories of 'otherness' according to the essential characteristics of our selves. What escapes us are other 'others' – those who we cannot categorize or pigeon-hole; those who surprise us and cannot be accommodated in our organization of knowledge.

Some 'othering' is presented to us by the multitude of images which intersect with the real experiences of people and places. In Chapter 6, Mike Crang discusses the charts, graphs, maps, pictures, films, and so on which provide ideas of bad image and good image in terms of specific people and places. In the travel agents' window, the wall posters, the output of news and entertainment media, and even the mental map of the journey to work, images can be deliberately promoted, managed or altered so as to suggest understandings of the world on which people act.

People inhabit these worlds of images, and we certainly need to understand what role images play in the construction of geographical knowledge.

The six chapters in this section of the book on *Foundations*, then, represent the very stuff of lively, interpretative, relevant and accessible human geographies. They help us to think through some of the recurring questions and issues involved in understanding the interconnections of people and places, and they help us to place ourselves in the picture as well. Far from being the 'boring theoretical stuff' they offer some keys with which to unlock thoughtful and nuanced accounts of everyday life. Enjoy!

CHAPTER I

Culture–nature

Sarah Whatmore

INTRODUCTION

Has anyone noticed how many television wildlife programmes seem to be scheduled around mealtimes? It's a mundane coincidence that illustrates just one of the ways in which we confront the tricky borders between culture and nature in our everyday lives. The feeding habits of the creatures on display and the food on the viewer's fork collide momentarily in millions of homes. In that moment, the cordon separating the things we call 'natural' from those we call 'cultural' loses its grip. Which is on the screen and which on the plate? At first glance, the big cat tearing into the flesh of its prey seems to embody nature at its most elemental – a world apart. But look again. This vision of nature 'red in tooth and claw' has been carefully framed by the hidden crews and technologies of film-making. They in turn are shaped by the conventions of science and television which establish our expectations of how a particular type of animal should eat and which aspects of feeding make good viewing. The meal in front of us, on the other hand, is more obviously of human making. But on closer inspection we cannot fail to be reminded that, however *haut* the cuisine or industrial the ingredients, we share the metabolic urges of our animal kin. Culture and nature, it seems, are not so easy to pin down.

Geography, as we are constantly being reminded, asserts itself as a subject uniquely concerned with this interface between human culture and natural environment. While the overt sexism of exploring 'Man's role in changing the face of the earth' (Thomas *et al.*, 1956) may have become outmoded (or at least better disguised), this classic description of the geographical project has lost none of its appeal (see, for example, Simmons, 1996). But it has also become shorthand for one of the underlying difficulties with the way the discipline is organized. The assumption that everything we encounter in the world already belongs either to 'culture' or to 'nature' has become entrenched in the division between 'human' and 'physical' geography and reinforced by the faltering conversation between them. As a result, even as geographers set about trafficking between culture and nature, a fundamental asymmetry in the treatment of the things assigned to these categories has been smuggled into the enterprise. Geography, like history, becomes the story of exclusively human activity and invention played out over, and through, an inert bedrock of matter and objects made up of everything else.

This division of the world into two all encompassing and mutually exclusive kinds of things, the so-called culture–nature *binary*, casts a long shadow over the way we imagine and live in it. It has not always been so and does not hold universal sway today. Rather, it can be traced to the European **Enlightenment** which, beginning in the fifteenth and sixteenth centuries, came to embrace the world through networks of commerce, empire and science. The geographical tradition of exploration and expedition played an important role in extending and mapping these networks and has left us with a thoroughly *modern* sense of nature as the world that lies beyond their reach (Livingstone, 1992). From this European vantage point, nature comes to be associated with the places most remote from where 'we' are – like jungles and wildernesses or, more recently, nature reserves and national parks. The trouble is that as the twentieth century draws to a close, 'we' seem to be everywhere – from the hole in the ozone to the cloning of sheep. Where is this pristine nature to be found now?

In this climate it is not suprising that geographers, like the rest of us, are having problems holding the line between the cultural and natural.

This chapter examines some of the ways in which contemporary Human Geography handles the relationship between culture and nature. The opening themes of food and wildlife are used to illustrate various approaches and interpretations, and to show the difference they make to the ways in which we understand and act in the world. The chapter focuses on two well-established kinds of accounts which explore different aspects of the ways in which human societies have refashioned natural environments over time. It concludes by looking at the growing dissatisfaction with such accounts and their assumption that we can best make sense of the world by first setting ourselves apart from everything else in it.

SOCIAL CONSTRUCTIONS OF NATURE

We begin then with established efforts by Human Geographers to make sense of the ways in which the ideas, activities and devices of human societies reshape the natural world. To put it another way, Human Geographers have treated nature first and foremost as a *social construction* although, as we shall see, they disagree over what this means. Two different, but in some ways complementary, traditions of academic work have been particularly influential over the last 25 years or so. The first is the **Marxist** tradition which has been concerned with the material transformation of nature as it is put to a variety of human uses under different conditions of production. The second is cultural geography which has focused on the changing idea of nature, what it means to different societies and how they go about representing it in words and images.

Producing nature

Writing at the height of the industrial revolution in Europe in the mid-nineteenth century, Karl Marx observed the ways in which plants and animals were being physically transformed by farmers using careful selection and breeding methods to produce commercially more valuable crops and livestock (Marx, 1976 (1867)). The lesson he drew from this observation was that with the rise of industrial capitalism, those things which we are accustomed to think of as natural were increasingly becoming refashioned as the products of human labour. This apparently contradictory idea of '*the production of nature*' has become a central theme for Human Geographers.

Noel Castree has identified three reasons for its geographical importance (1995). First, to acknowledge that nature is produced undermines the familiar, but misleading, idea that it is something fixed and unchanging. Instead we are forced to look at the specific ways in which human societies have interacted with natural environments in different times and places – from hunter–gatherer to post-industrial societies, from economies based on slave labour to those based on wage labour, for example. Second, it captures the double-edged sense in which the process of producing goods for human use and exchange simultaneously transforms the physical fabric of the natural world *and* people's relationship to it. For those of us whose idea of provisioning is stacking our shopping trolleys at the supermarket, it is difficult to imagine the intimate bonds that characterize societies in which the medicinal properties of plants or the seasonal habits of animals are part of everyday knowledge and practice. Third, it alerts us to the way in which capitalist production, in particular, seems to stop at nothing in its quest for profitability, turning landscapes, bodies and, these days, even the molecular structure of cells into marketable commodities.

Neil Smith's book *Uneven Development*, first published in 1983 and in a revised edition in 1990, has been one of the most influential elaborations of this analytical approach in contemporary Human Geography. Capitalism, he argues, for the first time in history puts human society in the driving seat, replacing God as the creative force fashioning the natural world. We can get more of the flavour of the argument from his own words.

> In its constant drive to accumulate larger and larger quantities of social wealth under its control, capital transforms the shape of the entire world. No god-given stone is left unturned, no original relation with nature is unaltered, no living thing unaffected. Uneven development is the concrete process and pattern of the production of nature under capitalism. With the development of capitalism, human society has put itself at the centre of nature.
>
> *(1990: xiv)*

This revolutionary social capacity to produce nature is termed *second nature* by Smith, and other Marxist geographers to distinguish it from nature in its 'god-given' or 'original' state, so-called *first nature*. These terms have an explicit historical dimension, marking off modern, or more particularly, capitalist societies from all those that have gone before in terms of their relationship with the natural world. In the same vein, a further transition is deemed to be going on today as we move towards postmodern social forms (see Chapter 2) accompanied by a *third nature* of computer-simulated and televisual landscapes and creatures.

The transition between first, second and third

nature also has significant geographical dimensions which are illustrated through the example of the potato in the three part sequence of Figure 1.1a–c. The potato arrived as an exotic curiosity from 'the new world' amongst the booty of fifteenth- and sixteenth-century explorers like Christopher Columbus and Walter Raleigh. Since then, the humble spud has become a staple of northern European diets and, in the guise of the McDonald's 'french fry', of a global fast-food cuisine. The image in Figure 1.1a dates from around 1600 and shows a drawing from Guaman Poma's encylopaedic survey of the ancient Inca state

of Tahuantinsuyu (in modern-day Peru) for the King of Spain. It shows sacks of potatoes transported by llamas being laid up for storage. It is the kind of image that from our own time seems to capture just what is meant by first nature – plants (and animals) in their 'original' state, remaining essentially unchanged by their encounter with a 'pre-modern' society. The Figure 1.1b is a photograph of potato harvesting in Brittany in the 1980s. The large featureless field, the monotony of the crop and the presence of the tractor tell us that this is an industrial agricultural landscape. A readily recognizable picture of second nature, wearing its human fabrication on its sleeve. The third image, Figure 1.1c is a cartoon illustrating some of the popular anxieties associated with the current transition to a third nature. Here, not only has the location and landscape of potato growing become a human artefact but the genetic structure of the potato plant itself has been mapped and engineered to enhance its commercial properties – in this case its size!

Figure 1.1a First nature. Seventeenth century drawing of the Inca state of Tahuantinsuyu. Potatoes being laid up for storage

Figure 1.1b Second nature. Industrial potato cultivation in Brittany, 1980s

Summary

- Nature is socially constructed in the sense that it is transformed through the labour process and fashioned by the technologies and values of human *production*.

"Who ordered the baked potato?"

Figure 1.1c Third nature. A genetically engineered potato. Credit: Pugh/*The Times,* 31/7/97

- From this perspective, nature–society relations are seen to have changed progressively over time from *first* (original) nature; to *second* (industrial) nature to today's *third* (virtual) nature.

Representing nature

A rather different interpretation of what is meant by the social construction of nature is that associated with the cultural tradition of Human Geography. In this geographical enterprise the natural world is understood to be shaped as powerfully by the human imagination as by any physical manipulation. This is because 'nature' does not come with handy labels naming its parts or making sense of itself, like a plant from the garden centre. Such naming and sense-making are the attributes of human cultures. The importance of this approach is that it forces us to recognise that our relationship with those aspects of the world we call natural is unavoidably filtered through the categories, technologies and conventions of human **representation** in particular times and places. As Alex Wilson, a Canadian landscape architect, puts it:

> Our experience of the natural world – whether touring the Canadian Rockies, watching an animal show on TV, or working in our own gardens – is always mediated. It is always shaped by rhetorical constructs like photography, industry, advertising, and aesthetics, as well as by institutions like religion, tourism and education.
>
> *(1992: 12)*

For cultural geographers, then, nature itself is first and foremost a category of the human imagination, and therefore best treated as a part of culture.

This can be a rather unnerving starting point for those who look to nature as the reassuring bedrock of a 'real' world that stubs your toe when you trip over it, regardless of any attempt to 'imagine' it otherwise. And one could be forgiven for not taking it very seriously if cultural geographers were arguing that nature is 'just a figment of our imaginations'. But, of course, they are not. What they are saying is that the relationship between the 'real' and the 'imagined' is no less slippery than that between nature and culture (see Chapter 6). These arguments are brought to a head in the concept of *landscape*. In everyday speech, landscape refers both to physical places in which we encounter the natural world and to artistic representations of such encounters and places. Cultural geography builds on these ambiguities to direct attention to the ways in which the relationship between the two – the 'real' and the 'represented' landscapes of nature – is far from straightforward.

In their influential book *The Iconography of Landscape* (1988), Stephen Daniels and Denis Cosgrove suggest that landscape is a way of seeing the world which can take a variety of forms:

> in paint on canvas, in writing on paper, in earth, stone, water and vegetation on the ground. A landscape park is more palpable but no more real, nor less imaginary, than a landscape painting or poem
>
> *(1988: 1)*

Whatever their form, these 'ways of seeing' the natural world share three common principles. The first of these is that the representation of nature is not a neutral process that simply produces a mirror image of a fixed external reality, like a photocopy. Rather, it is instrumental in constituting our sense of what the natural world is like. This is easy to accept for paintings or literature where we make allowance for 'artistic licence' in terms of the artist's vision and the technical qualities and stylistic conventions of their chosen medium, say, oils or poetry. But it holds equally well for natural history programme-making, or the geographical art of map-making, in which the nuts and bolts of the process of representation are less readily apparent or more actively hidden from view.

It follows, that the second principle of landscape is not to take representations of the natural world at face value, however much they seem, or claim, to be 'true to life'. The 'real' and the 'represented' cannot be so surely distinguished or firmly held apart in the practical business of 'seeing the world'. The work of the imagination, for example, has begun before a single brush stroke has been made on the canvas. What has brought the artist to this particular spot and made it a worthy subject for painting? As much as anything else, it is an established repertoire of cultural reference points for interpreting the natural world which repeat and ricochet off one another down the ages, like the biblical imagery of the 'wilderness' or the 'ark'. Likewise, representations of the natural world shift effortlessly from being understood as depictions of what it *is* like, to being used as blueprints of what it *should be* like in the guise, for example, of management plans for the conservation or restoration of historic landscapes.

The third principle of landscape is that there are many incompatible ways of seeing the same natural phenomenon, event or environment. For example, the drawings and accounts of eighteenth- and nineteenth-century European colonists depicted Australia as a 'waste and barbarous' land. Yet this representation could hardly have been more at odds with the 'dreamtime' landscapes of its aboriginal peoples whose communal stories and dances teem with plant and animal life. Such irreconcilable landscapes underline the importance of carefully situating different representations of the natural world, including our own, in the historical and social contexts which make them meaningful (see Chapter 23). They also alert us to the highly political nature of the representational process. British colonization and settlement of Australia were justified by treating it as an 'empty' continent. The prior claims and land rights of its aboriginal inhabitants went unrecognized in the country's constitution and legal process until an historic ruling in the Australian High Court in 1992.

We can illustrate these points by returning to the case of natural history progammes that opened this chapter. Television wildlife documentaries are widely taken to 'tell it like it is', providing us with a direct lens onto nature's creatures and landscapes. But as David Attenborough, one of the world's leading wildlife film-makers, made clear in an interview with geographers in the early 1980s, 'there is precious little that is natural . . . in any film'. He goes on to explain why.

> You distort speed if you want to show things like plants growing, or look in detail at the way an animal moves. You distort light levels. You distort distribution, in the sense that you see dozens of different species in a jungle within a few minutes, so that the places seem to be teeming with life. You distort size by using close-up lenses. And you distort sound.

> (Burgess and Unwin, 1984: 103)

Figure 1.2 illustrates this more complex relationship between the 'real' and 'represented' landscapes of wildlife. On the left we catch a rare glimpse of the people and equipment that mediate between an animal and its celluloid image. It gives us some sense of the discomfort and risk that film-makers face to 'get the shot'. But look closely at the lion. She is within spitting distance of the cameraman yet completely uninterested in him and his paraphanalia. The reason is that this photograph was taken in Serengeti National Park which has become a favoured location

for filming African wildlife precisely because, after decades of intensive management/planet and tourism, the animals have become habituated to human presence. Careful editing will be needed to make this animal and place live up to the standards of a documentary 'wildlife' landscape. Figure 1.3 adds a further twist. It shows the BBC Natural History Unit being caught out by a national newspaper over its filming of the Monarch butterfly for the series *Incredible Journeys*. In order to save money, the programme-makers decided not to travel to its native region in the Great Lakes of North America but to film captive-bred Monarchs which they had released into the comparable scenery of the English Lake District. Again, the film satisfies our sense of the wild aesthetically. But this time the representational process has left unknown consequences in its wake for both the butterflies and the regional ecology into which they were let loose.

Summary

- Nature is socially constructed in the sense that it is shaped as powerfully by the human imagination as by any physical manipulation. Our relationships with nature are unavoidably filtered through the categories and conventions of human *representation*.

- From this perspective, the *landscapes* of nature are understood as '*ways of seeing*' the world in which the 'real' and the 'imagined' are intricately interwoven.

ENLIVENING THE GEOGRAPHICAL LANDSCAPE

Whether their emphasis has been on its material transformation or on its changing meaning, Human Geographers have treated the natural world primarily as an object fashioned by the imperatives of human

Figure 1.2 Taking a close-up of a lion.
Credit: Roz Coward/Planet Earth Pictures

Figure 1.3   The Monarch butterfly's journey to Lake Windermere. Credit: Jonathan Anstee/*The Independent on Sunday,* 17/11/96

societies in particular times and places. Each perspective illuminates different aspects of the convoluted relationship between the things of human making (culture) and those that are not of our making (nature). But in different ways the creative energies of the earth itself, in rivers, soils, weather and oceans, and of the living plants and creatures assigned to 'nature', are eclipsed in both accounts. In their eagerness to stress the capitalist capacity for producing nature, the Marxist tradition, for example, too readily overlooks the active role of these natural entities and processes in making the geographies we inhabit. Likewise, the argument that our relationship to the natural world is always culturally mediated has tended to fix attention on the powers of the human imagination, ignoring the multitude of other lives and capacities bound up in the fashioning of landscapes.

In these marvellous worlds of exclusively human achievement nature appears destined to be relentlessly and comprehensively colonized by culture. Human Geography's long march from environmental determinism to social constructionism seems to have brought us, as the environmentalist Bill McKibben puts it, to 'the end of nature' (1990). Whatever their differences, both accounts of this triumph of human culture over the matter of nature are grounded in the assumption that the collective 'us' of human society is somehow removed from the rest of the world. Only

by first placing it at a distance can human society be (re)connected to everything else on such asymmetrical terms as those between producer and product, or viewer and view (see also Chapter 5). These are geographies whose only subjects, or active inhabitants, are people while everything consigned to nature becomes so much putty in our hands.

Such **_humanist_** geographies do not square with the anguish and infrastructure of environmental concern that characterizes the late twentieth century. In unimaginable and unforeseen ways the forcefulness of all manner of 'non-humans' has come to make itself felt in our social lives. From climate change to 'mad cow disease' there is a growing sense that our actions, and indifference to their consequences, are returning to haunt us. The popular face of this growing sensibility is illustrated by the image in Figure 1.4 which was circulating as a postcard at the Edinburgh Fringe Festival in 1995.

Likewise the pets and viruses, plants and wildlife that share the most urbanized of living spaces, and the peoples who over centuries have inhabited the deserts, jungles and swamplands where 'we' have seen only nature, make it apparent that 'the whole idea of nature as something separate from human experience is a lie' (Wilson, 1992: 13).

Over the last 5 years or so there has been mounting unease about the ways in which Geography has built

Figure 1.4 The cows come home. Credit: BSE II/Edinburgh Festival 1995

cal processes are woven together in the making of spaces and places. Three of the most important currents in this rethinking of the 'human' in Human Geography give a flavour of things to come.

The first of these currents is concerned with showing that the idea of nature as a pristine space 'outside society' is an historical fallacy. This idea is so pervasive today that it is difficult for many of us recognise it as a particular and contestable way of seeing the world. But historical geographers are helping to expose the ways in which the presence of native peoples was actively erased from the landscapes that came to be seen as wildernesses in colonial European eyes, and which are now revered by many environmentalists as remnants of 'pristine' nature (Cronon, 1995). Try looking again at the image in Figure 1.1a, in this light.

The second current extends this historical repudiation of the separation of human society and the natural world by paying close attention to the mixed-up, mobile lives of people, plants and animals in our own everyday lives. The place of animals has largely fallen off or, more accurately, between the agendas of contemporary human and physical geography. But a new focus on 'animal geographies' is emerging which seeks to demonstrate the ways in which they are caught up in all manner of social networks from the wildlife safari to the city zoo, the international pet trade to factory farming, which disconcert our assumptions about their 'natural' place in the world (Wolch and Emel, 1998).

Finally, and most provocatively, a third current of work against the grain of the nature–culture binary is trying to come to terms with the ways in which the seemingly hard and fast categories of human, animal and machine are becoming blurred. This blurring is achieved by technologies like genetic engineering and artificial intelligence which are seen to recombine the

this binary division between nature and culture into its descriptions and explanations of the changing world. This unease stems from several different concerns, not least the crippling effect this polarization had on the contribution that Geography can make to informing more sustainable living practices (Adams, 1996). In Human Geography, it centres on a growing recognition of the intricate and dynamic ways in which people, technologies, organisms and geophysi-

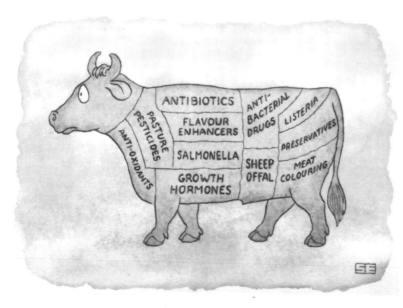

Figure 1.5 Mapping hybrid body spaces. Credit: Stan Eales

qualities associated with these categories in new forms, such as transgenic organisms, bionic enhancements and the like (Luke, 1997). In these 'cyborg geographies' the body is emerging as an important new site for geographical research. Returning again to the example of food, Figure 1.5 illustrates the ways in which body spaces are being reorganized at the most intimate of scales. Where does nature end and culture start for this cow?

Human geography has come a long way from defining itself as the study of 'man's [*sic*] role in changing the face of the earth'. The geographies now emerging challenge us to look again at how and where we draw the line between culture and nature and to recognize that this densely and diversely inhabited planet is a much more unruly place than these categories admit.

Further reading

• Anderson, K. (1997) A walk on the wildside. *Progress in Human Geography* **21** 4: 463–85.
A thorough review of geographical work on domestication as one of the longest-running processes connecting human, animal and plant life worlds in new ways and helping to put the novelty of genetic engineering into historical perspective.

• Cronon, B. (1995) The trouble with wilderness: or getting back to the wrong nature. In Cronon, W. (ed.) *Uncommon ground: towards reinventing nature*. New York: W.W. Norton, pp. 69–90.
A very readable piece by an eminent US environmental historian which reflects on both the intellectual and political problems of treating nature as 'outside' culture.

• Luke, T. (1997) At the end of nature: cyborgs, 'humachines' and environments in postmodernity. *Environment and Planning A*, **29**: 1367–80.
A difficult piece written with more than a hint of irony at the hyperbole with which some current writing about genetic engineering, artificial intelligence and the like are breaking down conventional boundaries between 'nature' and 'culture'. Worth the effort to get a sense of what the geographies of 'third nature' might look like.

• McKibben, B. (1990) *The end of nature*. Harmondsworth: Penguin.
This popular 'bestseller' argues that nature in its 'true' sense as a 'separate realm' has been eradicated by the relentless industrialization of human society. It is a passionate example of an environmental politics premised on maintaining the distinction between nature and culture. It's short and worth reading in its entirety, but the basic case is set out in Part 1.

• Wilson, A. (1992) *The culture of nature*. London: Routledge.
Written by a Canadian landscape designer, this book is a visual and literary feast concerned with the numerous ways in which nature and culture shape each other in post-war North American landscapes. The introduction and chapters on nature films (4) and nature parks (7) are particularly good.

Society–space

Susan J. Smith

INTRODUCTION

Each of us is unique: a distinctive blend of physical features, personality traits and social skills with our own particular biographies and life paths. Nevertheless, there are at least some characteristics that we share with other people. Sometimes these shared characteristics have little bearing on our sense of well-being or our access to resources. I am one of thousands of people who have red hair, blue eyes and write with their left hand. This makes no difference to my salary, my housing opportunities, my credit rating, my health or my likelihood of being burgled.

Nevertheless, there are a surprising number of shared characteristics which do affect our life chances. Whether we are sick or well, disabled or not, male or female, whether we are thought of as 'black' or 'white', young or old, what religion, nationality or sexual orientation we have, and even whether we are tall or short, can have a profound impact on what we get out of life. So although we think of ourselves as individuals, it is important to recognize that people are often grouped, or structured, into social categories. Their lives can be enhanced or impaired not by how hard they work, or by whether they are exceptionally lucky or peculiarly accident-prone, but simply because they are assigned to, or choose to identify with, one group rather than another.

These processes of social categorization do not occur 'naturally'. They are a product of how power and resources – which may be real (money, cars, homes) or symbolic (a question of how people think, and what they take for granted) – are struggled over and manipulated. Moreover, these struggles do not occur on the head of a pin. They take place. Just as history is relevant to an understanding of the present, so geography matters if we are to understand how people identify themselves, and how they label others. Geography matters if we are to grasp why certain social characteristics are salient and others are understated, in particular places and at certain times. To illustrate this, the following pages provide an overview of the different ways that Human Geographers have attempted to understand the links between society and space.

FROM SOCIETY TO SPACE

Perhaps the simplest way to conceptualize the interaction of society with space is to regard spatial arrangements as a straightforward reflection of social divisions. People from high-income groups, for example, can afford to pay more for housing than people from low-income groups. The cost and quality of housing vary over space, and it follows that those who can, will pay more for attractive homes in pleasant environments and well-serviced neighbourhoods. Because of this, people with high incomes tend to cluster in the same kinds of spaces, and these spaces tend to be separate from the neighbourhoods where poorer people live. Thus it is that income inequalities are expressed in spatial arrangements. Society is mapped onto space.

In practice, the picture is much more complex than this, because the straightforward relationship between income and ability to pay is cross-cut by other factors. For example, discrimination in the housing system has, in many countries, meant that for a given income band, 'white' households secure access to better goods, services and resources than 'black' households. This is true even in societies (like the north-west European welfare states) that rely on welfare transfers to compensate in cash or kind when groups lose out because they cannot compete in the marketplace. Women, gay people, sick people, people with physical impairments or learning difficulties, older people, and others may be disadvantaged (relative to male, heterosexual, healthy, younger people) in the same way. Nevertheless these 'complications' do not mean that the basic argument is compromised: all these social divisions do have some kind of spatial expression.

This first approach to the links between society and space has a great deal of common-sense appeal. This is probably why, under the label 'urban ecology' it became so popular, especially during geography's quantitative revolution in the 1960s. At its extremes it consisted of entering as many census variables as possible into a computer, producing clusters of statistically related population attributes, labelling them with terms like '**class**', 'family status' or '**ethnicity**', and mapping them. The idea was to show how particular combinations of socio-economic characteristics (particular arrangements of society) related to particular configurations of the urban environment (particular spatial structures).

In the end these complex techniques produced little that was of lasting value. They resulted in maps bearing labels like 'social disorganization' even though no-one seemed to know what this condition was, or what to do about it. This criticism of how the approach was put into practice does not, however, mean that *in principle* the ideas are wholly without merit. In fact, the notion that society should be considered in terms of space formed part of a rather exciting 'spatial turn' in sociological thought which occurred towards the end of the nineteenth century. The question on people's lips was how to handle the enormous complexity of the social world in the interests of developing social theory and devising public policy. The answer, voiced most explicitly by the philosopher Georg Simmel (b. 1855) in an eclectic series of books and articles, was to position society in space. Space provided a medium which 'fixed' social processes long enough for them to be scrutinized by scholars and policy-makers.

Simmel's ideas were enthusiastically taken up by his student Robert Park, who formalized the notion that spatial arrangements could be used to simplify the complexity of the social world. His argument, set out in an influential essay on 'The urban community as a spatial pattern and a moral order' (1926), was that that spatial patterns, unlike social processes, are tangible, visible and measurable phenomena. If the teeming chaos of society could (temporarily) be fixed in space, we might learn more about social problems, and this knowledge might help us devise appropriate solutions. If, as Park believed, 'human relations can always be reckoned, with more or less accuracy, in terms of distance', then it follows that 'society exhibits, in one of its aspects, characters that can be measured and described in mathematical formulas'. For Park, this pointed to 'the importance of location, position and mobility as indexes for measuring, describing and eventually explaining social phenomena'. Park's aim, then (unlike that of some of his successors), was not so much to reduce society to space, as to use space pragmatically, to pick out those aspects of social life relevant to the problem at hand.

Park's pragmatism inspired the branch of quantitative social science known as 'spatial sociology' – a term popularized by Ceri Peach in his collection of essays called *Urban Social Segregation* (1975). Here Peach describes Park's (1926) essay 'the fountainhead from which all else springs', and his own 'basic hypothesis' flows directly from it:

> the greater the degree of difference between the spatial distributions of groups within an urban area, the greater their social distance from each other . . . degrees of spatial similarity between socially defined groups are correlates . . . of the degree of social interaction between those groups . . .
>
> From the overall, spatial, residential mix of groups within urban areas, one can deduce the strengths of social divisions between those groups.

The basic thinking about society and space built into this approach is laid out in Table 2.1, column 1. It is a line of thinking which has been extremely influential, especially in relation to the idea of '**race**' and the politics of segregation. This, it must be said, is because the approach was largely **gender**-blind. It did not take into account inequalities in the social relations between men and women, because it focused on households, assumed they took the 'conventional' family form, and regarded this as unproblematic. Despite this obvious flaw, in the USA, the notion that spatial mixing is the key to 'racial' integration (and that 'racial' integration is the route to civil rights) has been enormously influential. It has underpinned a bitter debate over school bussing as well as a mountain of legislation and case law over the imposition of barriers to open housing. Moreover in places as far apart as Britain and the USA, not to mention the 'special' case of South Africa, analysts have argued persuasively that patterns of spatial separation contain and express a geography of racism (Cell 1982; Smith 1989). In this respect at least, the organization of residential space is an expression of deep-seated socio-economic and political inequalities in racially divided societies.

Spatial patterns do, then, express aspects of social

Table 2.1 Three ways of exploring the links between society and space

From society to space	The spatial construction of society	Thirdspace
Spaces are scientific and geometric, filled with an accumulation of social facts, providing an accurate but simplified representation of a more complex 'real' world.	*Spaces* have a material reality and a symbolic significance and can take on a life of their own. Spatial patterns express but also shape social relations.	*Spaces* which those marginalized by racism, **patriarchy**, **capitalism**, **colonialism** and other oppressions choose as a speaking position.
Geographies that are concrete, quantifiable and mappable.	*Geographies* that are negotiated and struggled over.	*Geographies* that were made for one purpose are apppropriate for another, redefined and occupied as a strategic (real or symbolic) location.
An explanatory framework which regards spatial patterns as an index and an outcome of social and political processes.	*An explanatory framework* which regards spatial patterns as informing and interacting with socio-economic processes.	*This is about being rather than explaining* – an approach which is emancipatory rather than predictive or interpretive.
Social categories and social identities are given. The social distances between groups are expressed in spatial separation; social interaction is signalled by spatial integration.	*Social categories* and identities are constructed through spatially discriminatory material practices (markets, institutions, systems of resource allocation) and cultural politics (struggles to control the imagination).	*Social categories* are resisted by those they are imposed on. Spaces on the margin provide a position from which to build open and flexible identities. Here commonalities are emphasized and differences tolerated.

difference. However, there are two kinds of qualifications to bear in mind. First, to be credible this approach cannot be content with simply describing the imprint of society on space. The real challenge here is to expose the *mechanisms* that translate social differences (categories and identities) into spatial patterns. Figure 2.1 points to some of the factors that come into play: the ideological norms in which the major markets and institutions are embedded; the ways these interact with, and interrupt, the fair operation of the housing system and the labour market; the way direct and indirect, personal and institutional discrimination open up spaces in which to live, work, learn and be creative to some people, and close them down to others. The challenge is vividly expressed in Norman Rockwell's painting (*see* Fig. 2.2). Research-based examples of how the mechanisms which lie behind such images may affect the spaces available to minority groups, people with health problems, and women in 'non-conventional' households are outlined in Smith and Mallinson (1996).

Second, although the logic of mapping society onto space continues to appeal, Livingston and Harrison's (1981) playful commentary on one of Lewis Carroll's allegorical poems contains a sobering thought for anyone tempted to push this line of reasoning too far. With apologies to both Lewis Carroll and Livingston and Harrison, I have drawn up a few

parallels between geographers' attempts to capture social relations in the container of space, and the fate of the band of sailors who embarked on the Hunting of the Snark (see page 15).

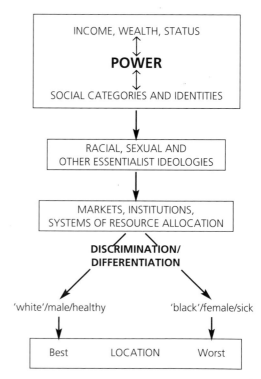

Figure 2.1 Exploring the mechanisms which translate social differences into spatial arrangements

Spatial sociology was a hunt for the best approximation of social processes that could be read off from spatial forms. Analysts knew they were in the right place and doing the right thing because they were following the wisdom of a previous generation of thinkers.

Human geographers knew what they were looking for because they worked with widely accepted social categories whose spatial distributions left their unmistakable marks on the patterning of urban space . . .

And they would know when they had found it, because they had a foolproof method in mind based on the new tools of quantitative social science

The hunt to specify the spatial structure of social life was on, and its success seemed inevitable. As long as we could think of space in geometric (or Cartesian) terms and take social categories for granted, patterns of residential segregation would provide clues to the structure of social relations, the processes of integration and the problems of ghettoization.

But there is a twist at the end of the story . . .

What happens if the Snark that we are so close to capturing and the Boojum, which we have only just heard about and are now trying desperately to avoid, turn out to be one and the same? This is what happened when Lewis Carroll's unfortunate Baker discovered the Sn———!

Source: Extracts from *The Hunting of the Snark* (1876) by Lewis Carroll

'Just the place for a snark'. I have said it twice,
That alone should encourage the crew.
Just the place for a snark, I have said it thrice:
what I tell you three times is true. . .

Come listen, my men, while I tell you again
the five unmistakable marks
By which you may know, wherever you go,
the warranted, genuine Snarks.

Taking 3 as the subject to reason about –
a convenient number to state –
We add seven, and ten, and then multiply out
By One Thousand diminished by Eight . . .

('That's exactly the method' the Bellman bold
in hasty parentheses cried.
'That's exactly the way I have always been told
That the capture of Snarks should be tried).

'But oh, beamish nephew, beware of the day,
If your Snark be a Boojum! for then
You will softly and suddenly vanish away,
And never be met with again!'

In the midst of the word he was trying to say,
In the midst of his laughter and glee,
He had softly and suddenly vanished away –
for the Snark *was* a Boojum, you see.

The Snark and the Boojum turned out to be one and the same, but the sailors' conception of the Snark could not accommodate the subtleties of the Boojum's identity. The Snark overspilled the conceptual category that the sailors had imposed on it. Despite clearly formulated goals and scientifically informed precision, their chosen means could never produce the ends they hoped for. They made an error of substance (the Snark was not what they thought), they made a practical, methodological error (they assumed the Snark was

Figure 2.2 School girl with US Marshals: 'the problem we all live with'. Credit: printed by permission of the Norman Rockwell Family Trust © 1964 The Norman Rockwell Family Trust

Table 2.2 Errors, limitations and reservations of human geographers' approaches to society and space

From society to space	The spatial construction of society	Thirdspace
ERROR OF SUBSTANCE: Socially constructed categories are depicted as 'real', fixed and mutually exclusive.	SUBSTANTIAL LIMITATION: Social categories are depicted as socially (and politically) constructed, but research continues to emphasize the binary divides between, e.g. 'black' and 'white', male and female, sick and well.	SUBSTANTIAL RESERVATION: Is the world as flexible and as open to people defining and redefining themselves as the concept of Thirdspace suggests?
METHODOLOGICAL ERROR: Spatial patterns are seen as independent of social processes (the spatial is set up as a measure of the social, but this could only work if the two really are separate). There are social relations that cannot be studied within the kind of spatial frameworks used (notably the social relations of gender difference).	METHODOLOGICAL LIMITATION: Spatial organization is implicated in the construction of social divisions, but in order to explore this, analysts are forced to work with concepts (such as the idea of race) or frameworks (the presumption of heterosexuality) whose potency they wish to challenge.	METHODOLOGICAL RESERVATION: Boundaries, borders, peripheries and other marginal spaces have become the fashionable positions from which to resist old social categories and create new identities. In practice, the radical potential of places on the margin may be open to question.
ETHICAL ERROR: Social categories are taken for granted. The reality of 'races', the division of society into economically unequal classes, and the conventional family form are regarded as the starting point of the analysis, not as the outcome of processes which remain to be explained.	ETHICAL LIMITATION: Continues to define social categories relative to one another (black in relation to white, women in relation to men, etc.). Can therefore imply that a particular, uneven, distribution of power is inevitable. Also criticized for remaining preoccupied with marginalized '**Others**' (e.g. black identities) while taking privileged selves (e.g. whiteness) for granted.	ETHICAL RESERVATION: Is it right to deny the potency of well-established points of difference just at the moment when previously powerless peoples want to put them to use?

something that could be captured with the equipment they had to hand), and they made an ethical error (that hunting was the way to deal with Snarks). Some spatial sociologists have fallen into the same trap. Their thinking also contains errors, as indicated in Table 2.2, column 1; and there is an extent to which they, too, have vanished – 'snarked' by a new spatial turn in social thought, which, as we shall see in the next section, refuses to take either social categories or spatial structures for granted. The very idea that social categories and identities can be fixed, captured and displayed on a spatial canvas at all, now seems remarkably naïve.

Summary

- One way of conceptualizing the relationship between society and space is to think of spatial patterns as a reflection of social structures, and to regard spatial processes as an index of social relations.

- This approach over-simplifies both society and space, but it can be useful when it focuses attention on the specific mechanisms that determine who gets what, where and why.

THE SPATIAL CONSTRUCTION OF SOCIETY

The novelty of quantitative social geography wore off in the 1970s, and the theoretical limitations of social area analysis, factorial ecology and measures of spatial dissimilarity became increasingly evident. This forced some reassessment of the relationships between society and space. The question was no longer 'how are particular social groups spread across physical space?' but rather 'how do spatial arrangements, how does place and position, *actively contribute* to the construction and reproduction of social identities?' So instead of talking about 'black' and 'white' and how they interact, or juxtaposing the West against the rest, analysts began to explore where those categories came from and how they are produced and reproduced. Human Geographers in particular became interested in the difference that space makes to these processes. Why, they ask, and how does geography matter for the construction of social life?

The nature of the difference in thinking that comes with this 'spatial constructionist' approach is illustrated in Figure 2.3, and summarized in Table 2.1, column 2. Instead of taking particular social cate-

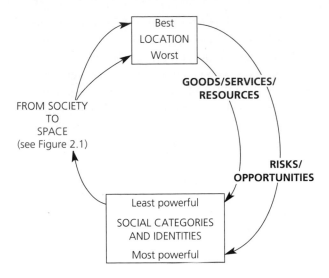

Figure 2.3   The spatial construction of society

gories for granted and mapping them onto space, analysts began to look at how socially divisive spatial arrangements mediate access to services and resources, underpin systematic differences in peoples exposure both to risks and to opportunities, and so feed back into the web of power relationships which influence how people are categorized by others, and how they identify themselves. This constructionist approach to the relationship between society and space is best illustrated by way of some examples.

Geography and the construction of gender difference

Pratt and Hanson (1994) begin their analysis of women in paid work in Worcester, Massachusetts in a fairly conventional way by developing the idea that the social characteristics of men and women help determine not only where they live and work, but also what these places are like. Society maps onto space. However, the study goes on to illustrate that once these social characteristics are rooted in space, they acquire a certain fixity. Spatial arrangements constrain social processes. In this example the characteristics acquired by different places so affect the employment prospects and practices within them that social divisions originating in the labour process are amplified through the organization of space. The result in Worcester is that social boundaries dividing men from women, and dividing different groups of women from each other, are reinforced, or hardened. Let's look at this more closely.

The case study neighbourhoods in Worcester are differentiated by the distinctive mix of 'race', class, gender and work traditions within them. Once this mix is established, employers make locational choices on the basis of it, in order to recruit from the labour pool they regard as most suitable for their businesses.

These locational decisions are partly based on stereotypes about the difference between men's and women's work, and partly on ideas about the suitability of different ethnic groups for particular kinds of tasks. As a result of these investment patterns, local labour markets with distinctive occupational niches develop. In this example we see how women workers in particular (because of the very local orientation of their lives) are positioned and segregated by the labour process. The most obvious social consequence of these spatial investment strategies is therefore a cementing of the gender boundary (through a spatial division of labour which plays on and feeds into the supposed differences between men and women). However, work practices are organized on class and race divisions as well, and a further outcome of this particular geography of placement is a hardening of the social boundaries between different groups of women.

Pratt and Hanson's (1994) study shows, in short, that the organization of space (in this case, as shaped by the investment decisions of employers) plays a role in constructing social difference. This can take several forms, not only enhancing existing social divisions (here, between men and women) but also creating divisions within groups (in this case women) whose interests might at some level be thought of as similar. Crucially, this example shows that space and place are not simply containers in which people's social lives develop. Rather, where people are placed has a direct bearing on those lives.

The spatial construction of 'race'

My own work on the politics of 'race' and residence in Britain follows a similar line of reasoning (Smith 1989; 1993). First, it shows that there is a material basis to residential segregation. South Asians, African-Caribbeans, and their descendents are statistically over-represented in the worst homes and neighbourhoods in the inner areas of the major cities. This is a consequence both of their relative exclusion from the most buoyant sectors of the economy (resulting in low incomes and vulnerability to unemployment), and their historically limited access to the institutions of the welfare state (particularly access to the better quality segments of the public rented housing stock). Society and social inequality are mapped onto space.

However, *where* marginalized groups live is not simply a reflection of a history of social exclusion. In addition, this positioning actively affects future employment and housing options. This is because the process of economic restructuring which has occurred in the last 20 years has been spatially selective. Jobs have been lost in the regions and sectors of cities which South Asians, African-Caribbeans, other black Britons and visible minorities are most likely to occupy. At the same time (and as part of the same process) these are the zones where property prices are lowest, house price appreciation is least, and where the options to move to take advantage of new employment opportunities are limited (the problem is compounded by spatial changes in the availability of public housing). So what was a spatial reflection of economic and social marginality becomes a spatial constraint on economic advance and social mobility (Figure 2.4).

This problem of 'spatial mismatch' is also being hotly debated in relation to North American cities. Entrapment in spaces of deprivation can prevent already disadvantaged groups from taking advantage of opportunities opened up by the restructuring of the economy. In the British example the organization of space contributes to a process of 'racialization' – of reiterating the salience of 'races' and making these social categories and the inequalities that divide them seem normal and inevitable. Space constructs society.

In this second example, the process goes one step further, because in addition to the material differences which underpin patterns of segregation, British political culture has also formed an *image* of 'racial segregation'. Over the last half century, this image has legitimized a range of public policies which – far from tackling the material inequalities that reproduce racial difference – has set the scene for things to get worse; for the divide between 'black' and 'white' Britain to harden, and for people's expectations of who should live, work and play where, to bear this in mind. (This process is not, of course, exclusive to Britain. The importance of ideas and images about the character of space for construction and conduct of social life is, for example, further illustrated by Anderson (1991 in her work on Canadian constructions of 'Chinatown'.)

The photographer Ingrid Pollard exposes the weight of this thinking in her exhibition 'Pastoral Interludes' (which is discussed at length by Kinsman (1995)). There are many ways in which the five photographs and their accompanying texts might be seen

and interpreted, but for me it is the placing of 'black' figures in the 'white' English countryside which is so striking. The juxtaposition is striking because it is not what we have been taught to expect. It is, after all, the image of the 'inner city' that most often accompanies publicly available photographs of black men and women. But Ingrid Pollard's photographs do more than simply illustrate the unexpected. They also challenge us to unpack the taken-for-granted assumptions that prompt us to see these images as in some way out of the ordinary. What has persuaded us – what has duped us – into thinking it is somehow 'normal' for black people to be absent from rural Britain?

The photographs further expose important injustices in the way we routinely conceptualize, and use, urban and rural spaces. Think, for example, about fear and anxiety. If you read newspapers or watch television, the spaces to fear – if you are 'white', which 'you' are often assumed to be – are those of the (implicitly 'black') inner city. Yet when Ingrid Pollard ventures into the countryside, she writes of an experience which 'is always accompanied by a feeling of dread'. There is another challenge here – to confront a division of space which is so socially divisive that many ordinary people feel uncomfortably out of place for much of the time.

In many ways, it could be argued that spatial constructionism is a more critical, sensitive and useful approach to the study of society and space than that adopted by the spatial sociologists. However, this approach, too, has its limitations, and these are summarized in Table 2.2, column 2. Perhaps the main reservation hinges around the tendency for academic research in this area to focus on the experiences of marginalized 'others' while taking privileged 'selves' for granted. This is especially notable in relation to work on race and racism, which has in the past been dominated by 'white' researchers studying 'black' people. In contrast, research dominated by men was, for many years either gender-blind, or simply failed to engage with feminist issues. Even then, a presumption of heterosexuality ran through the majority of pub-

Figure 2.4 Birmingham, England.
Credit: author's collection

'pastoral interlude'
. . . it's as if the Black experience is only ever lived within the urban environment.
I thought I liked the Lake District; where I wondered lonely as a Black face in
a sea of white. A visit to the countryside is always accompanied by a feeling
of unease; dread . . .

. . . feeling I don't belong. Walks through leafy glades with
a baseball bat by my side . . .

Figure 2.5   (a) and (b) 'Pastoral Interlude' (1988) Credit: Ingrid Pollard. Courtesy of the artist. Original in colour. Tinted silver print.
From the series of five

lished work. One response to these exclusions has been to turn attention to the construction of 'whiteness', the hidden agendas of masculinity, and the tricky terrains of sexuality and erotic preference.

Summary

- So far we have identified two quite distinct ways of looking at the relationship between society and space (cf. Table 2.1, columns 1 and 2) The second of these focuses on how social categories that were once taken-for-granted are in fact socially and politically constructed.

- These processes of social construction are mediated by the organization of space. Both material practices (the operation of markets, institutions and systems of resource allocation) and cultural politics (disputes over the meaning of space) play a role. The result is that geographies of placement reinforce social inequalities.

A THIRD WAY?

To overcome the limitations of both spatial sociology and spatial constructionism, feminist and postcolonial scholarship have developed a rather different way of thinking about society and space. This alternative approach encourages us to recognize not just that the social categories we once took for granted are in fact socially constructed, but also that this process of construction allows for the possibility of a redefinition and renegotiation of what the social world is like. People's identities may not be contained within one category or another: they may incorporate aspects of a number of categories; or they may develop in between or on the boundary of categories; and this may make a nonsense of the idea of working with categories at all. This suggests that we need a third way of approaching the constitution of society and space.

Soja (1996) explores this third option in a book which he calls *Thirdspace*. He writes from a conviction that 'the spatial dimension of our lives has never been of greater practical and political relevance than it is today' (p. 1). This in itself suggests that new ways of thinking about space as an element of society may be required. Indeed, Soja's book hints that the spatial turn at the end of the twentieth century may be even more radical and exciting than its nineteenth-century counterpart. At the very least, Soja argues, our old ways of thinking about space (what he calls first- and second-space thinking) can no longer accommodate the way the world works.

In practice, the third option has turned particular kinds of spaces into strategic locations. Space has become a source of community for those oppressed by the social categories that society has worked with for so long – by ideas about, and practices around – 'race', class, gender, sexuality, age, nation, region and colonial status. Experiences of marginalization and exclusion, and the spaces created by these processes, have been turned into a strategic resource. They have become a speaking position for voices which previously went unheard; a space in which old social categories are resisted by new processes of identification.

The ideas informing this third approach are summarized in Table 2.1, column 3. Perhaps the most important point to note here is that the spaces referred to are often metaphorical rather than geometric. Spaces on the margin *might* be material spaces – inner cities, spaces of unemployment, domestic spaces, segregated spaces – but they are also ways of thinking about how different groups are (and by implication how they should be) located relative to one another in the economic, political, social and cultural hierarchies. Thirdspace is created by those who reclaim these real and symbolic spaces of oppression, and make them into something else.

What does this mean for how we conceptualize society and its relationship with space? bell hooks, in her book *Yearning* (1991), provides a useful starting point. Whereas white Western scholars wrote about race and space, made black and white relational and understood black identities as a resistance to racism, bell hooks argues for African-American women that, 'We must deny the oppressive other ... *we* must determine how we will be.' To make this point, she argues that she is *choosing* marginality as the position from which to speak. She is not speaking as someone forced into a marginal position by her race and gender. She is inverting our idea of what it is to be marginal, and at the same time she is redefining what it is to be black and a woman.

---

bell hooks in Thirdspace

I was not speaking of a marginality one wishes to lose, to give up, but rather as a site one stays in, clings to even, because it nourishes one's capacity to resist. It offers the possibility of radical perspectives from which to see and create, to imagine alternatives, new worlds . . . [mine is] a message from that space in the margin which is a site of creativity and power, that inclusive space where we recover ourselves, where we move in solidarity to erase the category colonizer/colonized.

(1991: 149–52)

This new location – this Thirdspace – articulates, and is defined by, what Homi Bhabha (1994) has called hybrid identities. All the common approaches to managing social difference in the 'Western' world – the idea of 'race relations', the concept of 'multiculturalism' – are based on the assumption that when parties meet through relationships like colonialism, imperialism, and the process of world economic development, their identities remain intact. But Bhabha points out that after such a meeting, no-one can ever be the same. Each is changed by the other in the encounter, so all identities are hybrid.

Homi Bhabha in Thirdspace

All forms of culture are continually in a process of hybridity. But for me the importance of hybridity is not to be able to trace two original moments from which the third emerges, rather hybridity to me is the 'third space' which enables other positions to emerge. This third space displaces the histories that constitute it, and sets up new structures of authority, new political initiatives, which are inadequately understood through the received wisdom . . . The process of cultural hybridity gives rise to something different, something new and unrecognisable, a new area of negotiation of meaning and representation.

(1994: 211)

For Bhabha, the importance of this 'something new' is not just that it provides a different kind identity, but that that it leaves the world of fixed categories and identities behind, opening up 'the possibility of a cultural hybridity that entertains difference without an assumed or imposed hierarchy' (1994, p. 4). This produces what Paul Gilroy describes as 'a different view of culture, one which accentuates its plastic, syncretic qualities and which does not see culture flowing into neat ethnic parcels but as a radically unfinished social process of self-definition and transformation' (1993b, p. 61). Sometimes artists can be better than social scientists at getting this idea across. Rap music, for example, provides a space which can accommodate ideas about hybridity (Smith 1997), while the work of Jean-Michel Basquiat — a 'black' painter entangled with a 'white' art establishment exploring his Haitian and Puerto Rican roots, his American heritage and his masculinity — reminds us forcefully of bell hooks' exhortation: 'we must determine how we will be' (*see* Fig. 2.6).

Figure 2.6 Success, 1980 (acrylic, oil stick and gilt on wood) by Jean-Michel Basquiat (1960–1988). Credit: © ADAGP, Paris and DACS, London 1999/Bridgeman Art Library, London/New York

The relevance of this third way is not limited only to identities emerging from the dichotomies drawn between 'black' and 'white' or 'colonizer'/'colonized'. Gillian Rose (1993) talks about the 'paradoxical space' occupied by feminist thought (paradoxical because of the way that feminism is trapped within, yet seeks to transcend, masculinist discourse). This, she says, is 'not so much a space of resistance as an entirely different geometry through which we can think power, knowledge, space and identity in critical and, hopefully, liberatory, ways' (p. 159). Rose therefore talks about the construction by feminists of 'a different kind of space in which women need not be victims' (p. 159) and in which femininity is mediated by other social identities. In the end Rose leads us towards a feminist geography which is less concerned about the man/woman boundary, than about what feminism can be when it is not defined relative to masculinism or patriarchy; when it is defined in its own terms by those who identify with it, and who seek an alternative to the old gendered order.

> ### Gillian Rose in Thirdspace
>
> These studies interpret women's lives not through the categories of production and reproduction, but through another kind of sociality . . . [this work has] created a women's space, but one which does not depend on an essentialist understanding of women, and in this manner it escapes the terms through which masculinist geography interprets space . . . The subject of feminism must be positioned in relation to social relations other than gender . . . in order to displace masculinism . . . There is a notion of things that are not representable in masculinist discourse but which women themselves may sense if not articulate. Feminist critique depends on a desire for something else.
>
> *(1993: 136–8)*

Thirdspace seems to offer quite a different conceptualization of the constitution of society than the approaches discussed earlier. It turns our attention away from the givens of social categories and towards the strategic process of identification. It forces us to accept the complexity, ambiguity and multi-dimensionality of identity and captures the way that class, gender, and 'race', cross-cut and intersect in different ways at different times and places. The idea of Thirdspace may provide an opportunity to move beyond our historic preoccupation with social divisions — with what holds people apart — and think about what is to be gained from a discourse of belonging. Perhaps this is what Soja (1996) has in mind when he talks of exploring Thirdspace with a view to engaging in 'some form of potentially emancipatory praxis, the translation of knowledge into action in a conscious – and consciously spatial – effort to improve the world in some significant way' (p. 22).

Maybe the quest for belonging, the search for similarity, the hunt for commonality and community are what contain the next critical geography of society and space. But these are complicated times, and there are already some reservations about the nature of Thirdspace. Is it really as radical and open, as liberating and as politically correct as it might at first seem (Table 2.2, column 3)? As we move from Thirdspace into the next dimension, my recommendation is that whenever we think we have something definitive to say about the vexed questions of social categorization, cultural identification or the structure of society and space, we should think about the Snark and its ill-fated pursuers. It is always a mistake to take the status quo for granted. Perhaps, with Rose (1993) we need to conclude that the best, and the least, we can hope for is 'a geography that acknowledges that the grounds of its knowledge are unstable, shifting, uncertain and, above all, contested' (p. 160).

Summary

- Earlier social geographies are being challenged by groups who have historically been excluded from 'mainstream' societies, and who now claim the right to define themselves. Spaces on the margin provide a strategic location – a position of strength for those with new ideas about history, destiny, society and space.

- New processes of identification do not always coincide with the spatial, 'race', class and gender divisions that previously dominated the research agenda. This is forcing human geographers to think again about their presuppositions and their research methods.

- Thirdspace may hold the key to devising a geographically informed attempt to tackle long-standing social inequalities; but then again, it might not . . .

CONCLUSION

Like all attempts to classify knowledge into discrete bundles, the three approaches outlined in this chapter are to some extent caricatures. Every approach has some strengths; but they all come with problems and limitations. In every case there are analysts who use the approach wisely and constructively, mindful of its presuppositions and wary when interpreting results. In every case too, there are examples of bad practice, where ideas are applied uncritically or inappropriately, without adequate qualification, producing results which are theoretically and ethically suspect. It is important not to confuse the strengths and weak-

nesses of a particular approach with the merits or failings of particular applications.

It is tempting, nevertheless, to regard our three approaches to society and space as sequential and therefore progressive. Certainly, they developed chronologically and each new set of ideas grew out of well-founded dissatisfaction with the one that preceded it. But equally there are areas of overlap between, and debates within, these 'schools' of thought. As a result, none of the approaches is entirely redundant; and none is sufficient in itself to handle the links between society and space.

For example, labour markets and housing systems continue to segregate societies unequally with reference to ideas about race, class and gender. To address this we need to specify what this inequality looks like and how the mechanisms that produced it work. Likewise, we can only challenge the essentialism that lurks within social categories if we explore the way such categories are constructed and resisted. Finally, we can only move beyond the binary divisions between black and white, male and female, able and disabled if we recognise that old social categories can, in the end, be displaced by new forms of identification. What we want from human geography today is not a consensus that any one approach to the study of society and space is inherently superior to another, but rather an awareness of when and where a particular approach is appropriate, and a sensitivity to what it can and cannot achieve.

Further reading

For a short and accessible account of where studies of Whiteness fit into Human Geographers' interests in ideas about 'race' and space see:
- Bonnett, A. (1997) Geography, 'race' and Whiteness: invisible traditions and current challenges. *Area* **29**: 193–9.

A set of essays developing the 'spatial constructionist' approach to the study of social life.
- Jackson, P. and Penrose, J. (eds) (1993) *Constructions of race, place and nation*. London: UCL Press.

A classic collection of articles illustrating the various ways that 'spatial sociologists' have attempted to monitor social change using spatial measures is assembled in
- Peach, C. (ed.) (1975) *Urban social segregation*. London: Longman.

A well argued and wide-ranging critique of how geographers have thought about space and society, has been mounted from a feminist perspective by
- Rose, G. (1993) *Feminism and geography*. Cambridge: Polity Press.

This has become a key text introducing the most recent 'spatial turn' in sociological thought. It looks closely at the interweaving of real places and spatial metaphors.
- Shields, R. (1991) *Places on the margin*. London: Routledge.

An article which contains some reservations about the flexibility and openness that is supposed to characterize social life and Human Geography towards the end of the twentieth century.

- Smith, S. J. (1993) Social landscapes: continuity and change. In Johnston, R. (ed.) *A changing world: a changing discipline?* Oxford: Blackwell.

This is quite a complicated book and parts of it will be difficult to grasp if you are at the start of your studies in Human Geography. However, some sections, particularly Chapter 4. 'Increasing the openness of Thirdspace', provide an insightful account of the third way of exploring society and space.

- Soja, E. (1996) *Thirdspace*. Oxford: Blackwell.

CHAPTER 3

Local–global

Philip Crang

INTRODUCTION

Being a rather unimaginative soul, I always ask interviewees for the Geography course where I teach the same question: 'What would you say makes Geography an interesting subject?'. Unsurprisingly I also nearly always get pretty much the same four answers. That it has a lot of variety in the topics it covers (see the whole of this book for evidence of this). That it combines a concern with people and the natural world, especially through its contributions to environmental thought (see Chapter 1 for confirmation). That it deals with a reality that you can see when you look at the world around you, not abstract theories (see Chapter 6 for some doubts). And that it means you get to hear about, see slides of, and maybe even go to a lot of different places, exploring the local character of each and learning about areas of the world that otherwise one would be largely ignorant of. It is this last answer I am especially interested in here. For it rightly identifies, I think, a triumvirate of ideas that have long fostered Human Geography's understanding of itself as a distinctive intellectual endeavour. First, the idea of the 'local', stimulating an interest in the character of particular places. Second, the idea of the 'global', bound up with a desire to broaden horizons and foster a 'world awareness'. And third, mediating between the local and global, is an emphasis on 'difference' (both geographical and human). This chapter examines the relations between these three ideas. It will, I hope, give a sense of how productive they have been and can still be. However, it also argues for critical reflection. Notions of the local, the global and difference are not as simple and obvious as they might at first seem. It is important to think carefully about each of these ideas, and perhaps

even more so about how they relate to each other. If we do not, then we run the risk of simply reproducing conventional arguments about local–global relations, without even realizing that that is what we are doing, and unaware of other possible ways of thinking and acting. But before turning more directly to some different constructions of local–global relations, we will need briefly to outline how and why ideas of the local and the global have been so important to Human Geography.

LOCAL MATTERS, GLOBAL VISIONS

To start with the local, it, and associated notions such as place, locality and region, have long had a particular centrality in geographical imaginations, despite a tendency for their devaluation (at least until very recently) in other social and human sciences (Agnew, 1989). Many academic geographers have spent whole careers trying to document, understand and explain the individual 'personality' of an area (Dunbar, 1974; Gilbert, 1960). So, why is the local deemed so important to human geographical research and teaching? Nick Entrikin (1991, 1994) has argued that geographers have been interested in the local for three inter-related reasons. First, they have emphasized the actually existing variations in economy, society and culture between places; or what Entrikin terms the 'empirical significance of place'. Despite the homogenizing ambitions of the likes of McDonald's, everywhere is not the same. Landscapes vary. Life chances are materially affected by the lottery of location. Whether you happen to be born in Lagos or London or Los Angeles, or indeed in Compton or Beverley Hills, has an impact on the kind, and even length of life you can expect. And location is not just something we encounter and deal with. It is part of us. Where we are is part of who we are. Most obviously, this is the case through the spatial partitioning of the world into nationalities (see Chapter 21), imaginative constructions that are part of our identities, so powerful as to get people to kill and die in their name. So, places and the differences between them can be seen to exist and have real effects.

But the local also matters in a second way. Spatial variations do not only exist. They are valued, or seen as a good thing, not least by Human Geographers. There is, then, what Entrikin calls a 'normative significance to place'. Sometimes this is expressed as a celebration of difference: whether out of a suspicion of the power of global, homogenizing forces ('the media', 'American multinationals', and so on); or out of a pleasure gleaned from experiencing variety and the unexpected (see Chapter 31 for parallels to this in contemporary tourisms). Sometimes the local is cherished for its communal forms of social organization, for embodying an ideal of small and democratic organizations (for a critical and suggestive review see Young, 1990). And sometimes this social idealization goes hand in hand with an environmental utopia of self-supporting, environmentally sustainable local communities (Schumacher, 1973), or at least a worry about the environmental impacts of transnational trade and supply networks (as in, for example, calls in the UK to reduce the 'food miles', or the distance food travels to reach our plates, by localizing supply networks and reducing the market share of the large supermarkets). But whether culturally, socially or environmentally framed, in all such arguments the local does not just matter. It matters because it stands for good things.

There is a third importance attached to the local within Human Geography, according to Entrikin. This involves a concern with the impact of the local on the kinds of understanding or knowledges that geographers themselves produce; what he calls the **'epistemological** significance of place'. In part this involves a scepticism towards general theories that claim equal applicability everywhere. In equal measure it means a sensitivity to the importance of where knowledges come from (to their 'situatedness'). So, for example, the fact I live and work here in London has an affect on my ability to understand human geographies elsewhere in Britain, let alone anywhere else in the world.

It is not surprising, then, that at the same time as having a local fixation, Human Geography is also determinedly global in its scope, looking to break out of purely local knowledges. This interest in the global has been developed through combinations of ideas, institutions and practices (or what can be called '**discourses**' of the global). Let me draw out four such discourses which have been especially influential in constructing geographical understandings the world. Figure 3.1 displays a picture of the world from each.

First, we can identify a discourse of *exploration*, driven by a desire to 'know the world'. Exploration was central to Geography's early history – such that Geography's development as a science, from the sixteenth century onwards, went hand in glove with European explorations to the farthest corners of the earth (Livingstone, 1992b; Stoddart, 1986) – and still shapes some of the most popular parts of the subject – for example the student and other 'expeditions' sponsored by the Royal Geographical Society in Britain, or the mass-circulation *National Geographic*'s promotional claim to give American readers a 'window to the world of exotic peoples and places' (cited in Lutz and Collins, 1993: xi). Second, there is a discourse of *development*, with its hope of 'improving the world'. Here, a world vision matters not only in order to rectify ignorance of the world's diversity, but also to explain and act against inequalities between North and South (see Chapters 7 and 9). Third, and more recently, there is the discourse of global *environmentalism*, with its passion for 'saving the world' against planetary threats such as global warming or ozone depletion (see Chapters 13, 14 and 15). Here, thinking globally is essential not only to recognise the scale of these problems, but also to understand the true environmental impacts of our local actions (so, when I switch on my electric kettle I need to be aware of the impact of my domestic energy use on CO_2 emissions). Finally, and not unrelatedly, there is a discourse of global *compression*, with an emphasis on the 'shrinking of the world' (see Harvey, 1989: 240–307). Made familiar through the corporate boasts of the likes of IBM ('solutions for a small planet'), the emphasis here is on the increasingly dense interconnections between people and places on other sides of the world from each other, whether through telecommunications, global flows of money, or migrations and other forms of travel. '**Globalization**' is increasingly used as a label to describe such compression (for a very good critical review, see Allen, 1995), and in a globalized world our lives are led on a global scale. If you doubt this, just think about where the food you have eaten today comes from (you may have little idea and have to check, but the likelihood is that in your shopping basket, or on your plate, will be ingredients and foods from all over the world).

So, there are a host of good reasons why Human Geography should not myopically focus on the local. Because the global stands for important, 'big' issues

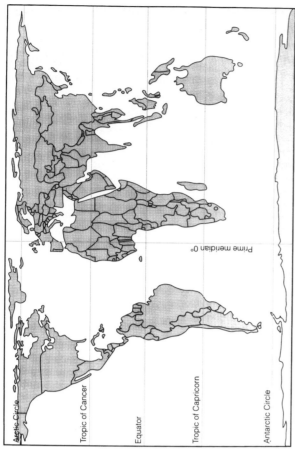

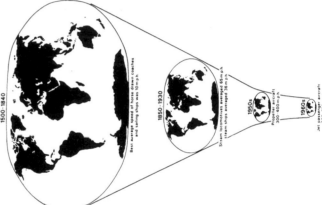

Figure 3.1 Four global visions. Top left: the conversion of the spherical globe into a flat map is here achieved through a Mercator projection. Developed in the seventeenth century, the Mercator world map is ideal for exploration as a constant bearing appears as a straight line, but this is achieved by distorting sizes, which makes tropical regions look far smaller than they actually are. Top right: the Peters projection, by contrast, is an equal area projection that distorts shape rather than size. First published in 1973, this projection was designed within development discourse to ensure the 'South' was given its proper global importance © Professor Arno Peters. Bottom left: 'spaceship earth' is an icon of contemporary environmentalism, portraying a living whole without apparent national boundaries or other political divisions. Bottom right: the shrinking earth of telecommunicational hype (reprinted by permission of Paul Chapman Publishing, from P. Dicken (1998) *Global Shift*.

and processes, as in this clarion call from David Stoddart:

> If as geographers we can get our heads out of the sands of our various minute concerns we will see the crisis which has overtaken us . . . The history of world population . . .: in 1750, . . .730 million; 1850, . . . 1200 million; . . . the end of the century, 6000 million . . Meanwhile environmental equilibrium is itself disturbed. The tropical rain forest destroyed at the rate of 1200 hectares an hour. Acid rain kills the temperate woodlands . . . Because of the greenhouse effect . . . sea-level around the world will rise . . . Quite frankly I have little patience with so-called geographers who ignore these challenges.
>
> *(Stoddart 1987: 334)*

Because these global scale processes impact on, and result from, our local places and lives. And because thinking globally allows us to compare, and even more usefully connect, our own lives and places to those of others.

Human Geography is therefore characterized by a concern with *both* the local *and* the global. At times, these can be understood as competing scales of interest; as when calls are made for geographers to escape local trivia and address the really important global issues; or, conversely, when global accounts are criticized for their ignoring of local differences. But, the local and the global can also be seen as two sides of the same coin. Explorers set out across the world to find new 'locals' to study (and all too often to exploit and conquer) (on the connections of exploration, geography and imperialism see Hudson, 1977; Driver, 1992; Smith & Godlewska, 1994). Environmentalists and multinational corporations both sloganize about 'thinking globally and acting locally'. So, how we understand and construct the global shapes our understanding of the local, and vice versa. In the second half of this chapter I therefore want to turn more directly to these relations between the local and the global. They can, I want to suggest, be thought of in a number of different ways. To illustrate, I will review three schematic accounts of local–global relations *(see* Fig. 3.2): the world as *mosaic,* the world as *system,* and the world as *networks.*

Summary

- Human Geography has fashioned itself as a distinctive intellectual endeavour both through emphasizing its interest in local places and specificities, and through stressing various global concerns (for example, with exploration, development, global environmental change and global 'compression').

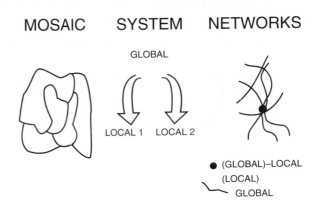

Figure 3.2 Figures of the local–global: mosaic, system and network

- Whilst the local and the global can be seen as alternative and competing scales of analysis, we need to recognize that they are always constructed in relationship to each other.

MOSAIC

One very popular way of thinking about human geographies is in terms of a mosaic, a collection of local peoples and places, each one being a piece in the broader global pattern. This pattern can be drawn out at a number of different scales, from neighbourhoods right up to whole continents. So, at the level of the city one could identify a patchwork of local areas, each characterized by different economies, residents and built environments. Think, for example, of portraits of the brutal contrasts in US cities between the 'hoods' and the 'subs', marked respectively by 'landscapes of decay and despair and landscapes of cash and comfort' (Riley, 1994: 151-2), so near to each other but separated, to use the title of Lawrence Kasdan's movie, by a 'grand canyon' experientially. Moving up a scale to the regional level, Geographers, and others such as regional novelists, have long evoked distinctive landscapes and connected them to distinct regional ways of life (Gilbert, 1972; Pocock, 1981). At the level of the nation the whole idea of nationalities depends upon constructing distinctive spaces; establishing borders and distinguishing between this country and that country, our people and those foreigners. And at the scale of the continent increasingly there are attempts to identify the boundaries and distinctiveness of continental economic, political and cultural units (for an example, see the critical analysis of Europe as constructed by the European Union in Chapter 28).

In many ways, this notion of the geographic mosaic has been so influential (see Gregory, 1994: 34-46) that it can be hard for us to see it as anything other than common-sense, a description of an obvious reality. Tourism, the world's largest industry, feeds off and actively constructs such an understanding, as it displays a world showcase of different destinations that the holidaymaker can visit. But the mosaic is only one possible way of framing local–global geographies, and it is a very particular framing, with its own preoccupations and blind spots. Three features are especially important. First, it puts an emphasis on boundaries. Geographical difference is seen in terms of distinct areas that can have lines drawn around them. Second, and relatedly, these areas are understood in terms of their unique characters, personalities or traditions. That is, each area is seen as having distinctive 'contents', whether that be its people, culture, economic activities and/or landscape, which cohere into some sort of unified geographical identity. Third, this means that any intrusions into this distinctive area tend to be seen as a threat to its unique character. For an example one could think of worries about how the global predominance of American popular culture, from fast food to TV programmes, is destroying local cultures and producing one Americanized global monoculture, where everybody wherever they are eats Big Macs, drinks Coca-Cola and watches *Baywatch* (see Peet, 1989).

All these features can be questioned factually. The world's geographical differences do not fit this neat model of the mosaic, and attempts to make them do so show just how rigid and unrealistic the model is. Apartheid in South Africa and ethnic cleansing in the former Yugoslavia would be examples of practices that have followed the mosaic, and its logic of each different thing in its own different place, to some brutal conclusions. Nor does the opening up of local places to global forces necessarily result in the destruction of difference. Take the example of American media products such as globally exported soap operas. Whilst living and researching in Trinidad the anthropologist Danny Miller was struck by the fact he had to stop his research for an hour a day whilst everyone watched the daytime US soap *The Young and the Restless* (Miller, 1992). But this was not the sign of an homogenizing Americanization. In fact, Miller argues, 'paradoxically an imported soap opera has become a key instrument for forging a highly specific sense of Trinidadian culture' (1992: 165). For in the extensive chat about this soap, what viewers identified was not an alien American world, to be aspired to or despised, but themes that resonated with deep existing structures of Trinidadian experience. In particular, viewers liked the way it dramatized what they called 'bacchanal', or the confusion and emergence of hidden truths produced through scandals, something also central to other Trinidadian

cultural forms such as Carnival. So, this globally distributed American soap actually helped produce a very distinctive, local Trinidadian sensibility. It was, as Miller concludes using a popular local expression, 'True True Trini'.

The problems with the figure of the mosaic are not only factual. They also stem from its political impulses and ramifications. To be fair, there are positive elements to the notion of the mosaic. Often underlying it is a desire both to recognize and respect differences; to appreciate, in both senses of the word, that everyone is not the same as you are, and that everywhere is not the same as here (an impulse to '**relativization**'). But it is not enough just to applaud this appreciation of difference. We have to think about how this idea of difference is used and constructed. To illustrate, let me take two examples of 'the world as mosaic' as constructed outside of the confines of academic Human Geography. The first is a piece of racism I overheard at a party not long ago. I live in the East End of London, an area that now has a significant British–Bengali population. At the party a white man, who had grown up in the East End, was complaining about the number of 'Asian' families on the housing estate we were on (the usual racist rubbish about the smell of curries, and so on). 'I'm not a racist,' he then said, 'but they're taking over the place.' So he explained his hostility by saying it was not directed at people *per se*, but at people in the wrong place. People, then, who were upsetting his view of the world's mosaic, one in which white (mostly working-class) British people lived in the East End, and Bengalis lived in Bengal. The very idea of British, East End Bengalis did not fit. Now this, fairly typical, piece of white racism is not only nasty but nonsense. The East End has a long history of immigration, especially due to its past proximity to London's docks. Indeed, Britain as a whole is a country defined by waves of immigration and 'people taking over the place' (Romans, Normans, Saxons, Jutes, as well as hundreds of years of history of Black Britains). What this appeal to a geographical mosaic does, then, is to fossilize difference and then use that fossilization as part of a defensive localism. It uses a particular way of thinking of local–global relations – recent waves of immigration upsetting a previously neat human geographic mosaic – to legitimate social and spatial exclusion.

A second example helps to explain how such fossilized ways of imagining the local–global came into being. It is the 1908 Franco–British Exhibition held at White City, London; an 'exhibition' of the world, its peoples and places, set up for both education and entertainment, and aimed at a mass market. The late nineneeth and early twentieth centuries saw a host of enormously popular exhibitions or world fairs in

North America and Europe, which, as they displayed to visitors scenes, objects and people from around the world also functioned to codify distinct national identities and legitimate national projects of imperialism. Fig. 3.3 shows some of the pieces of the mosaic constructed for the exhibition at White City in 1908.

As Annie Coombes (1994) argues, to make sense of the Sengalese Village, the Algerian Pavilion, the Indian Palace and the Irish Village at White City, one first must recognize the logics underpinning the geographical mosaic on display – one of distinct peoples and places differentiated along ethnic and racial lines – and then contextualize these logics in the times and places of their production – in a Western Europe where social elites were looking to consolidate popular affiliations to both the **nation–state** and to colonial empires. The appreciation of difference fostered by this exhibition involved, then, an active emphasis on the importance and interlinking of **race** and nationality (as echoed, nearly 90 years later, at a party in Whitechapel, East London). The exhibition did not just display a world of differences, but through displaying was actively shaping them into particular forms. Those forms in turn have to be understood within a broader historical–geographical context, one where racism and national jingoism were crucial legitimations of the imperial status and ambitions of European powers (see Chapter 29).

So, whilst not without its merits – in particular its recognition of difference – the mosaic as a way of thinking about local–global relations is deeply problematic – not least because of how it recognizes difference. We need to think, then, about whether Human Geography can combine the local and the global in other ways.

Summary

- A very common way of imagining local–global relations is to envision a world of many different local places and peoples, each being a piece in a wider human geographic mosaic.

- This constructs the local as a bounded area, made distinctive through the character of life and land within it. It also tends to construct global-scale processes as destructive to that local diversity.

- There are factual problems with this way of framing local–global relations. For example, local differences are not destroyed by global level processes; in fact they are often produced through them.

- There are also political dangers attached to it, in particular an impulse towards defensiveness and the exclusion of non-locals.

- The mosaic is only one way of imagining local–global relations, so rather than seeing it as a simple portrait of geographical reality the reasons for, and effects of, its use need to be analysed.

Figure 3.3 Displaying a Human Geographical mosaic at the Franco–British Exhibition of 1908: the Algerian Pavilion, the Indian Palace, the Sengalese Village and the Irish Village. Source: Coombes 1994. Credit: Hammersmith and Fulham Archives and Local History Centre.

SYSTEM

One alternative way of thinking about local–global relations is to see local differences as produced by a global system. That is, the differences between places are not seen as a consequence of their internal qualities but as a result of their location within the wider world. Perhaps the best examples of this argument come from within development studies.

To put it crudely, one way of thinking about the extreme differences, and inequalities, in wealth and life chances between different parts of the world would be to identify internal characteristics that explain these differences. So, we could say (and many do) that Europe and North America are so wealthy because of the economic innovation they have shown since the Industrial Revolution. And then we might argue that the Philippines, say, are comparatively so poor because of their lack of natural resources, or an inhospitable climate, or some deficiencies in their culture (endemic corruption or laziness). What this kind of explanation ignores, though, is the fact that Europe and the Philippines are not just separate places but places with long histories of interconnections through world political, economic and cultural systems (systems dominated by capitalism) (see Blaut, 1993). It excludes the possibility, then, that Europe and the Philippines are so different because of their relationships with each other rather than because of their internal qualities. To put it bluntly, maybe we need to think rather more about whether Europe is rich precisely because the Philippines are poor. That is a very simplistic statement but it has its virtues. It sensitizes us to the idea that there are a set of global relations between local places. And to the fact that these global relations actively produce differences between places; that differentiation is a process not a static fact. As Jim Blaut puts it, in arguing against the idea of a special European character that has led to its relative economic success:

> Capitalism arose as a world-scale process: as a world system. Capitalism became concentrated in Europe because colonialism gave Europeans the power both to develop their own society and to prevent development from occurring elsewhere. It is this dynamic of development and underdevelopment which mainly explains the modern world.
>
> *(1993: 206)*

Other chapters will discuss the merits of this 'world-systems' approach in more depth (see, for example, Chapters 7, 9 and 17). However, for our purposes here, a more concrete example may help to show the importance, and limits, of a systemic view of local–global relations. That example is the world coconut market as portrayed by James Boyce (1992). Boyce notes two main things about global coconut trade in the period 1960–85: first, 'the Philippines is king' with over 50 per cent of world exports; and second, that the Filipino producers of coconuts do not seem to be doing very well out of this dominant market position. Understanding either of these facts requires a global systemic focus. The prevalence of coconut production in the Philippines would have to be traced back to Spanish colonization (for example a 1642 edict for all 'indios' to plant coconut trees to supply caulk and rigging for the colonizers' galleons), to demand in the nineteenth century from European and North American soap and margarine manufacturers, and to US colonial control and post-colonial patronage in the twentieth century (which led to preferential tariff rates for Filipino coconut products in the US market until 1974). It reflects, then, an emergent international system in which the Philippines was positioned, by external powers, as a supplier of an agricultural commodity, whilst those powers used that commodity for their own purposes (for their ships or their own manufacturing industries). Low rewards for this agricultural production reflect declining global terms of trade, such that each barrel of coconut oil exported in 1985 would buy only half the imports it would have in 1962 (see Table 3.1). The explanation for this decline is complex, but principally stems from the success of manufacturers of potential substitutes in the developed world – both ground nut oil producers and petro-chemical companies – at getting subsidies and protection from their governments, thereby depressing world prices for all traded fats and oils. That is, it is the political and economic power of developed world producers and governments that means that the Filipino coconut industry gets an ever worse deal for its efforts. The world trading system not only differentiates through an international division of industries (you grow coconuts, we have petro-chemicals); it discriminates.

However, as well as stressing the global relations that have stimulated Filipino coconut production and worsened its terms of trade, Boyce's study also suggests some limits to purely global explanations. In particular, he stresses how the local trading relationships within the Philippines meant that whilst the majority of small growers reaped little reward, vast fortunes were made by a few powerful individuals. Under the guise of concern for small producers, the Marcos regime reorganized the industry to concentrate power in the hands of a single entity which controlled raw material purchases from farmers and marketing at home and overseas. This concentration was in turn used to reward a few close political associates, such as 'coconut king' Eduardo Cojuangco

Table 3.1 Terms of trade for Philippine coconut oil, 1962–85

Year	(1972 = 100)		
	Coconut oil export price index	Import price index	Terms of trade
1962	119.9	71.4	167.9
1963	133.7	76.2	175.5
1964	145.4	76.8	189.3
1965	159.8	78.1	204.6
1966	134.2	79.4	169.0
1967	142.3	81.2	175.2
1968	166.9	80.7	206.8
1969	140.1	82.7	169.4
1970	160.4	93.5	171.6
1971	143.9	95.5	150.7
1972	100.0	100.0	100.0
1973	198.8	128.8	154.3
1974	508.1	211.6	240.1
1975	208.7	219.6	95.0
1976	192.4	217.2	88.6
1977	296.8	241.2	123.1
1978	338.7	245.8	137.8
1979	512.6	270.1	189.8
1980	342.6	358.6	95.5
1981	284.3	398.6	71.3
1982	241.5	340.5	70.9
1983	286.8	342.4	83.8
1984	547.2	386.7	141.5
1985	295.7	363.8	81.3

Source: Boyce 1992

and the Defence Minister Juan Ponce Enrile, who siphoned off much of the dwindling national earnings from coconut trade. Thus, existing inequalities in economic and political power within the Philippines allowed actions that made these inequalities greater still. Declining global terms of trade were experienced particularly severely, and responded to in particularly unproductive ways, because of the distinctive (if not unique) political system in the Philippines. Local processes, as well as global processes, played their part in the impoverishment of coconut producers. Any attempt to rectify that impoverishment would have to deal with local and global trading relations and the political–economic structurings of each.

Summary

- Differences between places are not just the result of their 'internal' characteristics. They are produced by systems of global relations between places.

- Human Geography should therefore do more than document diversity. It should investigate the *processes of differentiation* through which diversity and inequality are produced.

- These processes of differentiation operate at both global and local scales.

NETWORKS

So far we have been looking at how to think of the relations between global and local scales. In the idea of the mosaic, the global is portrayed as a collection of smaller locals. It is, paradoxically, both an arena within which those locals can be recognized (world awareness allows a comparison between places and alerts us to their differences) and the site of forces that can destroy local uniqueness (through the invasion of non-local things and people). In the model of the system, the global is portrayed as a set of relations through which local differences are produced. Here the emphasis is less on collection and comparison than on connection. In this final section I want to take the idea of connection a little further. I want to suggest that we can see both the local and the global as made up of networks, or sets of connections that any one local place has to a host of other places the world over. In consequence, we may need to view the local and the global not as different scales but as two ways of approaching these same social and spatial networks. Networks in which the local is global, and the global is local. In which, to use a horrible piece of jargon, our human geographies are irresolvably 'glocal'.

Some examples may make this less opaque. Let's start with the (global) local. In fact, let's start by having a cup of tea. Having a cup of tea (in fact several) is often seen as a very English thing to do, whether in the setting of the upper-class afternoon tea party or the more working-class family or workplace 'cuppa'. As Stuart Hall observes, 'this is the symbolization of English identity – I mean, what does anybody in the world know about an English person except that they can't get through the day without a cup of tea?' (1991: 49). And yet, of course, tea is not simply English:

> Because they don't grow it in Lancashire, you know. Not a single tea plantation exists within the United Kingdom . . . Where does it come from? Ceylon – Sri Lanka, India. That is the outside history that is inside the history of the English. There is no English history without that other history . . . People like me who came to England in the 1950s [from the West Indies] have been there for centuries; symbolically, we have been there for centuries . . . I am the sugar at the bottom of the English cup of tea. I am the

sweet tooth, the sugar plantations that rotted generations of English children's teeth. There are thousands of others . . . that are . . . the cup of tea itself.

(Hall 1991: 48–9)

Indeed, Hall's point is that it is not just the cup of tea, but Englishness as a whole that is not simply English. The cup of tea is symptomatic of a wider condition; that the history of English culture, economy and society can only be understood if one analyses England's global, colonial networks (to the West Indies for sugar, to India for tea, and so on). England as a locally distinctive place, with locally distinctive features (like the ritualistic cuppa), is forged through a global web of connections, both past and present.

Doreen Massey on the global–local geographies of Kilburn and Cambridgeshire

Take a walk down Kilburn High Road, my local shopping centre. It is a pretty ordinary place, north west of the centre of London. Under the railway bridge the newspaper stand sells papers from every county of what my neighbours, many of whom come from there, still often call the Irish Free State . . . Thread your way through the often stationary traffic . . . and there's a shop which as long as I can remember has displayed saris in the window . . . On the door a notice announces a forthcoming concert at Wembley Arena: Anand Miland presents Rekha, live, with Aamir Khan, Salman Khan, Jahi Chawla and Raveena Tandon . . . This is just the beginnings of a sketch from immediate impressions but a proper analysis could be done, of the links between Kilburn and the world . . . It is (or ought to be) impossible even to begin thinking about Kilburn High Road without bringing into play half the world and a considerable amount of British imperialist history.

(1991: 28)

Think of the [seemingly isolated] Cambridgeshire village. Quite apart from its more recent history, integrated into a rich agricultural trade, it stands in an area which in its ancient past has been invaded by Celts and Belgae, which was part of a Roman Empire which stretched from Hadrian's Wall to Carthage . . . The village church itself links this quiet place into a religion which had its birth in the Middle East, and arrived here via Rome.

(1995: 64)

Doreen Massey has made a similar argument but starting from a more tightly defined local scale. Contra the model of a mosaic of bounded local places, she looks at how the distinctiveness of an urban neighbourhood or a rural village is not threatened by connections to the wider world, but actually comes from them (Massey, 1995). In consequence, she says, we need 'a global sense of place' (Massey, 1994): both in order to understand local places; and in order to appreciate them in a way that does not slip into a reactionary, defensive parochialism.

If we think of the world in terms of networks, then, we see how local places gain their different characters through their distinctive patterns of links to other places. In turn, we begin to see how the global is less some neat, all-embracing system with a single logic, than a mass of globally extensive yet locally routed fibres of connection. Not only do we globalize the local, but we localize the global. Again an illustration may help. You may remember from earlier on that one major '**discourse**' through which a sense of the global has been forged is that of 'shrinkage' or 'compression', of the world getting smaller. One key element of this shrinkage is often said to be new communication technologies. The Internet could be seen as exemplary, allowing as it does 'virtual communities' to meet, not face to face but in **cyberspace** (see Chapter 34). However, if we actually look at the facts of the Internet, rather than its hype, what we see is enormously unequal local access to it (look, for example, at Fig. 3.5, an attempt to picture just who is actually on the Internet and where they are located).

The Internet is global, sure, but it is also very local. What is more, the Internet is obviously not the only sort of global network, and its electronic messages are not the only thing being globally circulated (see Appadurai, 1990). There are also global networks and flows of money (often in an electronic form, routed through the major International Financial Centres in New York, Hong Kong, Tokyo and London). Global networks and flows of people (migrants, tourists, business travellers, even geographers), each with their own rather different patterns of movement. Global networks of cultural products (like music; see Lipsitz, 1994). And so on. Many networks, many flows. Sometimes interconnected, but equally often with very different geographies: so Jamaica is at best low-rise on our Internet traffic map, but in terms of world musics would be a skyscraper; and Mexico doesn't seem to e-mail the US much, but migrants often make that journey. What we have then is a global realm comprised of multi-directional, multi-fibred networks, the geographies of which are not mappable onto neat territories or neat systems. To grasp them, geographers will have to get inside these networks, go with the flows and look to connect.

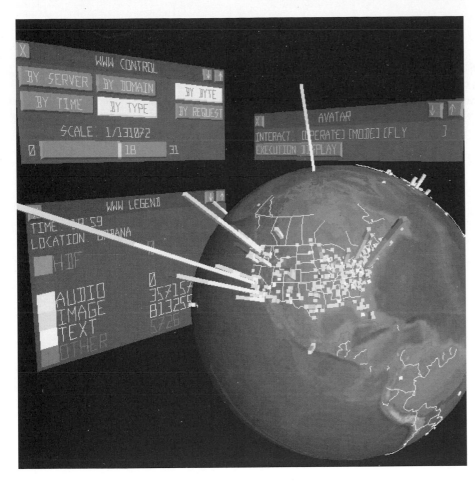

Figure 3.5 A global net? This is a printout from a VR simulation which shows the origin of all traffic destined for NCSA at Urbana – Champaign, the home of Web browsers Mosaic and Netscape. Note the skyscrapers in California and on the horizon in Europe, and the 'empty' spaces of much of Central and Latin America. Source: Lamm, Reed & Scullin 1996

Summary

- Local places get their distinctive characters from their past and present links to the rest of the world. In consequence, we need a 'global sense of the local'.

- All these global networks of links – with their flows of information, ideas, money, people and things – have locally routed geographies. In consequence, we need 'localized senses of the global'.

CONCLUSION

Human Geography is rightly interested in both the local – the specific place, with its distinctive qualities – and the global – the wider world, with its bigger picture. A crucial question that has always faced Human Geography is therefore how to conceptualize the relations between these two. Three general arguments have informed the discussion here. First, that appeals to ideas of diversity – a global collection of many locals – whilst commonsensical are deeply problematic: factually, politically and conceptually. Second, that rather than diversity the conceptual keystone of geographical work in this area should be 'differentiation', that is an

investigation of the ongoing productions of differences between peoples and places. Third, it is debatable (though quite rightly still debated) whether these processes of differentiation accord to a singular global logic (such as, 'developed countries make other countries underdeveloped as part of their own development'). Personally, I think tracing out the multiple networks that constitute both the local and the global offers a more convincing way of theorizing, and actually studying, local–global geographies.

Further reading

- Allen, J. (1995) Global worlds. In Allen, J. and Massey, D. (eds) *Geographical worlds*. Oxford: Oxford University Press, pp. 105–42.

Taking as its focus ideas about 'global compression', this student friendly reading deepens the discussion here on how to think of the global nature of contemporary human geographies.

- Blaut, J.M. (1993) The myth of the European miracle and After 1492, in *The colonizer's model of the world*. New York: The Guilford Press, pp. 50–151 and 179–213.

It is worth attempting a read of this for its powerful restatement of a world systemic approach. It is particularly strong on debunking the idea that European 'development' stems from qualities internal to Europe itself.

- Lipsitz, G. (1994) Kalfou Danjere. In *Dangerous crossroads: popular music, postmodernism and the poetics of place*. London: Verso, pp. 1–21.

A slightly unusual suggestion, written by a non-geographer, but this consideration of the geographies of 'world music' has some lovely examples of local – global networks of political and aesthetic affiliation.

- Massey, D. (1994; orig. 1991) A global sense of place. In *Space, place and gender*. Cambridge: Polity Press, pp. 146–56.

Already a classic article. There are lots of ideas here about globalizing the local and localizing the global but it is remarkably well written and really quite readable. It is also short!

- Peet, R. (1989) World capitalism and the destruction of regional cultures. In Johnston, R.J. and Taylor, P. (eds) *The world in crisis*. Second edition, Oxford: Blackwell, pp. 175–99.

This is one of the most explicit examples of an argument for seeing global level processes as destroying local distinctiveness. Written with real passion.

Structure–agency

Mark Goodwin

INTRODUCTION; STRUCTURE, AGENCY AND A NIGHT OUT

Think for a while about the last time you went for a night out. Think about where you went and what you did, think about who you were with and what they did. You probably chatted beforehand and decided to go to see a film or have a drink, or maybe go to a club. You decided who else to invite, and once you were there you decided what to drink and when to dance, and who to dance with. You decided when to go home and how to get home. You decided whether to get a taxi or take a bus. Filling in all these questions would give you an account of your last night out. It would also give you a very agency-centred view of your activities. It would place you at the centre of the night's decision-making, and would explain the events on the basis of your decisions. This view of events, in other words, would stress human agency by refering to the capabilities of particular human beings (in this case you and your friends) to pursue certain courses of action.

Now stand back for a bit and think through the evening again. In deciding where to go how much choice did you have? How many films were on in your local area, and had you seen some of them before? Did you want to see any of the others? Which clubs were open, and how much did they cost? Could you afford to go where you really wanted to, or did you have to settle for a slightly less expensive venue? Did the choice of venue for the evening depend on whether you are male or female, or whether you are black or white – are there certain places which are difficult for you to go, because of your gender or your race? Was the club gay or straight – did this affect where you went? Could you drink what you wanted

to, or was the selection available slightly limited? Could you dance with who you wanted when you wanted, and were the right records being played? Was there a bus home at the time you left the club or film, or had the last one just gone?

Introducing these issues into the picture of the evening raises the question of structure, and of the social contexts and constraints within which your actions were situated. These structures may be visible and material, like the numbers of cinemas or the time of the last bus, or they may be the sets of social and cultural rules which channel our behaviour in one way or another, like the 'choice' of fashionable clothes to wear or the appropriate way to dance. These structures may be borne by the actors themselves in terms of the position they occupy within broader sets of social relations – being a black woman or an unemployed man, for instance, or they may be imposed by others – a crowd dancing for instance, represents a structure to those stood watching. These issues are sketched out diagrammatically in Figure 4.1.

The actor in the top left of the diagram, the intentional agent, is you setting off for a night out. The structured social context in the top right are the sets of rules (usually immaterial) and resources (usually material) which operate to constrain your actions. This means that rather than being able to do anything you want to on that night out, you will be operating as a 'situated actor', located within the social context which defines your range of potential actions. Based on this range you then form an intention to act – deciding to go to this or that cinema, to see this or that film. Carrying out this intention by actually going to the film is understood as strategic action – doing something which is informed by your knowl-

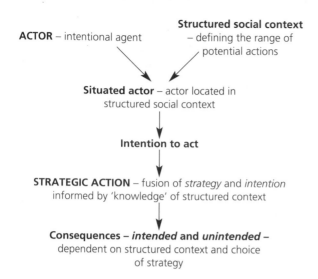

ACTOR – intentional agent

Structured social context – defining the range of potential actions

Situated actor – actor located in structured social context

Intention to act

STRATEGIC ACTION – fusion of *strategy* and *intention* informed by 'knowledge' of structured context

Consequences – *intended* and *unintended* – dependent on structured context and choice of strategy

Figure 4.1   Intention, strategy and action. Source: Hay, 1995 p.190

edge of the structured context. This may have a range of consequences, some intended, some unintended, which will then help to structure your future behaviour – such as bumping into an old friend in the cinema and inviting them to come for a drink.

The nub of the issue for the geographical researcher in all this lies in the explanatory weight given to either the structure or to the agency. In attempting to explain the creation of the human geographies which surround us all, do we stress agency or structure – or both? If you were attempting to understand the changing geographies of the leisure industry, for instance, which would you give primacy to in accounting for your night out? (*see* Fig. 4.2). These are enduring issues, which have troubled geographers to varying degrees ever since the foundation of the discipline Indeed, they have been problematic across most of the social sciences since the end of the nineteenth century, and are not just of concern to the

geographer. The puzzle over structure and agency was first set out succinctly by the German philosopher and economist, Karl Marx, when he wrote that whilst people 'make their own history' they do not do so 'just as they please nor under circumstances of their own choosing'. They also of course make their own geographies, as this book shows, but again they only do so under certain conditions and constraints imposed by social structures. The difficulty for those studying and seeking to explain these geographies is to tease out just how much of their making was contributed by human agency (people making their own geographies) and how much by broader social structures (the circumstances and conditions not of their choosing).

As ever, matters are a bit more complex than this and it is not enough simply to weigh up the relative contributions of structure and agency in causing any particular social process or event. This is because in practice it is remarkably difficult to separate the two. Indeed, most researchers would now accept that the broader structures are both the medium and the outcome of the practices which constitute social systems. In other words, we are operating within a set of rules and resources, yet our own actions created these. Moreover, structures are not to be conceived of purely as a barrier to action, but are essentially involved in helping to produce that action. Thus each element of structure and agency helps to create the other, and each provides the context through which the other works. This conception would see structure and agency as two sides of the same coin, rather than viewing them as two coins which occasionally bump up against one another (Hay, 1995: 197). However, for most of its history, human geography has steadfastly refused to see structure and agency as anything other than independent and analytically separate aspects of the social world, and its practitioners have usually chosen to stress one or the other in providing

Figure 4.2   Nightclub scene. Credit: Mirror Syndication International

accounts of human activity. We will briefly review some lessons which emerge from this history, before concluding with a summary of more recent research which views agency and structure as a duality in which each helps to constitute the other.

STRUCTURE AND AGENCY IN HUMAN GEOGRAPHY

One way of dealing with the competing claims of structure and agency is to focus exclusively on one or the other. This is basically what the discipline has done for most of its history, and for most of its past geography has tended to stress, almost uncritically, the role of structure. Perhaps because of its disciplinary beginnings in natural history, and later continuations in natural science, the human side to Human Geography was often absent. Instead there was a search for those broader processes which structured our geographies. Initially, an emphasis was placed on the structuring properties of climate and environment, in work which became known as environmental determinism. The contention in such work was that human activity, and hence human geography, was controlled (or 'determined') by the physical environment in which it took place. An example can be found in the writings of Griffith Taylor on the development of towns and cities in his 1949 book *Urban geography*. In this he sets out his basic tenet – that the key to the city is 'the dominant feature of the environment' (ibid.: 9) – before going on to look at the development of individual towns and cities. Figures 4.3 and 4.4 are graphs taken from his book which illustrate this search for the controlling factor of the

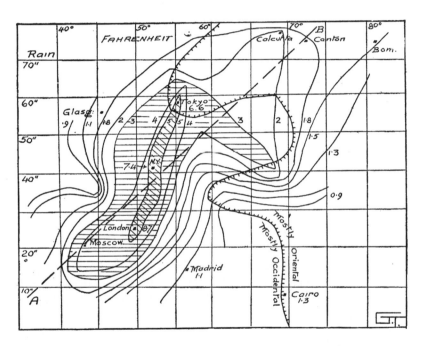

Figure 4.3 Cities, temperature and rainfall. A tentative isopract graph suggesting how great cities are controlled by temperature and rainfall. Figures on the isopleths indicate milions of city population. Source: Taylor, 1949 p.176

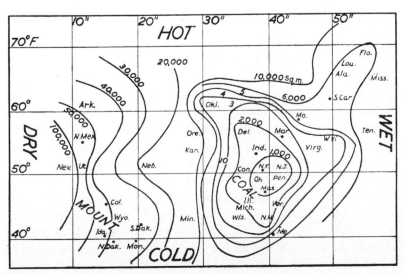

Figure 4.4 A tentative isopract graph showing how the distribution of towns (exceeding 17,000 folk) varies in the United States with climatic controls. Figures represent square miles allotted to each town. Source: Taylor, 1949 p.375

environment. Figure 4.3 'suggests how great cities are controlled by temperature and rainfall' (ibid.: 176), whilst Figure 4.4 shows 'how the distribution of towns varies in the United States with climatic controls' (ibid.: 375). Some mention is made in passing of economic and political matters, but in the main Taylor writes secure in the knowledge that 'No student of human geography will deny that in the broadest sense latitude is the variable which most controls human affairs' (ibid.: 15). Human agency is thus written out of the picture in favour of the structuring properties of the environment – and more specifically in this case, of latitude.

Although such **environmental determinism** lingered in human geography for a long time (my copy of Taylor's book was reprinted for the fourth time in 1964), the influence of latitude and temperature and rainfall on human activity was gradually downplayed. Yet this did little to reduce the emphasis on structure, and by the 1960s human geography had embraced a so-called 'scientific' approach, which had as its rationale the search for universally applicable laws of human behaviour. This work went under the label of **Spatial Science**, and adopted a positivist framework of enquiry. For the positivist, predictive and explanatory knowledge comes from the construction and testing of theories, which themselves consist of highly general statements expressing the regular relationships that are found in the world. In this process, scientific theories consist of universal statements whose truth or falsity can be assessed through observation and experiment (Keat and Urry, 1975). Those theories which were empirically verified would then assume the status of scientific laws. The full transfer of scientific method into human geography came in 1969 with the publication of Harvey's book *Explanation in geography* where he argued for the primacy of deductive theoretical forms of explanation. Under this form of explanation, the event or process to be explained is deduced from a set of initial or determining conditions and a set of general laws – given Law L, if initial conditions C and D are present, then event E will always occur. The event E is 'explained' by reference to the laws and the initial conditions.

Agency was again written out of the picture as people were reduced to little more than dots on a map or integers in an equation. We were all assumed to operate according to the same general laws – indeed, it was the very search for these laws which drove this entire approach. This kind of reasoning dominated human geography in the 1960s and most of the 1970s and generated the search for law-like statements of order and regularity which could be applied to spatial patterns and processes. Hence the succession of models which appeared in geography over this period – for

instance, Christaller's model of settlement hierarchy, Alonso's land use model, Zipf's rank size rule of urban populations, and Weber's model of industrial location. All were an attempt to use law-like statements in order to explain and predict the spatial outcomes of human activity.

One such model which human geographers used to explain patterns of flow between two or more centres was the so-called gravity model (*see* Fig. 4.5). This proposed that we can estimate the spatial interaction between two regions by multiplying together the mass of the two (equated conveniently with population size) and dividing it by some function of the distance separating them. The model was used to 'explain' all kinds of flows, from those of migration to passenger traffic, telephone conversations and commodity flows. Noticeable by their absence are any references to the actual motivations for the behaviour of the individuals who are migrating, or commuting, or speaking to each other on the phone, or purchasing the commodities. Human agency is given no space whatsoever, and people's behaviour is assumed to conform to a general pattern of behaviour – which itself is based on a model derived from a crude analogy with Newton's law of universal gravitation developed in 1687. Thus what was originally conceived as a way of accounting for the behaviour of distant bodies in the universe, was being used to explain a whole host of social, economic and cultural activites by reference to the two variables of population and distance.

Accounts which stressed human agency developed partly as a reaction to, and critique of, the determinism evident within spatial science. It was argued that the search by spatial science for cast-iron laws underlying human behaviour effectively dehumanized such behaviour, and robbed it of the very creativity, values and meanings which drive and sustain human activity (Cloke *et al.*, 1991: 69). Instead of the 'overly objective, narrow, mechanistic and deterministic view of [the human being] presented in much of the contemporary research', these critics advocated a '**humanistic** approach', studying 'the aspects of [people] which are most distinctively "human": meaning, value, goals and purposes' (Entrikin, 1976, 616, quoted in Cloke et al, ibid). As Gregory puts it in the *Dictionary of human geography*, such an approach is 'distinguished by the central and active role it gives to human awareness and human agency, human consciousness and creativity' (Johnston *et al.*, 1994: 263).

At last human agency had been placed centre stage, and the intentional human actor became the key focus of research work. A good example of the way such approaches informed research can be found in the work of Ley (1974, 1977) who sought to uncover

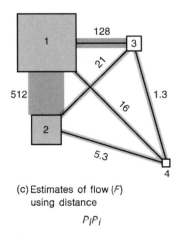

(a) Measure of population (P)

$P_1 = 64$
$P_3 = 4$
$P_2 = 16$
$P_4 = 1$

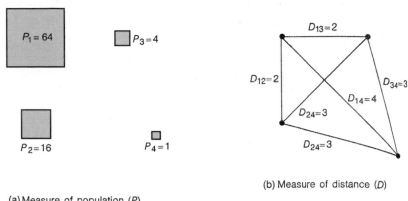

(b) Measure of distance (D)

$D_{13}=2$
$D_{12}=2$
$D_{34}=3$
$D_{14}=4$
$D_{24}=3$
$D_{24}=3$

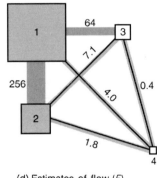

(c) Estimates of flow (F)
using distance

$P_i P_i$

128
21
512
1.3
16
5.3
1 3 2 4

(d) Estimates of flow (F)
using distance squared

$P_i P_i$

64
7.1
256
0.4
4.0
1.8
1 3 2 4

Figure 4.5 Elements in the estimation of flows between centres by using gravity-model formulation. (a) Population size and (b) distance elements are combined to estimate spatial interaction between four centres by the gravity model. Note that (c) the use of distance and (d) the use of the square of distance yield different estimates of flow. Source: Haggett, 1972 p.330

the usually 'taken-for-granted' meanings which informed the actions of street gangs in inner-city Philadelphia. In particular he argued that place was critical in shaping these meanings, and that to speak of a place is not to speak of an object alone, but is also to speak of an image and an intent. Thus in opposition to the formulations of the **gravity model** discussed above, he points out that

> the distant metropolis is never perceived in the perfect material terms that the gravity model with its economic determinism would have us believe. The metropolis has a meaning, it is . . . a state of mind, and it is always this meaning for the subject that precedes action: creative decision-making is not pre-empted by a mechanistic gravity field.

> *(1977: 507)*

He went on to show how the most mundane and everyday features of the urban environment can represent local societal values, as symbolized in the use of graffiti to help mark and form territorial space. As Ley puts it (1974: 218–19)

graffiti markings represent the language of space for members of the street gang culture. Where territories meet, space is most highly contested, and aggressive behaviour is most appropriate. With increasing proximity to the core of the turf, the meaning of space changes and there is an orderly decrease in assertive behaviour, until at the core security is perceived to be maximal. In this zone, where threat is regarded by gang members as unlikely, assertive behaviour against rivals becomes unnecessary, and graffiti obscenities are almost absent.

Through such research Ley sought to understand the everyday social world of black teenagers living in inner-city Philadelphia, and to explore what this social world means for these actors and what they meant by acting within it. Within such an approach graffiti is not just part of the garbage of inner-city dereliction, but instead becomes a meaningful part of the lives of the graffiti artists for whom it carries and expresses a whole set of shared values and shared meanings (*see* Fig. 4.6). However, even Ley with his emphasis on human agency, recognizes that the social

Figure 4.6 Mural with a message, West Houston Street, New York. Credit: Telegraph Colour Library

group (in this case the graffiti artists) is not autonomous in its decision-making. As he puts it 'for some men, the macro-social structure does not permit a wide range of action' (1977: 505). In work with Cybriwsky he exemplifies this through interviews with the graffiti artists. In the words of one,

> There isn't much choice of what to do . . . I did it because there was nothing else. I wasn't goin' to get involved with no gangs or shoot no dope, so I started writin' on buses. I just started with a magic marker an' worked up.
>
> *(Ley and Cybriwsky, 1974: 495)*

As Ley concludes (1977: 505), 'each individual has a history and geography which imposes constraints within his lifeworld: so begins the dialectic between creativity and determinism, charisma and institution' – and we might add so begins the dialectic between structure and agency.

Summary

- Accounts of human behaviour which stress the importance of wider structures have predominated within human geography for much of its history. This is the case whether these structures refer to the physical environment or to sets of abstract laws.

- Partly in opposition to these abstract accounts, a more humanistic geography developed during the 1970s which stressed the role of human agency and human consciousness in accounting for social behaviour.

BEYOND THE TWIN POLES OF STRUCTURE AND AGENCY

In recent years geographers have attempted to move beyond the twin poles of structure and agency in an effort to find and hold onto both the human agent and the social structure. Initial attempts to do so were made using the influential theory of **structuration**, developed by the sociologist Antony Giddens. Giddens takes seriously the notion of structure, but specifies it further in terms of sets of rules and resources. Social rules are those elements of interaction which individuals and institutions routinely implement, not as a set of codified laws but rather as a set of generalized encounters through which appropriate forms of behaviour are regularly followed and therefore reproduced. As well as these rules, social structures are also made up of resources which denote the power and influence which certain forms of authority and property are able to exert in order to control and constrain certain social interactions. The critical point, however, is that people are not seen as passive as these rules and resources structure their day-to-day activity. Indeed, they are seen as agents whose behaviour implements, reproduces and changes the rules and resources within which they operate. Thus, as the everyday social practices of human agents are produced through social interaction, so are the wider social structures reproduced. In this way structures enable as well as constrain behaviour, but behaviour can potentially influence and reconstitute structure (see Cloke *et al.*, 1991: 93–105).

What especially appealed to geographers about this formulation is that it is in and through specific spatial arenas that these engagements or interactions between the individual human agent and social structures are played out. As Thrift puts it:

> The region, initially at least, must not be seen as a *place* . . . Rather it must be seen as made up of a number of different but connected *settings for interaction* . . . Any region provides the opportunity for action and the constraints upon action.
>
> *(1996: 81, original emphasis)*

For Thrift then, the region, or locale, becomes conceptualized as a space where structure and agency interact – as literally the setting for structuration. Gregory (1982) takes these ideas further in his study of the historical transformation of the Yorkshire woollen industry. In this he presented an account of the continuous 'structuration' of this particular economic space, by analysing the interconnections between structure and agency as they were played out at a local, regional and a national scale. He looks at the ways in which the day-to-day flow of practical life was connected to broader changes in economy and politics through the outcomes of various situated social practices, and in this way he sets up the transformation from domestic to factory production as both the medium for, and the outcome of, these practices.

Despite these promising indications that structuration theory could be geographically sensitized to produce empirical accounts of the interaction between structure and agency, some human geographers have more recently attempted to move beyond the notions of structure and agency altogether rather than examine their interaction. As they point out, to examine their interaction requires one to work with the very objects one is seeking to transcend. Two influential steps along this route to transcendence have been taken by Pile (1993, 1996) and Thrift (1996). Their work, both separately and together (1995), provides an initial indication of the types of analyses that can be undertaken if we do wish to move beyond structure and agency rather than combine them. Thrift calls for an engagement with what he terms 'theories of practice' (1996: 6), those which see the human subject as primarily derived in practice. Such derivations emphasise the ways in which social agency is constructed *in* various sets of social processes, rather than being assumed to be a property *of* them. They also point to this process of construction as one that requires constant effort and adjustment, and as one where things other than human agency – for instance, texts, technologies, understandings, machines – are important (ibid.: 23–6). In such accounts 'understanding is not so much about unearthing something of which we might previously have been ignorant', whether this was to do with human agency or social structure, but instead is 'about discovering the options people have as to how to live' (ibid.: 8). Here structure and agency are dispensed with in favour of an emphasis on practice.

They are also dispensed with in Pile's account of the way 'the self' has been addressed in Human Geography. He concludes that for all their stress on human agency, humanistic geographies fail to problematize the construction of the human subject upon which they base their research. In other words, they work with relatively untroubled accounts of the human agent. Instead of this, Pile proposes that geographers should use **psychoanalytic** understandings of the human subject which emphasise the fragmented and partially 'unknowing' nature of the self. In this way, he contends, we will be able to re-conceptualize the relationship between the social and the self in ways which move beyond the 'cul-de-sac of the structure–agency dichotomy' and instead examine the 'intricate, dynamic and power-ridden relations between them' (1993: 137). For Pile, psychoanalysis does not provide the crossing point between structure and agency, or even a way of dissolving the distinction between them. Instead it can act as a means of

> re-interpreting both 'agency' and 'structure' in terms of each other . . . it can reveal the intricate inter-relationships between the personal and the social, which . . . provide the axes around which the self is organised, at the point of contact between the individual and the collective; it steps beyond the analysis of structure–agency.
>
> *(ibid.: 132)*

Summary

- Human Geographers have sought to move beyond the dichotomy of agency and structure, initially by adopting and developing structuration theory.

- More recently they have attempted to develop theories which move beyond the twin poles of structure and agency, rather than unify them. Indicative of these attempts are theories of practice and theories which stress the role of psychoanalysis.

CONCLUSIONS

We have then almost come full circle. From a position where Human Geography uncritically refused to engage with the dichotomy of structure and agency, preferring instead to emphasize one or the other, we have now reached a situation where researchers are advocating that we step beyond it altogether. As we have seen, however, in order to reach this stage, geography had to engage seriously with research elsewhere in the social sciences which not only problematized the dichotomy between structure and agency, but formalized its unification through the theory of structuration. One lesson shines through this story – whatever position the individual geographer might take in all this, it is nigh on impossible to neglect the twin concepts of structure and agency. Indeed, in common with the other social sciences, Human Geography has long grappled with their opposition. Sometimes it has resolved the problem by focusing on structure, sometimes it has focused on agency, sometimes it has attempted to unify the opposition and sometimes it has sought to step beyond it. What it has not been able to do for a long time is pretend that the opposition doesn't matter and that it is of little relevance to the discipline. On the contrary, all the signs are that a concern with structure and agency will continue to drive some of the more influential debates within the subject for a long time. Theories of practice and psychoanalysis may suggest directions in which these debates might move, but they will be far from the final words on the matter. In this area, as in so many others within human geography, the debates are both vibrant and exciting and promise to continue to be so.

Further reading

• Livingstone, D. (1992) *The geographical tradition*. Oxford: Blackwell.

This is an excellent book covering the history of geography as a socially constructed and contested enterprise. It contains very readable chapters on environmental determinism and spatial science, and smaller sections on humanism and structuration.

• Cloke, P. *et al.* (1991) *Approaching human geography*. London: Paul Chapman Publishing.

This contains very detailed accounts of both humanistic geography and structuration with smaller sections on spatial science and environmental determinism. The two chapters on humanism and structuration are still the best surveys of these approaches.

• Hay, C. (1995) Structure and agency. In Marsh, D. and Stoker, G. (eds) (1995) *Theory and methods in political science*. Basingstoke: Macmillan.

This is a very readable account of the problem of handling both structure and agency, and gives a broader social science view of the issue. It shows that the opposition between the two troubles other disciplines as well.

Self–other

Paul Cloke

INTRODUCTION: SELF-CENTRED GEOGRAPHIES?

Some people say that you should not judge a book by its cover. However, it is often interesting to pause and reflect on why books, organizations, or in this case subjects such as geography are represented by particular 'cover' images. Figure 5.1 shows the cover of the 1994 Annual Report of the Royal Geographical

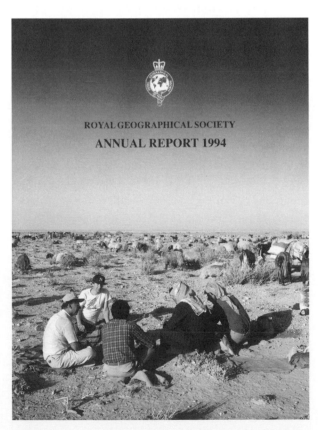

Figure 5.1   Annual report of the Royal Geographical Society, 1994. Credit: Chris Caldicott

Society, which is the organization representing academic and non-academic geographers in Britain. The image was designed to show geography in a positive light, as a subject which causes adventurous individuals to embark on exciting expeditions of learning in which they can discover the secrets of far-flung places and understand the lives of exotically different people. It is the 'us here' subjecting the 'them there' to serious geographical scrutiny.

This image, however, unintentionally poses other questions about 'us' and 'them'. The 'us' might suggest that Human Geographers can somehow be categorized as a homogeneous group of people, studying our geography in a somewhat standardized way – a bizarre supposition on a number of counts not least the 'maleness' of the encounter which is represented. The 'them' seems to have been selected on the grounds of their exotic difference to us. They too are in danger of being stereotyped. The strangeness of the place along with differences in skin colour, language, dress and 'culture' seem to be sufficient to mark out an appropriately 'other' subject of study. 'Us' encountering 'them' is on our terms. Exotic difference is defined by our mapping out of people and places in the world, and our assumptions about what is, and what is not, a normal view of life.

Perhaps these questions read too much from one particular image, especially since the RGS/IBG has subsequently sought to rectify in its output any previous perceptions of social or cultural insensitivity. However, these questions do reflect some of the most important themes to have arisen in Human Geography over recent years. The first is a highlighting and questioning of the geographical self. Not so many years ago, Human Geographers were taught to be 'objective' in their studies, so that anyone else tackling the same subject would come up with the same results. They were, in effect, being positioned as some kind of scientific automaton whose background, identity, experience, personality and worldview needed to be subjugated to the need for objectivity. The 'I' was personal pronoun *non grata* when it came to doing geography. However, the *self* does matter, and does influence the geography we practise. We do have different place- and people-experiences, different political and spiritual worldviews, different aspects to our identity and nature, and all of these factors will influence how we see the world, why our geographical imaginations are fired up by particular issues, and ultimately, what and how we choose to study.

The danger of *not* acknowledging and reflecting on the self is not only that we can unknowingly buy into other people's orthodoxies, but also that we can assume that everyone sees the same world as we do. We can, thereby, impose our 'sameness' onto others. The second set of questions, then, concerns a recognition of how we deal with 'others'. It is extraordinarily difficult sometimes to do anything but see things from our own perspective, however hard we try to escape from our self-centred geographies. Yet as soon as we move beyond the samenesses of self, we immediately begin to stylize and stereotype the differences of 'the other'. This has been the subject of much recent discussion across the range of human sciences, including Human Geography, under the rubric of debates on '**Otherness**' and 'Othering':

> as soon as we start to think about people who are not ourselves, we lapse into the language of 'Othering' and, as one urges oneself to consider 'Others' or to see the 'Other' side of the question, those who are not like 'me' can start to slide into a homogeneous mass of difference from 'me', *essentially* the same as each other. This is just as arrogant as the assumption that 'they' are essentially the same as 'me'. It is implicit in the terminology that the self is taken as prior. The Other is secondary, and the best that one can do for an other is to extend a liberal tolerance, a condescension flowing from a benign superiority.
>
> (Shurmer-Smith and Hannam, 1994: 89)

Dealing with others, and with questions of difference, is therefore fraught with difficulty, and requires considerable reflection rather than an uncritical assumption about the existence and obvious nature of otherness. The French anthropologist, Marc Augé, has suggested that we need to adopt a two-pronged approach to understanding otherness. First, we should seek *a sense for the other*. In the same way that we have a sense of direction, or family, or rhythm, he argues that we have a sense of otherness, and he sees this sense both disappearing and becoming more acute. It is being lost as our tolerance for others – for difference – disappears. Yet it is becoming more acute as that very intolerance itself creates and structures othernesses such as nationalism, regionalism and 'ethnic cleansing', which involve 'a kind of uncontrolled heating up of the processes that generate otherness' (Augé, 1998: xv). Second, we should seek *a sense of the other*, or a sense of what has meaning for others; that which they elaborate upon. This involves listening to 'other' voices and looking through 'other' windows onto the world so as to understand some of the social meanings which are instituted among and lived out by people within particular social or identity groups.

As the remainder of the chapter suggests, the attempt to inculcate the curricula and research of Human Geography with senses of the other, and with reflections on the self, has proved to be a complex and politicized process. Perhaps this reflects less the novelty of the ideas being worked with, than the way they speak to and critique an absolutely central concern of Human Geography: developing knowledge of people and places beyond those one already knows. This chapter argues that critique is worthwhile, and therefore discusses some of the delights, as well as difficulties, of bringing explicit reflections of self and other into our human geographies.

SELF-REFLECTIONS

In many ways, '**reflexivity**' has become one of the most significant passwords in Human Geography over recent years. To reflect on the self in relation to space and society has been seen as a key with which to open up new kinds of human geographies which relate to individuals more closely, and which individuals can relate to more closely. In particular, reflexivity has been used by **feminist** and **post-colonial** geographers in their respective political projects to persuade human geographers to reflect something other than male, white orthodoxies. A poem by Clare Madge urges geography to connect 'in here' rather than 'out there' by becoming a subject 'on my terms and in my terms'.

Clare Madge

An Ode to Geography

Geography,
What are you?
What makes you?
Whose knowledge do you represent?
Whose 'reality' do you reflect?

Geography,
You are not just space 'out there'
To be explored, mined, colonised.
You are also space 'in here'
The space within and between

That binds and defines and differentiates us as people.

Geography,
I want you to become a subject
On my terms and in my terms,
Delighting and exploring
The subtleties and inconsistencies
Of the world in which we live.

The world of pale moonlight and swaying trees in a
    bluebell wood.
The world of sand and bone and purple terror.
The world of bright lights flying past factory, iron and
    engine.
The world of jasmine scents and delicate breeze.
The world of subversion, ambiguity and resistance.
The world of head proud, shoulders defiant under the
    gaze of cold eyes laying bare the insecurity underlying
    prejudice.
The world of music, laughter and light,
Of torment and exploding violence
Of tar and steel strewn with hate
While the moon gently observes and heals.

Geography, could you be my world?
Will you ever have the words, concepts and theories
To encapsulate
The precarious, exhilarating, exquisite, unequal world in
    which we live?
I believe so.
By looking within and without, upside down and inside
    out,
Come alive geography, come alive!

(*Source: Women and Geography Study Group, 1997*)

Her frustration with the subject is echoed in the book *Feminist geographies* (Women and Geography Study Group, 1997) where the writing (usually by men) in geography is critiqued, but the problems of proposing alternative forms of writing (usually by women) are starkly acknowledged:

> Much academic writing . . . is characterised by a dispassionate, distant, disembodied narrative voice, one which is devoid of emotion and dislocated from the personal. In contrast to this, writing which is personal, emotional, angry or explicitly embodied is implicitly (and often explicitly) portrayed as its antithesis: something which (maybe) has a place in the world of fiction and/or creative writing, but which, quite definitely, is out of place in the academic world . . . to be masculine often means not to be emotional or passionate, not to be explicit about your values, your background, your own felt experiences. Feminist academics wishing to challenge those exclusions from the written voice of Geography find themselves in a dilemma, however, for if academic masculinity is dispassionately rational and neutral, writing which is overly emotional or explicitly coming from a particular personalised position is

often dismissed as irrational, as too emotional as too personal – as too feminine, in other words. Thus feminists who want to assert the importance of the emotional in their work, or feminists who want to acknowledge the personal particularities of their analysis, run the risk of being read as incapable of rational writing, of merely being emotional women whose work cannot be universally relevant.

(*Women and Geography Study Group, 1997: 23*)

It has therefore been important for Human Geographers not only to *theorize* the self in new ways, but also to position the self appropriately in the *practising* of human geography, such that knowledge is situated in the conscious and subconscious subjectivities of both the author/researcher and the subjects of writing and research. In terms of new ways of theorising the self, Steve Pile and Nigel Thrift (1995) 'discuss' four interconnected ideas which map out the territory of the human subject:

1. *The body*: which orders our access to and mobility in spaces and places; which interfaces with technology and machinery; which encapsulates our experiences of the world around us; which harbours unconscious desires, vulnerabilities, alienations and fragmented aspects of self, as well as expressions of sexuality and gender; and which is a site of cultural consumption where choices of food and clothing and jewellery, for example, will inscribe meanings about the person.

2. *The self*: which can be understood in a variety of ways, ranging from a personal identity formed by an ongoing series of experiences and relationships, but where there is no distinctive characteristic in these experiences and relationships to suppose an inner, fixed personality, to a personal identity in which self-awareness serves to characterize each experience as belonging to a distinct self.

3. *The person*: which is a description of the cultural framework of the self, and allows for different selves in different frameworks. For example, if you were born and brought up in Rwanda, or Albania, or Cuba, your person would reflect the cultual frameworks of life in those places.

4. *Identity*: where the person is located within social structures with which they identify. Traditionally this would have been seen to involve rigid structures such as class and family, but more recently identities have tended to be constructed reflexively and therefore often flexibly leading to new identity issues, for example focusing on alternative sexualities, ethnicities or resistance to local change.

The subject is therefore 'in some ways detachable, reversible and changeable', while in other ways it is 'fixed, solid and dependable'. It is certainly 'located in, with and by power, knowledge and social relationships' (1995: 12).

Some of these theoretical distinctions may at first be difficult to grasp, but they do serve to emphasize just how difficult it actually is to be reflexive about the self in our human geography. To what extent is it possible to know and to reflect on our selves, to appreciate fully how, precisely, the self is responsible for how we think, how our imaginations are prompted, how we interpret places, people and events, and so on? How much more difficult is it to understand the selves of others whom we might wish to study? These practices, which I have identified earlier as being important political and personal projects in Human Geography, are perhaps more difficult than they first appear. The multidimensionality of the body, the relationally dependent and often subconscious nature of the self, the culturally framed (and therefore flexible) person, and the changeable and overlapping influences of identity render reflexivity a most complex, and some would say impossible, task.

Nevertheless, the breaking down of detached and personally irrelevant orthodoxes in Human Geography has remained a task which many continue to consider sufficiently worthwhile to warrant attempts to bring reflexivity into a prominent position in the practice of Human Geography. Three interconnected and often overlapping strategies are briefly outlined here. First, a strategy of **positionality** can be identified in which 'telling where you are coming from' can be employed tactically as a contextualization of the interpretations which are to follow. Sometimes this involves the identification of key political aspects of the self, for example, a **feminist** positioning, which will self-evidently influence what occurs subsequently and which provide us with new positions from which to speak. On other occasions, particular spatial or social experiences will be described which are used to claim expertise or insight into particular situations. Take, for example, George Carney's autobiographical preface to *Baseball, barns and bluegrass*, his book on the geography of American folklife. Here, he describes his childhood in the foothills of the Ozarks, and how the folk knowledge accumulated during that time has been translated into a scholarly pursuit of cultural tactics of American folk-life more generally. Not only does his folk heritage equip him for this work, but it also punctuates what he writes and how he writes it.

George Carney's autobiographical preface

The first eighteen years of my life were spent on a 320-acre farm in Deer Creek Township, Henry County, Missouri, some six miles south of Calhoun (population 350), ten miles northwest of Tightwad (population 50), and five miles west of Thrush (population 4). My parents, Josh and Aubertine, inherited the acreage and farmstead buildings from my grandpa and grandma Carney, who retired and moved to Calhoun. The eighty acres to the north of the farmstead consisted of hardwood timber (walnut, hickory, and oak), Minor Creek, which flowed in an easterly direction as a tributary to Tebo Creek, and some patches of grazing land. The remaining 240 acres, south of the farmstead, were relatively rich farmland where my Dad planted and harvested a variety of crops ranging from corn and soybeans to alfalfa and oats. Classified as a diversified farmer, he also raised beef and dairy cattle, hogs, sheep, and chickens. Thus, my roots lay in a rural, agrarian way of life in the foothills of the Ozarks.

My early years fit the description that is often used to define the *folk* – a rural people who live a simple way of life, largely unaffected by changes in society, and who retain traditional customs and beliefs developed within a strong family structure. I was experiencing the *folklife* of the Ozarks. Folklife includes objects that we can see and touch (tangible items), such as food (Mom's home-made yeast rolls) and buildings (Dad's smokehouse). It also consists of other traditions that we cannot see or touch (intangible elements), such as beliefs and customs (Grandpa Whitlow's chaw of tobacco poultice used to ease the pain of a honeybee sting). Both aspects of folklife, often referred to as material and nonmaterial culture, are learned orally as they are passed down from one generation to the next – such as Grandpa Carney teaching me to use a broad axe – or they may be learned from a friend or neighbor – for example, Everett Monday, a neighbor, instructing me on the techniques of playing a harmonica.

Through this oral process, I learned many of the traditional ways from the folk who surrounded my everyday life – parents, relatives, friends, neighbors, teachers, preachers, and merchants. The most vivid memories associated with my early life among the Ozark folk are the six folklife traits selected for this anthology – architecture, food and drink, religion, music, sports, and medicine.

Since leaving the Ozarks for the Oklahoma plains some thirty-five years ago, I have developed a greater awareness and deeper appreciation for American folklife and all its spatial manifestations. My teaching and research interests have been strongly influenced by those folk experiences of yesteryear. Students in my introductory culture geography classes are annually given a heavy dose of lectures and slides on the folklife traits covered in this reader. My research has increasingly focused on two of these traits – music and architecture. Clearly, my roots have made a lasting impression – one that I have converted into a scholarly pursuit.

Carney (1998: xv–xxii)

Second, a more radical strategy of **autoethnography** involves interpreting people, places and events through the perspective of your own involvement. An influential figure here has been Elspeth Probyn, a sociologist from Montreal, whose use of autoethnography is carefully and critically reviewed in her book called *Sexing the self* (1993). Probyn focuses on some very personalized passages in her own life in order to discuss gendered positions in cultural studies:

> My project here is to rethink what the self might be in and for feminism. I want to reconstitute its force and reveal the material forces behind its motion. In previous articles I have implicitly worked through my own particular self in relation to distinct concerns: anorexia; the gendering of the local; the death of my mother; and to body in general. Yet something central always seemed to evade me. It somehow seemed unseemly to speak of what drew me to these subjects: as I described these matters, I made them into objects separated from myself . . . However, my idiosyncrasies aside, I must state that fundamentally I am drawn to these subjects both near and dear because I am convinced that gender must be represented as processes that proceed through experience.

(1994: 3)

Autoethnography opens up intriguing possibilities for studying, for example, our gender, race/ethnicity, sexuality, sense of place, and also our work, leisure, tourism, and other activity geographies through the medium of our personal involvement. At one and the same time, there are opportunities to practise the geographies which we understand most – our own – and yet dangers of becoming so self-obsessed that nobody else's geographies matter. Even with autoethnography, then, there are strong arguments for including 'other' voices in our own stories.

A third strategy therefore, is to acknowledge *intertextuality* in our practice of human geography, by finding ways of recognizing the significance of our selves as important influences which shape our geographies, whilst at the same time seeking to listen to other voices. The texts which result from such encounters are complex dialogues. The Human Geographer will shape the conversation, both by the individuality of their own subject-experience and by the questions which are asked of the 'other'. In turn, other individuals will have different, changing and even competing experiences, and will represent themselves differently to different people. The 'results' of the encounter will usually be 'interpreted' by the Human Geographer in the light of their self-positioning. This may involve a process of 'finding new places to speak from', and bringing them into the conversation, or it may involve a *tactic* of 'letting people speak for themselves' and seeking for a plurality of voices (a 'polyphony') to emerge. Interpretations are then usually written down, often using quoted extracts of other voices, but almost always with the author in control, exerting power over what is included and excluded, what is contextualized and how, and what story-lines are used to shape the narrative of the 'findings'. In all these processes and practices, the need to recognise the interconnections between the powerful self and the 'subjected to' other is paramount.

The increasing use of ethnographic strategies and qualitative methods in human geography (see Cloke *et al.*, 2000) has certainly helped to provide research practices with which we can be more reflexive about our selves, and the relationships between our selves and others. In the end, however, we have to realize just how 'easy' it can become to think and write about ourselves, and how difficult it is to know enough about our selves to be reflexive in our geographies. Delvings into psychoanalysis (Sibley, 1995) have began to help our understandings here but there still seems to be an inbuilt desire to empower the self over the other, however much a many-voiced, polyphonic geography is being aimed at. In the more general context of the problems in the world, such preoccupations with the self might be regarded as inappropriate if not positively dangerous!

Summary

- Reflexivity – reflecting on the self in relation to society and space – is an essential process in recognizing how our individualities contribute to all aspects of our practice of Human Geography. It also gives us grounds on which to challenge seemingly 'orthodox' geographies and to make our human geography more relevant to us and to others.

- The difficulties involved in understanding the self are often underestimated. The human subject is a complex mix of body, self, person and identity and, for some, spirit and soul will also be important considerations.

- There is an interconnected range of strategies by which the self can consciously be included in the practice of human geography.

- The dangers of exaggerating the self in our reading, thinking, researching and writing about Human Geography are very real, and can divert us from important issues relating to others.

SENSING THE OTHER

There is currently what Chris Philo (1997: 22) calls 'an exciting swirl of interest' in Human Geography about the need to take serious notice of different kinds of people who are situated in different kinds of spaces and places, and who experience, mould and negotiate these spaces and places in a different way to ourselves. This interest in the differences of the other has implications for the ways in which we conceptualize and practise our human geographies, and also for the ways in which these geographies are politicized. Dealing with the 'other' is of course linked to dealing with the 'self'. To reiterate, the arrogance of the self is often manifest in an assumption that others must see the world in the same way as we do. Alternatively, we will often place ourselves in the centre of some 'mainstream' identity which is defined not only around our self-characteristics but also in opposition to others who are not the same as us. Think, for example, about the way white people often assume that only 'non-white' people have an ethnicity, and find their own whiteness unremarkable. As Philo further suggests, then, we are often 'locked into the thought-prison of "the same"' (ibid.: 22) which makes it impossible for us to appreciate the workings of the other. Indeed we will often seek either to *incorporate* the other into our sameness, or to *exclude* the other from our sameness, in order to cope with the threat that difference seems to present to the perceived mainstream nature of our identity (see Sibley, 1995). Both incorporation, and exclusion are highly political acts which trap the other in the logic of the same.

The interest in recognizing 'other' human geographies focuses attention not only on that which is remote to us, but also should make us re-think what is close to home. Two examples serve here to highlight some of the principal themes in the recognition of otherness in proximal and remote situations. The first relates to the neglect of 'other' geographies close to home and focuses on rural geographies (see Chapter 27) although the principles involved relate to a wide range of human geography contexts. Philo's (1992) review of 'other' rural geographies emphasized that most accounts of rural life have viewed the mainstream interconnections between culture and rurality through the lens of typically white, male, middle-class narratives:

> there remains a danger of portraying British rural people ... as all being 'Mr Averages', as being men in employment, earning enough to live, white and probably English, straight and somehow without sexuality, able in body and sound in mind, and devoid of any other quirks of (say) religious belief or political affiliation.
>
> *(1992: 200)*

Such a list is important in its highlighting of neglect for others, but also runs the risk of immediately producing a formulaic view of what is other. Thus, we can recognize that individuals and groups of people can be marginalized from a sense of belonging to, and in, the rural on the grounds of their gender, age, class, sexuality, disability, and so on. However, as David Bell and Gill Valentine (1995) remind us, the mere listing of socio-cultural variables represents neither a commitment to deal seriously with the issues involved nor a complete sense of the *range* of other geographies. Indeed, our very recognition of *these others* serves to 'other' *different others* and exclude them from view.

A specific illustration within this rural context is offered in Chapter 2, which presents a well-known self-portrait by the photographer Ingrid Pollard (*see* Figure 2.5, and also Kinsman, 1995). Her autobiographical notes suggest that the photograph is a self-aware comment on race, representation and the British landscape. She sets herself in the countryside, and through juxtaposing her identity as a 'black photographer' with the cultural construction of landscape and rurality as an idyll-ized space of white heartland, she graphically expresses a sense of her own unease, dread, non-belonging – of other. The black presence in 'our' green and pleasant land says much about whiteness = sameness in this content. However, as the Women and Geography Study Group (1997) point out, the otherness in this representation is by no means a uni-dimensional matter of race. They suggest that:

> Pollard is claiming a different position from which to look at and enjoy English landscapes (albeit an uneasy pleasure); a right to be there and a right to be represented and make representations. She challenges, disrupts and complicates the notion of a generalisable set of shared ideas about England and the implicitly white and masculinised position from which it is usually viewed.
>
> *(1997: 185–6)*

Ingrid Pollard the 'black *woman* photographer', then, exposes another critical edge of otherness in this content and clearly the multi-dimensional nature of identity is by no means exhausted by these labels. In our seemingly known worlds, therefore, we make assumptions about the nature of people and places; about who belongs where, and who doesn't fit into the sameness of our mainstream; about who, what, where and when is other (see Chapter 23).

The second illustration is even better known within human geography, having achieved almost cult status in attempts to formulate **postcolonial** approaches to the subject. Edward Said is Professor of English and Comparative Literature at Columbia University in the USA (*see* Fig. 5.3). He is a Palestinian, born in Jerusalem and educated in Egypt and America, who is most famous for his analysis of the way the West imagines the Orient or East (including the Arabic Middle East) as different to itself (for a

Figure 5.2 A guard with a zither player in an interior, by Ludwig Deutsch (1855–1935). The illustration was used on the cover of Edward Said's *Orientalism*, 1995. Credit: © Christie's Images, London, UK/Bridgeman Art Library, London/New York

review of these and other 'imaginative geographies' see Chapter 24). In his classic book *Orientalism* (1978; 1995) Said traces how the Arab world has come to be imagined, represented and constructed in terms of its otherness to Europe:

> the French and the British – less so the Germans, Russians, Spanish, Portuguese, Italians and Swiss – have had a long tradition of what I shall be calling Orientalism, a way of coming to terms with the Orient that is based on the Orient's special place in European Western experience. The Orient is not only adjacent to Europe; it is also the place of Europe's greatest and richest and oldest colonies, the source of its civilisation and languages, its cultural contestant, and one of its deepest and most recurring images of the other.
>
> *(1995: 1)*

Representations of the romantic, mystical Orient, he argues, act as a container for Western desires and fantasies which cannot be accommodated within the boundaries of what is normal in the West. Yet at the same time, representations of the cruel, detached and money-grabbing nature of the Oriental Arab serve to underline the assumed hegemony of the West over political–economic and socio-cultural norms:

> Arabs, for example, are thought of as camel-riding, terroristic, hook-nosed, venal lechers whose undeserved wealth is an affront to real civilisation. Always there lurks the assumption that although the Western consumer belongs to a numerical minority, he is entitled either to own or to expend (or both) the majority of the world's resources ... a white middle-class westerner believes it his human prerogative not only to manage the non-white world but also to own it, just because by definition 'it' is not quite as human as 'we' are.
>
> *(Said, 1995: 108)*

Through the process of Orientalism, the societies and cultures concerned are marginalized, devalued and insulted, while the imperialism and moral superiority of the West are legitimized. Said's contestation of the othering of Orientalism points the way for wide-ranging inquiry by Human Geographers into how different people and places are similarly othered. It also shows us that at the heart of what we take to be familiar, natural, at home, actually lurk all kinds of relations and positionings to that which is unfamiliar, strange and uncanny (Bernstein, 1992).

From these illustrations it becomes clear that whether otherness is close to home, or positioned in some far-off exotic space, it is often difficult to detach ourselves, both conceptually and empirically from a frame of study which validates the self, the same and the familiar as waymarkers for the understanding of others. Two sets of issues arise from this conclusion. First, there is a need to think through much more deeply about what constitutes otherness in human geographical study, otherwise our main contribution may only be to further emphasize the othernesses which are *reinforced by* such study. At one level, this requires a grasp of the multidimensional nature of identity. As Mike Crang (1998) puts it:

> very few people are the 'same' as others – everyone is different in some respects. The most we could say is that certain groups share certain things in common, so who is counted as part of a group or excluded from it will depend on which things are chosen as being significant … Belonging in a group depends on which of all the possible characteristics are chosen as 'defining' membership. The characteristics that have been treated as definitive vary over space and time with significant political consequences attached to deciding what defines belonging.
>
> *(1998: 60)*

We need to recognize, therefore, that 'same' and 'other' identities are:

1. *Contingent* – in that differences which define them are a part of an open and ongoing series of social processes.

2. *Differentiated* – in that individuals and groups of people will occupy positions along many separate lines of difference at the same time; and

3. *Relational* – in that the social construction of difference is always in terms of the presence of some opposing movement.

(Jones and Moss, 1995)

Even with greater sensitivity for other identities, we are usually still trapped in a concern for what Marcus Doel (1994) calls 'the Other of the Same' – that is, we translate othernesses into our language, our conceptual frameworks, our categories of thought, and thereby effectively obscure the other with the familiarities of the samenesses of our self. The real difficulty then is to find ways of accessing 'the Other of the Other' – that is the unfamiliar, unexpected, unexplainable other which defies our predictive, analytical and interpretive powers, and our socio-cultural positionings.

The second set of issues relates to the methods we employ in order to encounter 'others'. As with our self-reflections, the increasing use of **ethnographic** and qualitative methods is important to this project. However, researching the other through ethnography takes a long time. Drawing lessons from Anthropology, we would have to conclude that to carry out appropriate studies of unknown peoples and worlds can take several years. Consider, for example, the account of French anthropologist Pierre Clastres (1998) who spent two years with so-called 'savage' tribes of Indians in Paraguay in the 1960s. He acknowledges that even 'being there' with his research subjects did not break down the very considerable barriers of communication and cross-referenced understanding, until circumstances changed many months into his research. Even over this timescale it proved difficult to form a bridge between himself (and here we might wonder whether his concept of 'savages' got in the way of effective communication) and the mythologies, embodiments and social practices which lay at the heart of the very existence of the Guayaki Indians (see opposite).

We need to acknowledge just how difficult it is to form a bridge between ourselves and the complicated essential existences of others, whether far off or close to home. It can be argued that the pressure to publish in the contemporary academy has run the risk of too many 'quickie' ethnographies of othered subjects. As with the Guayaki, an appreciation of the other geographies and experiences of, say, homeless people in a city like Bristol require long-term commitment rather than brief encounters. Only by reconceptualizing otherness, and reviewing the quality of our encounters with it, are Human Geographers likely to become any more attuned to a sense for the other and a sense of the other as suggested by Augé at the beginning of this chapter.

Summary

- Sensing the other is inextricably linked with understanding the self. By assuming that others are somehow the same as us, we can be locked into the 'thought prison' of the same, which makes it impossible to sense the other appropriately.

Pierre Clastres
Chronicle of the Guayaki Indians

Figure 5.3 Jyvukugi, chief of the Atchei Gatu. Source: *Chronicle of the Guayaki Indians* by Pierre Clastres, 1998 pp. 57–9

They really were savages, especially the *Iroiangi*. They had only been in contact with the white man's world for a few months, and that contact had for the most part been limited to dealings with one Paraguayan. What made them seem like savages? It was not the strangeness of their appearance – their nudity, the length of their hair, their necklaces of teeth – nor the chanting of the men at night, for I was charmed by all this; it was just what I had come for. What made them seem like savages was the difficulty I had in getting through to them: my timid and undoubtedly naive efforts to bridge the enormous gap I felt to exist between us were met by the Atchei with total, discouraging indifference, which made it seem impossible for us ever to understand one another. For example, I offered a machete to a man sitting under his shelter of palm leaves sharpening an arrow. He hardly raised his eyes; he took it calmly without showing the least surprise, examined the blade, felt the edge, which was rather dull since the tool was brand-new, and then laid it down beside him and went on with his work. There were other Indians nearby; no one said a word. Disappointed, almost irritated, I went away, and only then did I hear some brief murmuring: no doubt they were commenting on the present. It would certainly have been presumptuous of me to expect a bow in exchange, the recitation of a myth, or status as a relative! Several times I tried out the little Guayaki I knew on the *Iroiangi*. I had noticed that, although their language was the same as that of the Atchei Gatu, they spoke it differently: their delivery seemed much faster, and their consonants tended to disappear in the flow of the vowels, so that I could not recognize even the words I knew – I therefore did not understand much of what they said.

But it also seemed to me that they were intentionally disagreeable. For example, I asked a young man a question that I knew was not indiscreet, since the Atchei Gatu had already answered it freely: "*Ava ro nde apa?* Who is your father?" He looked at me. He could not have been amazed by the absurdity of the question, and he must have understood me (I had been careful to articulate clearly and slowly). He simply looked at me with a slightly bored expression and did not answer. I wanted to be sure I had pronounced everything correctly. I ran off to look for an Atchei Gatu and asked him to repeat the question; he formulated it exactly the way I had a few minutes earlier, and yet the *Iroiangi* answered him. What could I do? Then I remembered what Alfred Métraux had said to me not long before: "For us to be able to study a primitive society, it must already be starting to disintegrate."

I was faced with a society that was still green, so to speak, at least in the case of the *Iroiangi*, even though circumstances had obliged the tribe to live in a "Western" area (but in some sense, wasn't their recent move to Arroyo Moroti more a result of a voluntary collective decision than a reaction to intolerable outside pressure?). Hardly touched, hardly contaminated by the breezes of our civilization – which were fatal for them – the Atchei could keep the freshness and tranquility of their life in the forest intact: this freedom was temporary and doomed not to last much longer, but it was quite sufficient for the moment; it had not been damaged, and so the Atchei's culture would not insidiously and rapidly decompose. The society of the Atchei *Iroiangi* was so healthy that it could not enter into a dialogue with me, with another world. And for this reason the Atchei accepted gifts that they had not asked for and rejected my attempts at conversation because they were strong enough not to need it: we would begin to talk only when they became sick.

Old Paivagi died in June 1963; he certainly believed that he had no more reason to remain in the world of the living. In any case, he was the oldest of the Atchei Gatu, and because of his age (he must have been over seventy) I was often eager to ask him about the past. He was usually quite willing to engage in these conversations but only for short periods, after which he would grow tired and shut himself up in his thoughts again. One evening when he was getting ready to go to sleep beside his fire, I went and sat down next to him. Evidently he did not welcome my visit at all, because he murmured softly and unanswerably: "*Cho ro tuja praru. Nde ro mita kyri wyte.* I am a weak old man. You are still a soft head, you are still a baby." He had said enough; I left Paivagi to poke his fire and went back to my own, somewhat upset, as one always is when faced with the truth.

This was what made the Atchei savages: their savagery was formed of silence; it was a distressing sign of their last freedom, and I too wanted to deprive them of it. I had to bargain with death; with patience and cunning, using a little bribery (offers of presents and food, all sorts of friendly gestures, and gentle, even unctuous language), I had to break through the Strangers' passive resistance, interfere with their freedom, and make them talk. It took me about five months to do it, with the help of the Atchei Gatu.

- Geographies of other people and places can be close to home or in far-off exotic worlds. In either case, human geographers should see themselves as observers who are situated *within* the objects and worlds of their observation.

- At the heart of what we take to be familiar, natural and belonging lurk all kinds of relations and positionings with that other which is unfamiliar, strange and uncanny.

- There is a need to think through much more deeply what constitutes otherness in Human Geography. It is usually very difficult to bridge over between self and other.

- There is also a need to avoid methodological shortcuts in encounters with others.

CONCLUSION

This discussion of the interconnections of self and other raises a number of important issues about our human geographies. First, there is the risk that in acknowledging our selves in our work we become too self-centred and too little concerned with political and other priorities in the world around us. Second, there is the potential for losing our sense of otherness. Third, there is the conceptual and methodological complexity involved in encountering the other of the same, let alone the other of the other. Finally, there a concern over the way in which we can sometimes privilege certain kinds of otherness without giving due attention the need for sustained, empathetic and contextualized research under appropriate ethical conditions. There can be a tendency to 'flit in and flit out' of intellectually groovy subjects, with the danger that research becomes mere tourism or voyeurism of the subjects concerned.

When we have negotiated these tricky questions,

there is one further important issue of self-other inter-relations to resolve. In the words of Derek Gregory (1994): 'By what right and on whose authority does one claim to speak for those "others"? On whose terms is a space created in which "they" are called upon to speak? How are they (and we) interpellated?' (p. 205).

In seeking to encounter the stories of other people and worlds, is it inevitable that we become mere tourists, burdened by the authority of our selves and the power of our authorship? Or are there ways in which we can be sufficiently sensitive about the positionality and intertextuality of our authorship that we can legitimately seek to understand and write about the stories of others, without polluting them with our voyeuristic or touristic tendencies, the exclusionary power of which are so graphically illustrated in Figure 5.4? I believe that in this we can learn much from Gregory's emphatic and optimistic answer:

> Most of us have not been very good at listening to others and learning from them, but the present challenge is surely to find ways of comprehending those other worlds – including our relations with them and our responsibilities toward them – without being invasive, colonising and violent . . . we need to learn how to reach beyond particularities, to speak of larger questions without diminishing the significance of the places and the people to which they are accountable. In so doing, in enlarging and examining our geographical imaginations, we might come to realise not only that our lives are 'radically entwined with the lives of distant strangers' but also that we bear a continuing and unavoidable responsibility for their needs in times of distress more.

> *(Gregory, 1994: 205)*

In this agenda lies a pathway towards more sensitive and meaningful engagements of self and other in Human Geography.

Figure 5.4 The power to exclude when engaging in touristic or voyeuristic geographies. Credit: Earthscan/D. Dibbs

Further reading

For comprehensive discussions of subjectivity and self, and ethnography and self respectively, read:

- Pile, S. and Thrift, N. (eds) (1995) *Mapping the subject: geographies of cultural transformation*. London: Routledge.
- Probyn, E. (1993) *Sexing the self: gendered positions in cultural studies*. London: Routledge.

For further commentary on ethnography and qualitative methodologies in human geography, read:

- Cloke, P., Crang, P., Goodwin, M., Painter, J. and Philo, C. (1999) Practising human geography. London: Sage.
- Cook, I. and Crang, M. (1995) *Doing ethnographies*. IBG CATMOG No. 58, Norwich: Environmental Publications.

For an excellent case study of the conceptual and practical outworking of self–other in ethnography, see:

- Crang, P. (1994) It's showtime: on the workplace geographies of display in a restaurant in Southeast England. *Environment and Planning D. Society and Space* **12**, 675–704.

To continue with the illustrations of rural otherness and Orientalism read, respectively:

- Cloke, P. and Little, J. (eds) (1997) *Contested countryside cultures: otherness, marginality and rurality*. London: Routledge.
- Said, E. (1995) *Orientalism*. London: Penguin (reprint with Afterword).

CHAPTER 6

Image–reality

Mike Crang

INTRODUCTION

Geographers spend a lot of time working with images. They do not, however, generally call them 'images'; they are called charts, graphs, and even, maps. The term 'images' tends to be used in a negative way. The word image is used to imply superficial, not factual, obscuring and covering up reality, conveying a biased impression. There are, it seems, 'bad' images that are things out there in the world that get in the way and mislead people (poor dears) and 'good' images that allow the geographer to grasp what is really going on. Nor is this confined to geography, Clifford Geertz comments of scientific writing in general 'that "symbolic" opposes to "real" as fanciful to sober, figurative to literal, obscure to plain, aesthetic to practical, mystical to mundane, and decorative to substantial' (in Baker, 1993: 10). Image is taken to imply the opposite of real, in a series of binary pairs where two terms are opposed and we are trained through years of education to value the second terms on Geertz's list. Geography has seen many variants on this pattern, some of which are explored in more detail later. Geographers have studied '**mental maps**' to see how these diverge from 'reality', the 'perception' of risk as opposed to statistical likelihood, tourist images as glamourising real places, facades and regenerated areas as images concealing real economic processes or literature as a 'subjective' representation of a region. The discipline has often implied that images obscure or deviate from a reality revealed by careful geographical study.

This chapter will suggest the relationship of image and reality is rather more complex than this. We might, yes, look at how images refract, reflect and alter the world. Images have impacts on the world in terms of how they shape action by people. Images can be deliberately promoted, massaged or altered to achieve desired ends, but they all go into forming the ideas and understandings of the world, based on which people make choices and act. There are geographies of images in terms of what areas they do, or do not, show and how they move through society. Moreover, geographers produce images of the world, so we need to re-evaluate what role images play in geographical knowledge. Changes in how the world is seen can tell us something about those doing the seeing.

PROCESSES OF PERCEPTION

People do not take in the whole world as they go about their business. Everyone selects and filters what they see and what they make of it. We might look at this through three ideas; biology, (physical and social) position and cultural frames. The first highlights that human senses connect us to the world in particular ways. With the philosopher of science Donna Haraway, I would suggest walking a dog shows how specific is human experience. The dog's world is visually poor but full of odours carrying information. Our dependence on the visual, our sense of scale and our sense of location are all based on the human body. **Perception** does not start as a free-floating moment, but is grounded in frailties and adaptations of the body. We are disposed to make sense of the world and make order from the sensory stimuli our body gets – think of the little parlour games of optical illusions that play on that tendency (*see* Fig. 6.1). Not that this is bad, just think how useful it is that we can 'see' three dimensions on a flat piece of paper.

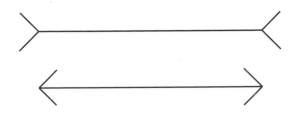

Figure 6.1 Things are not always as they appear

Our orientation also plays an important part in ordering experience. The world comes to us in terms of high and low, near and far, present and absent. We understand ourselves spatially, as we recall the world in mental maps to situate ourselves. These are egocentric maps, where our life and experiences are centre stage, and the world fades off into the distance around us. We do not know every part of the planet equally, nor our country, our city, even neighbourhood. We have areas of concern or interest, and different ways of understanding these. As Edward Relph (1976: 5) put it, consciousness is always consciousness of something, it is not free-floating. We are inescapably immersed in the world. Thus our immediate surroundings are grasped as left and right, while more distant places are subjects of abstract images. These images may be our own experiences as they are remembered or they may be memories of images produced by others – blurring the boundaries of sensed and imagined worlds. These remembered spaces are not based on the geometrical spaces of latitude and longitude. They are shaped by our experiences, travels and the tasks we have at hand (*see* Figs 6.2, 6.3). Our perception of the world is spatial in the way we define objects of interest as a foreground set off against a background and in relation to our viewing position. Images create a relationship between three terms – the perceiving subject, the viewed object and the relationship between them.

Our images of the world are not simply our own but are derived from social sources. Different cultures have different ways of seeing the world and representing it. For example, twisted and notched sticks formed maps for travelling in the Arctic for the Inuit. Even colours can shift between cultures, so that we read Homer's 'wine dark sea', because he did not clearly distinguish blue and green (Eco, 1985: 159). Colours are described by reference to other objects ('red like fire', 'white as snow', 'yellow as cornfields'). Our descriptions refer to other things, not direct experience. Images of the world are formed, understood and communicated through materials at hand to that culture, not by a universal sensation.

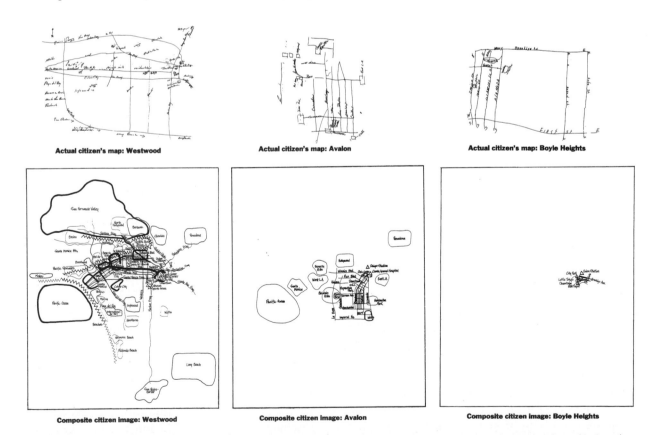

Figure 6.2 Cognitive maps of Los Angeles as perceived by predominantly Anglo American residents of Westwood, predominantly African American residents of Avalon, and predominantly Latino residents of Boyle Heights. Source: *The visual environment of Los Angeles*, Los Angeles Department of City Planning, April 1971, pp 9–10

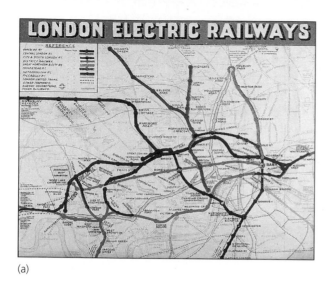

(a)

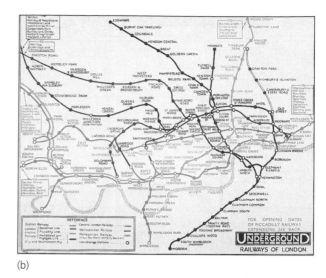

(b)

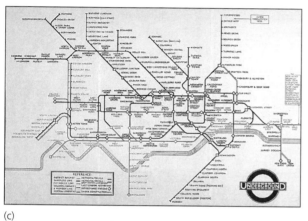

(c)

Figure 6.3   (a) The first London Underground Railways folding pocket map, issued free in 1908. An accurate reproduction in which lines were related to a central London map. (b) The 1932 London Underground map designed by F.H. Stingemore. This map recorded the expansion of the network into the suburbs. (c) The 1933 Underground map by Harry Beck. A diagrammatic approach that offered clarity rather than a close relationship to actual directions and locations. Credit: © London Regional Transport. Reproduced by kind permission of the London Transport Museum

MAKING PICTURES

The philosopher Martin Heidegger suggested that a crucial shift in how Western people experienced the world was when it became conceived as a picture. The world became seen as separate and detached from the viewer. Up until then, Heidegger argued, people had seen themselves as part of the world. This change can be linked to the rise of new techniques for producing images such as the *camera obscura*. At its simplest, this is a darkened room with a hole in one wall, while the one opposite forms a screen on which an image of the outside world appears – like a large pinhole camera. Observers could draw directly from life. These images could be assessed by their direct correspondence to the outside world: a *correspondence theory* of truthfulness. It produced this image through the seclusion and detachment of the observer, separated from the world. This way of producing images thus became a model of truth that saw a world that could be known and represented through a detached observer.

The viewing position affects the knowledge created. For instance, landscape painting is a specific rela-

tionship of the viewer to the world, looking on the scene for pleasure, as an aesthetic object, not as an immersive workaday environment, reflecting how paintings were often commissioned by owners to show their estate. A famous example is Gainsborough's *Mr and Mrs Andrews* (*see* Fig. 6.4) which depicts Mr Andrews standing by his seated wife in the left-hand corner and overlooking the house and estate. The composition is of Mr Andrews showing off his belongings – with his wife included in that category. Thus not just the content tells us about rural life but also the composition of these images reveals a class-based and gendered way of looking at the world. This image relates to scenic country parks created by excluding the rural poor, removing traditional rights and privatizing the land. This was a period where villages and public rights of way were moved to create views for the owners of country houses and where new laws criminalized intrusions. Many rural people had traditionally hunted game which was redefined as poaching. At the start of the nineteenth century over a quarter of all prosecutions in England and Wales were for breaches of these 'game laws'. The image of beautiful and charming rural scenery reflects real

Figure 6.4 *Mr and Mrs Andrews* by Gainsborough. Credit: © National Gallery, London

struggles over access to the land. The detachment from the scene is also linked to **patriarchal** power and a masculine viewing position. Landscapes and women tend to be the objects of a male viewer separated from what is depicted, tending to link the natural and the feminine (Rose, 1997). The art critic John Berger (1972) suggests that in most art, men act or are viewing subjects while women appear as objects of that view.

A vantage point from which the world can be made sense of is often called an 'Archimidean point'. One example is the viewpoint of maps, often suggesting an observer outside what is depicted seeing the true picture. Although apparently factual, maps can also have hidden implications of who is looking at whom from where. Thus British students are familiar with a map centred on Europe and the Atlantic, not one centred on the Americas or on the southern hemisphere (*see* Fig. 6.5). What implicit assumptions are there in the familiar Mercator map about who is of central importance and who is not? The projection downplays the size of places near the equator and exaggerates those near the poles – making the 'West' disproportionately large. An alternative projection – the Peter's Projection – equalizes for area but is less useful for navigation. Each emphasizes certain features of the world and not others.

Not all images rely on detached vantage points,

and we can usefully consider why they may not. In art, the rise of Cubism, after World War 1, apart from using rectilinear outlines, also did away with perspective and notions of realistic correspondence. What does the emergence of these images tell us about the world? Stephen Kern (1983) argues that they express changing experiences of space and time. This was a period seeing the climax of rapid changes in transport (the expansion of railways, steam ships and mass transit systems) and communication technologies (the telegraph, telephone, then radio). People, goods and information were circulating more rapidly, in greater numbers and over greater distances than ever before, offering no stable vantage point at which the whole picture came together or from where it could all be controlled. The great cities, modern communications and transport created a fragmented experience not a coherent whole. The world could no longer be depicted in the same way. All these changes in images suggest novel ways of organizing and understanding the world and shifts going on in society at large.

Summary

- The perceptual world may be more 'real' and truthful to our experiences, and thus as important in orchestrating actions, as any scientifically defined reality. As the world and our experience of it change, the images that are produced may also change.

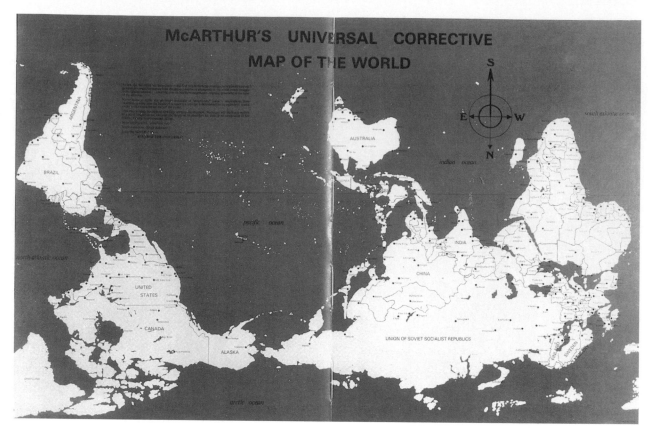

Figure 6.5 McArthur's corrective map of the World, by Stuart McArthur. (Further information available from Stuart McArthur, 208 Queen's Parade, North Fitzroy, 3086, Tel: 9842 1055)

- Geographers need to be aware of 'from what angle', literally and metaphorically, people perceive the world. Particular ways of creating images imply relationships between the viewer and the world. The detached point of observation from which a coherent picture can be made may depend on relationships of power and control more than its 'realistic' depiction of the world.

GEOGRAPHIES OF IMAGES

There is a basic geography of making images; production facilities concentrated in Hollywood dominate the flows of images coming to Europe, while there are vibrant movie-making districts in Hong Kong or 'Bollywood' in India. In an era of satellite pictures beamed around the world, of global news corporations and media events, the conventional geography of national broadcasters breaks up, and we need to think about the circulation of images. Images circulate over increased distances at an increasing rate and in vast numbers. Much of this is controlled by multinational business but it also offers opportunities for minority cultures to use media technology to link groups dispersed over wide distances, as in Inuit soap opera in Canada, or the circulation of videos of

Australian aboriginal rituals (Ginsburg, 1991, 1993; Perlmutter, 1993). In a world of global transmission it has become an important issue who can control the representations of communities and their appearance through the media. For many minority groups becoming visible to the world at large is an important political tool – but one which also changes the internal dynamics of groups irrevocably by altering the power and status of members who control those images (Moore, 1992, Turner, 1991).

Issues of power and control are highlighted when we remember that technologies of vision have been an important part of military development over the last century. Control of vision has been a vital stake – who can see and not be seen. Reconnaissance flights, high altitude spy planes, night sights, spy satellites – all create new images. The Gulf War saw Western forces deploying advanced image-producing technology. An issue was control of pictures. The US military blamed pictures from the Vietnam conflict for undermining support, so this time reporters were tightly controlled. Although portable satellite links meant instant transmission, and the military timetable paid heed to American network news slots, live footage did not mean free reporting. The military provided pictures, from cameras in 'smart bombs' and war planes, ready for broadcast. Unsurprisingly they recorded successful missions, emphasizing what the military wanted and not the carpet bombing of conscripted troops. It is revealing that the commanders had to remind viewers that these pictures were not 'video games'. The images were similar – with troops only seeing their enemy through their screens and describing missions like arcade games. It becomes less clear where the division of image and reality might be.

We might also look at the geography of what is made visible through popular pictures. Some of the most ubiquitous images are tourist snapshots. Estimates suggest something over a billion are taken every year (Stallybras, 1996). We could imagine a map of the world which just recorded these as points of light. Think of the dense clusters around the great sights – the Taj Mahal, Grand Canyon, Tower of London, and so on – tailing off into darkened hinterlands – in the slums of Mumbai, into Harlem, former mining villages around Durham. So we have a geography of what is regarded as photogenic. This geography would have to include images in brochures and postcards. These emphasize the good points of a place; the weather is sunny, the scenery picturesque and the beaches clean. We can think of the types of places they market and depict – so we might think of how they show a world where ethnic stereotypes pervade. Thus there are Amerindians in 'tribal' clothing, next to 'totem poles'; exotic cultures of Asia represented by attractive women; stereotypes of all sorts (Albers and James, 1988; Dann, 1996; Edwards, 1996). We know these images will be selective in what they portray. They illustrate and help shape the desires of tourists, what they want to see, and thus play a vital role shaping flows of tourists even if 'inaccurate'. We must begin to ask questions about whether images do not shape reality as much as reflect it.

GEOGRAPHIES SHAPED BY IMAGES

An example of how images shape reality comes from tourism when places are altered to conform to expected images. More subtly, the experience of those places is shaped through images. So in, Don DeLillo's novel, *White Noise* the narrator visits the most photographed barn in America:

> We counted five signs before we reached the site. There were forty cars and a tour bus in the makeshift lot. We walked along a cowpath to the slightly elevated spot set aside for viewing and photographing. All the people had cameras; some had tripods, telephoto lenses, filter kits. A man in a booth sold postcards and slides – pictures of the barn taken from the elevated spot. We stood near a grove of trees and watched the photographers. Murray maintained a prolonged silence, occasionally scrawling some notes in a little book.
>
> "No one sees the barn," he said finally . . . "Once you've seen the signs about the barn, it becomes impossible to see the barn . . . We're not here to capture an image, we're here to maintain one . . . They are taking pictures of taking pictures . . . We can't get outside the aura. We're part of the aura."

(Cited in Frow, 1991: 126, and Nye 1991)

The imageworthy element of the barn is not its inherent qualities but that it appears in so many images. We have all seen so many images of wild animals filmed with advanced equipment that actual encounters can be disappointing. Umberto Eco has called this '**hyperreality**' where the copies are more important, and realistic, than their originals. Going one step further images become **simulacra**, that is copies for which there is no original. They become entirely self-sustaining without referring to any exterior reality. Jean Baudrillard argues this has been a long-running trend where gradually images have come to stand for and then replace things, calling the whole category of reality into question. Examples might include themed shopping malls that have facades conjuring up images of Parisian cafés which have never existed, or TV series like *Heartbeat* where signs for Aidensfield, the fictional village in the series, stay up all year in the real setting of Goathland.

The impacts of images shaping the world can be more direct. We can consider the global flows of news pictures and how this brought pictures of the Tiananmen Square massacre to the whole world. We could compare the effect of globally circulated images with others that have not emerged – such as the few pictures of Indonesia's invasion of East Timor. One example of how powerful images can be is the Ethiopian famine of 1984. Here striking news footage of starving people in relief camps produced an enormous charitable effort, global fund-raising events and a massive relief effort. The images also suggested helpless victims, fuelling stereotypes about Africa, and said nothing about the way southern Ethiopia remained a food exporter. The cause of the disaster was simplistically presented as a lack of rain and consequent decline in food availability (*see* Fig. 6.6). The images can be criticized for inaccurately depicting the causes yet we have to acknowledge their enormous real world impacts.

Summary

- Images selectively portray peoples and places. We need to consider who in the current global economy decides

Figure 6.6 World weatherman

what is shown and controls how it is circulated. How peoples are represented and whether they represent themselves can be the subject of political struggles.

- Images do not just reflect reality but shape actions, experiences and beliefs. Intuitively we think reality comes first and images second. However, the relationship can be more circular. It is possible to suggest that in some cases this could be a closed circle of image referring to image without needing to refer to an external reality.

THE TRUTH IS OUT THERE?

These issues do not only affect people 'out there' but also 'academic' images. Traditionally, geographers have seen themselves as making progressively more accurate images of the world. However, activities such as map making have played a more active part in shaping the world, for instance being promoted by imperial states such as Britain in order to administer conquered territories. The images produced by geographers are not exempt from the sorts of processes outlined above. We spend most of our time working with **metaphors**, images, models and so forth. How do we assume they relate to reality? Quite often the answer is in terms of some 'correspondence' theory. Yet a perfect correspondence would lead to some bizarre conclusions. The novelist Jorge Luis Borges discussed the perfect map, on a one to one scale, that was thus as large as the territory it depicted. More imaginatively, even if you managed to shrink this map, if it shows all the activities and features in the territory, then somewhere

on it there would have to be an image of you holding it. And on that image there would have to be an even smaller image of you holding it. This is known as a problem of *infinite regress*. It may be then that trying to find some underlying reality behind the images is the wrong approach. It reminds me of a story.

A young pupil approached a Lama and asked:
  'Oh wise one, what is the world on which we dwell?'
The Lama paused, then replied, 'The world is a great disk with seven mountains and nine seas surrounded by the outer ocean.'
The pupil thought, then asked, 'Oh wise one, on what does the disk of the world sit?'
  'It sits on the backs of four celestial elephants, and it shudders as they shift its weight on their shoulders.'
The pupil thought for a minute in silence.
  'Oh wise one, on what do the four celestial elephants stand?'
  'They stand, facing out towards the stars, upon the vast carapace of the world turtle.'
The pupil hesitated.
  'Oh wise one . . .'
  'Forget it', said the Lama, 'after that it's turtles all the way down.'

A silly example, but we have to face the possibility that we may only be able to understand reality through words or images. We need to think then, not of how they may distort reality, but what effects and meanings they have for their beholders. In this way we need to see images as actively creating the world rather than simply transmitting a prior reality.

Further reading

- Berger, J. (1972) *Ways of seeing*. Harmondsworth: Penguin. An old book but a classic about the hidden implications of particular ways of looking in art.

- Cosgrove, D. (1997) Prospect, perspective and the evolution of the landscape idea. In Barnes, T. and Gregory, D. *Reading human geography*. London: Arnold, pp. 324–41.
A careful historical survey looking especially at the invention of realistic perspective in the Renaissance.

- Eco, U. (1987) *Travels in hyper-reality*. Picador: London. Written by an Italian Professor of semiotics (a study of images and communication) who is also a novelist – so it is readable. It documents what he sees as a trend for images replacing the originals they are supposed to depict.

- Rose, G. (1997) Looking at landscape: the uneasy pleasures of power. In Barnes, T. and Gregory, D. *Reading human geography*. London: Arnold, pp. 342–55.
A trenchant argument that many conventional ideas of images and landscape in geography are founded on an assumed masculine viewer.

- Woods, D. (1994) *The power of maps*. New York: The Guilford Press.
A nice summary of how what may appear to be factual images corresponding to reality can have many hidden assumptions.

PART II

Themes

INTRODUCTION

One of the things which we have always loved about geography is the sheer diversity of its subject matter. Looking back, we're sure that what first attracted us to take Human Geography at university was the fact that we could legitimately study a huge range of human behaviour and social activity, and we're equally sure that this has played a large part in holding and stimulating our interest ever since. We always felt slightly smug at university when in conversation with other social scientists, as their subjects seemed so narrow. The economists stuck to the study of economics and the historians to the study of history. The political scientists examined politics and the planners looked at planning. This seemed a bit of a shame, especially when by contrast we were able to study all of these areas, as well as many more. Among the hundreds of lectures we have attended, picked almost at random, were those on revolutions in Latin America and Asia; on welfare provision in Sweden; on urban politics in the USA; and on industrial development in Scotland. Among the thousands we have given include those on the geographies of famine and hunger in Africa; rock and roll music in the USA; poverty and deprivation in urban Britain; and theme parks and shopping centres in Western Europe. Other geographers could come up with similar experiences - you probably can yourself. The point lies not in the particular subjects but in their breadth.

Geography's abiding concern with space and spatial organization has long given it a wonderful licence to study just about anything it wants - as just about everything has some kind of spatial outcome or component. However, in order to make sense of this breadth, and to make it more manageable, geographers tend to work in and around particular specialist areas, commonly called sub-disciplines. These sub-disciplines are the different 'themes' which we explore in this section of the book. In Section 1 we looked at some of the critical dualisms which have structured and shaped the development of Geography as a whole. Now we begin to break that discipline down into its constituent parts. Those we have chosen to highlight are amongst the most vibrant areas in today's Human Geography. In each case we have selected three or four key aspects of these themes which we present as individual chapters. Each is written by an acknowledged expert in their particular field who is at the forefront of contemporary geographic research. This should enable you to get some flavour of the developments taking place in the geographical literatures. Any less and we would have been trying to cover too much in each chapter; any more and the book would have become unwieldy and (even more) expensive.

We begin with three chapters devoted to development geographies, before moving on in turn to look at economic, environmental, historical, political and social and cultural geographies. We have separated these themes in this section, although as you read through you will begin to pick out the connections - between, for instance development geography and geopolitics; between the economy and sustainability; between the historical geographies of memory and heritage and social and cultural geographies of the imagination. Indeed, one of the most exciting developments in Human Geography in recent years has been the willingness to explore the connections and interfaces between the sub-disciplines, which has raised a whole number of new avenues to study. This issue is taken up in more detail in the final section of the book, where these interconnections are explored in a number of different contexts.

For now though, we will hold these interfaces somewhat apart in order to explore some of the key developments which have recently taken place in each sub-discipline. Each 'theme' has its own editorial introduction, briefly summarizing the key concerns of the sub-discipline and introducing the chapters in relation to these. We are not attempting to be comprehensive here, and each section is not meant to present all the work going on within any one theme. There will inevitably be arguments about the areas of each sub-discipline that we have either included or excluded. Hopefully these arguments will make interesting debates in their own right. In making the choice of content, the idea is to present you with a picture of, and a feel for, some of the breadth and excitement of contemporary Human Geography. We cannot provide an account of the whole discipline in the limited space available in this section, but we have tried to present an interesting and novel set of chapters which individually and collectively begin to push against the more traditional sub-disciplinary boundaries. In doing so they open up a whole series of new and vibrant issues for the Human Geographer to examine in the new century.

Development geographies

INTRODUCTION

Issues of 'development' are an inescapable part of our everyday lives. The imagined geographies of our world, as fuelled by the graphic pictures and reporting of television news and documentary, are regularly topped up with images of the latest famine, warfare, impoverishment or refugee migrations in some far off 'Third World' country. Mega charity-spectacles such as Children In Need and Comic Relief regularly pull at our heartstrings and purse-strings in response to the victims of 'underdevelopment'. Our environmental concerns, for example, over the chopping down of rain forests or the erosion of biodiversity, bring us into direct contact with issues of how the commercial conduct of economic development (in this case of agricultural production) is often at odds with global ecological objectives. The more discerning of us may even allow the components of our everyday diet to remind us of the global geographies of food, and of the likelihood that what we are eating is directly connected with unfair trade and production conditions which benefit big international firms but impoverish those whose labour has been directly involved in food production.

Geographies of development, then, are a vital part of bringing the global into the local – of forcing us to understand that our plenty is directly related to others' poverty. With that in mind, some of the early practice of development geography tended to be overly tied to detached indicators of the state of development in different countries. Thus, details of production, consumption, investment, demographic characteristics, health, education, income, and so on come to dominate the agenda. However, geographers have gradually spread the news that development is diverse, complex and often contradictory, and the real-life experiences of people in the countries and regions concerned are made trivial by being reduced to a series of such indicators. Nevertheless, as Stuart Corbridge points out in Chapter 7, there is still more work to be done by geographers interested in development in order to unpack the in-built assumptions on which our studies are often founded. He argues, for example, that development studies have led to generations of students and policy-makers believing that a combination of national economic planning in the Third World, and foreign aid and direct foreign investment from the First World, will make traditional societies modern, and poor people affluent.

The unpacking of development certainly involves new perspectives on the development process itself, and in particular on the ways in which discourses of development are produced and circulated. The work of the **post-developmentalist**, Arturo Escobar, is discussed in Chapter 7 in this respect. The unpacking will also, however, involve listening to 'other' voices in developing countries. Paul Routledge takes us in this direction in Chapter 8. He shows how many development practices have resulted in the '**pauperization**' and marginalization of indigenous people, with peasants, tribal people, women and children usually being viewed as impediments to progress and thereby being excluded from participation in the development process. It is important, therefore, to hear the voices of *these* people, who will sometimes organize themselves into place-specific social movements so as to resist the threats to their economic and social survival.

In Chapter 9, Sarah Radcliffe helps us to think through some of the unanswered questions about geographies of development. In addition to the recognition of geographies of resistance, we need to address issues of how colonial pasts still pervade contemporary development practices, and studies of those practices. Postcolonial theory will therefore help to put nuanced geographies and histories back into essentialized labels such as First/Third Worlds, and Traditional/Modern ways of life. Equally, we need to be aware that conventional scales and arenas of development studies need to be transformed. Maps of development should not be confined to nation–states. There are now richer spaces in poor countries, as well as poorer spaces in rich countries. Also, new institutional structures have emerged at the interface of the market, the state and non-governmental organizations. All of these spaces, arenas, processes and practices are shot through with gendered power relations, and we still have a great deal to learn about women's agency and the diversity of masculinities involved.

Development geographies are therefore becoming more attuned to many of the wider theoretical considerations of human geography. Sensitivities to resistance and to gender are being allied with perspectives drawn from postcolonial and post-developmental theory. The key question is how such postcolonial and post-developmental geographies might be implemented in practice. As the socialist alternative has largely melted away, how far is it possible to engender wider access to, and participation in, capitalist development without that development merely being a vehicle for yet more pseudo-colonialism and exploitation?

Further reading

An excellent 'first stop' for further reading on development geographies is Stuart Corbridge's informative and challenging reader:

• Corbridge, S. (ed) (1995) *Development studies – A reader*. London: Arnold.

For *the* key contribution to postcolonial and post-developmental approaches, refer to:

• Escobar, A. (1995) *Encountering development – the making and unmaking of the Third World*. Princeton: Princeton University Press.

Otherwise journals such as *Third World Quarterly* are worth browsing through and regular reviews of contemporary themes in development appear in *Progress in Human Geography*.

Development, post-development and the global political economy

Stuart Corbridge

The true function of social science is to render problematic
that which is conventionally self-evident.

*(after Max Weber, and as commended to the author by Derek
Gregory many moons ago)*

INTRODUCTION

All subjects are blessed with a number of able teachers, and economics is no exception to this rule. In
Cambridge, during the 1980s, one of the most
respected teachers of economics was given responsib-
lity for a course on development studies. For several
years in succession the first lecture on the course took
the same form. The economist spoke for 50 minutes
about the asymmetries that structure the world economy. The underdevelopment of 'Third World' countries was blamed upon their subordinate relationships
with 'developed' countries, and students were quickly
exposed to dependency theories, unequal exchange
models and a critique of the role played by multinational corporations in the South. The future for these
benighted countries, our economist claimed, was to
delink from the capitalist world economy; the choice
lay between barbarism and socialism. And then our
economist bundled up his notes and acetates. Moving
to the door he paused only to say: 'Everything I have
told you today is at best misleading or simplistic, and

at worst just plain wrong. In the remainder of the
course, I will explain why.'

It is not my purpose in this chapter to be as polem-
ical as my friend in the Cambridge economics faculty
(he now teaches in London). But I do want to use the
first part of the chapter to question – to render problematic – three claims that are put forward as self-evident propositions in some parts of social science: (a)
that intentional (post-1950) development has failed;
(b) that this 'failure' exposes the shortcomings and
iniquities of a global political economy that is run by
and for a handful of richer countries and corporations; and (c) that true 'development' can only be
secured by strategies of survival and resistance that
take shape outside the modern world economy and at
the local level. In the rest of the chapter I develop
some more constructive points about the global political economy. I recognize that 'development' has not
been secured evenly or in a stable fashion throughout
the international system since 1950, nor has it lived
up to the hype that surrounded the modern invention
of '**developmentalism**' at about this time; of course
not. I also acknowledge that Human Geographers
have something to learn from what is being called
'**post-developmentalism**' (which is associated with
the three propositions listed above). Even so, I argue
for accounts of development and the global political
economy that are properly sceptical of the utopian
and dystopian visions of arch-modernizers and com-

mitted anti-developmentalists. In my view, these accounts should have close regard for questions of market access in a changing world economy, and for questions of risk avoidance and empowerment at different spatial scales. Such accounts should also recognize the very real achievements made by many developing countries, and should acknowledge the key role played by states and international agencies in creating landscapes of empowerment or disempowerment.

DISCOURSES, DEVELOPMENTALISMS AND THE GLOBAL POLITICAL ECONOMY

One of the most amazing things that has happened since about 1950 is that most of us have come to accept that poorer countries can and should be developed or are developing. It was not ever so. In the first half of the twentieth century it was widely believed that poor countries were meant to be poor. Countries in Asia and Africa were not considered candidates for directed economic growth or 'progress', either because of trying climates or because they were populated by racial groups lacking the supposed advantages (cerebral and other) of white Caucasians. The rise of Japan, together with the post-1945 rise to ascendency of the USA and the beginnings of decolonization, changed all this. The 1960s was designated the Development Decade by the United Nations, and a number of economic models were put into play which claimed to show how 'latecomer' countries could very quickly become rich by imitating the development trajectories pursued by 'pioneer' countries like the UK and USA. In sum, an ideology of developmentalism emerged in the period between about 1950 and 1970. Development studies emerged as an academic and practical discipline, and a generation of students and policy-makers was brought up to believe that a combination of national economic planning in the Third World (this is what I mean by intentional development), plus foreign aid and direct foreign investment from the First World, would rapidly make traditional societies modern, and poor men and women more affluent. In so far as this ideology was challenged, the challenge came from the political Left. In the 1960s and 1970s, academics led by Gunder Frank and Samir Amin (amongst others) argued that the industrial development of the South would only be effected as and when the periphery severed its ties with the capitalist First World – when it broke with rapacious multinational corporations and their main protectors, the USA, the World Bank and the International Monetary Fund.

But now times are changing again. Following the apparent collapse of socialism as an alternative to capitalism, many critics of capitalist development are minded to criticize not so much capitalism as the ideology of development itself. The work of Arturo Escobar is cited admiringly in this context. In his book *Encountering Development: The Making and Unmaking of the Third World* (Escobar, 1995), Escobar moves far beyond the critiques of capitalist development and underdevelopment that were advanced by neo-Marxists in the 1960s and 1970s. Where critics like Frank or Amin once commended socialism as the true path to development (in the sense of removing people from poverty and providing them with access to power and the means of production: Frank, 1967; Amin, 1990), Escobar and his fellow post-developmentalists condemn the development project itself for 'discovering' (creating, producing) mass poverty and for promoting the illusion of social improvement by means of Eurocentric trajectories of economic change and advancement. Escobar contends that the dream of development has:

> progressively turned into a nightmare. For instead of the kingdom of abundance promised by theorists and politicians in the 1950s, the discourse and strategy of development produced its opposite: massive underdevelopment and impoverishment, untold exploitation and oppression. The debt crisis, Sahelian famine, increasing poverty, malnutrition, and violence are only the most pathetic signs of the failure of forty years of development.
>
> *(Escobar, 1995: 4)*

Escobar's arguments are more nuanced than this quotation suggests, and his book is one that I would urge all students of development to read. A particular strength of Escobar's work, and of post-developmentalism more generally, is its willingness to examine the origins of words or concepts like 'development' and 'mass poverty', and its insistence that keywords are put into play to serve particular interests or constellations of power. Escobar's hero is the French social theorist Michel Foucault, and he makes use of Foucault's work on knowledge–power relations and intellectual histories (*genealogies*) to argue that the Third World was called into existence and defined in a speech delivered by the President of the United States, Harry Truman, on 20 January 1949. In this speech, Truman announced his 'fair deal for the whole world'. In less than one hundred words, Truman described the 'underdeveloped areas' of the globe as suffering from disease and inadequate food, and proposed that 'a program of development based on the concepts of democratic fair dealing' should be put in place and funded by the USA and its richer allies to guarantee 'prosperity and peace [through] greater production' (Truman, 1949, quoted in Escobar, 1995, p.3).

What makes Escobar's work exciting is his refusal to take the humanism and apparent internationalism of Truman's fair deal at face value. His first chapter questions why and how the USA came to define the Third World in these terms, and how and why the 'developed countries' came to define themselves in opposition to the 'underdeveloped countries' *and* demanded that the latter be remade in the image of the former. The remaining chapters of Escobar's book are set up to deal with these questions. Having discussed the origins of what he calls the discourse of development, Escobar considers how the Western mind-set defines the Third World as a broken world of poverty, disease, pathology and deprivation which has to be mended by infusions of Western capital and know-how. These infusions, he goes on to claim, are organized first by the aid industry and later by Western private capital, and in almost all cases they prove to be inattentive to the expressed needs and knowledges of local people. The result, Escobar contends, is a nightmare world of bureaucratically organized development 'projects' and an explosion of ill-conceived urban and industrial ventures which remove men and women from their local resource bases and well-formed agro-ecological traditions and practices. So-called development ends in deforestation and desertification as fragile ecosystems are bent to Western conceptions of commercial and industrial development. By the same token, further integration into the world economy results in the accumulation of huge debts to Western financiers, the loss of remitted profits to **transnational corporations** (TNCs), a diminution of local food production capacities to serve the appetites of Western consumers, and a progressive loss of local and national sovereignties as economic and other powers are leached to the TNCs, the World Bank and the IMF. Real or alternative (even anti-) development, Escobar concludes, depends upon breaking these ingrained attachments to the discourse of development and an associated discourse of globalization. In the final chapters of his book Escobar tries to 'imagine a post-development era' and commends various 'defenses of the local' as one precondition for breaking the holds of Westernization and an ideology of 'modernizing development' (p. 226). Among the more effective such defenses of the local, Escobar maintains, are those mounted by women's, ecological and people's organizations, including farmers' movements in India (*see* Fig. 7.1) and the Chiapas rebellion in Mexico.

Summary

- Post-developmentalism offers a radical critique of the development project 'itself', and not just its capitalist

Karnataka
Rajya Ryota Sangha

(Karnataka State Farmer's Association)

NATIONAL "SEED SATYAGRAHA"
March 3rd & 4th 1993
"SIEGE OF DELHI" BY FARMERS

The Second Salt Satyagrha has started to-day. This "Seed Satyagraha" is to protect the right of the farmers to produce, modify & sell seeds. "Seed freedom" is freedom of the Nation

- Our genetic resources are our national property.
- We oppose patents or any form of intellectual Property protection on plants & genes.
- To produce modify & sell seeds is the right of the farmers.
- We want the preservation of the Indian Patent Act 1970, which excludes patents on all life forms.
- We want a ban on the entry of Multi-National Corporations in the Seed sector.
- We oppose total control of global trade by Multi National Corporations.

Let us Banish Multi National Corporations From The Country

Figure 7.1 Poster distributed by an Indian farmers' movement

versions. Many post-developmentalists believe that development has failed and that it has wrought havoc in previously well-adjusted communities and countries.

- Escobar dates the invention of modern developmentalism to 1949. He pays close attention to the discursive strategies by which the so-called developed world defines itself in opposition to a so-called underdeveloped world and demands that the latter is made like the former.

'POST-DEVELOPMENTALISM' IN CONTEXT

There is much to be said for Escobar's work, and his book is a welcome alternative to texts and plans which treat development as no more than a technical issue, more or less emptied of ethical considerations and cut off from an engagement with the voices of those who are marked out for 'development' whether they like it or not. At it best, work in this genre is intellectually liberating, and promises the liberation of bodies, locales and regions caught up in globalized power relations over which they exert little control. Such work helps us to make sense of a world economy that should on occasions strike the sane among us (and who might that be?) as insane. The cartoonist Steve Bell captured this insanity very well in a cartoon he drew for *The Guardian* newspaper on the occasion of the United Nation's Social Development Summit in Copenhagen in March 1995; a Summit that coincided with a news story breaking from Singapore which featured Nick Leeson as the man who broke the back of the British merchant bank, Barings (*see* Fig. 7.2). Leeson, it transpired, had been gambling with electronic representations of other people's money in the futures and derivatives markets of east and south-east Asia. The fact that Leeson gambled away almost £700 million was proof for many commentators of the sickness of the modern world economy. Bell took this diagnosis a stage further by juxtaposing the 'unreal' worlds of currency trading and the real worlds of future poverty.

Yet even here things aren't what they seem. For a start, Bell is not concerned to deconstruct 'poverty'. Poverty, for Bell, is a reality and not just a concept put into play by the West to (mis-)represent the Third World. Some post-developmentalists seem unwilling to acknowledge this reality (Yapa, 1996). Second, the collapse of Barings Bank did not lead to a more general crisis of confidence in the world's financial system. The Crash of 1929 was not replayed as the Crash of 1994. New systems of financial insurance had long since been put in place to guard against this possibility. Third, if the Barings story begins and ends in

London, most of the action takes place in the Asian Tiger economies. Notwithstanding the difficulties that many of the Tigers experienced in 1997–8 (and the severity of the difficulties should not be underestimated: South Korea may be the world's eleventh largest economy measured in conventional GDP terms, but it is hugely in debt to the IMF and other creditors), the rise to economic prominence of large parts of east and south-east Asia since 1960 suggests that 'development' has not been a failure, pure and simple (see Amsden, 1989; Wade, 1990). Some countries have done better than others and we need to ask why.

These points can all be generalized. One objection to 'post-developmentalism' (and to Escobar's work) is that it collapses a set of very different, and constantly changing, observations on the processes of development into a singular discourse of developmentalism. It is ironic, perhaps, that in texts which demand a sensitivity to different voices and different experiences of social life, we sometimes find an author reducing 'the' ideology of developmentalism to phrases delivered in a speech in 1949, or to the cruder models of modernization published by Lewis and Rostow in the 1950s and early 1960s (Lewis, 1955; Rostow, 1960). But this will not do. It is misleading to suggest that a singular discourse of development has continued unchanged over the past 30 or 40 years, or that mainstream development theorists and practitioners haven't come to

Figure 7.2 It's a mad world. Credit: © Steve Bell 1995

terms with gender or environmental issues, or with those new configurations of the global political economy which make terms like First World, Second World and Third World unhelpful or even redundant. The current President of Brazil, Fernando Henrique Cardoso, was himself a noted theorist of development in the 1960s and 1970s. In 1977 he took issue with the pessimistic views of those who maintained that the capitalist industrialization of the periphery of the world system was logically impossible (because the development of some few countries depended on the active underdevelopment of many more countries). Cardoso turned his eyes to the emerging Asian Tiger economies, and declared that: 'history is preparing a trap for the pessimists'. To read the work of some anti- or post-developmentalists today one would think that this history lesson has still not been learned.

But it is not only theory that should concern us. In a recent review of *Encountering Development*, David Lehmann has explored how Escobar mines the historical record of 'developmentalism' to paint development in the blackest of terms (Lehmann, 1997). Development, in Escobar's work, is reduced to the Sahelian famine, the debt crisis, gigantic dams and other failed development projects. In less sophisticated texts development becomes synonomous with the Mexican default of 1982 or the Bhopal disaster of 1984. Such an approach has the virtue of 'keeping the raw nerve of outrage alive' (as E.P. Thompson once put it with regard to the build-up of nuclear weapons), but it also draws the eye from the very real changes and improvements that have occurred in many poor coun-

tries since 1950. Singular and disturbing incidents deflect attention from more continuous and broad-ranging data sets; data sets which record, for example, that 'More people today live longer, healthier, and more productive lives than at any time in history' (World Bank, 1992, p. 24). And this is not just World Bank propaganda. The World Development Report that I have just quoted from goes on to acknowledge that these 'gains have been inadequate and uneven. More than 1 billion people still live in abject poverty' (ibid.). The point, though, is that further reductions in poverty are unlikely to be secured without 'sustained and equitable economic growth' (ibid.). The point, too, is that many such reductions in poverty, or improvements in life choices and capabilities, have been secured since 1950, for women as well as for men (*see* Fig. 7.3). In India, in 1990, life expectancy at birth for men was 60 years, for women it was 58 years; in 1965 the corresponding figures were 46 and 44 years. In China the relevant figures would be 71 and 69 years (1990) and 57 and 53 years (1965). India and China are home to just over one-third of the world's population. Such improvements – even in countries beset by continuing and appalling problems of female infanticide – hardly suggest that development has 'failed'; nor, for that matter, do the recent economic or epidemiological histories of Botswana or Chile, Malaysia or Kenya suggest anything so one-dimensional. If they are suggestive of anything, they are suggestive of what John Toye calls the dilemmas of development, or the failure of sustained economic growth to take hold in many poor regions (Toye, 1993).

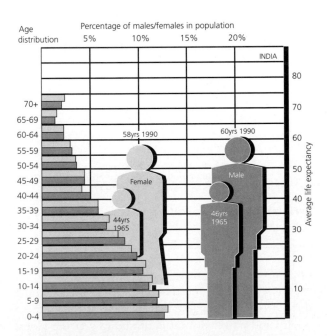

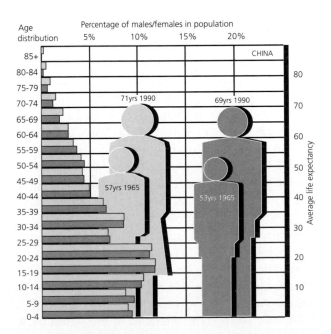

Figure 7.3 Life and death in India and China

Such distinctions matter when it comes to politics and to broader characterizations of the global political economy. Two dangers implicit in the romanticized claims of post-developmentalism are as follows: (a) such claims suggest that sustained economic growth will not significantly affect rates of poverty in poorer countries; and (b) they suggest that people can *only* be empowered or (post-)'developed' on a local scale and in opposition to the forces of growth, development and globalization. The first of these suggestions is not borne out by the cross-country data sets that are available to us. Sustained economic growth is not by itself a guarantee of poverty reduction (and far less a panacea), but it is difficult to secure reductions in absolute poverty over the longer term without the benefits of economic growth (Department for International Development, 1997). These benefits would include cheaper food, more employment and higher wages. The second suggestion is misleading in so far as it sets up a local-as-good versus global-as-bad dichotomy. Men and women can be empowered at the local level, and on very many occasions they need to defend themselves against the forces of brute modernity at this level. But political actions at this level *only* are unlikely to be successful, and to the extent that they do succeed they might cut people off from the pleasures of development. I have yet to meet a man or a woman in the village where I have been working for several years in Bihar, India, who wouldn't want electricity to be brought to the village. We should be wary of assuming that 'local' people want very different things from 'developers' or Westerners. In addition, this second suggestion constructs the extra-local as a threat, and not as a possible threat *and* site of empowerment. Multinational corporations get demonized for what they take out, rather than for what they put in and take out. Foreign aid is stigmatized as 'neo-imperialist', rather than worked on and through as a means of forcing richer countries to come to terms with at least some of their obligations to the needs and rights of distant strangers in poorer countries (possibly with the help of local non-governmental organizations). And the World Bank and the International Monetary Fund are chastised for imposing stabilization and adjustment programmes on indebted poor countries, when the harder issues to face up to include the following: (a) what is the alternative?; (b) how can people be empowered locally in a country like Bolivia in 1985 which had an annual rate of inflation of 5,000 per cent: might economic stabilization at the macro-scale be a precondition for effective local actions?; and (c) how realistic is it to assume that countries like India, or even Uganda or Indonesia, can be bossed around by the World Bank and the IMF (isn't this to rob such countries 'in writ-ing' of those few powers they might have accumulated in practice since decolonization)? More generally still, we should ask whether human resource development is (or might be) positively related to rates of global integration and private sector development, as the World Bank likes to maintain (*see* Figure 7.4)?

Development diamonds for Ghana, Thailand, and Malaysia, 1989

Writing some years before the Asian financial crisis of 1997, the World Bank maintained that:

> The recent dynamic growth in East Asia shows what can be achieved by pragmatic government policies and a disciplined hard-working population that responds to the right incentives. Malaysia and Thailand, which were poorer than Ghana in the 1960s, managed to double per capita income and reduce poverty dramatically in about ten years. They have now eclipsed Ghana in other measures of development as well [*see* Fig. 7.4]. Ghana can profit from East Asia's experience by emphasizing three areas: education and health, openness to international markets, and partnership between government and the private sector.

(World Bank, 1994: 40)

This view – this policy agenda – is far removed from the viewpoints and agendas being pressed by post-developmentalists. What do you make of the World Bank's perspective? What further evidence would you need to make an informed assessment of the Bank's policy agendas?

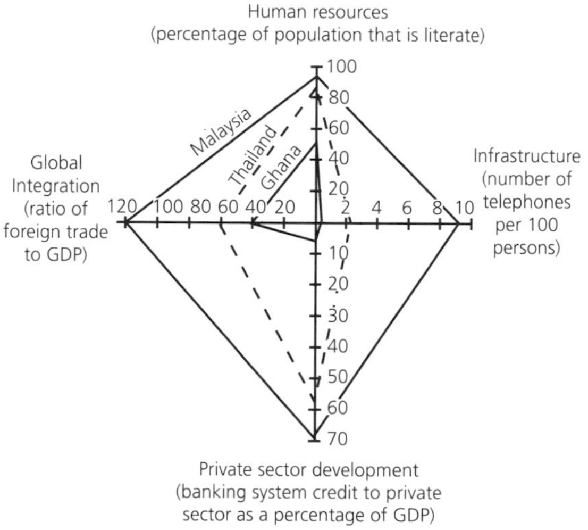

Figure 7.4 Development diamonds for Ghana, Thailand and Malaysia, 1989

Questions such as these may be misunderstood or even resented in some quarters. My point in raising them is not to suggest that massive inequalities of wealth and power do not continue to scar the global political economy at all spatial levels; my purpose, rather, is to 'render problematic that which can be taken to be self-evident' in some parts of Human Geography, and to set the scene for some closing remarks on the changing shape of the global political economy.

Summary

- Some accounts of post-developmentalism oversimplify the processes of development and reduce them to a one-dimensional and negative stereotype.

- Excessive concentration on the discursive aspects of development can direct attention away from the materiality of many social problems (including high and persisting rates of absolute poverty in some countries), and from the very real successes that we might associate with intentional (willed, planned) and immanent development since about 1950.

- A comparison of the development trajectories of Malaysia and Ghana does not suggest that integration into the global economy is detrimental to human development. To the contrary: efficient and equitable use of private and public capitals (and social capital formation) is worth commending.

ACCESS, RISK AND EMPOWERMENT IN THE GLOBAL POLITICAL ECONOMY

How, then, might we characterize the global political economy at the end of the twentieth century, and how best can we get to grips with the problems and opportunities facing poorer countries in our changing world? I have been trying to sketch an answer to the first of these questions throughout this chapter. In my view, we need to develop characterizations of the global political economy: (a) that recognize the enduring legacies of imperialism and colonialism in the production of rich and poor countries (to wish away the 'Third World' plays into the hands of those who refuse to acknowledge these legacies and the continuing inequalities in wealth and power that divide most postcolonial countries from their colonial oppressors: *see* Fig. 7.5); (b) that recognize, at the same time, that some developing countries have been able to access the global market-place more successfully than some other developing countries over the past 40 years, so much so that it no longer makes sense to describe some of the Southern Cone countries in Latin America or the Asian Tigers as 'developing countries'; and (c) that recognize, further, that recent developments in communications and information technologies are encouraging well-situated regions, classes and ethnic groups in some poor countries to access the global market-place on not unfavourable terms – regions that would include southern China, the Delhi–Mumbai corridor (and Bangalore) in India, and Mexico City and its hinterland. Such characterizations allow us to move beyond maps of the global political economy that are in tow to a mosaic model of a world system made up of nation–states alone (see Chapter 3 by Philip Crang). It allows us to take account of 'First Worlds in the Third World' (south New Delhi would be a good example), and of 'Third Worlds in the First World' (like south-central Los Angeles, or Albania and Bosnia-Herzegovina in Europe). At the same time, and to repeat, it encourages us to grasp these complexities and interconnections without losing sight of the broader picture, which remains one of disturbingly solid asymmetries in wealth and power between the OECD world economy and what is sometimes called a Fourth World of low-income countries.

But even these characterizations are only a start, and they remain rooted in economic considerations.

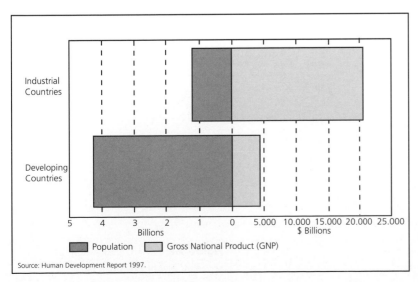

Figure 7.5 Our divided world – shares of population and income 1994

When it comes to power relations within the global political economy we see similar fractures, divisions, shifts and renegotiations. The major fractures and divisions exist between the G8 powers and the rest of the world. Only a fool or a charlatan would deny that countries like the USA, Germany and Japan enjoy economic and political powers enjoyed by few other countries, or that the USA, in particular, exerts considerable influence on institutions like the World Bank and the IMF, and on 'the market' through Wall Street and the Federal Reserve System. Nevertheless, many students of international relations would also accept that power in the world economy is much less centred on states in general, and the USA in particular, than it was in the heyday of the Bretton Woods system (roughly 1944–71). The USA shares power today with countries like Germany and Japan, regional blocs like the European Union, and major market-makers like financial and industrial corporations and the Bretton Woods institutions. Figure 7.6 offers one representation of these new configurations of power. It combines several of the observations made above, and is meant to invoke a **deterritorializing** world economy based on flows of capital and information as well as fixed production systems. It also makes reference to the idea of First World–Third World collisions or hybrids, and to the possibility of economic power (hegemony) being exercised by agencies other than nation states.

What these new landscapes threaten or promise is much debated. Optimists like Frances Cairncross see limitless opportunities for economic growth throughout the world economy, based on the diffusion of new information and digital technologies (Cairncross, 1997). Manuel Castells, in contrast, foresees a world economy which will divide ever more sharply between an expanding and geographically disparate group of individuals and countries who command

various 'spaces of flows', and those who are poor in information and capital (Castells, 1996). Castells' map of the world in the twenty-first century is not unlike the world I have described in Figure 7.6.

Summary

- Modern maps of 'development', or wealth and poverty, cannot reasonably be confined to countries or so-called nation–states. New industrial and informational possibilities are adding to existing richer spaces in poor countries, and poorer spaces in rich countries. This trend can be expected to continue.

- Although richer countries and corporations continue to dominate the global political economy, most developing countries and regions are not powerless and new opportunities for power sharing are beginning to emerge. It is part of the task of development geography to explore the conditions under which economic and political agents in different geographical areas are or might be empowered to access different markets and different state structures.

CONCLUSION

If Castells is right, as I think he might be, it is a fair bet that questions of market-access, empowerment and risk management will be at the heart of studies of development and the global political economy in the decades ahead. How best to opt in might be the question, not how to opt out. So, how can people, regions and countries be empowered (at all spatial levels) to *access* the different market-places and information opportunities that might be made available to them? In most poor regions and countries, men, women and children first need to be *empowered* by gaining access

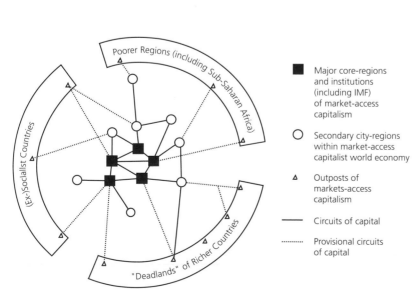

Major core-regions and institutions (including IMF) of market-access capitalism

○ Secondary city-regions within market-access capitalist world economy

△ Outposts of markets-access capitalism

—— Circuits of capital

········· Provisional circuits of capital

Figure 7.6 New configurations of economic plenty, power and poverty

to publicly provided systems of education and health-care – systems that presuppose effective governmental and non-governmental institutions. States matter; governments make a difference. Development studies is coming to terms with this agenda through studies of governance, institutions and social capital formation (see Fox, 1996; Harriss *et al.*, 1995; Leftwich, 1995; Platteau, 1994, Tendler, 1997). Poor regions and countries also need to win access to national and international market-places, and this will require further debt reductions, continued actions to secure fairer trading systems world-wide, better rights for workers employed by foreign and domestic firms, less exploitation of rural groups by urban groups (not least in sub-Saharan Africa), and so on. Development studies is coming to terms with this in part by means of major comparative studies of regional growth trajectories, and through considered studies of international trade and capital flows. As regards social and environmental *risk*: how best can individuals, (gendered) households and groups, regions, countries and indeed planet Earth be insured against the instabilities that dog any large-scale economic system but which are expanded (and made less even or equal) when the engines of economic expansion are removed from state control and entrusted to 'the market' and market-makers? How, too, can we be protected from the threat of ecological catastophe that some critics believe will accompany a rapid and uncontrolled expansion of the industrial bases of a large number of countries?

In my view, there are (provisional) answers to these questions. Scholars and practioners are experimenting with international regimes to coordinate reductions in CO_2 emissions, for example, or a possible tax on foreign exchange transactions (to dampen down the global casino so alarmingly stoked up by Leeson and his pals). In my view, too, these provisional solutions fall a long way short of the 'delinking' and anti-development arguments that are gaining ground in some parts of social science and in some people's movements. Development is about dilemmas and contradictions, and we should not expect to smooth out these contradictions in the name of some idealized end-state, be it the 'free market', communism, or post-development. But this is only my view (and my view right now). It is your job, and mine, to answer back; to render this chapter (and others) problematic and to suspend its claims to be self-evident.

Further reading

- Escobar, A. (1995) *Encountering development: the making and unmaking of the third world*. Princeton: Princeton University Press.
The key academic text so far in the pantheon of post-developmentalism. Demanding at points, but generally well written. See Lehmann (1997) for a coruscating review.

- Ferguson, J. (1994) *The anti-politics machine: 'development', depoliticization and bureaucratic power in Lesotho*. Minneapolis: University of Minnesota Press.
A disturbing account of the power of the aid industry and various national and international bureaucrats to capture and depoliticize local development agendas. Again, read it critically: does Ferguson's critique of livestock development projects in Lesotho add up to a sustained critique of immanent and intentional development (as he seems to imagine)?

- Brohman, J. (1996) *Popular development: rethinking the theory and practice of development*. Oxford: Blackwell.
A thoughtful contribution from a geographer who mixes and tries to match ideas emanating from radical–political economy and post-developmentalism.

- Toye, J. (1993) *Dilemmas of development: reflections on the counter-revolution in development theory and policy*. (2nd edition), Oxford: Blackwell.
A sustained but fair critique of New Right ideas in development studies. The New Right made the running in development policy in the 1980s and early 1990s, although you wouldn't always think so reading the work of Human Geographers. It is important that we engage with these ideas.

- Agnew, J. and Corbridge, S. (1995) *Mastering space: hegemony, territory, and international political economy*. London: Routledge.
Chapter 7, on 'Transnational Liberalism', expands several of the points made in this chapter.

- Stallings, B. (ed.) (1995) *Global change, regional response: the new international context for development*. Cambridge: CUP.
Hits the target.

CHAPTER 8

Survival and resistance

Paul Routledge

INTRODUCTION: THE THREATS TO PEOPLE'S SURVIVAL

On 27 September 1987, in the Tibetan capital, Lhasa, Buddhist monks stage a peaceful demonstration at Jokhang Temple carrying forbidden Tibetan flags and demanding independence. They are arrested and charged with fomenting counter-revolution.

In late December 1990, on the road that joins the Indian states of Madhya Pradesh and Gujarat, a procession of several thousand predominantly tribal peoples – who have marched 250 km across central India to protest the construction of the Narmada Dam – are confronted by armed police and prevented from crossing the state line.

On New Year's Day 1994, ski-masked Mayan guerrillas emerge from the Lacandon jungle, capture the town of San Cristobal de las Casas in Chiapas, Mexico, and declare war on the Mexican state.

In late April 1997, on the horizon before Brasilia, the hoes and machetes of thousands of *campesinos* glint in the sunshine as this movement of Brazil's landless peasants approaches the country's capital to demand land from the government.

While very different, these four moments are examples of conflicts that are occurring in both the developed and developing countries. These conflicts represent mobilizations of the dispossessed, the poor, the marginalized: those threatened with, or experiencing, displacement, **cultural ethnocide** and the grinding ravages of poverty. They speak of struggles over the allocation of resources, over self-determina-

tion and over rights of economic and cultural survival, and it is to a consideration of these struggles that this chapter is devoted.

Within the global economy, economic development has had a variety of results. Certain locally-based development practices have improved health and education services, enhanced environmental quality, and generated employment opportunities that foster equity and self-sufficiency. However, many of the development practices that have been enacted throughout developing countries have resulted in the **pauperization** and marginalization of indigenous peoples, women, peasant farmers, and industrial workers, and the deterioration of labour, social, and environmental conditions.

Such development, which has been state-directed, and/or financed by private corporations and international institutions such as the World Bank and International Monetary Fund, has emphasized economic growth, modernization and industrialization as the panacea for poverty (see Chapter 7 by Stuart Corbridge). Capital-intensive schemes have displaced traditional and subsistence economies which are labour-intensive (often resulting in unemployment); and Western values (of capitalist production, economic growth) have been emphasized at the expense of indigenous and traditional systems of knowledge, economy and culture. As a result, peasants, women, children and tribal peoples have been viewed as impediments to progress and modernity and excluded from decision-making and participation in the development process. Moreover, such development has facilitated both the state's and transnational corporations' securing control over natural and financial resources and consolidated the power of those directing and benefiting from the development apparatus – national ruling elites, and international institutions (Nandy, 1984). In the process, traditional subsistence economies and their associated cultures are destroyed, people face displacement from their homes and lands, lose access to their resources, and become economically marginalized.

In addition to the threats posed by development, are those posed by the practices and policies of particular states. These may be aimed at enabling development practices to take place, or they may be aimed at securing the state's political (and economic) control over resources and territory. These territories are frequently inhabited by groups who perceive such state policies as an intrusion on their political and cultural rights. The assault upon the lifestyles of these groups – which include indigenous peoples and peasants – has led to the emergence of myriad social movements who articulate struggles for political autonomy, and cultural, ecological and economic survival. This chapter considers the four examples of contemporary popular resistance that opened this chapter: the Tibetan Independence Movement; the Save Narmada Movement of India; the Zapatistas of Mexico; and the Landless Movement of Brazil. It investigates how geography – and geographers – can lend important insights into these struggles.

GEOGRAPHY AND RESISTANCE

Human Geography can lend some important understandings to people's resistance, providing valuable insights into the place-specific character of struggles, explaining why these conflicts arise, and why they emerge where they do. This is because different social groups endow space (and its associated resources) with a variety of different meanings, uses and values. Such differences can give rise to various tensions and conflicts within society over the uses of space for individual and social purposes, and the control of space by the state and other forms of economic and cultural power such as transnational corporations. As a result, particular places frequently become sites of conflict between different groups within society, which reflect concerns of ecology (e.g. struggles to prevent deforestation and pollution), economy (e.g. peasant struggles to secure land on which to grow food), culture (e.g. struggles to protect the integrity of indigenous people's communities), and politics (e.g. struggles for increased local autonomy). These concerns are also associated with what Gedicks (1993) terms the 'resource wars': the struggle over the remaining natural resources between indigenous and traditional

peoples, state and national governments and transnational corporations.

In response to these different concerns, people frequently organize themselves into social movements, which are ongoing collective efforts aimed at bringing about particular changes in a social order. Many of these social movements are place-specific actively affirming local identity, culture and systems of knowledge as an integral part of their resistance. In doing so, these movements articulate localized '**terrains of resistance**' with their own place-specific idioms of protest, which motivate and inform their struggles. However, resistance can also be global in character, spanning both national and international space. This is because with the impacts of globalization – where localities are increasingly influenced by non-local economic and cultural forces – many social movements feel obliged to focus their resistance both within and beyond the confines of their immediate locality, in order to attract as wide a support for their struggle as possible. Such 'globalized resistance' (Brecher and Costello, 1994) often involves coalitions of different social movements and **non-government organizations** who co-ordinate their struggles across a variety of levels in response to the emerging global economy and the actions of particular governments.

Summary

- Different social groups endow space and its resources with different meanings, uses and values arising in conflicts over the uses and control of particular places.

- Certain development practices have emphasized economic growth and industrialization as the solution to poverty, and viewed indigenous and traditional systems of knowledge as impediments to progress.

- As states and transnational corporations seek control over resources and territory for development, so the inhabitants of the areas concerned are frequently displaced and economically marginalized.

- In response to such development processes, people often organize themselves into place-specific social movements to resist threats to their economic and cultural survival.

- Geography can provide insights into the place-specific character of resistances, explaining why these conflicts arise, and why they emerge where they do.

SOCIAL MOVEMENT STRUGGLES: RESISTANCE FOR SURVIVAL

Social movements operate on a number of interrelated 'levels' within society. At the level of the economy, they articulate conflicts over access to productive natural resources such as forests and water, as well as conducting struggles in the workplace. The economic demands of social movements are not only concerned with a more equitable distribution of resources between competing groups, but are also involved in the creation of new services such as health and education in rural areas (Guha, 1989). Indeed, social movements have emerged in many areas, including civil liberties, women's rights, and science and health, that are themselves often related to problems caused by development. At the level of culture, social movement identities and solidarities are formed, for example, around issues of class, kinship, neighbourhood, and the social networks of everyday life. Movement struggles are frequently cultural struggles over material conditions and needs, and over the practices and meanings of everyday life (Escobar, 1992).

At the level of politics, social movements are frequently autonomous of political parties (although some have formed working alliances with trade unions, voluntary organizations and non-government organizations). Their goals frequently articulate alternatives to the political process, political parties, the state, and the capture of state power. By articulating concerns of justice and 'quality of life', these movements enlarge the conception of politics to include issues of gender, ethnicity, and the autonomy and dignity of diverse individuals and groups (Guha, 1989).

At the level of the environment, social movements are involved in struggles to protect local ecological niches – e.g. forests, rivers, and ocean shorelines – from the threats to their environmental integrity through such processes as deforestation (e.g. for logging or cattle grazing purposes) and pollution (e.g. from industrial enterprises). Many of these social movements are also multidimensional, simultaneously addressing, for example, issues of poverty, ecology, gender, and culture.

Social movements are by no means homogenous. A multiplicity of groups including squatter movements, neighbourhood groups, human rights organizations, women's associations, indigenous rights groups, self-help movements amongst the poor and unemployed, youth groups, educational and health associations and artists' movements are involved in various types of struggle. Many of these struggles take place within the realm of civil society, i.e. those areas of society that are neither part of the processes of material production in the economy nor part of state-funded organizations, and can be either violent or non-violent in character.

While social movements in the developed and developing countries share some of the broad charac-

teristics mentioned above – e.g. they articulate such issues as ecology, gender and ethnicity – there are also important differences between them. In the developed countries, social movements have often concentrated upon 'quality of life' issues, whereas in developing countries, movements have often focused upon the access to economic resources. An example of this difference is represented by the issues faced by ecological movements. In the developed countries, the ecology movement has taken much of the industrial economy and consumer society for granted, working to preserve nature as an item of 'consumption', as a haven from the world of work. In the developing countries, however, those affected by environmental degradation – poor and landless peasants, women and tribals – are involved in struggles for economic and cultural survival rather than the quality of life. Such groups articulate an 'environmentalism of the poor' (Martinez-Allier, 1990), whose fundamental concerns are the defence of livelihoods and of communal access to resources threatened by commodification, state take-overs, and private appropriation (e.g. by national or transnational corporations), and with emancipation from material want and domination by others.

Social movements rarely articulate their demands on only one level. As we shall see below, economic struggles may also contain political dimensions, political struggles may also contain cultural elements, and so on. Moreover, the responses of state authorities to social movements vary, according to the type of movement resistance, and the character of the government involved. When faced with social movement challenges, governmental responses include repression, cooption, co-operation, and accommodation. Repression can range from harassment and physical beatings, to imprisonment, torture and the killing of activists.

At the level of the economy, social movements articulate conflicts over the productive resources in society such as forest and water resources, involving demands for a more equitable distribution of resources, the creation of new services, and the integrity of local, traditional forms of economic practice. The *Movimento Sem Terra* (MST), or Landless Movement, in Brazil provides an interesting example of the organized struggle for access to land resources (*see* Fig. 8.1). The MST is a mass social movement of some 220,000 members which has developed during the past 17 years, and is made up of Brazil's dispossessed – the croppers, casual pickers, farm labourers and people displaced from the land by mechanization and by land clearances. Many of those involved are homeless, or live in roadside tents, and earn less than 60 pence a day. Of Brazil's population of 165 million

Figure 8.1   Sem Terra activists hold aloft machetes and hoes which have become symbols of their struggle for land. Credit: ©
Sebastiao Selgado/Network Photographers

people, fewer than 50,000 own most of the land, while 4 million peasants share less than 3 per cent of the land. While approximately 32 million people in Brazil are malnourished, over 42 per cent of all privately-owned land in Brazil lies unused. Hence the principal demand among the dispossessed has been for land. The MST targets Brazil's vast estates that lie unused. First, groups of people illegally squat on the uncultivated land in 'land invasions'. After the area has been 'secured' the MST resettles massive numbers of people on the squatted sites, who then construct houses and schools, and commence farming. Since 1991, the MST has occupied 518 large ranches and resettled approximately 600,000 people. The process has been far from peaceful, as the large landowners and their private armies have attacked and killed the squatters. For example, in one incident in 1996, in the state of Para, 19 MST activists were shot dead by police in the pay of local landowners (Vidal, 1997).

While the right of the government to redistribute land that is not being farmed is enshrined in the Brazilian constitution, successive regimes have failed to exercise this right, due in part to the political power wielded by the country's landowners. However, the Brazilian government has begun to tentatively initiate agrarian reforms, providing credit for new settlements, and confiscating some of the ranches

in the state of Para to settle some of the families of those MST members killed in 1996. This is due, in large part, to the success of the MST in mobilizing popular support for its cause, and its ability to develop alliances with trade unions and other grassroots organizations. Evidence of this support was dramatically shown in the mass demonstration in Brasilia, in April 1997, mentioned at the beginning of this chapter, where over 120,000 landless people lined the streets to demand land reform (Vidal, 1997). In attempting to change the government's agrarian policy, the MST is also operating within the field of political action.

At the ecological level, social movements struggle to protect remaining environments from further destruction, and to ensure the economic (and cultural) survival of peasant and tribal populations. An example of such a struggle is that of the resistance against the Narmada River valley project in India. This river, which is regarded as sacred by the Hindu and tribal populations of India, spans the states of Madhya Pradesh, Maharashtra, and Gujarat, and provides water resources for thousands of communities. The project envisages the construction of 30 major dams along the Narmada and its tributaries, as well as an additional 135 medium-sized and 3,000 minor dams. When completed, the project is expected to flood 33,947 acres of forest land, and submerge an estimated 248 towns and villages. According to official estimates (based on the outdated 1981 census), over 100,000 people – 60 per cent of whom are tribal – will be forcibly evicted from their homes and lands. With two of the major dams already built, opposition to the project has been focused on the Sardar Sarovar reservoir, the largest of the project's individual schemes. The resistance to the project has been coordinated by the *Narmada Bachao Andolan* (Save Narmada Movement) – a network of groups, organizations and individuals from various parts of India who have demanded the curtailment of the scheme (*see* Fig. 8.2).

The movement's repertoire of protest has included mass demonstrations, road blockades, fasts, public meetings, and disruption of construction activities. While localized protests have occurred along the entire Narmada valley, wider public attention has been drawn to spectacular events such as mass rallies, and the protest march mentioned at the beginning of the chapter. While the movement has been almost completely non-violent, its leaders and participants have been harassed, assaulted, and jailed by police. In one tragic incident, police opened fire on a demonstration in the Dhule district of Maharashtra, killing a 15-year-old boy. The movement has attracted widespread global support from various environmental groups and non-government organizations such as Survival International. However, despite the resistance, construction of the dams continues. In representing a threat to the ecology of the area surrounding the Narmada river, the construction of the dams also threatens the economic survival of the tribal and peasant peoples who will be evicted from their homes and lands – from which they earn their livelihoods – when the land is submerged. Moreover, these inhabitants have a profound religious connection to the landscape around the Narmada River. This spiritual connection to place – which eviction threatens to sever –

intimately informs their customs and practices of everyday life. Hence opposition to the dam also articulates the inhabitants' desire for cultural survival. In addition, many of the villages that border the Narmada are demanding a level of regional autonomy, seeking 'our rule in our villages', thereby articulating political demands as well (Gadgil and Guha, 1995).

At the political level, social movements challenge the state-centred character of the political process, articulating critiques of **neoliberal** development ideology and of the role of the state. One of the most prominent recent examples has been that of the *Ejercito Zapatista Liberación National* – the EZLN or the Zapatistas – in Chiapas, Mexico, which has articulated resistance to the North American Free Trade Agreement (NAFTA) and the Mexican state. The Zapatistas, a predominantly indigenous (Mayan) guerrilla movement, have emerged in Chiapas due to several factors. First, the state of Chiapas is rich in petroleum and lumber resources which have been ruthlessly exploited causing deforestation and pollution. Second, the increasing orientation of capital-intensive agriculture for the international market has led to the creation of a class of elite wealthy farmers, and forced Indian communities to become peasant labour for the extraction and exploitation of resources, the wealth of which accrues to others. In addition, large landowners and ranchers control private armies who are used to force peasants off their land, and to terrorize those with the temerity to resist. Third, although it is resource-rich, Chiapas is amongst the poorest states in Mexico with 30 per cent of the population illiterate, and 75 per cent of the population malnourished. Fourth, the production of two of the main crops from which *campesinos* (peasants) earn a living in Chiapas – coffee and corn – has undergone severe economic problems in recent years, and will be further damaged by NAFTA. Finally, government reforms in 1991 enabled previously protected individual and communal peasant land-holdings to be put up for sale to powerful cattle ranching, logging, mining and petroleum interests.

The Zapatistas initially engaged in a guerrilla insurgency by occupying the capital of Chiapas and several other prominent towns in the state, as noted at the beginning of the chapter. However, they staged their uprising in a spectacular manner to ensure maximum media coverage, and thus gain the attention of a variety of audiences – including civil society, the state, the national and international media, and international finance markets. For, although their guerrilla bases were in the Lacandon jungle in Chiapas (*see* Fig. 8.3), the Zapatistas were particularly concerned to globalize their resistance. The appearance of an armed insurgency, at a moment when the Mexican economy

Figure 8.2 Indian activist Medhar Patkar addresses Narmada demonstrators. Credit: © Patrick McCully/Survival

Figure 8.3 Although based in the Lacandon Jungle, the Zapatistas have been able to globalize their resistance through creative use of the internet. Credit: © Antonio Turok

was entering into a free trade agreement, enabled the Zapatistas to attract national and international media attention. Through their spokesperson, Sub-commandante Marcos, the Zapatistas engaged in a 'war of words' with the Mexican government, fought primarily with rebel communiqués (via newspapers and the Internet), rather than bullets. Through their guerrilla insurgency and their war of words the Zapatistas have attempted to raise awareness concerning the unequal distribution of land, and economic and political power in Chiapas; to challenge the neoliberal economic policies of the Mexican government; to articulate an indigenous worldview which promotes Indian political autonomy; and to articulate a call for the democratization of civil society. The success of the Zapatista struggle has lain in its ability, with limited resources and personnel, to disrupt international financial markets, and their investments within Mexico, while exposing the inequities on which development and transnational liberalism are predicated (Harvey, 1995; Ross, 1995).

However, despite certain successes, the movement has been faced with repression from the Mexican government. Over 15,000 army personnel have been deployed in Chiapas; villages suspected of being sympathetic to the Zapatistas have been bombed; and peasants suspected of being Zapatistas have been arrested and tortured. At present an uneasy cease-fire is in place between the Zapatistas and the government and peace-talks between them are stalled. Since its emergence in 1994, the Zapatistas have posed more than just a political challenge to the Mexican state. In their demands for equitable distribution of land, their calls for indigenous rights and ecological preservation (i.e. an end to logging, a programme of reforestation, an end to water contamination of the jungle, preservation of remaining virgin forest), they also articulate an economic, ecological and cultural struggle.

At the cultural level, social movements frequently affirm and regenerate local (place-specific) identity, knowledge and practices, which at times are expressed in the language and character of the struggles. Local resistance may incorporate local linguistic expressions, such as songs, poems, and dramas that imbue and affirm local experiences, beliefs, and cultural practices. An example of such struggles would be that of the Tibetan people's resistance against Chinese occupation of their country.

In October 1950, the People's Liberation Army of China invaded Tibet and annexed the country. During the next 44 years the Chinese government has attempted to integrate the political, economic, and cultural life of Tibet into the People's Republic of China. The consequences of this policy have been a systematic assault upon Tibetan culture including the destruction of approximately 90 per cent of the Buddhist monasteries that existed before 1950; the population transfer of 6 million Chinese into Tibet in order to make Tibetans a minority in their own country; and the death of between 150,000–1,000,000 Tibetans due to disappearances, famine, executions and detentions.

Faced with suppression of their culture and civil liberties, Tibetans have articulated myriad forms of resistance against the Chinese occupation, including a periodic, localized armed struggle between 1951–59. Subsequently, resistance has taken a predominantly non-violent character, including demonstrations in Lhasa – the Tibetan capital – particularly since 1987, by Buddhist monks and nuns (such as that mentioned at the beginning of this chapter), appeals for international assistance, and the submission of petitions to the Chinese Army requesting that they leave Tibet. Between 1987 and 1993 there have been 178 peaceful independence demonstrations in Tibet. Non-violent resistance has also been conducted amongst the Tibetan diaspora, particularly in India, where the Tibetan government-in-exile is located. Since he fled Tibet in 1959, the Dalai Lama has pursued a policy of soliciting international support for Tibetan self-determination. In October 1959, 1961, and 1965, the United Nations have passed non-binding resolutions concerning the upholding of human rights in Tibet, which have been largely ignored by the Chinese government. Indeed, government repression of Tibetan resistance has included arrests, detention without trial, torture, and firing upon peaceful demonstrators resulting in fatalities (*see* Fig. 8.4).

Alongside openly declared demands for independence, Tibetans have also articulated myriad cultural expressions of resistance to Chinese occupation. Traditional hand-carved wood-blocks (traditionally used to depict Tibetan scriptures), have been produced

Figure 8.4 Peaceful demonstrations for Tibetan independence have frequently been met with Chinese repression such as arrest and detention without trial. Credit: The Tibet Information Network

by monks demanding independence. These woodblocks have then been used to run off leaflets, pamphlets, and manifestos for distribution both inside and outside of Tibet. Songs and poems have also been composed articulating resistance to the Chinese. They speak of prison conditions, and repression under the Chinese occupation, but also celebrate the Tibetan religion, the Dalai Lama, and articulate the desire for independence. Tibetans have also used forms of smuggling as a non-violent sanction. For example, pregnant Tibetan women are smuggled across the border with India so that they can have their babies born and educated in exile. Also miniature versions of the UN Declaration of Human Rights, translated into Tibetan, have been smuggled into Tibet to be distributed amongst the Tibetan population (Bennett and Routledge, 1997).

Under conditions of attempted cultural ethnocide, the practices of everyday life – the wearing of traditional dress; the use of the Tibetan language; the use of religious artefacts such as prayer flags; and the participation in religious rituals (such as prayer meetings) – have all become cultural expressions of resistance for Tibetans attempting to maintain their cultural identity. Moreover, in their struggle for independence, the Tibetan resistance also articulates a political struggle against Chinese occupation.

Summary

- Social movements articulate resistance within society in the realms of economics, politics, culture, and the environment.

- At the economic level, social movements articulate conflicts over the productive resources in society, involving demands for a more equitable distribution of resources, the creation of new services, and the integrity of local, traditional forms of economic practice.

- At the political level, social movements challenge the state-centred character of the political process, articulating critiques of development ideology and of the role of the state.

- At the cultural level, social movements frequently affirm and regenerate local (place-specific) identity, knowledge and practices, which at times are expressed in the language and character of the struggles.

- At the ecological level, social movements struggle to protect remaining environments from further destruction, and to ensure the economic (and cultural) survival of peasant and tribal populations.

CONCLUSION: RESISTANCE AND GEOGRAPHY

The four examples described above are but a few of the resistances to development projects and repressive governments, that have proliferated across the world, during the past 15 years. These have involved leftist guerrillas, social movements, non-government organizations, human rights groups, environmental organizations and indigenous peoples movements. Frequently coalitions have formed across national borders and across different political ideologies in attempts to revitalize democratic practices and public institutions, promote economic and environmental sustainability, encourage grassroots economic development, and hold transnational corporations accountable to enforceable codes of conduct. Such resistances are frequently responses to local conditions that are in part the product of global forces, and resistance to these conditions has taken place at both the local and the global level. In contrast to official political discourse about the global economy, these challenges articulate a 'globalization from below' that comprises an evolving international network of groups, organizations and social movements.

Human geography can provide valuable insights into the place-specific character of these different forms of resistance, explaining not only why conflicts emerge, but why they arise where they do. Hence, in our examples, the landless peasants of the MST have focused their struggle in those areas of Brazil that have large estates with unused land; the Narmada movement has been most active in those areas threatened by submergence by the construction of dams; the Zapatistas have emerged in Mexico's poorest and most economically and ecologically exploited state; and Tibetan resistance has been focused in Tibet's capital of Lhasa – a symbol of Tibetan political autonomy – and in Dharamsala in North India where the Tibetan government-in-exile is located. As geographers, then, we can contribute to the understanding of struggles for survival in different cultural contexts. Given that we live within an increasingly interdependent world, we might also consider ways of attempting to contribute towards these struggles, to make our contribution towards an environmentally sustainable and socially just world.

Further reading

For a fine collection of essays on the themes of development and social movements read:

- Peet, R. and Watts, M. (eds) (1996) *Liberation ecologies: environment, development, social movements*. London: Routledge.

For a discussion of globalization and resistance read:

- Brecher, J. and Costello, T. (1994) *Global village or global pillage*. Boston: South End Press.

Recent overviews of grassroots social movements around the world can be found in:

- Ekins, P. (1992) *A new world order*. New York: Routledge.
- Ghai, D. and Vivian, J. (1992) *Grassroots environmental action*. London: Routledge.

CHAPTER 9

Re-thinking development

Sarah A. Radcliffe

INTRODUCTION

In the 1990s, in what appears to be an integrated world, 'the global village', people in practice are living with highly differentiated experiences of, and stakes in, development. Cosmopolitan jet-setters in São Paulo live one kind of development while women in sub-Saharan Africa walking for hours to collect water experience a completely different kind of development. How do we listen to this difference? What analysis best suits this geographical diversity, while also linking our concerns to the obvious gender and class-based inequalities we see? Rethinking development means reconsidering the categories we use in development geography, and unpacking the power relations that shape them.

Since the widely perceived crisis, or 'impasse' in development studies in the mid-1980s, there has been a lot of rethinking of development going on. Never a month goes by without a new article or book proclaiming a new critique or way forward in what appears from the outside to be a morass of contradictory arguments between advocates of distinct approaches depending on new interpretations of social structures, liberation movements or human–environment relations. As can be seen from Chapter 7 and Chapter 8, the insertion of a critical sensibility into development studies has generated much heat – and hopefully light – on the questions about how to write and analyse this complex thing called development. To a novice of the development literature, there is even a question about whether development still exists as a condition of certain populations, or as a coherent interdisciplinary field, in which geography has made exciting inroads. Yet to read the latest publications on development (e.g. Booth, 1994;

Lehmann, 1997; Peet and Watts, 1996; Simon, 1997; Rowlands, 1997) is to be reassured that certain principles remain, including a moral commitment to eradicate poverty in practice and analyse its foundations in theory. While once overwhelmingly economic, development studies today comprises various theor*ies* in which the specificity and interrelationships of development's political, cultural and institutional facets are recognized (Booth, 1994: 305). Another principle is that the causes and effects of development are multifactoral, whether one sympathizes with 'post'-development perspectives or not (see Chapter 7). Finally, rethinking development is about power – its operations, its geographies, its highly uneven distribution, contradictions, and strategies for getting it, so the analysis of power is central to our post-'impasse' ideas. In the remainder of the chapter, three windows on to development's power will be examined, namely the power of writing, the power of gender and the powers of the state.

GEOGRAPHY AND POSTCOLONIALISM

Postcolonialism has attracted considerable attention in geography in recent years; the latest edition of the *Dictionary of human geography* (Johnston, 1994) – always a barometer of trends in contemporary geography – remarks on how postcolonialism will become increasingly important in geography in the future. But what does '**postcolonialism**' mean and how does it relate to development geographies? Postcolonialism refers to a way of criticizing the legacies of colonialism in the South (see Chapter 29), which are material (found in the state, and social organization as well as urban structures) as well as related to ideas (how people think about themselves and their relationship with the developed world). Postcolonialism thus critiques the ways in which the West has 'made' knowledge about the South, the Third World; postcolonialism also tries to work towards what Ngugi wa Thiong'o has called 'decolonizing the mind'. This 'decolonization of the mind' might involve new ways of learning, by including a wider range of people in the collection of information about development experiences. Postcolonialism thus does *not* refer to the status of independence achieved by most countries of the South by the 1960s. The Spanish American colonies gained independence between the 1820s and 1898, while British and French colonies in Africa and Asia became independent during the middle years of the twentieth century (*see* Fig. 9.1). Although the

term postcolonialism suggests a chronological break between the two statuses of colony and independence, we are now more aware of continuities rather than breaks. The continuing effects of industrial countries' economies and policies on the South, and the continued importance of models that see development through Western eyes illustrate this (McClintock, 1995).

As a form of critique, does postcolonialism help us re-think development? David Simon argues that postcolonialism takes us away from a static view of 'tradition' – seeing the Third World as unchanging – while removing the Northern labels on peoples and places which have done so much to create negative stereotypes about 'basket-case African countries' or tin-pot dictators in South America (Simon, 1997). By pulling

(a)

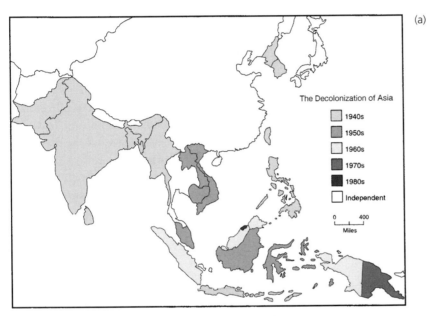

(b)

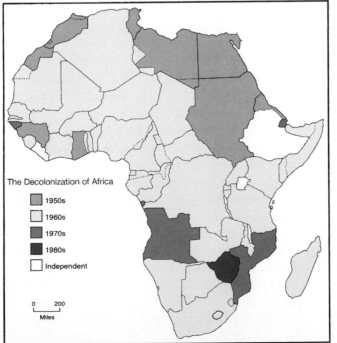

Figure 9.1 Decades of independence for former colonies in the South. (a) Asia (b) Africa

apart the history and geographies to categories such as 'Third World' and 'Third World women', their power can be taken apart, to be replaced by an understanding of how location, economic role, social dimensions of identity and the global political economy differentiate between groups and their opportunities for development. Moreover, postcolonialism has important implications for development geography as it challenges notions of a single 'path' to development.

While recognizing the need to overcome such gulfs in income and well-being that characterize North–South relations (*see* Fig. 7.5), postcolonial approaches acknowledge the diversity of perspectives to priorities in development. Needs, however basic, are not pre-given but are determined by social identities and local–regional cultures, within highly politicized contexts. Practical (basic) needs policy – such as the Basic Needs Approach (BNA) of the 1980s – addresses the concerns of 1.2 billion people in absolute poverty. Nevertheless, it risks obscuring their 'strategic' needs (that are more liberatory and small-p political). The politics of defining and satisfying needs is thus a crucial dimension of current development thought, getting us to look again at *who* is voicing a development concern; is it a Northern development expert? A First World feminist? An older male religious leader of a village? How do these participants' identities and structural roles in local and global society shape their priorities, and who is silenced as a consequence?

Postcolonialism also encourages us to consider that we – sitting in our libraries, far from the women collecting water in Nigeria – have a role to play in all of this, and that our identities and our worlds are made in that same global economy, as it works its uneven way around the world in ever-faster circles. Development is thus not only about economic relations and income differentials, but also about the politics of needs, about a political economy of racism (how colonialism and its relations underpin our attitudes to issues as varied as AIDS, famine relief and population growth), and about how tradition can be re-invented. Having outlined the powers entangled in the writing about development, we now turn to look at two dimensions of the development debate, namely issues of **gender** and **feminism**, and second, the state.

Summary

- Postcolonialism is a critique of the ways in which European colonialism and **neocolonialism** shape the experiences of, and the writing about, development. It also hopes to contribute to new ways of writing about 'knowing' development.

- Postcolonial perspectives encourage us to put the geography back into a world divided between First and Third Worlds, North and South, and rethink what is meant by traditional vs. modern.

CONTEXTUALIZING DEVELOPMENT: FEMINISM AND GENDER

In a world where women do two-thirds of the world's work, earn 10 per cent of the world's income and own less than 1 per cent of the world's property, the promise of 'postcolonialism' has been a history of hopes postponed.

(McClintock, 1995: 19)

In her critique of development, Anne McClintock highlights the particular ways in which postcolonialism celebrates too much, too soon when it comes to the position of women. National independence and the new international division of labour result in a world where women and men often experience widely differing lives, whether in terms of citizenship rights, work, or kin and family relationships. To take a global view, these differences reduce women's power, visibility and access to resources.

Yet while McClintock points to women's continued marginalization, she represents women as a homogeneous group in which differences are glossed over. This quote from McClintock galvanizes our political solidarity with women and human rights, yet it also reiterates what the feminist film-maker Trinh Minh-ha calls the 'third world difference', that is, a way of identifying all women in the South in a distinct category to Western women. As Chandra Mohanty has argued, the third world difference portrays Third World women as powerless, burdened, cloistered and oppressed by their (male) kin (Mohanty, 1991). Found in writings by colonial officials, development experts and geographers, this 'third world difference' is an example of a colonial legacy. A postcolonial perspective therefore attempts to examine Third World women as agents of change, and as highly differentiated. As feminist geographers have pointed out, gender divisions of labour and gendered distribution of resources and income depend on a large number of factors which differentiate women along lines of class, location, religion, age/generation, race/ethnicity, and sexuality.

In contrast to their passive stereotype, women in the South have diverse strategies for dealing with a combination of forces, including capitalist economies that see them as the cheapest labour; racism; colonialism, and male power in families and states. The gulf in living standards between women of different classes often translates into highly diverse strategies, with poor women facing the most difficulties in securing income, health and a voice for themselves and their communities. Social movements draw women together to struggle for livelihood, often alongside men (Chapter 8). The Chipko movement to protect forests in the Indian Himalaya was one movement in which men and women organized together to protect their meagre income and well-being. As well as public protests, women are engaged in small-scale, everyday forms of resistance, in their households and daily encounters. As Mohanty says, 'Agency is thus figured in the minute, day-to-day practices and struggles of third world women' (Mohanty, 1991: 38). Women are increasingly differentiated between those coping or not under conditions of structural adjustment and impoverishment; between female heads of households (accounting for an estimated one in three of the South's households) and women who are maintaining their households; and between women using public and private forms of resistance (Radcliffe and Westwood, 1993).

Ways of 'being female' are intimately bound up with the nature of production relations, the practices of the nation–state, ideologies and political culture as well as kinship. In this context, Third World women see their female condition resulting from the intersection of these processes; their actions are not driven by 'just' gender concerns, but by gender *in association* with struggles over class, race and ethnicity, colonial legacies and global capitalism. Feminism in the South is thus a problematic term; Third World women criticize First World feminists for their perceived overemphasis on separate 'women's issues'.

With Western development interventions, similar conflicts have arisen. In policy, early planners viewed Third World women as falling outside development and advocated their incorporation into the development process. Later, socialist feminists argued that incorporation had already occurred but was profoundly shaped by the relations of production. A focus on class reminds us that the costs of development fall particularly on low-income women, and men, while for many urban wealthy or middle-class women and men development benefits are extensive (Sen and Grown, 1987) (*see* Fig. 9.2). Since the 1980s, these ideas have been superceded by the Gender and Development model, or GAD. In this approach, male–female power relations are seen as central to an understanding of women's position in development, which involves women and men in different combinations of reproduction, production and community management work (Moser, 1993). The GAD approach highlights how gendered power results in women's unequal treatment in education, land distribution, training and credit provision, all key aspects of formal development.

However, not all the questions about why and how low-income women are marginalized systematically have been answered. Commentators have suggested that the GAD approach, for all its strengths, tends not to recognize the *diverse* forms of gendered power in different areas. Again, a recognition of complex geographies is called for. Additionally, the GAD approach tends not to incorporate a sense of how women *do* have capacity for action (although this is often constrained), and that female empowerment needs to be brought into analysis.

In the focus on gender, the discussion about men and masculinities in development has hardly begun. Overall, men and women in households, settlements and the state negotiate the ways in which work, value

Figure 9.2   Credit: © Angela Martin

and income are allocated between individuals and groups. Yet by emphasizing women's frequent disempowerment in these arrangments, the diverse positions and attitudes of men in this process are often forgotten. 'Patriarchy' or male power can no longer be viewed as if it were a universal male characteristic; rather, different masculinities (ways of being male) are at work. Many development projects are having to recognize that the views of men influence programme outcomes. Whether development projects are targeted at women-only groups (gender-specific projects) or at mixed groups, the attitudes of men in the community and their assistance or objections affect the project's outcome as well as women's confidence, as a case study from peasant associations in Southern Peru illustrates.

Masculinities and development: the case of Southern Peru

Indigenous (Indian) small farmers in the Peruvian Andes are engaged in a multiplicity of livelihood activities that include subsistence farming, cash crop production, temporary and permanent migration, as well as local wage labour. Gender divisions of labour in agriculture are clear-cut, with men ploughing and acting as the 'main' farmers, while women sow, weed and harvest. Development assistance, primarily in the form of credit and technical training largely goes to men as head of households. Organization into regional and national peasant confederations allows small farmers to address their concerns to government and external agencies, as well as co-ordinate local development initiatives. Peasant leaders are overwhelmingly men, although a small number of female activists have promoted women's concerns. Male attitudes towards women's organization and demands are crucial in determining the outcome of negotiations, whether in a household between partners, or in the unions. If men demonstrate *comprensión* (understanding) of their wives and union colleagues, women can travel around the country carrying out promotion and organizational work, and gain support in confederation structures for gender-specific programmes. Supportive masculinities contrast with obstructive male attitudes that block women's appointment to leadership positions, marginalize their concerns in union agendas, and prevent female mobility. Where this happens, peasant women set up their own parallel organization for their counterparts in the Andes.

(Source: Personal fieldnotes, 1988–90)

In the development field, men as heads of households were for a long time assumed to be equitable decision-makers on behalf of family members. The failure of redistribution within Indian (and other areas') households, resulting in illness, malnutrition and even death for young female household members made an examination of household relations an urgent priority. Opening up the 'black box' of the household to closer scrutiny means, in this respect, asking specific and detailed questions about the ways in which masculinities are organized and reproduced in public and private spaces.

Summary

- Gender relations in the South are diverse, with different types of opportunities and marginalization associated with ways of being male (masculinities) and female (femininities).

- Third World women are active agents in development, yet have been stereotyped by much writing from the North.

- Women, especially those in low-income groups or in younger age groups, are – in comparison with menfolk in their society – more likely to work longer hours, earn less money for their work, and have less secure access to land or property title, credit or development assistance.

- Development policy needs to address gendered power relations, women's agency and the diversity of masculinities found in regional cultures.

DEVELOPMENT, POWER AND THE STATE

Another institution whose powers have been opened up for scrutiny in the South is the state. Long seen by developing countries' rulers and development practitioners as *the* institution carrying out development, the state was presumed to have the interests of its population in mind, and the organizational capacity to administer and manage development. Using a version of the liberal nation–state model, developing countries hoped to achieve the practical means of statehood as well as the promise of modernity (Chatterjee, 1986). However, the colonial legacy could not be forgotten. Colonization in many areas put into place a heavily militarized command hierarchy, a centralized executive administrative system, with limited legislative powers and representative procedures. These 'overdeveloped states' satisfied narrow elite or sectoral interests and remained unchallenged by generally weak civil societies, particularly in Pakistan and sub-Saharan Africa. Where a dynamic civil society was complemented by a strong legislative structure, such as in Sri Lanka, more inclusive and equitable development outcomes could be envisaged (Martinussen, 1997).

The negotiation and implementation of development priorities may thus be affected by the nature of the state structure. The inheritance of specific colonial state forms may result in selective and limited development policies. By contrast a 'developmentalist state', a successful state, combines an ability to organize a productive agricultural sector (through land redistribution, and taxation), together with publicly held principles of law-making, guaranteed territorial sovereignty, and independent judiciaries and administrators. Such states are most likely to bring about well-being for their populations, by providing a widely agreed set of priorities as well as the mechanisms to achieve them.

During the 1980s, neoliberal development policy began to question technical aspects of the state's role. With market liberalization and the removal of international trade barriers, the state was to concede its developmentalist role to the market through 'rolling back'. Under pressure from the international financial institutions interested in post-debt stability, the Third World state 'shrank' by laying off employees from administrative and manufacturing jobs, reducing subsidies on family food baskets, and generally 'freeing up' constraints on capitalist economies (Toye, 1993). Under these far-reaching structural reforms, restructuring of developing countries' economies *and* states occurred.

In a formal sense, democratic participation in the South has increased as civilian regimes running regular elections have increased in number. While the formal aspects of democracy are not a guarantee of citizenship rights, numbers of civilian governments have risen dramatically from 25 per cent globally in 1973 to 68 per cent in 1992 (Leftwich, 1993: 614). Moreover, governance, that is improving the efficiency and effectiveness of state interventions, entered the core of World Bank's concerns. Yet while authoritarian (and sometimes violent) patterns of rule are being overturned, more diffuse limits on citizens' participation continue. Political cultures that marginalize or silence 'other' social groups in comparison to the 'standard' citizen (white, male, conversant with Western culture) remain in place, although they are challenged by diverse disempowered groups and by political parties.

Summary

- Colonialism and narrow elite interests have made the Third World state in the South an important, but often inequitable, institution for development.

- During economic globalization and 'roll back' policies, the state's role in development was reconfigured, changing what it can do as an institution of development.

NEW INSTITUTIONAL FORMS

Alongside 'roll-back' and democratization, state decentralization is promoted as a means to create closer (and more effective) ties between government and the governed. Increased tax-raising powers and decision-making rights for municipalities and the designation of regional authorities in several Latin American states arguably work towards a greater degree of participation in government, reduced levels of corruption and the designation of funds for local development initiatives. The Law of Popular Participation in Bolivia from 1994, for example, redirected power to local organizations – including Indian communities – in a programme to provide 'growth with equity', combining a neoliberal economic environment with access to decision-making for the mostly rural, marginalized poor populations. In other cases, joint ventures of local and municipal governments together with **non-governmental organizations** (NGOs) are emerging to deal with housing, health care, waste disposal and even education, drawing on international funds.

In development, the institutional framework has been radically transformed. NGOs have stepped into welfare provision, development extension work, credit and job-creation programmes that would, at one time, have fallen automatically to the state. In Latin America currently, it is estimated that around 11,000 NGOs are working in multiple activities, from popular education to agricultural development, through to human rights work. Often welcomed as bearers of appropriate technology, participation and sustainable development, NGOs are a highly diverse group of institutions, with divergent aims, practices and degrees of success. Nevertheless, NGOs are a well-established part of the development scene, raising questions for both participants and observers about the extent to which they can – or should – ameliorate broader structural inequalities.

At another level, innovative 'hybrid' institutional forms are being created in many countries to deal with the imperative of development, resulting from complex new geographies, where global development agencies work alongside local associations and reformed state agencies. In this multi-levelled space, new institutional forms draw on transnational resources and networks to promote local alternatives to state-led development and top-down measures. Drawing on the experiences of grassroots groups, regional NGOs and rural federations, the Ecuadorian coalition PRODEPINE, for example, aims precisely to re-draw the spaces of decision-making, power and territory (*see* Box). By-passing current administrative districts, PRODEPINE targets 288 parishes across the country that contain the highest concentrations of black and indigenous groups, poverty, social deprivation and a degree of grassroots organization. With international funds from international agencies, PRODEPINE organizes poverty reduction and training projects not at a national level, but rather through scattered, 'local' geographies mapped out in recognition of development's spatial and social unevenness.

Summary

- New spatial forms of organization are emerging, with decentralization legislation and non-governmental organizations working at sub-national scales.

- New institutional structures have emerged at the interface of the market, the state and NGOs, and are shaping the responses to development's social and spatial unevenness.

Organizations involved in PRODEPINE: Project for the Development of Indigenous and Black People in Ecuador

- National Council of Planning for Indigenous and Black Populations (CONPLADEIN): a state agency.

- Committees of Local and Regional Management.

- Second Grade Organizations (Spanish acronym OSGs), around 160 local associations and federations for irrigation, education, production, as well as sports, gender issues and the elderly.

- Indigenous nationalities and their organizations, including CONAIE (National Confederation of Indigenous Nationalities of Ecuador), FENOCI and FEINE, as well as FEI, FENACLE and FENOC.

- Afroecuadorian organizations, such as ASONE (Association of Ecuadorian Afroecuadorians)

(Source: Consejo Nacional de Planificación de los Pueblos indígenas y negros del Ecuador (CONPLADEIN), Quito, 1997)

CONAIE   http://conaie.nativeweb.org/

The Confederation of Indigenous Nationalities of Ecuador (CONAIE)

Since its formation in 1986, CONAIE has led the Indigenous peoples of Ecuador from relative isolation to a position at center stage of Ecuadorian society. CONAIE is the representative body that guarantees Indigenous people the political voice that has too long been denied them, and that expresses their needs and goals within a rapidly changing world.

- Introduction to CONAIE (In English)
- Introdución a la CONAIE (en español)

Important: National Constituent Assembly / Asamblea Nacional Constituyente

- Capítulo 5: De los derechos colectivos (De los pueblos indígenas y negros o afroecuatorianos) (agosto 1998)
- Derechos Colectivos de los Pueblos Indígenas y Negros (Mayo 1998)
- The Plurinational State (Jan. 1998)
- The Project of Forming The Political Constitution of Ecuador (Jan. 1998)
- Urgent request for help with the Promotion of the National Constituent Assembly (November 1997)
- Promotion of the National Constituent Assembly (Sept. 1997)

Information on CONAIE / Información sobre la CONAIE

- A Brief History of CONAIE (Dec. 1992)
- The Present Situation (Dec. 1992)
- Achievements of the Indigenous Movement (Dec. 1992)
- Indigenous Plurinational Mandate (Nov. 1992)
- Political Declaration of Ecuador's Indigenous Peoples (Dec. 1993)
- Interview with Luis Macas (Dec. 93)
- Indians want "A state within a state," say landowners (May 1991)
- Assassination of Indigenous Leader and Continuing Threats of Violence against Indian Communities (Sept. 1991)
- Indian Uprising In Ecuador Protester Killed by Army (Oct. 92)

News from CONAIE / Noticias de la CONAIE

- CONAIE's view on some key concepts (Jan. 1998)

Indigenous Culture / Cultura Indígena

- La Música de las Nacionalidades Indígenas del Ecuador
- El Festival de Cine y Video (Junio 1999)
- Segundo Festival de Cine y Video (Oct. 1996)

Mapa de Nacionalidades Indígenas del Ecuador
Map of Indigenous Nationalities in Ecuador

Related web pages / Otras páginas web

- Indigenous Peoples in Ecuador
- Abya Yala Net (Indigenous peoples in Latin America)
- Red Intercultural Tinku Andino Nórdico (Algunas páginas están en Español pero la mayor parte se encuentra en Finés)

Email CONAIE at conaie@ecuanex.net.ec

Web design by Marc Becker, marc@nativeweb.org

23/09/98 14:40

Figure 9.3 Web site of CONAIE. Source: www.nativeweb.org/abyayala/conaie

CONCLUSIONS

Development geographies are not contained within neat postcolonial boundaries of independent nation–states, each individually addressing questions of rural–urban inequalities or what products to sell on the international market. Development operates in a number of overlapping, and in many cases conflicting, institutions at a number of spatial levels. For example, Andean local associations fighting the state for recognition of their land titles, or squatter settlements in Nairobi lobbying Oxfam for funds, are both engaged in a web of economic flows as well as networks of ideas, information and politics. Institutions in development include the diverse massed ranks of NGOs, local associations, 'slimmed-down' states and their agencies, international financial institutions, as well as neighbourhood associations, and cyber-groups on the Internet (*see* Fig. 9.3). These institutions are hybrids, arising from power-filled relationships of different points in space and time. Under the umbrella of increased global integration, the complex geographies of development remain to be mapped, by participants and students of development alike.

Further reading

A clear summary of the arguments about why democracy is such a contested idea in the developing world, and why the issue is bound up with global financial institutions, political changes in the West, and social struggles is found in:

- Leftwich, A. (1993) Governance, democracy and development in the Third World. *Third World Quarterly* **14**(3), 605–24.

An impressive and easy-to-read summary and discussion of the main theories about development, covering economics, sociology and politics is:

- Martinussen, J. (1997) *Society, state and market: a guide to competing theories of development*. London: Zed.

Two classic studies of the development concerns of Third World women, alongside the major critiques of the ways in which these concerns have been marginalized by governments, liberation struggles, and Western feminists are:

- Mohanty, C. (1991) Cartographies of struggle: Third World women and the politics of feminism. In Mohanty, C., Parker, A. and Russo, A. (eds) *Third World women and the politics of feminism*. London: Routledge.
- Sen, G. and Grown, C., (1987) *Development crises and alternative visions: Third World women's perspectives*. New York: Monthly Review Press.

Economic geographies

INTRODUCTION

The contemporary economy is an extraordinarily complex set of processes, operating in and around a huge variety of institutions and activities. It embraces everything from a teenager receiving and spending pocket money to the most advanced manufacturing technologies in the world being employed by global corporations. It touches most of our daily lives, and directly affects what we eat, how we dress and where we sleep. We are surrounded and confronted by advertisers extolling us to purchase their products; we spend ages agonizing over which ones to buy; and huge swathes of our towns and countryside are devoted to the production of goods and services. Even the most peaceful rural scene is riven by economic relations and processes which connect the small village in the UK to a global food industry.

The usual way in which geography has dealt with such complexity is to break it down by economic activity or sector – agricultural geography, transport geography, industrial geography and the geographies of trade and services, for instance. This section introduces to you a different way of breaking the economic world up into manageable chunks – not by sectors of economic activity but into three coherent parts (production, money and finance) of a single unified process. Within the contemporary economy, the dominant sets of relations are **capitalist** in nature, and thus the unified process which represents most of the global economy is known as the circuit of capital – set out diagramatically in Figure I.i.

Money (M) is placed into the circuit at the top of the circle by those who wish to invest. This largely takes place in and through financial centres like the City of London or the New York Stock Exchange on Wall Street. Moving clockwise around our diagram-

matic circle, this money, fictitious or real, is then used to purchase **commodities** (C) in the form of labour power (LP) – say, car workers – and the means of production (MP) – say, an assembly line and bits of steel, rubber and various other metals. These are then combined in a production process (P) – say, in a car factory – which produced further commodities (C′) – say, in the form of a car. This new commodity is then sold, for more money (M′) than was originally invested (our initial M). The difference between M and M′ is known as surplus value, or more usually in everyday language as profit. This amount is ready to re-invest in a further round of production. The realization of a surplus is the rationale behind the capital-

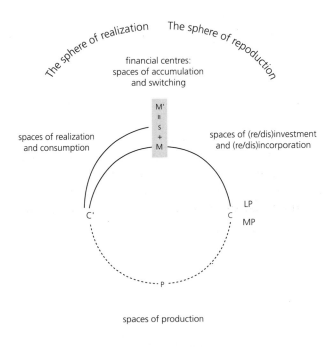

Figure I.i The circuit of capital. Source: Lee, 1999

ist economy – those firms, organizations and individuals who do not manage to do this will quite simply go bust. Those who manage it on a regular basis will thrive and prosper. Under capitalism therefore, production, consumption and exchange are all combined as the means to an end of making a profit – or generating surplus.

The geographies of all this are highly dynamic and mobile since one of the defining features of capitalism is that its key component parts – capital and labour – are notionally free. The creation of value, or a surplus, takes place in and through specific economic spaces. We can break these spaces down into three main kinds – those of production, those of investment and those of consumption. The geographies of each of these are considered in this section. Initially, in Chapter 10, Roger Lee explores the geographies of production. He points out that production is not just confined to the formal economic spaces of the factory and the office, but instead takes place in the home around the issue of **social reproduction**. He also stresses how the spaces of production are increasingly dynamic and flexible, yet at the same time are vulnerable and subject to almost instant demise. In Chapter 11, Adam Tickell looks at what stimulates production in the first place by interrogating the notion of money. He shows how money, although footloose and highly mobile, has a very particular geography. A small number of world financial centres, which are the major source of investment capital, form a very tight core within the global finance industry. He then shows how this global situation is mirrored at the national level, where financial services tend to again be concentrated and how this in turn leads to increased social exclusion and highly unequal patterns of wealth and poverty.

By contrast, in Chapter 12 Jon Goss discusses the pervasive nature of consumption and seeks to understand the urge inside us all to consume particular products. He shows how this urge is the product of a desire to make ourselves distinctive – yet in the act of consuming we become just like millions of others. He goes on to discuss not the geographies *of* consumption, but the ways in which geography *is* consumed. By this he doesn't mean how you as students consume

the discipline, but rather focuses on the ways in which spaces of consumption are increasingly based on a number of archetypal spatial settings, such as the festive market-place.

Through the act of consumption we close one circuit of capital and help to contribute to the beginnings of the next one. But the chapters all show how this is not just an economic process, and is instead one which is rooted in key sets of social and cultural processes. The key question for geographers to explore is how these processes combine to produce and work through a whole host of very precise spatial settings. In this sense the economy only becomes realized in particular spaces. It never takes place in the abstract, as in Figure I.i, but different sections of the circuit are constantly being played out to different degees in different places. Yet as all the chapters indicate, this is not a simple mapping of economic activities onto space – the spaces are themselves constitutive of the economies and are crucial to their success or failure.

Further reading

Although published over a decade ago and concentrating primarily on the UK, the following two books still offer the best introductory material on economic geography. There are other more recent texts which concentrate on one or other aspect of the economy, but these are the most informative general introductions to the topic:

- Allen, J. and Massey, D. (eds) (1988) *The economy in question*. London: Sage.
- Massey, D. and Allen, J. (eds) (1988) *Uneven re-development: cities and regions in transition*. London: Hodder and Stoughton.

More recent books which also cover the broad sphere of economic geography, but which are not so introductory in nature, are:

- Lee, R. and Wills, J. (eds) (1997) *Geographies of economies*. London: Arnold.
- Storper, M. (1997) *The regional world: territorial development in a global economy*. New York: Guilford.

For the most up to date writings on economic geography you should scan recent editions of the journal *Economic Geography*. This will give you a sense of the current topics being pursued in this field.

Production

Roger Lee

STORY ONE: OF CAPITAL, MOTOR RACING AND ECONOMIC REGIONS

Just before Christmas 1997, one of many articles on the controversy surrounding the British government's attempt to exempt formula one (F1) motor racing from the EU's proposed ban on the sponsorship of sporting events by tobacco companies appeared on the front page of the *Financial Times*. Like most articles in the *FT*, it was about economic geography but, by questioning the veracity of the threat by *Formula One* (the company which organizes grand prix events) to relocate its grand prix motor racing activities outside Europe if the advertising ban went ahead, it showed how *geography is a powerful formative influence upon and not merely an outcome of economic activity*.

The threat of relocation was taken seriously by the government which assumed, simplistically, that the location of *production* of racing cars is linked merely to their *consumption* (at grand prix circuits). If F1 racing was relocated, the production of F1 cars could easily follow. This would be a matter of some concern because the production of F1 and Indy racing cars is a high-tech industry contributing skill, knowledge and innovations to the avionics and motor vehicle engineering industries from which it derives many of its inputs. It generates around 50,000 jobs, many of which are highly skilled, and so a threat to reduce or even discontinue production in Britain extended in its effects well beyond the geographical and organizational boundaries of the production of racing cars.

The source used by the *Financial Times*' journalist, John Griffiths (1997), to counter the threats of relocation was research from two economic geographers – Nick Henry and Steven Pinch (1997) – on the production of racing cars in 'motor sport valley' (*see* Fig. 10.1).

This is a regional cluster of motor sport firms centred around mid-Oxfordshire, which dominates the world in the production of racing cars. The research asked why, in an age of globalisation and telecommunications, should most of the world's leading motor sport firms choose to be located in a relatively small part of England?

(ibid: i)

Its conclusions were that the world dominance of the region in the production of motor sport vehicles 'lies

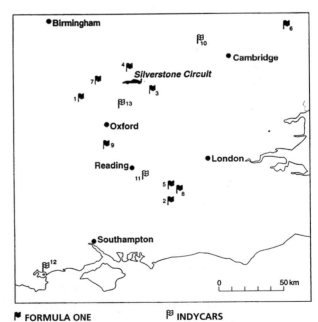

Figure 10.1 Motor sport valley. Source: Henry, Pinch and Russell, 1996

⚑ FORMULA ONE
1 Benetton Formula
2 Ferrari UK Design Centre
3 Footwork Grand Prix International
4 Jordan Grand Prix
5 McLaren International Ltd.
6 Pacific Grand Prix
7 Simtek Grand Prix
8 Tyrrell Racing Organization
9 Williams Grand Prix Engineering

⚐ INDYCARS
10 Lola Cars Ltd.
11 March Cars Ltd.
12 Penske Cars Ltd.
13 Reynard Racing Cars Ltd.

in its features as a geographically-concentrated knowledge-based community'. A highly flexible mode of industrial organization. The small and medium-sized firms characteristic of the industry produce specialized technical knowledge and outputs which are exchanged amongst themselves across dense networks of interaction within motor sport valley. At the same time, they draw upon the knowledge and outputs of some of the world's largest engineering and avionics companies operating elsewhere, creating and exchanging knowledge along a steep learning trajectory and connected to some of the largest companies in the world.

So, 'to be outside . . . the Valley is to risk your position within the knowledge community' and to reduce or even deny the possibility of engaging effectively in production within the industry. Thus *Formula One*'s threat to relocate was severely compromised. The firms comprising the racing car industry were so embedded within the geographies that they themselves had produced within motor sport valley that relocation was hardly an option in the short or even the medium term. This was because 'the unit of competitiveness is not the company, nor the sector, but the network and the region and its flows of knowledge; it is an *innovative region*.' In short, the socially constructed *geography* of production in this industry was what mattered and this geography could not, therefore, so easily be reproduced elsewhere. The ignorance of the UK government on this matter exposed it to embarassing and damaging criticisms not only of having been bought by large donations of money[1] but also of failing to understand the profound significance of geography in economy.

GEOGRAPHIES OF PRODUCTION

This story illustrates a range of issues critical to the understanding of production and its geographies:

- Production itself has to be produced. The creation of a geography of production – here the highly networked knowledge community – is an essential condition of production itself. *The geography of production is not merely an outcome of the business of production but is rather the means through which production takes place and it is a formative and determinant influence on the effectiveness of production.*

- Furthermore, people (Bernie Ecclestone, the boss of *Formula One* for example) matter in this process.

> Production is a social process and . . . its development, including its geographical development, is not a mechanistic outcome of external forces. [However,] . . . it is not only the personal idiosyncracies of managers

which are the issue . . . but the fact that historical change comes about through social processes and social conflict. [People] . . . certainly can . . . make a difference . . . though only within the broader context in which they are operating.

(Massey 1995: 15)

- *Production is only a moment in a wider set of circuits* with which it is closely integrated: investment, involving the flow of capital into (and out of) production activities; the hiring and firing of labour and the organization of often highly complex processes of production at particular locations; the sale and distribution of the finished product; the wider processes of production within related sectors; and the possibility of continuing to invest in further rounds of production. *The point of all this activity in capitalist society is the generation of profit which may be reinvested to begin further rounds of accumulation or profit-making.* These may be quite different in scale, geography, technology or sector from those which preceded them.

- Production is subject to prevailing norms of evaluation (of profitability in capitalist economic geographies) within the circuit. The ways in which capital and labour – but especially highly mobile capital (an example is the ability of *Formula One* to move the staging (another kind of production) of grand prix racing from Silverstone) – circulate in response to geographies of profitability and risk (here affected by lucrative advertising revenues) exert a profound influence on the trajectory and geographies of production. Thus *production cannot be understood unless its links with the prior and formative processes of capital investment and subsequent processes of the returns to capital in production are also specified.*

- At the same time, production is a highly complex and dynamic process constantly changing in response to technological change and the demand for its products. *It is, therefore, very difficult to define 'industries' and to draw classificatory boundaries around them as they flow into each other and are perpetually reshaped.* In the case of motor sport valley, its 'industries' are difficult to separate from avionics and advanced motor vehicle engineering.

- Just as production itself is a complex social product, shaped by a range of contextual influences, so too are the stories told about it. The account of motor sport valley outlined above has been challenged (Lilley and DeFranco, 1997) on the grounds that it over-stresses *geographies of supply* as influences on production:

It is the success of *Formula One* that induces world-wide investments in racing cars and teams. It is no accident that the Silverstone Grand Prix race facility is at the epicentre of the fertile valley. If something happened to *Formula One* as a European-wide economic phenomenon, then the presence of the derivative business would be at risk in the UK.

It is not difficult to discern here an argument lobbying for the exemption of the ban on tobacco advertising at circuits like Silverstone whence comes the demand for the output of the F1 industry – grand prix motor racing. If tobacco advertising is prevented, then the returns from race meetings at such circuits would fall and so be made less attractive. The implication is that *Formula One* may then decide to relocate the site of the revenue-generating site of consumption from Silverstone to another circuit – maybe outside Europe – where the threat to tobacco advertising is less potent. Thus a major locational influence on the geography of motor sport production would be removed and production in motor sport valley undermined, thereby threatening the future of this high tech industry within Europe. This interpretation tells a different story – engages in an alternative **discourse** – *in which production is shaped and structured by the demand for its products* rather than the geographical conditions under which it takes place.

- This alternative discourse raises a wider question concerning the social construction of the conditions in and through which production takes place. 'You can't', a former Prime Minister of Britain was fond of remarking, 'buck the markets'.[2] The implications of this assertion are that economic geographies are somehow automatic and that market processes are mechanistic driving forces – beyond social control – of such geographies. The economic geography of motor sport valley suggests otherwise.

The argument between the two accounts outlined above centres on the relative significance of demand and supply influences in the sustenance of motor sport valley as a location of production. And yet *both demand and supply conditions are socially produced*. Demand emanates from the media-enhanced spectacle and **iconography** of F1/Indy car racing and is affected by legislation controlling broadcasting and the ways in which the spectacle may be represented (e.g. through advertising) while supply (made manifest in geographies of production of racing cars) may be encouraged by polices directed at, for example, the development of public–private partnerships to foster the high

levels of co-operation and interdependence which characterize innovative regions like motor sport valley and enhance it as an increasingly effective motor of production. So, far from being 'unbuckable', markets themselves are social constructs which can be shaped in various ways to achieve a range of objectives.

These observations suggest that, in order to begin to understand production, we should reflect a little on the wider contexts and linkages – the circuits – within and through which it takes place.

PRODUCTION AND REPRODUCTION

No matter how revelatory it may be, one example cannot tell the whole story about production. A glaring omission from the account discussed above is one which is so often absent from stories of economic geographies of production. Most production takes place not in specialized spaces and places of paid and formal work – like the factories, test sites, offices and networks of communication which make up motor sport valley – but in spaces of unpaid work – most notably the home.

Homes are the geographies within which the production of household services – the 'vast bulk of production in households' (Murgatroyd and Neuburger, 1997: 63) – takes place. Estimates – based on alternative assumptions of valuation – of the value of household production in the UK vary from around 40 per cent to around 120 per cent of GDP. These figures are staggering. But they are significant not merely because of their sheer size but for a number of other reasons too. Just three will be mentioned here.

Despite the continuing convergence of the proportion of women and men in the paid labour force, it is women who are primarily responsible for domestic production. While the proportion of time spent in paid work by UK males is fifteen per cent,[3] for women the equivalent figure is 9 per cent. By contrast, nearly a quarter (23 per cent) of the time of women is spent in unpaid, overwhelmingly domestic, work whereas only 16 per cent of men's time is so spent. Thus women in the UK spend 32 per cent of their time working in paid and domestic production while men so spend 31 per cent of their time. Given the rapidly growing (relative to men and in absolute terms) participation of women both in the formal economy and in non-passive activities outside it (Lee, 1998), this **gender**-differentiated engagement in production is a profound influence on the complex of changing relations between men and women in contemporary society and the (often pathological) social consequences that flow from them.

A second reason why household production matters is that 'much of what is presented in the national accounts as final consumption is in fact intermediate consumption by the household production industries' (Murgatroyd and Neuberger, 1997: 64). At the same time the relationship between household production and paid labour is far from constant – varying from activity to activity, place to place (the ratio of paid to unpaid work in Denmark, for instance is around 2:1, in the USA around 1:1 and in the Netherlands around 1:2 (ibid.)) and from time to time in the same place. Furthermore, the two sectors of production 'are as often substitutes as complements' (ibid.). The changing technology of domestic production may itself be a factor in the growth of the paid female labour force in many countries and may help to shift certain forms of domestic work like cleaning, for example, into the formal economy.

Figure 10.2 Domestic labour and the sale of financial products: 1950s ad for personal loans. Credit: © HSBC Holdings Plc

Such links and circuits of productive work point to a third feature of household production: its significance for the circuits of **social reproduction** which sustain (or fail to sustain) human life. Economic geographies are not just about the moments of production, consumption and exchange in the sports grounds, offices, shops, factories, concert venues, art galleries, markets and all the other places in and through which they take place. Rather, economic geographies involve the social struggle to construct sustainable circuits of social reproduction – through which the production, consumption and exchange of values fundamental to the continuance of social life may be maintained across space and time – and to establish the geographically and culturally constituted and socially directed conditions through which such circuits may be created, sustained and transformed.[4]

Circuits of social reproduction demonstrate the close interdependence and mutual constitution of work and home. The production and consumption of value, often based in and around the home, is necessary to reproduce the work force and to enable its continued participation in production at work while the consumption of the productive value of means of production within workplaces (e.g. machines, communications networks, places of production) is necessary to sustain both consumption and production. The enormous quantitative contribution of domestic production and the gendered relations which sustain it are, therefore, vital components in sustaining circuits of social reproduction. Thus the men who make up the majority of workers in motor sport valley rely on the consumption of a wide range of services produced at home to enable their continued participation in formal work. To a very large extent, domestic services are not paid for and so the firms employing these workers are supported by a hidden subsidy of massive proportions.

The circuit of capital as a circuit of social reproduction

The establishment of a specific and at least temporarily acceptable set of social relations sustainable across geographical space is an essential condition of existence for any circuit of social reproduction. Social relations provide the bases for social communication and understanding by establishing norms of social action and indicators of progress and regress – what is good/bad, better/worse. Thus the form taken by a circuit of social reproduction and its objectives, meanings and purposes are defined and shaped by prevailing sets of social relations.

Within the contemporary world economic geography, the dominant and even determinant set of social relations are capitalist – the essential feature of which is the separation of the ownership and control of capital (by capitalists) and of labour power – the power to engage in productive work (by labour). Although such a representation of a binary divide between labour and capital is a gross over-simplification of contemporary **capitalism**, and much effort has been put into trying to reduce its ideological and political significance, the social relationships between capital and labour remain a tremendously powerful determinant of the trajectory of capitalist circuits of social reproduction and of the ways in which they work in practice.

As labour and capital are 'free' of each other under capitalist social relations, the only reason for advancing capital into the circuit is the prospect of increasing its value. Furthermore, the mutual 'freedom' of capital and labour creates highly competitive conditions under which each must seek the other out and must combine within productive units at the highest possible level of productivity in order to sustain their competitive edge. So, under capitalism, production, consumption and exchange are merely the means to the end of the production of a surplus – or profit – which can be accumulated by capital and set to work again in search of a further surplus. And the mutual freedom of labour and capital endows capital especially with great mobility in space and time. *Geographies of capitalist circuits of social reproduction are, therefore, highly dynamic and expansionary.* It is within this context of a mobile, dynamic and highly directed circuit of reproduction that capitalist production takes place and geographies of production are constructed.

Summary

- Production is merely a moment in circuits of social reproduction which sustain networks of consumption, production and exchange across space and time.

- Production takes place throughout these circuits – most notably in the home where it is largely unpaid.

- Circuits of reproduction are dependent upon the establishment of social relations of communication, meaning and direction which guide participation in and understanding and evaluation of economic geographies.

- The social relations which shape production in the contemporary world are those of capitalism which bring particularly powerful and formative objectives and criteria to bear upon social reproduction.

SPACES OF PRODUCTION UNDER CAPITALISM

It is in and through spaces of production (*see* Fig I.i on p. 93), including the home, that value is created or,

more precisely, in which value is transformed with the intention of its expansion. Material and mental inputs are combined within the labour process in ways which are directed at the enhancement of value by producing outputs, the value of which exceeds that of the inputs used to produce them. A haircut, for example, involves the productive combination of a place of production (the home, the street or a hairdressers, for example), the means of production (scissors, shampoos, water, electricity) and skilled labour brought together to produce the product – the haircut. But that is far from the end of the story. The intention of the transaction is that the value of the haircut to the consumer (appearance, convenience, status) – its exchange value – should exceed the value of the inputs used up in the process of production and so generate a material surplus as well as enhancing the reputation of the haircutter – thereby increasing further the potential exchange value of her/his labour power – and, most important in terms of the evaluation of this productive activity, increase the value of capital embodied in the business. As suggested above, objectives of the production of value in this way are its realization as profit and its accumulation as capital which may be used in further rounds of accumulation in the circuit of capital. The production (and consumption) of the haircut are simply the means to this end.

The production (and integration) of production (systems)

This does not imply that production itself is a simple or an incidental process; far from it. In order for production to take place and value to be created, three sets of processes have to be set in motion:

- Sources of finance and investment have to be identified and put in place. These processes take place – are themselves produced – within the sphere of reproduction (*see* Fig. I.i) in the circuit of capital.

- A location must be established and serviced in such a way that the means and forces of production may be assembled and combined in productive work, the most effectively to produce value. The commodities produced must then be circulated to realize their (surplus) value.

- These activities must be organized and dynamized through the exercise of corporate power.

Studies of production within economic geography have tended to focus on the second of these sets of processes despite the fact that it can make little sense without the formative context of the first and the third. Thus economic geographies of production have tended to neglect the crucial importance of what Michael Storper and Richard Walker refer to as the 'integration of production systems' (Storper and Walker, 1989: 125).

Such integration tends to occur in spaces of **hegemony** within the sphere of reproduction (Lee, 1999) Activities which take place within the sphere of reproduction involve the evaluation of the effectiveness (measured though various forms of profitability) of the circuit of capital. *But these evaluations have themselves to be produced.*[5] Information is collected and processed by experts, such as financial analysts, specialized in particular fields of reproduction (like oil production, for example, or grocery retailing) working in financial institutions concerned to invest their own capital most effectively and to sell their expertise at a profit to other investors. Using their 'expert' knowledge (or belief) analysts look back across earlier rounds of accumulation to assess, amongst other things, the effectiveness of different locations, organizations and forms of production against the goal of expected profitability. At the same time, they look forward to assess the risks and potential profitability of future rounds of accumulation. On the basis of such assessments, capital may be switched into certain spaces of production, withdrawn from them or withheld and invested in other activities that may enable more profitable surplus extraction and optimize the relationship between risk and profitability.

Such work requires access to large amounts of (often up to the second) data from around the world, the ability to translate them into meaningful information and to assess their significance. It is hardly surprising then that the geographies of evaluative production take place within spaces of hegemony (Lee, 1999) located predominantly in highly localized clusters of financial activities within major financial centres such as London, New York and Tokyo and within smaller centres such as Paris, Amsterdam and Frankfurt (*see* Fig. 10.3) with a more restricted geographical reach. It is in such centres that the data needed for evaluations of the circuit are readily available. But that alone would not be a sufficient reason for localization as the wide and relatively cheap availability of information technology which can disseminate information quickly and easily across geographical space in real time creates the potential for the geographical decentralization of such activities.

However, if they are to be useful, data have to be converted to information and information must be assessed. These requirements necessitate contact with knowledgeable others within and beyond the workplace vital. The value of face-to-face communication and access to the multifarious uses and users of information facilitate its evaluation and so promote the clustering of the productive activities engaged in this task. Even those, like branch bank managers apparently operating at the local level within the sphere of reproduction (*see* Fig I.i), increasingly use standard assessments of risk and return based on centrally administered credit rating and scoring systems rather than local or personal knowledge. Thus local discretion is removed and displaced upwards to spaces of hegemony where evaluative criteria reflecting less local concerns and more those of international financial assessments are brought powerfully to bear.

The production chain

One way in which the complex of activities involved in production may be represented is through the notion of the production chain (Dicken, 1998). Effective production involves the integration of the complex range of activities within and between a range of production units (firms, industries, produc-

Figure 10.3 Banking centre in Frankfurt.
Credit: German Information Centre, London

tion systems), each of which contributes to the creation of value. The question is what is it that ties some activities more closely together (so creating 'industries' and 'production chains') and sets others further apart? Two sets of influences are thought to be especially important in this integration.

Workplaces and firms

The boundaries of industries are extremely difficult, if not impossible, to define. Storper and Walker (1989: 128; see also Massey, 1995: Chapter 2) rightly point out that industries cannot be fixed as they are but 'moments in a dynamic process of division and integration of labour'. It is, therefore, necessary to build from the smaller units – workplaces and firms – of which industries are physically and legally constituted to begin to appreciate the dynamic constitution of industries and production systems. However, workplaces – homes, workshops, factories, offices, shops, infrastructural nodes, cars – are immensely varied in scale and form and in the range of activities carried out within them, whilst their form changes dramatically over time and space. The size and range of their activities are shaped by economies of scale (the effects of size of output/means of production on unit costs) and scope (the effects on unit costs of the range (scope) of activities encompassed by a firm.)[6]

Transactions

The combination of economies of scale and scope, along with associated **transactions costs** within or between firms, shape the degree of integration or disintegration within industries. Firms have a choice as to whether to internalize production within a workplace, or series of workplaces or whether to externalize production by purchasing inputs from another firm. At the extreme of externalization, each activity may be undertaken in individual firms, with the transactions between them governed by market or network relationships, or through subcontracting based on contracts which may allow the subcontractor some freedom and legal independence from the lead firm. At the other extreme of internalization, the whole chain may be located within an individual firm and transactions organized hierarchically or through a series of internal markets and profit centres.

Production chains are, then, constituted of sequences of technologically linked and socially directed activities, dynamically delimited by economies of scale and scope along with consequent transaction costs. They are co-ordinated in a variety of ways and made up of 'complex networks of small and large firms, big and little workplaces (whose boundaries do not always correspond with the industry's boundaries)' (Storper and Walker, 1989: 136). As well as offering a helpful perspective on the dynamics and complexity of production in modern economic geographies with highly developed social divisions of labour, the production chain gets away from more formal and static definitions of 'industries'. It reflects the notion of economic geographies as circuits of *r*eproduction through which production of value is constantly evaluated before decisions are taken to engage in or disengage from further rounds of accumulation.

And *production chains are not simply unidirectional flows of value moving from production to consumption.* Consider, for example whether, as you read this book, you are creating value or destroying it. Opening and reopening the book, turning its pages and, maybe, marking them you are, as a consumer, destroying its physical value. But in thinking about what it has to say you are changing yourself, extending (we hope!) your abilities and level of understanding. In such ways, you are not merely consuming value but creating it: you are not merely a consumer but a producer. And, as motor sport valley demonstrates, the same is true of many activities along the production chain too. Information acquired by one participant may be embodied into its own activities and output and fed back to the originators of information in the form, for example, of improved products.

Territorial production complexes

Territorial production complexes are extensive work sites that bring

> disparate production activities into advantageous relation with each other, at a larger scale and scope than the individual workplace, firm or even, ... industry. Territorial formations are fundamental to the operation of industry because they offer the means of integrating production systems above and beyond those found in organization charts, market institutions, or the laws of ownership and contract.
>
> *(Storper and Walker: 138)*

An excellent example of a territorial or regional production complex would be that of motor sport valley; the City of London would be another. Spatial propinquity within such complexes reduces the cost of links, movements and the sharing of information, the locational fixity of infrastructure establishes a resource base for effective interaction and their boundaries limit the extent of external movement.

Such transactions contribute towards the generation of external economies of scale[7] which may be intensified through agglomeration as a result of reductions in the costs of exchanges, increases in their frequency and form, and acceleration of the circulation of capital through the system. Thus spatial clustering may engender further economies of scale and may lead to a further widening of the social division of labour through horizontal and vertical disintegration[8] as a result of the operation of (dis)economies of scope.

Further economies arising from the reproduction of a specialized labour force (via local training and educational systems), the provision of (specialized) knowledge and infrastructure, improvements in the exchange process as a result of the ease and frequency of ready conversation and comparisons, and in the development of localized means of communication trust, understandings, vocabularies, forms of knowledge and performance-related structures of governance, may also follow from agglomeration. In short, a dynamic and flexible territorially based structure of production, based on intense interaction between and within firms and workplaces capable of responding especially well to short-term changes within the geography of the circuit of capital may be enabled within territorial production complexes, thereby enhancing the productive power of the firms located within them.

But there are limits to the efficacy of such geographies of production. Diseconomies of scale such as congestion and pollution may operate and spatial clustering is more common in industries with unpredictable and unstable systems of short-term interactions, where face-to-face communication is essential and where shared knowledge and facilities are especially important. At the same time, just as external economies within territorial production complexes may lock out competitors they may also lock in participants. Such rigidities limit the ability of territorial complexes to cope with severe structural shocks. An example of such a shock would be the relocation of F1 motor racing from Silverstone which, according to one account of motor sport valley would, as we have seen, serve to undo this territorial production complex. Furthermore — and perhaps most significantly — the geographically expansive and intrusive circuit of capital may create conditions in which local producers increase their links over time with the world beyond the production complex, so disrupting its internal coherence. This may well be achieved through the dynamic corporate geographies of large trans-national firms with their strong internal corporate links and relatively high levels of economic power (see Dicken, 1998).

The organization of production and the production of organization

Production is merely a moment — a complex, messy, dramatic, depressing and uplifting moment but a moment nevertheless — in the lengthier and wider circuits of social reproduction. Individual capitals must organize and dynamize their activities in as effective (profitable) way as possible. Thus firms are not only legal entities but the framework within which capital can operate, disperse its means of production across geographical space, hire and fire labour, store surplus capital and administer the complex processes of production. Above all, perhaps, firms are the major entities through which capital engages in competition. Although the relationships between firms and the territorial production complexes in which they take place are highly formative in this respect, Peter Dicken and Nigel Thrift (1992: 284) argue that this is not the whole story:

> capitalist social relations — and their geographies — are articulated in particular ways by business enterprises through their adoption of specific product and process technologies, through their procurement and investment decisions, through their influence on employment relations, the labour process and the division and integration of labour. Such processes do not occur in a general abstract form; they take on specific cognitive, cultural, social and political forms in an environment which is shaped very largely (although not exclusively) by business enterprises, especially large business enterprises which are able to wield more social power.

Such organizations cannot be understood simply as organizational charts or spatial structures of production but

> as the way people in firms do things, using particular kinds of rationality, applying particular beliefs, ideologies and collective understanding, participating in particular social networks and influencing particular legal and political institutions. In other words forms are cognitively, culturally, socially and politically embedded.

> (ibid.: 283)

And the doing of such things in firms involves the wielding of power — the shaping of environments, both internal and external (other firms, states, cities, regions, markets) — in ways which generate competitive advantage. Of course, the effectiveness of the organizational dynamics of firms is judged within the sphere of reproduction (see Fig I.i) and may be closely shaped by its active participation within the division of labour of a territorial production complex. Nevertheless, the organization of production and the webs of power through which it is undertaken are not only inherent within but highly significant influences upon spaces of production.

The possibilities open to firms within a chain of production — within which, as we have seen, value may flow in both directions — must be recognized and exploited by those firms. The Swedish home furnishings and household goods manufacturer and retailer —

IKEA – serves to demonstrate the point. IKEA is an innovative production organization not only in terms of the distinctive nature of its retail outlets (*see* Fig. 10.4) but in its relationships with suppliers with which it enters into productive partnerships – selling expertise, advice and finance – and so both consuming and producing the value generated by them. At the same time, the customers assembling the furniture bought from IKEA become part of the company's (unpaid) productive workforce and, incidentally, open up possibilities for further developments in the chain of production for specialist firms of self-assemblers to respond to the needs of those challenged by the DIY ethos.

Summary

- Production creates value through the consumption and transformation of values embodied in other goods (e.g. labour power, books, machines, knowledge, finance).

- This complex process is regulated, evaluated and integrated within the sphere of reproduction.

- The production of value flows forwards and backwards along chains of production.

- This interlinking of production, consumption and exchanges of value underlies the formation and emergence of territorial production complexes within which transactions between firms and workplaces may be facilitated.

- Nevertheless, the success or otherwise of individual firms cannot be deduced entirely from their place within a territorial production complex but from the interaction between the possibilities of such complexes, which may be more or less localized, and the organizational capacity of individual firms.

Figure 10.4 IKEA: a space for the production of retail services located on a large site, well served by road. The overcrowded car park and congested approach roads are not shown.

STORY TWO: OF CAPITAL, FINANCE AND GLOBAL ECONOMIC GEOGRAPHIES

> If this bank does not have 30 per cent of its earnings coming from Asia in the next three to five years, it will not be a global bank but a European bank.
>
> *(Simon Murray, group executive chairman of Deutsche Bank Asia-Pacific (Fisher, 1996))*

Deutsche Bank is one of the largest banks in the world. Its business is the production of financial services and financial instruments. Measured by its tier one capital base[9] it ranks sixth and, by total assets, second in the world. It ranks first in Europe on both counts. In 1996, it announced that it was substantially to increase its operations in the Asia-Pacific (A-P) region as a means of transforming itself from a regional bank in Europe to a global bank. The aim is to become one of the top three investment banks in Asia: to grow from earning a negligible proportion of its total earnings from the A-P region in 1993 to at least 30 per cent by 2001; in 1996, the proportion was about 10 per cent. This would finally transform a bank founded in the nineteenth century as a national institution, designed to reduce German dependence on British finance, into a global operation.

Why did it adopt this strategy of increasing its productive capacity by expanding into a new region? One answer is that it had little choice. The increasing **globalization** of the world economic geography (manifest, for example in the increasing proportions of world trade, portfolio and direct investment, flows of knowledge and global standards of evaluation exerted by interest rates on financial instruments such as government bonds) presented the bank with the need to make itself capable of producing financial services for clients who require global rather than regional advice and products. Furthermore, if it was to retain an ability to compete with its major globalizing competitors, it had to go with the flow and make itself global too before losing clients, themselves increasingly global, to them.

Although Deutsche Bank has a presence in 28 emerging markets (outside the major regions of economic development in the contemporary world economic geography) many of its competitors are more widely based.[10] Staying as a predominantly national bank is not an option as even domestic customers demand globally-informed financial products, including advice and financial services. The bank's former chairman, Hilmar Kopper, who retired in May 1997, feels the bank is 'not strong enough in Europe, outside Germany' and should be 'among the top five' in global investment banking while recognizing 'that we will never be able to push aside the big Americans' (Fisher, 1997).

The strategy adopted was to expand into A-P

through the expenditure of DM6 bn and by establishing a physical presence in 18 countries within the region and employing over 5 000 people to produce its financial services. This is a form of direct investment which expands the bank's capacity for production by putting fixed investment at risk and by exposing a large amount of finance capital to the risks of investing in a region in which the bank has only relatively little previous experience.

So why choose this geography? For one thing, it has proved notoriously difficult for aspiring European banks to compete with US banks in the USA.[11] For another, the Asia-Pacific region has a number of attractive features relating to its local characteristics and its place within the world economic geography. Rapid economic growth (until the late 1990s) which was shifting the global geo-economic balance of power from the Atlantic to the Asia–Pacific region had created the demand for huge investments in water, transport and telecommunications infrastructure. At the same time, high savings ratios in many parts of the region created a pool of capital not especially well managed by existing banks and without alternative outlets in well-developed financial markets. Economic development had also stimulated rapid growth in internal economic integration. By 1997 almost 40 per cent of total international trade was conducted within the A-P region, so strengthening the triadic structure of the global economic geography (Castells, 1996; Dicken, 1998). Furthermore, the presence in the region of the network of the immensely economically powerful overseas Chinese offered a possible stepping stone to China itself.

But this was a risky venture – not least because of the fragile nature of the local financial systems in economies like South Korea, Thailand, Malaysia and the Philippines, to say nothing of Japan, as the crisis which broke and spread throughout the region during late 1997 and early 1998 demonstrated. There is no guarantee of success and, big though it is, Deutsche Bank is still beholden to the financial markets for their assessments of its performance as a global bank. The evaluative criterion set by the bank for its venture into the region is a return of 25 per cent on its investments. Currently the return is barely more than half that and the crisis in south east Asia made its achievement far more difficult. Thus in January 1998

Deutsche Bank shocked the stock market . . . [in Frankfurt] . . . by announcing that profits for 1997 would be sharply lower as a result of the Asian financial crisis . . .

Germany's largest bank said it was making extra provisions of DM1.4 bn ($783m) to cover all possible risks in Asia, where it has been building its commercial and investment banking operations. This would depress operating profits by a third. In 1996, these totalled DM5.8 bn – a rise of 37 per cent.

(Fisher, 1998)

That a bank as large as Deutsche Bank needs to set aside provisions amounting to around a quarter of the previous year's profits gives some indication of the severity of the crisis for banks operating in south east Asia and the risks that even the biggest run.[12] It also demonstrates the point that production is but one, constantly evaluated moment, in the wider circuits of social reproduction.

At the same time Deutsche Bank is up against strong competition from the big US investment banks such as J P Morgan, Merrill Lynch, Goldman Sachs and, in commercial business, one of the largest, Citibank, as well as a number of European competitors. If its strategy is successful, the bank will secure for itself new geographies of production which will enable it to operate effectively (profitably) in a globalizing economic geography. If it fails, however, its own value will diminish and its devalued geographies may be taken over by stronger competitors, who will either add Deutsche Bank's geographies to their own, and so increase their global power, or rationalize them to avoid costly duplication and overlap. As well as marking the demise of a financial corporation and a corporate geography of production, this would also serve if the new corporate strategists are up to the task, to enhance profitability within the rest of the global investment banking industry. Thus a corporate geography of production created at great expense of capital, commitment and effort, would be wiped out, in whole or in part, the circuits of social reproduction in the localities and amongst the workers most directly affected challenged, and the geographies of financial production transformed.

CONCLUSION

Production is a complex but intensely practical process subject to evaluations of effectiveness made by financial analysts working in financial markets with direct and indirect power to influence the geographical trajectories of highly mobile capital. It can take place in a very wide range of environments from within networks centred on the home to those articulated through the most sophisticated corporations or territorial production complexes. It is directed above all at the production of values and thereby at the sustenance of social reproduction. But the nature of 'value' or 'worth' is defined by the social relations in and through which production operates. Within capitalist societies value is defined in terms of profitability and so the phrase 'shareholder value' represents an objective towards which its neoliberal defenders would wish it to progress.

In this project of the creation of value, production is a vigorous creator of landscapes, landscapes capable both of enhancing or restricting the capacity of production and of transforming the appearance and ways of life of those places and people caught up within it. Equally the switching of capital away from certain landscapes of production and towards others is capable of wreaking enormous damage on these same people and places judged/valued as irrelevant to the continued profitable circulation of capital and so left, with abandoned identities (*see* Fig. 10.5a and b).

Production can, simultaneously, endow and strip away identity through its capacity to integrate individuals within a network of relations of production while demanding – more, or less, insistently – conformity with corporate objectives. The absence of production, or of access to production, for a substantial and growing minority of unemployed people often congregated together in particular localities and amongst particular age groups – in many Western societies, amongst young and old men rather more than women – separated from an effective circuit of social reproduction, is increasingly recognized as a major social problem. It feeds a sense of **alienation** and disconnectedness in which social norms are seen both as personally irrelevant and as the cause of the plight of the unemployed.

At a global level, the two-thirds or so of the world's population left outside effective circuits of social reproduction are – in the face of separation from such circuits and in the midst of social conflict generated not least by destitution and desecrated landscapes of social organization bestowed by colonizing powers – often too weary trying to sustain life to mount serious resistance. Nevertheless, the powerful amongst the dispossessed are capable of engaging in economies of crime, based, for example on the (large-scale) production, processing and sale of illicit drugs. In one sense,

(a)

(b)

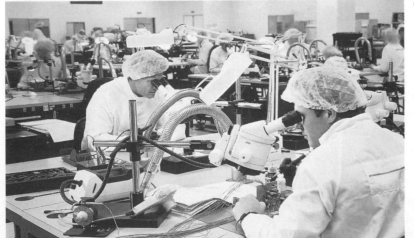

Figure 10.5 (a) A view of the Stanton blast furnaces in Nottinghamshire. Credit: Popperfoto (b) Main micro-circuitry assembly room for communications satellites. Credit: Telegraph Colour Library

dreadfully enough, this is not an unreasonable response. Production is neither geographically nor socially autonomous. It is shaped in its objectives, technology, internal and organizational networks and in its outputs by circumstances socially constructed within circuits of social reproduction.

NOTES

1 Bernie Ecclestone, the promoter of F1, donated £1 million to the UK Labour Party during the run up to the general election in May 1997. In the wake of subsequent controversy over tobacco-based sponsorship at F1 events, the newly-elected Labour government was forced to return the donation.

2 A quote attributed to Margaret Thatcher in responding to financial or industrial crises which, following neoliberal discourses of economics, she interpreted as somehow natural and so beyond social construction and control thereby making (state) intervention not only pointless but harmful.

3 This and the figures which follow refer to ALL males and females aged 16 and over.

4 These ideas are explored by Lee (1999).

5 Production takes place within the spheres of reproduction and realization as well as within the sphere of production. Although it is convenient to distinguish a sphere of production within the circuit of capital, the production of value is central to the functioning of the sphere of reproduction, the sphere of realization as well as within the sphere of production. Production of value within the sphere of reproduction takes the form primarily of financial services and financial instruments; within the sphere of realization, it takes the form of trade, retail and logistical services.

6 Economies of scale are generated when the unit cost of output falls with the increasing scale of output; diseconomies occur when unit costs rise with increases in the level of output. Economies of scope exist when two or more activities can be carried out together within a single firm or unit of production more cheaply than if they were carried out separately; diseconomies arise when it is cheaper to disintegrate the tasks to specialized production units.

7 Reductions in unit costs of production with increased levels of output as a result, for example, of the development of links with other firms supplying inputs and demanding outputs, the development of a skilled labour force, or of specialized knowledge or support services.

8 Vertical integration refers to the combination within a firm of several stages of the production chain; disintegration involves the splitting of these stages asunder into separate locii of control. Horizontal integration involves the merging of activities and/or firms at the same point in the production chain; disintegration involves de-merging.

9 A measure of size based on capital that may be accessed at short notice.

10 In 1996 Citibank was present in 77, HKSB in 53, ABN Amro in 46 and ING–Barings in 37.

11 However, the USA is the most highly developed market for investment banking and, in late 1998, Deutsche Bank announced the purchase of Bankers Trust, the eighth largest US bank.

12 It is, of course, appropriately ironic that a widely held view of the cause of the crisis in Asia was ill-advised lending policies indulged in by the major banks operating in the region.

Further reading

The many ways in which geography exerts a decisive influence over production has been explored most fully by:

- Massey, D. (1995) *Spatial divisions of labour: social structures and the geography of production*. 2nd edition. Basingstoke and London: Macmillan.

A brief review of the ways in which geographers have examined production is provided by:

- Dicken, P. and Lloyd, P. (1990) *Location in space: theoretical perspectives in economic geography*. 3rd edition, New York and London: Harper and Row.

Part I of this book provides an outline of the locational–analytic approach to industrial location which, by isolating the geography of production diminishes its significance. Part II attempts to re-establish the geography of production within the social and political–economic context in which it takes place.

The significance of territorial production complexes is explored by:

- Storper, M. and Walker, R. (1989) *The capitalist imperative*. New York and Oxford: Basil Blackwell.

Contemporary tendencies towards the globalization of production are explored both theoretically and empirically through a wide range of case studies by:

- Dicken, P. (1998) *Global shift*. 3rd edition, London: Paul Chapman.

Some insight into the significance of production in sustaining social reproduction and giving identity to localities and the struggles involved in trying to establish and sustain alternative bases of decision-making to those founded merely in concepts of corporate rationality in the face of industrial restructuring are explored in diverse ways by the essays in:

- Hayter, T. and Harvey, D. (eds) (1993) *The factory and the city: the story of Cowley automobile workers in Oxford*. London and New York: Mansell.

CHAPTER II

Money and finance

Adam Tickell

INTRODUCTION

At the end of August 1994, police on the remote Scottish island of Jura were called out by a local fisherman who had discovered the charred remains of numerous £50 notes on the sea shore. Shortly afterwards *The Observer* newspaper reported that the K Foundation had unceremoniously burnt £1 million as part of their campaign of 'art terrorism' (*see* Fig. 11.1). Previously the Foundation, which was made up of the two members of the band KLF, had planned an exhibition called 'A major body of cash' which would consist of a million pounds hammered into a frame. As news of the K Foundation's actions on Jura seeped out it was met by disbelief, bewilderment and anger. Burning such a large amount of money is in many ways incomprehensible. Even if the destroyers had no need for the money themselves, surely they could have found some constructive use for it. The poor are, after all, always with us.

And yet it was in many senses a creative destruction. The act of burning £1 million forces us to think about money. As Bill Drumond, one of the K Foundation, said in a rare interview, 'We created something there. That thing there now exists. The fact that that one million pounds exists in the mind . . . it'll gnaw away at you' (thee data base, 1996). Perhaps, compared to the billions that are spent on developing mechanisms to kill people more effectively or on genetically engineering crops, the symbolic burning was money well spent. Perhaps. But whatever we think of it, it does remind us that although money has become fundamental to modern life, and it is worth remembering that the extent of this is both a recent development and not yet globally pervasive, few of us stop to think about what money actually is.

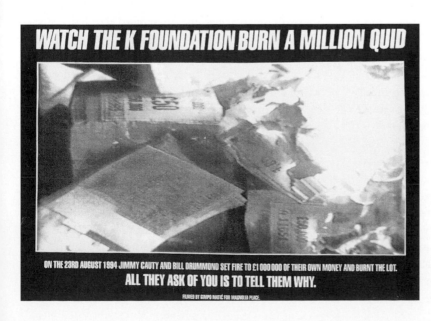

Figure 11.1 KLF Poster. Source: http://jumper.mcc.ac.uk/~ttl/klf/images/ kfposter.jpg

In some respects, the K Foundation's activities remind us that money is more than simply dollars, pounds, yen and euros, notes and coins, cheques and credit card slips, and digits on a screen whizzing around the world at the click of a financier's mouse. It is a special commodity which obscures the nature of production. When the Roman emperor Vespasian, who built the Colosseum, introduced a tax on the contents of the city's urinals his son, Titus, objected. Vespasian replied, 'Money does not smell.' In much the same way, our use of money today hides the social and geographical processes which have gone into getting something on to a supermarket's shelves. Take an everyday item like chocolate, for example. The production of chocolate may involve paying the growers of cocoa beans and sugar very low prices; spraying pesticides and insecticides on plants; using hormones to stimulate overproduction of milk in cows; wasteful use of energy to bring the beans to processing centres; low wages for shop workers, and so on. Yet none of this is important when we buy chocolate – all that we need is the right amount of money and the entire production process from raw materials to finished product is both mobilized for us and obscured. Money is able to achieve this because it is a *store of value*, we exchange things for money because it allows us to buy other things with it later.[1]

In theory, money is not really geographical. Unlike most other products, money does not rot or rust (although inflation and variations in the value of different currencies do affect its 'value') and a pound is the same in Aberdeen as in London, while a dollar is the same in San Francisco as in New York – or even in Bogota. Furthermore, since the middle of the 1970s, there has been a rapid growth in the extent to which financial markets in different countries have become connected with one another, as governments have increasingly pursued 'liberal' economic policies and deregulated financial markets at the same time as technological advances in the telecommunications industry have facilitated global trading. These developments have led some commentators to believe that money no longer 'respects borders' and that the whole financial industry will cease to be a geographically variable one. In a highly influential book, for example, Richard O'Brien (1992: 1–2) claimed that the:

> end of geography, as a concept applied to international financial relationships, refers to a state of economic development where geographical location no longer matters in finance, or much less than hitherto. In this state, financial market regulators no longer hold full sway over their regulatory territory . . . For financial firms, this means that the choice of geographical location can be greatly widened, provided that an appropriate investment in information and computer systems can be made . . .

> There will be forces seeking to maintain geographical control . . . Yet, as markets and rules become integrated, the relevance of geography and the need to base decisions on geography will alter and often diminish.

However, this chapter shows that finance has not reached an 'end of geography' state. Not only is finance an industry with a very particular geography, but it has marked geographic impacts. In this chapter I focus on the continued importance of international centres for the financial industry and on the ways in which geographical restructuring of the financial industry unwittingly contributes to the decline of inner cities.

THE GEOGRAPHY OF INTERNATIONAL FINANCE

With its intangible inputs and heavy reliance upon the 'liberating' telecommunications technologies, the financial industry should be free of the traditional constraints of location. Paradoxically, however, finance is highly spatially concentrated within a small number of international financial centres, and prominent among these are London, New York and Tokyo which form the core of the international financial system. The overwhelming majority of international financial transactions take place in these cities and in secondary centres such as Singapore, Chicago and Frankfurt. In 1993, for example, there were 342 foreign banks in New York, 312 in London and 97 in Tokyo; the three cities accounted for over 40 per cent of international bank lending and headquartered 23 out of the 25 largest securities firms (Choi *et al.*: 1996).

International financial centres are able to maintain their dominance because their very size has become self-reinforcing. Part of the reason for this is that globally integrated financial markets which are effectively open 24 hours a day require a physical location in which to operate and, according to Saskia Sassen (1990), co-ordinating centres which ensure that the markets retain an overall coherence. The historical and economic strengths of London, New York and Tokyo gave the cities an initial competitive advantage in becoming the 'regional representatives' of the international financial system, while their continued strength rests in part because they are located in different time zones which allow around-the-clock trading. On the London stock exchange, for example, the trading of foreign equities dwarfed trading in domestic stocks for some time.

However, it is not simply that international finan-

Figure 11.2 London International Financial Futures and Options Exchange (LIFFE) trading floor, Canon Bridge. Credit: LIFFE

cial centres are conveniently located that explains their continued dominance. Within Europe, the authorities in Paris and Frankfurt are continually attempting to grow their financial service infrastructure at the expense of London, while similar processes are underway – with more success – in Asia as Singapore and Hong Kong position themselves as major participants in the international financial system. Yet finance remains highly concentrated in a relatively small number of financial centres because the very nature of the industry positively promotes geographical concentration. Economically, financial centres benefit from 'agglomeration' in terms of technology and labour. With a large number of potential clients, established centres such as London and New York have been at the cutting edge of innovation in information and telecommunication technologies. Similarly, they develop a pool of highly skilled – and highly paid – workers which helps to draw in new financial institutions requiring skilled workers. As Reed has argued,

> an institutional infrastructure that permits it to coordinate its sourcing . . . and marketing . . . of capital and information to achieve desired degrees of customisation for an individual country, region, institution, or individual without sacrificing economies of scale and specialisation of its sourcing and marketing activities. Computer and data banks central to the highly integrated global systems of financial and commodity markets, investments, trade, production, and economic and political intelligence are pre-eminent in this centre. In addition, many of the centre's activities transcend the authority and control of any single nation-state, including its home country.
>
> (Reed, 1981: 61)

Furthermore, increasing numbers of banks in a financial centre not only increases the number of competi-

tors for business, it also increases the total size of the market.

Less tangible factors also help to explain why technological change is unlikely to lead to the end of geography in international finance. Lenders of money want to be able to have access to their money quickly if they need it and large, established financial markets theoretically provide this. However, as Nigel Thrift (1994) has demonstrated, large markets are often made up of networks of smaller markets where prices can change quickly on unreliable information. Rather than undermining larger markets, however, this volatility underpins them, because social networks allow traders to interpret market movements reliably (*see* Fig. 11.2). In this sense, firms need to be 'sociable', because 'who you know' helps develop business relationships and profits, and developing relationships is easier if people are physically close to one another. Furthermore, as the financial sector grows larger and more complex, financial centres become more important as proving grounds for new products and as centres of trust.

The development of integrated global financial markets has speeded up a tendency towards the 'flattening' of financial space, where time horizons have shortened and geographical distance has become less important in determining the price of goods. Although this process has been underway since at least the nineteenth century, the development of new financial products in the global markets means that although buyers and sellers 'may be very distant from one another in geographical space, the time–space distance between them may be negligible' (Leyshon, 1996: 70). Arguably, the most significant development has been the growth of the 'derivatives' market. Derivatives were developed in order to let companies manage risk and volatility in financial markets. The most common are 'futures' and 'swaps' and these

Table 11.1 Market for selected financial derivatives (billions of US dollars)

	1986	**1987**	**1988**	**1989**	**1990**	**1991**	**1992**	**1993**	**1994**	**1995**
Futures	583	724	935	1259	1541	2251	3019	5103	5945	6074
Swaps	500	867	1330	1952	3450	4449	5346	8475	11303	17713
Options with futures			371	953	750	1268	1615	2668	2918	3112
Swap related		180	300	450	561	577	635	1398	1573	3705
Total	1083	1771	2936	4614	6302	8546	10615	17643	21739	30602

Note: Data for 1995 are not fully comparable with earlier periods due to a broadening of the reporting population.
Source: BIS (1996) *International Banking and Financial Market Developments*, 26 August, Basle, p. 35 and BIS (1997) *International Banking and Financial Market Developments*, 28 February, p. 33.

essentially allow firms to know what rate of exchange or interest they will be paying at a given date in the future, in much the same way as individuals borrow money at fixed rates in order to be sure that they will not be subject to interest rate rises on loans.

Since the early 1970s, and particularly during the closing two decades of the twentieth century, the use of derivatives has mushroomed and derivatives have become a ubiquitous feature of business life. As Table 11.1 shows, the value of derivatives contracts outstanding in 1995 was estimated at $30,602 *billion*, representing a growth of over 2,800 per cent over a 10-year period. However, although cautious use of derivatives allows firms to manage risk, they have also been implicated in a series of high profile losses. The most striking collapse occurred in 1995 when a trader in the Singapore branch of Barings Bank used derivatives disastrously to back up his belief that the Japanese stock markets would stay relatively stable when, in fact, they rapidly fell leading Barings to lose £900 million and its independence. The Barings crisis is instructive about the changes which have occurred in the international financial system during the past century. In 1890 Barings had previously got into severe financial problems after lending in Argentina went sour, but the leisurely pace of life then meant that the crisis took years to come to a head, allowing the Bank of England to organize a rescue. The 1995 crisis took days to blow up and once the Bank of England became aware of the problems it had less than 24 hours.

Tragic though the failures of institutions like Barings are for the people involved, a wider concern about the internationalization of financial markets and the development of new financial products is that they may have the potential to increase the levels of a widespread collapse of global financial markets. In 1994 the General Accounting Office (GAO), the investigative arm of the US Congress, released a report which argued that the growth of derivatives markets had led to substantially increased levels of risk in the international financial system. If a large financial institution suffered major losses using derivatives, the GAO argued that the highly interconnected nature of firms and markets could result in losses being transmitted throughout the world. Furthermore, during the 1990s it has become clear that two further factors have increased risk in the international financial system. First, there has been the emergence of large, integrated financial conglomerates with highly complex financial and corporate structures (one of the contributory factors in the collapse of Barings, for example, was the complex nature of the bank's corporate structure). Second, an increasing share of international financial transactions is dominated by a small number of institutions from a small number of countries, if one of these got into difficulty the contagion effect would be more serious than in a less concentrated market while at the same time putting pressure of smaller companies to cut corners to develop their market share.

Summary

- There has been a globalization of financial markets.

- Global markets are concentrated in a small number of major financial centres.

- **Globalization** of financial markets has contributed to levels of risk in the international financial system.

NATIONAL GEOGRAPHIES OF FINANCE

If financial services have internationalized over the past two decades, during the past two centuries, financial systems *within* capitalist countries have evolved from localized to being increasingly centralized. This has perhaps been most marked in the United Kingdom whose governments, unlike those in countries such as Italy, Germany or the United States, have appeared to be unconcerned about the existence of a regional financial infrastructure. The nineteenth century and early twentieth century saw the gradual erosion of regional financial systems which had developed their own banks and stock exchanges, as financial institutions merged with rivals based in London.

That process left Britain with one of the most highly concentrated financial systems in the capitalist world – concentrated in terms of geography in London and concentrated in terms of power in the hands of a small set of banks and insurance companies. This centralization of control has been reinvigorated during the 1990s – despite a relative decentralization of employment – as mutually owned building societies have become private companies and merged with London-based banks. In terms of employment, in 1995 nearly 300,000 people worked in the financial sector in London and together with the rest of the South East region this accounts for over half of all workers in the sector. Furthermore, London is the only place in Britain where more men work in finance than women, reflecting the relatively high status and maintenance of male power in the industry (see McDowell, 1997).[2]

Ron Martin and Richard Minns (1995) have analysed the spatial structure of the UK pension fund system and show that although contributions to pension funds come from all regions, their management and control are overwhelmingly concentrated in London and the South East. This concentration occurs for similar reasons as concentration in international financial centres. As fund management is a well-paid occupation, such geographical concentration is a direct contributor to the regional economy of the South East. However, Martin and Minns believe that the principal geographical distortion from the pension fund industry lies in the way that the money being managed is actually invested. Taking investment in UK shares, which accounted for just over half the total, they argue that over two-thirds is invested in companies which are themselves based in the South East of England and little money trickles back to the regions to stimulate investment or business expansion (*see* Fig. 11.3). As Martin and Minns depressingly argue,

the private occupational pensions system . . . undermines regions by extracting savings from all over the UK and centralising their administration, management and investment in one region, where fees accrue, control over investment policy is concentrated and where tax subsidies are skewed to benefit relatively high income contributors helping to support an investment regime which has little to do with the promotion of capital investment throughout the UK.

(1995: 139)

Pension fund investment is not the only aspect of the financial sector which has reproduced geographical patterns of wealth and poverty. Wrigley (1998) has recently demonstrated that the ways that American retailers organize their borrowing is directly reflected in the geography of American retail openings and closures. Similarly, the **venture capital** industry, which invests in relatively high risk new and growing businesses with the prospect of very high returns, has been very strongly concentrated in a small number of growth regions. In the United States, for example, Florida and Smith (1993) have shown that investment is overwhelmingly centred in Silicon Valley and New England. Further, approximately 60 per cent of venture capital investments made in the United Kingdom during the 1980s went to companies in the South

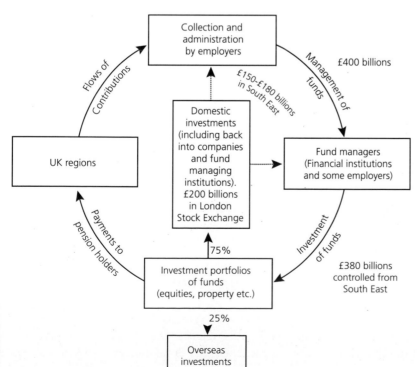

Figure 11.3 The pivotal role of the South East region in the circuit of UK pension fund finance

East region, while every other part of the country had a share smaller than its share of the UK economy. However, such patterns are not set in stone and Colin Mason and Richard Harrison (1998) have recently shown that as the venture capital industry in the United Kingdom has developed, it has become more adventurous in its pattern of investment and by the mid-1990s only three regions (Northern Ireland, Wales and the South West) received less than their 'share' of venture capital investment.

If finance has a national geography which tends to reinforce existing patterns of relative wealth and poverty, perhaps of more significance to many people are the ways in which the geography of the financial system affects their daily lives. Until very recently, the extensive branch networks of the banks acted as highly effective barriers to entry by new competitors. Branches allowed banks to offer a range of services across a range of areas and provided them with access to information about their clients which was then used to assess creditworthiness. Geographically, firms responded to this by ensuring that they had an extensive network and success was partially measured in terms of the number of customers with accounts. However, the growth of cash machines has meant that customers no longer have to go near their branch for money, the development of generic financial products and sophisticated databases has undermined the intelligence gathering role of the branch and banks have stopped cross-subsidizing customers with profits made from more lucrative clients.

Yet at the same time as the branches have become less important for the banks, the new technologies have allowed new entrants into an already overcrowded market. On the one hand, brand names with little history in finance, such as Virgin or British Gas in the United Kingdom or General Motors in the United States, are targeting particular segments of the financial market, while on the other hand, established companies are establishing direct sales subsidiaries which rely primarily on telephone communication. These organizations have far lower operating costs, greater levels of accessibility and, on the whole, a much more affluent set of customers. Relatively well off customers receive better rates of interest but another outcome is that banks are increasingly 'rationalizing' their networks. Between 1989 and 1995 the big four banks in Britain closed over 2,000 branches, contracting their branch networks by more than 20 per cent (Pratt *et al.*, 1996). In the United States, poor neighbourhoods are increasingly becoming stripped of banks at the same time as more affluent areas are seeing growth. In Los Angeles, for example, the two largest banks closed 51 branches during the 1980s and over two-thirds of these were in low-income neigh-

bourhoods. At the same time, the total number of bank branches in Los Angeles county grew by 44 per cent (Pollard, 1996).

There are costs to these processes and – somewhat predictably – the costs are being felt by the poor. As insurance companies are able to price risk more accurately, whole areas of the inner cities face crippling premiums for house and car insurance or simply cannot get quotes at all. Leyshon and Thrift (1997) have convincingly demonstrated that the restructuring of the banking sector is resulting in poorer people getting *reduced* access to the formal financial system at precisely the time that it is becoming more important (in terms of discounts on utility bills or in terms of pay). These processes have been going on for longest in the United States and the withdrawal of the 'financial infrastructure' from poor areas has resulted in the growth in the 'second tier' financial sector: pawnbrokers, money-lenders, cheque-cashing firms and hire-purchase shops. Commenting on this, Gary Dymski and John Veitch have pointed out that:

> Financial structures . . . amplify growth and decay in urban neighbourhoods. Second-tier financial firms which service lower income communities provide no way to pool community savings to finance community investment in new human or physical assets that enhance future economic growth. These new non-banking financial firms meet the financial needs of a community in strictly limited ways. They charge higher fees for services and facilitate households' decumulation of their stock of assets to meet current income crises. In consequence, lower income, higher minority population communities are increasingly isolated from more prosperous communities. Social polarisation is the result.
>
> *(1996: 1257)*

Summary

- At the national level, financial services tend to be geographically concentrated.

- Financial service investment reproduces existing patterns of wealth and poverty.

- During the 1980s and 1990s, less affluent people have increasingly been excluded from the formal financial system in the USA and, to a lesser extent, elsewhere.

CONCLUSION: A ROLE FOR GOVERNMENT?

This chapter has explored some of the ways in which geography affects, and is affected by, the financial sector. In particular, the development of risk in the system and the contribution of finance to uneven regional development have been highlighted. These raise fundamental questions about the ways in which governments should respond to financial sector restructuring. Should, for example, governments attempt to regulate international finance more closely? Any such moves would go against the thrust of policy over the past twenty years and would face a number of potentially insuperable barriers (not least that financial institutions would be likely to move to countries with lower levels of regulation). However, not to do so may result in a widespread financial meltdown which could plunge the world into a 1930s' style recession and this explains why governments seem increasingly prepared to cooperate in framing very limited controls at the international level. Similarly, should governments attempt to control the local activities of financial institutions to ensure, for example, that they provide equal access to the poor or invest in inner city areas? Such measures may meet resistance from the banks, who would almost certainly lose money from them. However, as access to financial services is increasingly a badge of citizenship, such measures may increase a sense of belonging (*see* Chapter 20) and begin to redress the trend towards social exclusion.

ACKNOWLEDGEMENT

I would like to acknowledge the support of ESRC for the support for the research fellowship 'Regulating finance: the political geography of financial services' (H52427001394).

NOTES

1 Money also has other qualities which set it apart. It is not simply a means of exchange, it actually creates value through what Marxists term the circuit of capital (*see* Fig. 1.i, p. 93).
2 The data were obtained by the author from the Office for National Statistics' 1995 Census of Employment via NOMIS.

Further reading

Research on the geography of finance remains in relative infancy. Although there are now a growing number of articles in academic journals (see, particularly, *Environment and Planning A*, *Economic Geography*, *Geoforum*; *Regional Studies*; and *Annals, Association of American Geographers*), there is a limited number of books which deal with the issue. Three collections of essays on the geography of finance have recently been published:

- Leyshon, A. and Thrift, N.J. (1997) *Money/Space*. London: Routledge.
- Martin, R. (ed.) (1998) *Money and the space economy*. Chichester: John Wiley.
- Corbridge, S., Thrift, N.J. and Martin, R. (1996) *Money, power and space*. Oxford: Blackwell.

A more specialist book which examines the essentially masculine nature of much of the City of London is also to be recommended:

- McDowell, L. (1997) *Capital culture*. Oxford: Blackwell.

CHAPTER 12

Consumption

Jon Goss

INTRODUCTION

Geography's engagement with consumption has been a fairly recent one. This is surprising given the importance to us all of the act of consumption. Consumption is one of those aspects of social life which by its very nature is ubiquitous. In this chapter I set out to ask where this ubiquity comes from and what it might mean. I then move on to look at the ways in which geography itself is actively consumed, showing a concern not just with the particular geography of consumption, but also with the peculiar consumption of geography. Geography has hitherto paid far more attention to the other two spheres of the circuit of capital, finance and production. For a long time consumption was something which was left to the disciplines of sociology or cultural studies. Recently, however, there has been an appreciation that consumption is an important facet of the everyday geographies which surround us all. The objects we have in our home help to define that space and give it meaning, and the clothes we wear help to give us our identity. Moreover, our cities and our countrysides are increasingly being shaped as places of consumption rather than production. The urban environment is now dominated by retail parks and shopping malls as much as by factories and offices. Crucially, though, this is more than a geography of consumption – as we shall see, in each place there is also occuring a very particular consumption of geography.

THE CHANGING FACE OF CONSUMPTION: FROM NEGATIVE MORALITY TO CONSUMPTION REDEEMED

In its historical sense consumption is literally a dirty word. Its original meaning is to exhaust or destroy, and in the vernacular it described pulmonary tuberculosis, a fatal wasting disease. The term was partly rehabilitated in the specialist discourse of eighteenth-century political economy, from which it got its neutral meaning as the utilization of products of human labour. Even then, however, consumption was conceived as secondary to production. Adam Smith, for example, dismissed consumption as unproductive investment of resources necessary to realize production, while Karl Marx argued that production simultaneously produced objects for consumption (commodities) and social subjects (classes) to consume them.

The view of consumption as destructive and subordinate to production persisted among modern cultural critics who lamented manipulation of consumers by the 'consciousness industries'– mass media marketing, advertising, and ancilliary activities. There are almost endless variations on the theme, but the main points are that 'false needs' corrupt basic human needs, hedonistic 'communities of consumption' replace real community, public life gives way to organized commercial spectacles, anonymous transactions displace interpersonal interaction, private solutions are sought to social problems, and democratic politics degenerates into electoral role-playing. Commodity relations have penetrated into all spheres and whether citizens or students, parents or patients, we adopt a 'consumer attitude', experiencing life as challenges to be overcome by the acquisition of appropriate goods and services (Bauman, 1990: 204). It is almost impossible to escape, and even dissent is commodified as marginal subcultures and principled resistance become merely other 'styles' (Frank and Weiland, 1997). It seems that we are all consumers now. But perhaps we are not even really that, and have ourselves become objects of consumption: not only are our consumer identities and practices sold as products – for example, as television audiences or mass marketing lists – but objects actively possess us with their characteristics as much we passively possess them (Baudrillard, 1988).

There has recently been a backlash against this negative view. Research shows that consumers are not manipulated dupes of forces of production but are in many ways unmanageable (Gabriel and Lang, 1996). The vast majority of new products fail, and the consciousness industries do not create markets as much as identify and meet latent desires.

The 'cool' consumer

In a recent article in *The New Yorker*, Malcolm Gladwell shows how marketers identify emerging trends by employing 'coolhunters' to search out 'cool' consumers, who pioneer style. There are three rules of cool: 1) cool cannot be defined, only felt; 2) cool can only be observed by those who are cool; and 3) cool cannot be manufactured. Cool consumers cannot be told what to consume by advertising campaigns, but they set the trends, so marketers, who themselves are said to possess a sixth sense for 'coolness', visit the inner cities of North America to consult with selected street kids. There is an interesting geographical component to cool, and marketers talk in terms of the conventional innovation diffusion model, recognizing innovators, early adopters, and late adopters within the urban hierarchy and across the United States. In the market for Reebok shoes, for example, the main source of innovation is Philadelphia and trends move along the East Coast, such that if Chicago, New York and Detroit adopt the innovation, then it spreads out from these major cities and is a guaranteed hit. Even within cities, some neighborhoods are more cool than others. Within New York, for example, Harlem is most sophisticated and cool in terms of sneaker markers, while The Bronx is colorful and glitzy, and Brooklyn tends to be more 'preppy'.

(Gladwell, 1997: 84)

We therefore make our own meaning from products, displaying considerable knowledge and discriminating skilfully between different offerings. Consumption is not merely about using up stuff, but about creatively exploiting products to construct lifestyles expressive of individual and collective identity.

Women still dominate everyday consumption, of course, and misogyny may partly explain the contempt in which predominantly male critics have held it, but the housewife is now hailed as the 'global dictator' given the effect of her decisions on the international economy relations, and as 'hero' due to her thrifty management of the moral economy of the home (Miller, 1995). Recent feminist histories have shown how consumer desire created public spaces and employment opportunities in department stores, transforming **gender** relations in the late nineteenth century. Today, consumers collectively use the power of markets to influence corporate behaviour, government legislation and foreign policy, and the politics of consumption played a significant role in the 'collapse' of communism in the late 1980s. Shopping, once condemned as a frivolous female activity, is now the metaphor for citizenship. Academics acknowledging this power of the consumer, argue that 'consumption, and not production, is the central motor of society' or 'the vanguard of history' (Corrigan, 1997: 1; Miller, 1995) and it is now even fashionable for intellectuals to admit their own pleasure in commercial culture (Cowen, 1998).

Consumption depends upon 'The Gap' between what we know and what we wish to believe, which is why advertising, for example, moves between rational and fantastic motivations for acquisition of products, and why we attach fantastic meanings to mundane material things, as well as judge their value and efficiency and our need. This depends partly upon the verbal rhetoric of advertising, but particularly upon the real or imaginary settings in which objects are presented, what the geographer Robert Sack (1992) calls the **'context of the commodity'**. The context helps us to imagine something other than merely material activity is going on, and objects are invested with lives of their own. Have you ever noticed, for example, how advertising tries to persuade you that you are not merely like the rest of us – a modern shopper buying mass-produced objects – but that you are particularly discriminating of real meaning and value? Or that places you shop evoke somewhere else in space and time, where natural value of objects putatively inheres and endures? You cannot have escaped observation that commodities are supposed to possess qualities that in turn possess us: for example, the brand of cigarette brands the rugged man, and designer label labels the sophisticated woman. Of course, we know that this not really true, but we often act as if it were. We also now understand the general environmental and social consequences of consumption, but we still believe in the purity of the product and innocence of our possession. Some marketing campaigns even play ironically on our self-conscious cleverness in containing this contradiction (*see* Fig. 12.1). This capacity for simultaneously knowing and not-knowing, and for conflating the material and symbolic, both so vital to consumption, is captured by the difficult concept of **commodity fetishism**. It describes how inanimate things are possessed with life of their own, and helps explain our desire for objects.

If you need a watch to make you look cool, you need more than a watch to make you look cool.

T·O·U·R·N·E·A·U
1-800-348-3332
From the world's premier pen maker, a line of functionally elegant Swiss watches.
Presenting the new Swiss made Sonoma™ Series for him & her. 1-800-ATCROSS.

CROSS
SINCE 1846

Figure 12.1 Advertising irony: the choice of watch that says that you are cool because you know that the watch you choose is not what makes you cool! Source: Cross Watches USA

Summary

- The view of consumption has shifted from that which emphasized the negative moralities involved in the process, to that which celebrates the creativity and the culture of consumption.

- Despite this shift, the process is still seen as contradictory and one that is based on the objects of consumption being invested with 'lives' of their own.

THE DESIRE FOR OBJECTS: SOCIAL DISTINCTION AND MASS CONSUMPTION

Although the acquisition of objects realizes universal human need for meaning in the world, things are quantitatively and qualitatively distinct in modern consumer society. We voraciously consume the very latest products in order to realize timeless values. We consume things that money was never supposed to buy. We accumulate objects that compensate for **alienation**, but their obviously accelerating obsolescence only intensifies our sense of loss. We are caught in a maelstrom of progress and can only look back. We love things but are not sure we should. In short, consumption is contradictory. Contemporary consumption purposefully blurs boundaries between art and commerce, and adopts practices functionally equivalent to magic in simpler societies – it is no coincidence that glamour originally referred to casting of spells. It asks us to believe in the transformative powers of material objects, such that the perfect slipper makes the princess, or, perhaps, in the modern version of the story, the perfect sportwalker makes the perfect estate attorney (*see* Fig. 12.2). Rationality tells us that the 'real' origin of power lies in those social relations that legitimate the object, not in the object itself, but we nevertheless want to believe its fetish character.

The term fetish originally applied to objects of traditional religion that were believed to possess animate powers. Karl Marx used it in the nineteenth century to describe our failure to see that the value of **commodities** lies not in their inherent nature, but in the human labour they embody. His labour theory of value is controversial; nevertheless, the prices you pay

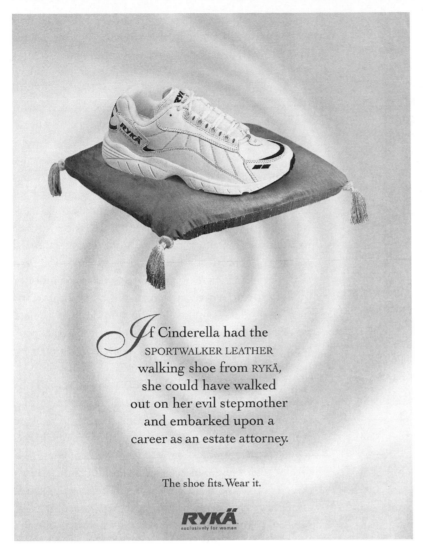

If Cinderella had the
SPORTWALKER LEATHER
walking shoe from RYKÄ,
she could have walked
out on her evil stepmother
and embarked upon a
career as an estate attorney.

The shoe fits. Wear it.

RYKÄ
exclusively for women

Figure 12.2 The magic object of the modern fairy-tale. Source: Rykä

at the supermarket are not determined by social relations, relative need, or capacity to pay, but more or less by the costs of production. Moreover, the material origins of commodities are generally obscured, in the sense that they appear for sale with little evidence of geographical and social relations of production, and so occupy their own perfect worlds.

Jean Baudrillard (1988), a French philosopher, in turn problematizes the concept of fetishism. For him, it is not so much the usefulness or price of objects that clinches sales, but the image of usefulness and price, neither of which are properties of objects *per se*. How else could we explain such phenomena as the wildly popular 'Pet Rocks', more or less valueless lumps of coal that sold several years ago as lovable low-maintenance concept-pets? It is clear that consumption is as much an act of imagination as the using up of things and we are thus advised to 'forget that commodities are good for eating, clothing and shelter; forget their usefulness and try instead the idea that commodities are good for thinking; treat them as a nonverbal medium for the human creative faculty' (Douglas and Isherwood, 1979: 62).

In this sense commoditics communicate both metaphorically and metonymically: they open up a whole world of associations. American automobiles, for example, are utilitarian objects quite similar in style and function from one manufacturer and model to another. Their names, however – from Colt to Cougar, and Falcon to Phoenix – are metaphors for powerful animistic forces associated with speed, freedom of the wild and mythical transformations. The logic is simple, but in case you do not get it, a Chrysler advertisement instructs: 'You are what you drive. Fantasize.' The car, like other commodities, is a prop in the performance of our cultural and personal fantasies. Also, the context in which the automobile is presented evokes other geographies which are more desirable: it might be parked on a rocky outcrop overlooking the Grand Canyon or in the forecourt of a posh hotel, and it will take you anywhere and everywhere. The car, like other commodities, transports us from the here and now. Most of our expenditures are for more mundane household needs, but even everyday products possess intangible qualities, and the 'cool' of clothing styles or 'carefree' of toiletries transform us

---

Commodity magic

While doing fieldwork in Mall of America, I found two excellent examples of fetishism by which commodities are stripped of their real material origins and placed in an imaginary world. In the 'Love From Minnesota' store there was a $155 stuffed bear for sale: the store's tag prominent on its right ear said 'Minnesotans who live deep in the northwoods among the loons, wolves and scented pine trees, listen to the gentle lapping of the waves of the shoreline while they handcraft unique mementos of our homeland, like this one, to share with you'; the manufacturer's label hidden behind its other ear said, 'Bear made by Mary Meyer Corp, Townshend, Vermont . . . Made in Indonesia.' In COLORADO, an outdoor outfitter, I picked up a coat made in Sri Lanka and imported by Columbia Sportswear of Oregon and asked the assistant about the connection, I was told, 'It's just a name. Most people associate it with the outdoors. I mean, if we called it Iowa or something . . . who'd wanna come?' The store is headquartered in New York and has no outlet in Colorado itself, but the slogan of the store explains the connection: 'Remember COLORADO isn't only a store: it's a state . . . of mind.'

---

in consumption, taking us to places of our dreams. It needs stressing that however tenuous the connection, one is not purchasing merely an abstract sign, or the image rather than the substance. The point about commodities is that one always buys both.

One way in which image and substance combine is in the promotion of social distinction. Distinction of style, taste and 'class', are important in urban societies which require regular interaction among strangers and in capitalist society more generally where social status is determined by wealth, rather than, say, physical attributes which are more immediately apparent. Conspicuous consumption thus provides us with an immediately interpretable system of signs of social identity and status – think what you can tell about someone from the clothes they wear or the newspaper they read. Pierre Bourdieu (1984: 136), a sociologist, says that consumer choices are never merely personal judgements, but constitute 'position-taking' within a hierarchy of competencies distributed largely according to education. Like Karl Marx, he sees consumption as ultimately determined by the status of consumers as producers. Consumption reproduces social distinctions, as suggested by popular expressions of contempt, 'I wouldn't be caught dead in that', or 'that's so common/tacky', which serve to differentiate those 'in the know' from others, and marketing sometimes trades explicitly on distinction by poking fun at those who have to ask what it means (see Fig. 12.1).

Still, consumption is more than striving to 'keep up with the Joneses'. Partly because of the expansion of consumption choices, there is much greater horizontal differentiation of social groups, or segmentation of markets. Large and rigid status categories have fragmented into specialist 'consumer tribes' (Maffesoli, 1996). Although I am aware of some of the dangers of the analogy, I particularly like this concept because it describes some key features of contemporary consumption: for example, that it is a communal activity defining coherent groupings and primary loyalties; that membership is conditional upon acquisition of knowledge, often quite arcane, and increasingly outside of formal institutions; that identity is performed in ritualized activities; that 'primitive' traits, captured by the concepts of the fetish and totem are conserved (a totem is a revered object that serves as a symbol of identity for a collectivity that shares its mystical qualities); and that contexts of consumption are where familiarity with the landscape and its meanings provide for a sense of belonging. Each consumer tribe has its totem and its territory.

In a segmented consumer society, we think of consumption behaviour in terms of individual choice and that collective good results from private pursuit of well-being. The fact is, however, that governments regulate exchange, define minimum standards for products and levels of consumption, and subsidize or directly organize health, housing, education and recreation services. There was perhaps a key 'consumer revolution' in the early to mid twentieth century with the development of '**Fordism**', a form of industrial capitalism based on mass production and mass consumption mediated by the state. Consider, for example, the massive suburbanization of capitalist economies since the Second World War, which could not have occurred without public investment in infrastructure, financial incentives, facilitating legislation and regulation. Suburbanization dramatically expanded production and capital accumulation, not only in construction but also consumer durable industries, since each home required an automobile, a washing machine and lawnmower. At the same time, it afforded independent lifestyles and pleasures of consumption to workers, compensating for increased overall productivity and intensified discipline in the workplace (Schor, 1998).

Similar arguments can be made about the role of government in promoting the development of contemporary consumer landscapes, from shopping centres and festival market-places, to sports arenas and cultural complexes, which both facilitate capital accumulation and publicly subsidize consumption. There is, of course, a very particular geography to these subsidies which disproportionately enhance the consumer lifestyles of the middle classes: the countryside becomes their residential dormitories and the city their urban playgrounds. If there is a particular geography of consumption, however, there is also a peculiar consumption of geography.

Summary

- Consumption is fed from our desires, and these in turn depend as much upon our images of the objects to be consumed as on the objects themselves.

- Distinctions in taste and style are important in shaping these desires, and hence in shaping consumption.

- Despite these distinctions, a form of mass consumption has emerged, increasingly regulated by the state.

THE CONSUMPTION OF GEOGRAPHY

The nostalgic narrative tells of progressive alienation from the natural world and our true nature: rationality subordinating imagination, technology severing culture from nature, the cult of newness delivering the present from the past, and mass media undermining genuine sociality. The retail-built environment exploits this sad tale and creates idealized settings that stage restoration of lost innocence and authenticity through the redemptive powers of commodities. These settings include rather specific spatio-temporal archetypes, including Public Space, Market-place and Festival Setting, and Nature, Primitiveness, Heritage and Childhood. The spatial themes evoke the possibility of community among strangers in an authentic public realm, social interaction and transaction in a free market, and spontaneous drama of festivity. The temporal themes evoke memories of human origins in nature and primitive society, of cultural origins of modernity and national heritage in the nineteenth century, and of personal origins in childhood. Transport to these spatio-temporal realms occurs through imagination and memory and via various vehicles: transport to the past in restored sailing vessels and trains; memory evoked in old photographs and 'authentic reproduction' antiques and handicrafts; and fantasy stimulated by myth and magic objects. Most shopping centres and many shops evoke these realms in multiple complex ways.

Public space

Genuine urban life is thought to have disintegrated under the assault of the automobile, urban redevelopment, and violent crime, and it is restored in forms reminiscent of pedestrian streets, avenues, parks, gardens, and train stations, often from places distant in time and space – nineteenth-century European themes, for example, dominate North American

malls, where nodal points quote plazas and piazzas, and food courts resemble street cafés. Despite private ownership and control, malls evoke spaces of social aggregation, people-watching, and public performance. They often host public institutions such as day care, health clinics, religious centres, and government offices, sometimes even police stations and high schools; they provide handicap access and family restrooms; and they raise funds for charity, sponsor community events, and pioneer recylcing programmes (Goss, 1993; 1999). The aim is to reconcile commerce and community, restoring trust in our business institutions which are generally perceived to have abandoned civic responsibility in the pursuit of wealth.

Market-place

The modern economy purportedly undermines traditional markets characterized by unmediated exchanges borne of custom, reciprocity and need. The retail-built environment, therefore, exploits what I playfully call agorafilia, a nostalgic desire for the lost spirit of the market-place, evoking exotic and historic market-places in designs reminiscent of bazaars, stylized historic shop fronts and detailing, numerous variations on the 'Ye Olde Shoppe' theme, and costermonger carts (see Fig. 12.3). It highlights local products and small business, recalling the values of petty enterprise and the regional basis of traditional trade. Again, consumption blurs boundaries: even as cultural critics complain that the exploitation of nature and art commercializes aesthetics, the retail built environment aestheticizes commerce.

Festivity

According to the narrative of decline, we no longer assemble for real public rituals, but submit ourselves to spectacular **simulacra** of festival orchestrated by

Figure 12.3 The costermonger barrow evokes traditional marketplaces. Credit: Jon Goss

mass media. The contemporary retail-built environment provides settings for its restoration in spaces conducive to circulation, aggregation and observation of people, on stages for public performances (*see* Fig. 12.4) and in decor that changes according to ritual calendars. It mobilizes the excitement of the crowd, the aesthetics of performance, and the erotics of the body to tempt you treat yourself, 'discover a new you', try a new look, or to 'try it on', through the transformative potential of commodities, reminiscent of the function of costume in carnival.

Nature

In contemporary consumption nature stands for an original purity and contemporary scarcity. The retail-built environment exploits a pastoral or edenic aesthetic and evokes fugitive nature, often focusing on fragile ecosystems and endangered species (*see* Fig. 12.5). It seems to me that nature products are allegories of the fate of the commodity under contemporary capitalism, where we are exhorted to 'buy while stocks last'. Our predilection for stuffed animals, precious stones, and dinosaurs, naturalizes rapid obsolescence of form and function, but also promises potential restoration provided we have faith in commodity aesthetics and values that never go out of style. They are nature's equivalent of the antiques and collectibles that decorate shopping centres and stores.

Primitiveness

In the modern imagination pre-industrial peoples 'naturally' exhibit values and behaviours that modern civilization has long repressed, enjoying stewardship of nature, simple economies, and innocent faith in animate powers in the material world. Like nature, primitiveness is endangered, which of course enhances its value, and consumers like tourists seek signs of its authentic presence even as it ceases to exist. Primitiveness is evoked in exotic settings, often incorporating themes of historic travel and colonial adventure so that native products are souvenirs of disappearing way of being in the world. The promise, literally, of 'a bit of the Other'.

Figure 12.4   Shopping centres stage traditional public life.
Credit: Jon Goss

Figure 12.5   The nature theme often
focuses on endangered species.
Credit: Jon Goss

Heritage

According to the nostalgic narrative a steep cultural decline has occurred since the late nineteenth century, a time associated with the simple virtues of craft labour and local community, stable identity and public purpose. Contexts of consumption evoke early industrial cities, small towns, and villages, in forms such as the commercial waterfront reproduced in festival market-places, main streets in shopping malls, and village squares in speciality centres. They are decorated with antique tools and bric-a-brac, souvenirs of obsolete forms of production. The Museum Store in your local mall, or the gift store in your local museum, again blur conventional boundaries, both forms incorporating the display aesthetic to narrate how these respective institutions re-store the lost meaning and values of objects.

Childhood

As Western civilization has 'grown up', modern rationality and instrumentality have repressed our childlike impulses. Like primitives, however, our children are not yet fully modern and partially inhabit the time of myth and realm of fairy tale. While typically acknowledging adult values in the form of price, efficiency and educational value, the retail-built environment is organized around the experience of the child, appealing directly to imaginative capacities for transcendence and instant gratification. As an adult, I pooh-pooh fairy tales and the trickery of animation, gigantism and miniaturization mobilized in the mall; I smile knowingly at games of dressing up, anthropomorphization of animals, and superstition of magic objects; and I certainly do not believe in Santa Claus or fairies. But I believe in the innocent belief of children, and I believe through them that belief is possible. Without our faith in children we might not be able to ourselves believe in the genius of things: in magic slippers, amazing dreamcoats, elixirs of everlasting health, potions of supernatural power, the wizardry of laundry soaps, or even the alchemy of the credit card.

Summary

- The particular geography of consumption partly operates through a specific consumption of geography.

- The geographies which are consumed invoke very specific archetypes, ranging from public space and marketplace to nature, heritage and the spaces of childhood.

CONCLUSION

This chapter has argued that consumption provides compensation for the felt **alienations** of modernity and the retail-built environment seeks to restore our sense of well-being by creating an alternative reality from geographical and historical archetypes of an orignal, authentic and immanent world. In these contexts, it seems as if something other than mere consumption is going on, and fetishized commodities appear as other than merely material objects of human labour. There is a certain undeniable magic. Ironically, however, while offering images of authentic public life, market-places, and festival settings, the increasingly privatized and socially controlled retail-built environment undermines genuine urbanity, and while incorporating images of nature, primitiveness, heritage and childhood, its technical virtuosity seems to make increasingly redundant myth, memory and imagination. It seems to present the world of **commodities** innocent of the **commodification** of the world, and magic turns out to be an illusion. Only by acknowledging that in our material society our dreams and desires are transformed into wants for commodities will we be able to materialize our visions of authentic life in social form rather than project them onto objects. By embracing and understanding our dreams, we might, like the subject of psychotherapy, come to understand the 'true' object of our repressed desires.

Further reading

A basic, but also very contemporary introduction to consumption and everyday life is provided by:
- Mackay, H. (ed.) (1997) *Consumption and everyday life*. London: Sage.

For a review of contemporary research in consumption from a variety of disciplinary and theoretical perspectives, including geography, see:
- Miller, D. (ed.) (1995) Consumption as the vanguard of history. In Miller, D. (ed.) *Acknowledging consumption: a review of new studies*. London: Routledge.

Another multidisciplinary and diverse collection of essays is:
- Shields, R. (ed.) (1992) *Lifestyle shopping: the subject of consumption*. London: Routledge.

For a sustained geographical approach to consumption and consumption landscapes, see:
- Sack, R. (1992) *Place, modernity and the consumer's world: a relational framework for geographical analysis*. Baltimore: Johns Hopkins University Press.

Environmental geographies

INTRODUCTION

In many ways, environmental issues are a 'natural' area of concern for geographers. Indeed, traditional defin-itions of the subject have tended to emphasize the relationships between people and their environment, and geographers have for a long time been interested in the connections between human activity and envi-ronmental degradation, or environmental hazards. For many years, geographers tended to adopt rather for-mal approaches to studying these interconnections. For example, a 'science' of 'environmental impact assessment' has grown up which suggests methods either for cataloguing the environmental conse-quences of a particular project, or for a more struc-tured quantitative accounting of the costs and benefits to the environment of that project.

More recently, however, human geographers have turned their attention to environmentalism as a focus of study. Environmentalism collectively describes a wide range of ideas and practices which demonstrate a concern for nature–society relations. Many of these ideas and practices can be brought much closer to the geographies of our everyday life than the previous evaluative foci. For example, Human Geographers have been interested in different forms of environ-mentalism, from the deep ecology movement, which has exposed the principle of living in harmony with nature, to the different elements of the 'green' move-ment, which have used popular protest to highlight key environmental concerns, and the New Age movement which has brought a range of personal and spiritual attachments to nature into sharp focus.

As we move into the new millennium, the global and local considerations of different environmen-talisms will become increasingly important. Moreover, as the three chapters in this section demonstrate, there

have been further changes in the way in which these issues have been discussed in the public arena. Most notably, 'environmentalism' has been transformed into 'sustainability' which Bill Adams notes has been the subject of a meteoric rise to the status of global buzz-word. In Chapter 13, he clearly shows how sustain-ability has become a powerful concept *because of* the lack of precision involved in its definition. Put simply, sustainability is defined in different ways by different interests, even though these interests might have opposing objectives! It follows that the introduction of sustainability ideas into practical policy, such as in the principles of Agenda 21, means that every seemingly firm commitment to environmental action is likely to be accompanied by a get-out clause. Thus a potentially exciting set of concepts is often rendered important because they can mean all things to all people.

The rise of environmental policy is detailed by Andrew Jordan and Tim O'Riordan in Chapter 14. They trace how long-term shifts in public opinion towards quality-of-life issues, and the increasingly active and effective green pressure groups, have resulted in a range of policy changes. These include green accounting, environmental auditing, strategic planning and some greater acceptance of intergenera-tional equity. However, they suggest that technocen-trism – that is the view that the management of growth permits the exploitation of the environment by humans for their own utility – still holds sway over its ecocentric alternative.

Jacquie Burgess helps us to see why. In Chapter 15, she shows how environmental issues and crises are socially constructed. The ways in which knowledge about the environment is presented to us are greatly influenced by the media and by 'experts' who help to frame the issues involved, and to 'spindoctor' public perception of environment crisis. It is in this production of environmental knowledge that the communication of environmental problems is variously effective in persuading people to adopt sustainable lifestyles. Information is often globalized, but human response is often based in the home or work locality – hence 'think global, act local'.

The rise of sustainability has placed environmentalism in a broader field of coalition – building across a wide range of policy communities which were previously unconnected. Health and social cares, crime prevention, family solidarity, and environmental stewardship by business have become entwined across national borders and all levels of government. In this flux, outdated environmental outlooks are beginning to atrophy. The 'old guard' of environmental leaders is slowly and painfully being replaced by a more opportunist, managerialist class. As environmentalism tackles sustainability, however, the political stakes have become much more resistant to the demands of tax reform, economic and political redistribution, and

spatial reorganization that the new agenda of sustainability is making. The nerve centres of power have recognized that sustainability needs to be tamed, and this process is now the centre of the struggle.

Further reading

Geographies of environment and environmentalism are both vast and multifaceted, yet there are relatively few texts which succeed in pulling the various strands together. For a lively introductory text read:

- Park, C. (1997) *The environment: principles and applications.* London: Routledge.

For more focused, but also more discursive accounts, you might like to look at:

- Bell, M. (1998) *An invitation to environmental sociology.* Thousand Oaks: Pine Forge Press.
- O'Riordan, T. (1998) *Environmental science for environmental management.* (2nd Edn) Harlow: Longman.
- Pepper, D. (1996) *Modern environmentalism: an introduction.* London: Routledge.

CHAPTER 13

Sustainability

W.M. Adams

INTRODUCTION

I have long wanted to complain to Bill Gates that my word processor's spell-checker does not recognize the word 'sustainability'. At first this surprised me, for the word has become indispensable in any discussion of human impacts on the environment. It is hardly one of those secret words only used by academics and other toilers in the depths of libraries, for it peppers the speeches of politicians, teachers, business leaders and environmentalists and, of course, Geographers.

But perhaps Microsoft's dictionary-compilers are not so behind hand, for in fact sustainability has only recently become so widely used. The ideas for which it is so convenient (and, as we shall see, so slippery) a label have been around for many decades, but sustainability is very much a word of the 1990s. Specifically, it owes its global reach to the vast media roadshow surrounding the United Nations Conferences on Environment and Development (UNCED, or the 'Earth Summit') at Rio de Janeiro in Brazil in June 1992.

That event brought together 128 Heads of State and 178 governments and their attendant lobbyists, as well as a host of non-governmental organizations (Chatterjee and Finger, 1994). Sustainability was UNCED's 'Big Idea'. Thus launched upon the world, its meteoric rise to global buzzword began. But what does sustainability mean? Why did it emerge at Rio? What use is it? These questions are rather harder to answer than they look.

WHAT IS SUSTAINABILITY?

What does sustainability mean? Irritatingly, we could answer, 'everything and nothing'. It is a word that promises much. A dictionary definition offers a range of meanings, each of which captures something of the meaning of sustainability. Longman's *Dictionary of the English Language* (1991) gives the following definitions of the verb 'to sustain':

- to give support or relief to;

- to supply with sustenance, nourish;

- to cause to continue, prolong;

- to support the weight of;

- to bear up under, to endure (to suffer, to undergo).

They are all to do with continuity ('prolonging', 'nourishing', 'supporting', 'enduring'). They are also all basically positive, all things that might be thought of as broadly desirable or admirable. In public debate about environment and development, use of the word sustainability suggests that change can be allowed to happen (or made to happen), that the best of what has done before is maintained, whether that change is in an economy or society ('sustainable development') or in an **ecosystem** ('sustainable environmental management').

Internationally, the dominant definition of sustainable development has undoubtedly been that of the Brundtland Report, in *Our Common Future*: 'development that meets the needs of the present without compromising the ability of future generations to meet their own needs' (Brundtland, 1987: 43). This definition is both rhetorical and vague (Lélé, 1991), but it proved compelling as a way of pulling to together concerns about environmental degradation and present and future poverty (often spoken of as **inter-** and **intra-generational equity**).

The word sustainability first appeared in British legislation in 1991, in the Act establishing the conservation organization Scottish Natural Heritage (SNH). SNH was charged with achieving and promoting the conservation and enjoyment of landscapes and wildlife 'in a manner that is sustainable'. This apparent coup for environmental thinking caused some scratching of heads in the new agency, because while the word might have sounded wonderful in Parliament, its meaning was far from clear. Eventually, SNH suggested that sustainability should mean 'the ability of an activity or development to continue in the long term without undermining that part of the environment which sustains it' (SNH, 1993). While heart-warming (and arguably very wise) as a general principle, this is hardly a sharply focused definition. When other UK national conservation agencies set up in 1991 got in on the act, their own definitions were no more specific. Their definitions of sustainability and sustainable development were:

- The ability of an activity or development to continue in the long term without undermining that part of the environment which sustains it (Scottish Natural Heritage, 1993).

- Sustainable development seeks to improve the quality of human life without undermining the quality of our natural environment (English Nature, 1993).

- Sustainability implies that human use of or enjoyment of the world's natural and cultural resources should not, in overall terms, diminish or destroy them (Countryside Commission, 1993).

Summary

- Sustainability is an emotive word, with complex meanings.

- The concept of sustainability began to be used internationally in the 1980s (notably in the report of the Brundtland Commission in 1987), and first appeared in British law in 1993.

SUSTAINABILITY: MORE THAN A BUZZWORD?

The Council for the Protection of Rural England describe sustainable development as 'the latest buzzphrase to hit the planning profession' (Jacobs, 1993:

8); indeed, it has become almost impossible to avoid using it. Geography has certainly not been free of its influence, as a glance round any mainstream publisher's catalogue, or a swift perusal of geography textbooks (including this one, of course!) would prove. Sustainability is therefore quite obviously important, if one measures something's importance by the number of people talking about it. What is less clear, however, is whether anything much lies behind the glittery promise of the word.

Superficially, the concept of sustainability seems very simple, yet it can have a wide range of meanings attached to it. Rio Tinto Zinc, or Shell, might speak of sustainability in the context of mineral extraction, but mean something very different to a Friends of the Earth campaigner or someone from Nigerian Ogoniland (see Fig. 13.1); the Chancellor of the Exchequer might (indeed, has) spoken of 'sustainable' national economic management, and mean something very different to a proponent of a zero-growth economy (see Box).

Figure 13.1 Ken Saro-Wiwa, Nigerian Ogoniland protester and playwright, killed by the Nigerian State in 1996 following protests about environmental pollution and the oil industry in the Niger Delta (much of it led by Shell International).
Credit: Popperfoto/Reuter

Michael Jacobs on sustainable development

Governments often speak of aiming for 'sustainable growth': they mean economic growth without inflation rather than without environmental degradation, and the usual interpretation of 'sustainable' is lasting about four years, or until the next election, whichever is the sooner.

(Jacobs, 1993: 9)

It needs to be remembered after all that sustainable development and sustainability were not originally intended as 'economic' terms. They were, and remain, essentially ethico-political objectives, more like 'social justice' and 'democracy' than 'economic growth'. And as such, their purpose or 'use' is mainly to express key ideas about how society – including the economy – should be governed.

(Jacobs, 1995: 65)

Both environmentalists and conventional economic and political planners use the word sustainability to express their own vision of how economy and environment should be managed. The word does not end the debate about how society should exploit non-human nature, it simply re-labels it. Indeed, such is the power of sustainability to allow different ideas to be smuggled forward in its ample conceptual folds that it effectively delays debate and pushes it underground. Radical opponents of roads and other infrastructure have literally taken to the ground (or the trees) in opposition (*see* Figure 13.2). Their concerns have not been met by the growing debate about sustainability and transport in the UK. For a single neat word, sustainability hides a theoretical maze of great complexity (Daly, 1990; Lélé, 1991). It offers a verbal flourish, but at its core is a theoretical black hole (Redclift, 1984, 1987).

Of course, it is not strictly fair to say that sustain-ability has no theoretical core. Its intellectual roots lie in population biology, ecology and economics. Through the 1920s and 1930s, biologists were developing simple mathematical models of population growth and competition, from which, in time, grew the notion of maximum sustainable yield, that populations of organisms (initially fish, but the point was generally true) could be harvested at a rate that allowed the population to reproduce itself.

These scientific ideas about how animal populations fluctuated, and what happens when people start to harvest them, comprise one stream of biological ideas feeding into sustainability. A second is in ecology, particularly in the concept of the ecosystem (proposed in the 1930s), and in ideas about plant succession. As ecology became influenced by systems thinking in the 1960s, ideas of equilibrium in ecosystems provided a further natural science basis for ideas of sustainability. The science of ecology seemed to show the vulnerability of the environment to human impacts, and the need for those impacts to be moderated. Meanwhile, from economics came concepts of renewable (flow) and non-renewable (stock) resources. These are diverse enough roots, but onto them many other ideas were grafted from the emerging worldview of environmentalism, particularly about population growth, resource exhaustion and the toxic and shocking effects of industrialization and urbanization (Adams, 1990).

Summary

- Sustainability tends to be defined in different ways by different interests, for example, by environmental organizations and big business.

- The concept of sustainability draw on scientific studies of the dynamics of animal populations and ecosystem equilibrium, and ideas about the economics of renewable resource exploitation.

Figure 13.2 Swampy protesting over Manchester Airport, 1997. Credit: Popperfoto/Reuter

SUSTAINABILITY AND THIRD WORLD DEVELOPMENT

The concept of sustainability first emerged at the United Nations Conference on the Human Environment, held at Stockholm in 1972 (Adams, 1990; McCormick, 1992). This meeting was the direct forerunner of the Rio Conference 20 years later. Like Rio, it saw profound divisions between industrialized countries and the Third World. The poorer non-aligned countries saw the First World's concerns about pollution and technology as the worries of an exclusive club of wealthy countries, and a potential threat to their ability to industrialize effectively. They also feared and resented the obsession of environmentalists in the First World with population growth (see Box). The concept of sustainable development was coined explicitly to argue that an option existed that would allow appropriate (i.e. rapid) economic growth and industrialization without environmental damage. This happy outcome has been the target of all subsequent calls for sustainability.

Since Stockholm, different interests have emphasized different aspects of sustainability, and sought to claim the concept for their own. In 1980 the World Conservation Strategy (IUCN, 1980) took a strongly conservation-oriented position. It defined conservation as sustained resource use, and suggested three objectives for global conservation. The objectives of the World Conservation Strategy are:

1. to maintain essential ecological processes and life support systems (such as soil regeneration and protection, the recycling of nutrients, and the cleansing of waters);

2. to preserve genetic diversity (the range of genetic material found in the world's organisms);

3. to ensure the sustainable utilization of species and ecosystems (notably fish and other wildlife, forests and grazing lands).

Six years later, the report of the World Commission on Environment and Development, *Our Common Future*, (called the Brundtland Report after its Chair), had a very different emphasis. It deliberately broadened the debate, locating environmental issues within the economic and political context of international development debates. It therefore linked basic development needs and environmental degradation, arguing that one could not be solved without the other. The way forward, it suggested, was through global multilateral co-operation between rich and poor countries to achieve development: sustainability achieved through careful economic growth.

The Brundtland Report, published in 1986, led directly to UNCED, in two ways. First, it was debate of the report in the General Assembly of the UN that led to the resolution to hold what became the Rio Conference. Second, it was this message of adapted or 'green' growth that provided the carrot to persuade both rich and poor countries to come to the negotiating table. None the less, the task of finding common ground was Herculean, and lasted through a full five years of preparatory meetings before the conference itself (Chatterjee and Finger, 1994). The documents produced by this 'Rio process' were the fruit of wearying debate far into the night by government delegations determined to produce a form of words that gave least away in terms of their own national interests.

Inevitably, divisions opened up between the distinction between countries in the industrialized 'North' and the underdeveloped 'South'. They disagreed over what the main global problems were (global atmospheric change, **biodiversity** loss and tropical deforestation in the industrialized countries, poverty and the environmental problems associated with it in unindustrialized countries), and they disagreed over who should pay for any action needed. Third World countries feared that their development would be stifled by restrictive international agreements on atmospheric emissions (just as at Stockholm in 1972), and they were jealous of their right to use the natural resources within their boundaries for development (notably tropical forests) without restriction by environmentalists in the First World (whose environmental concerns, arguably, were only possible because of a wealth itself created by polluting freely and consuming forests and other forests).

Eventually, some kind of agreement was patched together, and the conference agreed a slightly rambling set of 29 principles in the 'Rio Declaration', a much watered-down set of principles for forest management (unexcitingly titled a *'Non-Legally Binding Authoritative Statement of Principles for a Global Consensus on the Management, Conservation and Sustainable Development of all Types of Forests'*), and the vast compendium of good intentions in Agenda 21. Samuel Johnson, the essayist and first compiler of an English Dictionary, is said to have apologized for the length of a letter, saying 'I did not have time to make it short.' So it was with Agenda 21, which contains more than 600 pages of text in 40 separate chapters. These were divided into four sections, covering socio-economic and environmental aspects of sustainable development, the actors who could make it happen, and the means of implementation (*see* Box p. 130).

Agenda 21 has become an icon of sustainable devel-

Population, environment and sustainability

In the 1970s, First World environmentalists laid particular emphasis on the problem of population growth, arguing that as populations rose resource exploitation would inevitably become unsustainable. This argument was particularly made about rural populations in the Third World. Books like *The Population Bomb* by Paul Ehrlich (1972), and papers like Garrett Hardin's 'The tragedy of the commons' (published in *Science* in 1968), started a new and apocalyptic 'neo-Malthusian' debate about people and environment. Through the 1970s and 1980s, the drylands of Sub-Saharan Africa were singled out in particular as a place where rapid population growth was leading to environmental degradation. Drought and famine were both significant problems in this decade, but attention focused in particular on the problem of 'desertification', and human-made deserts. It was widely held (not least by geographers) that population growth inevitably led to desertification, as farmers and pastoralists pushed semi-arid ecosystems past some natural limit.

However, research in recent years has begun to show that in several parts of Africa agricultural systems appear to have coped with significant levels of population growth without loss of sustainability. The most important study of this kind was conducted in the Machakos District in Kenya (Tiffen *et al.*, 1994). In the 1930s, government officials despaired of Machakos, which was thought to be on the verge of ecological collapse due to over-population. Fifty years later, the changes have been remarkable. The population has soared (from 0.24m in 1930 to 1.4m in 1990), but far from destroying the environment, farmers have developed it. Terracing is extensive, cattle are stall-fed and their manure applied to the land (*see* Figs 13.4 and 13.5), and with the advent of cash crops (particularly coffee), the volume and value of output have increased to match population growth. It is an astonishing story, and while the experience of this area might not be a good model for the whole of Africa (among other things, the international city of Nairobi is not far from the borders of Machakos, and the main road to the coast passes through it), it is clearly unwise to make the automatic assumption that rural population growth is unsustainable.

Figure 13.3 Smoke plume from UK power station: concern about acid rain was one of the environmental problems that led to the Stockholm Conference. Credit: Mike Read/Planet Earth Pictures

opment, worshipped but not much read. Because of the way it was written through negotiation, every commitment has a get-out clause somewhere nearby. It contains therefore within it most possible arguments, and is readily mined for nuggets of text that can be used to legitimate any given point of view.

However, the Rio Conference was a watershed. From then on, governments and international agencies began to re-interpret their normal work of economic planning within the new, internationally agreed, terminology. There was a substantial shift in political and bureaucratic rhetoric, but the new language sometimes lay wafer-thin across old and not obviously 'sustainable' policies. A shift in the language of policy of this magnitude, of course, is no small matter. It was due to two related forces. First, it was a straightforward response by politicians (particularly in Europe and North America) to the surge of environmentalism that took place within Western societies in the early 1990s. Behind that pragmatic (perhaps sometimes cynical) politics lay a perception of environmental limits, which had itself driven that rise of environmental concern. As Bill McKibben wrote in his best-selling book *The End of Nature*, 'The greenhouse effect is the first environmental problem we can't escape by moving to the woods' (McKibben, 1990: 188). Sustainability and sustainable development were the words people in the 1990s came to use to express that thought, and on which they tried to build arguments for reform.

The contents of Agenda 21

Section 1   *Social and Economic Dimensions*   Eight chapters, covering international co-operation, combating poverty, consumption patterns, population, health, settlements and integrated environment and development decision-making.

Section 2   *Conservation and Management of Resources for Development*   Fourteen chapters on the environment. These covered the atmosphere, oceans, freshwaters and water resources, land resource management, deforestation, desertification, mountain environments, sustainable agriculture and rural development. They also covered the conservation of biological diversity and biotechnology, toxic, hazardous, solid and radioactive wastes.

Section 3   *Strengthening the Role of Major Groups* Ten chapters discussing the role of women, young people and indigenous people in sustainable development; the role of non-governmental organizations, local authorities, trade unions, business and scientists and farmers.

Section 4   *Means of Implementation* Eight chapters, exploring how to pay for sustainable development, the need to transfer environmentally sound technology and science; the role of education, international capacity-building; international legal instruments and information flows.

(Robinson, 1993)

Figure 13.4   Terraced farmland in Machakos. Credit: Bill Adams

Figure 13.5   Stall-fed cow, Machakos. Credit: Bill Adams

Figure 13.6 Purple Saxifrage, a rare arctic-alpine plant in the UK, potentially threatened by global warming. Credit: Bill Adams

Summary

- The key event in the development of thinking about sustainability was the Rio Conference in 1992.
- At Rio a series of agreements were signed, the most important of which was the massive Agenda 21.

SUSTAINABILITY AND BIODIVERSITY

The fundamental element in thinking about sustainability is the state of the environment. Some views see the environment as consisting of particular critical elements that must be maintained as economies and societies develop. Others see the most important issue as the maintenance of the flow of economic benefits from the environment over time. The first of these approaches was recognized at Rio in those parts of the Convention on Biological Diversity that dealt with the traditional approach to the conservation of species and ecosystems, which emphasized the preservation of nature over its utilization. In the UK this concern relates directly to conventional nature conservation activity, for example the fate of relict arctic-alpine species on British mountains in the face of global warming (*see* Figure 13.6). The second approach (about the environment as a source of benefits for people) emphasizes the need for the environment to provide safe and productive livelihoods for the poor.

Of course, sometimes preservation and economic needs can be closely aligned, for example in the management of a lake so that both fishery and fish biodiversity are sustained, or in the need of communities in the rural Third World for continued supplies of products from a forest (and hence maintenance of its ecosystem). The poor also need un-polluted environments, and protection against ruthless short-term profit-oriented businesses that might see pollution (and vestigial employee care) as a way of cutting costs. In these kinds of cases, sustainable development may indeed optimize both environmental protection and gains in human welfare.

Traditionally, conservation has tended to make severe demands on local communities, particularly in the Third World, when protected areas (such as National Parks) have been created. However, from the 1980s onwards, conservation strategies based on excluding local people have been replaced by attempts to integrate conservation and economic development. Strategies include Integrated Conservation and Development Projects, which attempt to promote sustainable resource management by targeting development aid to communities in or around National Parks. Another approach is to develop wildlife tourism activities that yield significant benefit to local people. A third approach is the so-called CAMPFIRE model developed in Zimbabwe (*see* Fig. 13.7), where

Figure 13.7 Community grinding mill funded out of CAMPFIRE funds in South East Zimbabwe. Credit: Bill Adams

legal rights in wildlife can be granted to District Councils, who can then allow communities to charge safari hunters for shooting rights.

Where economic development is on a larger scale, opposition between the maintenance of biodiversity and the maximization of economic returns can be more difficult to reconcile. As we have seen above, the concept of sustainable development was precisely devised as a means of suggesting that this need not be the case. One way to do this is to place economic values on nature (Pearce *et al.*, 1988; Barbier *et al.*, 1994). A distinction may be drawn between two kinds of capital, 'conventional' ('human-made') capital and that represented by natural resources and nature's services (which could be anything from rainfall to the dilution of pollution by a river), called 'natural' capital. Sustainability can then be defined in terms of the maintenance of constant stocks of natural capital over time. One organization that has adopted this approach is English Nature. They have suggested treating vital conservation areas as 'critical natural capital' (*see* Fig. 13.6), because they cannot be replaced if lost (or at least, not within feasible time frames), and cannot therefore be substituted for human capital or compensated for by positive projects elsewhere (English Nature, 1993).

However, behind this lies a potentially exciting set of concepts. Although to date, the radical potential of sustainability as a political and policy tool has tended to be smothered, it has not yet been wholly lost. A critical question for Geographers must be what happens when the over-inflated balloon of green rhetoric comes down to earth. Specifically, how will the challenge of sustainability posed at Rio be taken up, and the fine words turned into policy by central and local government, businesses and non-governmental organizations? This question is addressed by Andrew Jordan and Tim O'Riordan in Chapter 14 of this book.

Although it has exciting potential in policy terms, the concept of sustainability is less satisfactory in theoretical terms. It has little rigour to offer Geographers anxious to get an intellectual grip on issues of environment and development. Here too, however, sustainability cannot be ignored, for it has few equals as a challenge to Geographers to develop their theoretical ideas, and then to use them to engage in understanding the relations between society and nature, and perhaps even changing them for the better.

Summary

- One aspect of sustainability emphasizes the importance of preserving the quality of the environment, particularly biodiversity. Another aspect emphasizes the flow of economic benefits through time, particularly the livelihoods of the poor. The concept of sustainability is attractive because it suggests that both these objectives can be achieved together.

- Biodiversity conservation strategies (such as National Parks) are increasingly being designed to have a positive and not a negative impact on local people.

- One approach to integrating environmental and welfare aspects of sustainability is through identifying the economic values of the environment, and treating biodiversity as 'natural capital'.

CONCLUSION: LIVING WITH SUSTAINABILITY

Clearly, at Rio and everywhere else, debates about sustainability have involved a large amount of hot air.

Further reading

- Chatterjee, P. and Finger, M. (1994) *The earth brokers: power, politics and world development*. London: Routledge.
This book provides very useful insights on the Rio Conference and the meetings leading up to it; it is thought-provokingly critical about the failings of the 'Rio Process'.

- Holdgate, M. (1996) *From care to action: making a sustainable world*. London: Earthscan.
An excellent summary of environment and development problems, and what might be done about them. This book updates the 1990 follow-up to the 1980 World Conservation Strategy in the light of the Rio Conference: sane, practical, readable, unradical.

- McCormick, J.S. (1992) *The global environmental movement: reclaiming Paradise*. London: Belhaven.
- Redclift, M. (1996) *Wasted: counting the costs of global consumption*. London: Earthscan.
Michael Redclift's latest book on sustainability offers a typically readable critique both of the concept itself, and of the unsustainability of global 'business-as-usual'. John McCormick's book gives an excellent introduction to the history of the environmental movement.

Environmental problems and management

Andrew Jordan and Tim O'Riordan

THE ORIGINS OF ENVIRONMENTAL POLICY

Recognition of the need to both transform and adjust to nature is a fundamental aspect of the human condition. While we may think of 'the environment' as a modern political issue that gained popular appeal in the 1960s, the roots of environmentalist thinking stretch back far into the past (O'Riordan, 1976). The natural environment provides humanity with the material resources for economic growth and consumer satisfaction. But throughout history there have always been social critics and philosophers who have felt that humans also need nature for spiritual nourishment and aesthetic satisfaction. John Muir, the redoubtable founder of the Sierra Club in the USA (*see* Fig 14.1), felt that without wild places to go to humanity was lost:

> Thousands of tired, nerve-shaken over civilised people are beginning to find out that going to the mountains is going home; that wilderness is a necessity and that mountain parks and reservations are fountains not only of timber and irrigating rivers, but as fountains of life. Awakening

Figure 14.1 The Yosemite National Park has inspired many generations of modern Americans including the redoubtable John Muir, a pioneer of the American environmental movement. Credit: Dave Lyon/Planet Earth Pictures

from the stupefying effects of over-industry and the deadly apathy of luxury, they are trying as best they can to mix their own little ongoings with those of Nature, and to get rid of rust and disease . . . some are washing off sins and cobwebs of the devil's spinning in all-day storms on mountains.

(quoted in Pepper, 1984: 83)

Environmental protection is justified in remarkably similar terms today. What is dramatically different is the *extent* of popular concern. The critical question which needs to be asked is *why did modern* **environmentalism** *blossom as a broad social movement spanning different continents in the late 1960s and not before?* There is strong evidence that environmental problems like acidification and pesticide pollution materially worsened and became more widespread in the public mind in the 1960s and 1970s. The American sociologist Ronald Inglehart (1977), however, believes that we also have to look to society for an explanation. On the basis of careful and intensive public opinion analysis he argues that modern environmentalism is the visible expression of a set of 'new political' values held by a generation of '**post-materialists**' raised in the wealthy welfare states of the West. This liberated class no longer had to toil to supply their material needs and set out to satisfy what the psychologist Maslow (1970) terms its 'higher order' or quality of life wants like peace, tranquillity, intellectual and aesthetic satisfaction. This was surely a 'post-materialist' sensibility, but at first it was confined to a vociferous minority that tried to push their values onto the majority who steadfastly regarded themselves more as consumers than as citizens.

Other commentators, however, highlight the tendency for environmental concern to exhibit a cyclical pattern over time, with particularly pronounced peaks in the late 1960s and late 1980s. Closer scrutiny reveals that these short-term 'pulses' coincided with periods of economic growth and social instability, which at first blush seems consistent with Inglehart's thesis. Other sociologists have also observed that materially richer and better educated sections of society tend to give much higher priority to environmental protection than poorer ones, with the highest rates among those working in the 'non-productive' sectors of the economy such as education, health and social care (Cotgrove and Duff, 1980). Conversely, concern tends to tail off during periods of economic recession (Downs, 1972), and is not normally as pronounced in poorer sections of Western society or in developing countries. The birth of the modern environmental movement in the late 1960s certainly coincided with a period of economic prosperity and societal introspection. Whether this led to or was caused by the accumulating evidence of environmental decay is open to interpretation.

Be that as it may, landmark books like Rachel Carson's *Silent spring* (1964) and *A blueprint for survival* (The Ecologist, 1972) gave the impression that a global crisis was underway and found a keen and receptive audience in the USA and Western Europe. It is often said that the first evocative images transmitted from space of Earth floating precariously in the inky blackness of deep space, catalysed and intensified public concern. They emphasized much more clearly than words the finiteness of the planet and did much to encourage a *global* perspective. Before long, though, the economy slipped into recession and the clamour for change died away. But environmental policy weathered that storm as it has done the painful economic recession of the early 1990s. Studying environmental politics through recent history demonstrates that, ratchet-like, the laws and public institutions put in place to protect the environment during each upsurge, fix environmental values in place and provide the platform for the next burst of attention. For example, the European Union's (EU) environmental division, DG XI, the UK Department of the Environment (DoE) and the United Nations Environment Programme (UNEP) all date from the early 1970s. Since being established, each has worked to drive green principles into its institutional settings, 'locking' environmental values in place. The broader point being made, however, is that modern environmentalism needs to be seen against the backdrop of long-term shifts in the economy and society. In other words – and this is important – 'the environment' is as much a *social* and thus intensely subjective and political phenomena as it is a set of objectively defined natural resources.

We can summarize the endurance of the environmental movement since the mid-1960s on the basis of the following points. First, a steady diet of varied but symbolic *environmental crises* from *Amoco Cadiz*, Seveso, *Exxon Valdez*, Three Mile Island, Chernobyl, the Spotted Owl, Indonesian forest fires and Amazonian rain forest destruction, which were popularised by environmental pressure groups (*see* Fig. 14.2). Second, the emergence of national and international *data of environmental change* based on satellite imagery and more comprehensive surveys and integrated ecosystem analysis at all spatial levels. Third, the concurrent arrival of *environmental protection agencies* backed by national and international laws, regulations, assessment procedures and consultative mechanisms, all of which alerted business, government, pressure groups and academics to the political demand for environmental protection. Fourth, a string of *international conferences* from Stockholm to Rio and beyond, resulting in a suite of international regulatory protocols that forced a higher order of public and private awareness, much greater financial

Figure 14.2 Modern environmental politics is stimulated by institutional crises such as the nuclear disaster at Chernobyl nuclear plant in 1986. Credit: Popperfoto

investment, and corporate planning. Finally, a new era of *environmentally concerned citizens* moving into the late 1990s' economy and society with this outlook and the emergence of tools of coping with complex relationships between economy, society and ecology.

The consequences of all this is the politicization of environmentalism in a number fascinating ways:

- *Integration of policy* across departments, budgets and borders. The EU is mandated to promote environmental policy integration, through which environmental considerations are to be factored into all aspects of policy-making and the evaluation of policy options.

- Creation of *green accounts* through which indices of sustainable economic welfare are being assessed by imaginative means and incorporated into national measures of economic growth.

- Publication of ***environmental audits***. Many governments publish regular reports on their environmental performance – the EU has developed a series of environmental Action Plans and hundreds of local authorities across the world are developing Local Agenda 21 policy documents explaining how they will implement sustainable development at the sub-national level.

- *Strategic planning*. The days of specialized, 'end of pipe' regulation are fast disappearing. In their place is a fresh look at whole company audits, green technology and integrated 'whole plant' planning, to ensure that environmental resources are safeguarded.

- The *precautionary principle* is based on the presumption of inadequate knowledge, avoidance of possible dangers that cannot fully be calculated, and a weighting of decision-making towards the well-being interests of the most vulnerable to future social and economic change. In effect, it lays a duty of environmental care on all actors. As yet, that duty is not fully codified in law. But increasingly its outlines are clear in the form of codes of practice and the kind of responsibilities Bill Adams outlines in Chapter 13 are now facing the custodial agencies in the UK.

Part of the precautionary principle is the notion of **inter-generational equity**, a concept that is fraught with difficulty. As yet there is no institutional device to reduce it, let alone remove it. Neither the most advanced multidisciplinary science, nor the most sophisticated intergovernmental agreements show any sign of addressing this notion convincingly. Inter-generational inequity is the non-sustainability 'Cinderella ethics' of the modern age. Its institution-alized continuation spells trouble, but no-one can say how serious that deliberate neglect will prove to be. So it remains institutionally guaranteed and insulated against serious scrutiny. Sustainability is particularly radical because it involves 'holding back' from certain types of development. It also requires policy co-ordi-nation across a range of different spheres of activity to an unprecedented extent. However, individual departments' eagerness to preserve their own admin-istrative 'turf', creates strong inertial pressures in the administrative system which prevent radical or sudden changes in the direction of policy (see Box).

Summary

- Environmentalism is a broad social movement which reflects a growing public demand for a better environment, untainted by pollution and resource extraction. It is reflected in long-term shifts in public opinion towards 'quality of life' issues such as peace and tranquillity, and an increasingly active and diverse range of 'green' pressure groups.

- Although modern environmentalism blossomed in the late 1960s in the industrialized countries of the North, it is firmly rooted in an age-old conflict between the perceived need both to exploit and to conserve natural resources.

- The recent upsurge in environmental concern is as much a symptom of societal change as a long-term decline in the material quality of the environment. That said, the rate and scale of human development increased to such an extent in the twentieth century that only a very few 'wilderness' areas of the world remain entirely untouched by human intervention.

A case study of policy conflict: road aggregates

Aggregates (i.e. sand, gravel and crushed rock) are required for building and road construction. What would a sustain-able strategy for aggregates look like? At present about 15 per cent are re-used in road beds or building foundations; the rest are disposed of. The demand for aggregates is pro-jected in terms of economic growth rates. It is an article of faith that the need for fresh aggregates is a sign of a healthy economy. The UK Department of the Environment, for example, forecasts that demand will virtually double by 2011. As yet, the idea that demand should be curtailed is extremely controversial because it strikes at the very heart of modern consumer societies (Owens, 1997). Providing for the predicted level of demand raises two very important questions.

First and foremost, where would all this material come from? As the scope for winning aggregates in Southern England or off the coast is reduced due to local protest and the need to nourish beaches, fresh aggregates are being sought from the hard rock 'peripheries', notably the outer Scottish islands. If Scottish superquarries create jobs and provide a restoration bond (which in no way will compen-sate for the loss of regional amenity and wildlife), is this socially and environmentally acceptable? Those near the superquarries certainly do not think so (Barton, 1996). The answer may be yes on very weak sustainability grounds, but less clear-cut on strong sustainability grounds.

Second, which department would put in place a sustain-able policy for aggregates? The DoE really only handles planning and it comes in at a relatively late stage in the pol-icy process. The demand-led modelling is rooted in the Departments of Transport and Trade and Industry. Any attempt at amenity policy would have to pass through these two departments as well as the Treasury, who would be very unwilling to see any earmarking of funds for compen-sation or restoration. Any coherent policy on aggregates would have to connect demand reduction, recycling, **plan-ning betterment** and pricing in a highly sophisticated manner: not easy.

THE NATURE OF ENVIRONMENTAL POLICY

At the root of most environmental conflicts is a clash between two modes of thought or ideologies: **tech-nocentricism** and **ecocentricism** (O'Riordan, 1976) (*see* Table 14.1). *Technocentrists* have faith in the ability of humans to *manage* nature and live in harmony through the application of science and technology. They believe that sustainable development requires a judicious mix of regulations and market-based instru-ments such as green taxes to correct market failures and ensure that the environment is fully considered in

Table 14.1 Environmental worldviews

	Technocentric		Ecocentric	
	Cornucopian	**Accommodation**	**Communalist**	**Deep ecologist**
Green label	Resource exploitative	Resource conservationist	Resource preservationist	Extreme preservationist
Type of economy	Anti-green: unfettered markets	Green: markets guided by market instruments	Deep green: markets regulated by macro-standards	Very deep green: markets heavily regulated to reduce 'resource take'
Management strategy	Maximization of Gross National Product (GNP): assumes human–environmental resources are infinitely substitutable	Modified economic growth: infinite substitution rejected (i.e. some 'critical' capital)	Zero-economic growth: complete protection of 'critical' natural capital	Smaller national economy: localized production (bio-regionalism)
Ethical position	Instrumental (man over nature)	Extension of moral considerability: inter- and intra-generational equity	Further extension of moral considerability to non-human entities (bioethics)	Ethical equality (man in nature)
Sustainability label	Very weak sustainability	Weak sustainability	Strong sustainability	Very strong sustainability

Source: Pearce *et al.* (1993, 18–19)

decision-making. Technocentrists do not believe there are limits to growth *per se*: human and natural capital is infinitely substitutable, so the diminution of one is acceptable if it is offset by an equivalent increase in the other. *Ecocentrists*, on the other hand, see science and technology very much as part of the problem rather than the solution. They tend to view humans as one small part of nature rather than a superior and omnipotent resource 'manager'. Accordingly, humans need to find ways to live *with* nature, rather than over it. For ecocentrists, the overall *scale* of economic activity has to be reduced if humanity is truly to live within its environmental means, and radical changes made to the economy and society. Certain 'critical' aspects of the environment such as keystone species or rare habitats are regarded as inherently non-substitutable and must be saved regardless of the economic cost. The preservation of such resources effectively becomes a constraint or 'limit' upon economic growth.

Environmental policy has been defined as 'public policy concerned with governing the relationship between people and their natural environment' (McCormick, 1991: 7). Environmental politics is the process through which the two worldviews described above shape policies. However, groups of individuals differ greatly in their view of what kind of policies should be adopted. This in turn generates political conflict, argument and dispute. The ubiquity of conflict in society partly explains why policies sometimes differ markedly from the recommendations of scientists or other technical experts. 'Poor' policies emerge and are implemented not because the policy process is irrational or dysfunctional, but because it has to accommodate the conflicting needs of the different actors.

Seven factors differentiate the environment from other policy arenas. First, the environment needs to be seen in terms of public 'goods' and public 'bads'. Environmental quality is a classic example of a public good in that large parts – the atmosphere and large areas of the ocean – are common property and nobody can be excluded from using them. This means that while benefits of resource use tend to be concentrated, the costs are typically dispersed across myriad individuals. Think of a smoking factory chimney pouring ash on to the houses surrounding it, for example. One implication is that resource users have an incentive to 'free ride' – that is consume resources in the knowledge that the costs, or 'externalities', will be shared by others. Meanwhile, polluters have an incentive to fight to protect their rights while the losers are usually too widely dispersed to mobilize into a coherent group. It is often the case that the preferences of the concentrated interests fighting to exploit the environment prevail over those of more diffuse or disorganized interests fighting for the 'public' interest or the welfare of non-human entities. Such conflicts involve a mixture of motivations and are often as much a demand for more 'democracy' (i.e. local involvement in decision-making) as for the preservation of the environment *per se*. Of course nowadays, these politics can be global in scale, as in climate change, with conflicts between the industrialized nations of the 'North' grouped on one side and the poorer states in the 'South' on the other. The consequence of this is the activities that have to be regulated are increasingly diverse, heterogeneous, ubiquitous and operating beyond the realm of formal regulation. One way forward is to apply financial instruments which harness the power of the market for environmental ends (Pearce *et al.*, 1993).

The second characteristic is the need for state intervention to regulate human behaviour. Environmental damage originates from otherwise socially legitimate activities such as energy generation and food production. Even those on the far Right of the political spectrum accept the need for government intervention to regulate to keep the level of damage to a socially acceptable level. Economists sometimes assume that governments act even-handedly in setting the level of regulation to produce outcomes that are socially 'optimal'. In reality, some groups have more of a say than others and can 'capture' the regulatory process and maximize the externalities they impose on society. Political power is by no means only a matter of observable manipulation or mobilization. It is also exercised structurally, meaning that rules, procedures and beliefs support the interests of the powerful without the powerful having to decide on every occasion what should be allowed onto the political agenda for action.

Third, environmental politics have an important temporal and spatial dimension. The tendency for costs and benefits to be distributed unequally in society immediately raises important questions of fairness and equity. Balancing these different needs becomes a central matter determined by political dispute. The language of consensus suggested by sustainability overlooks the fact that while growth and environmental conservation may be compatible at the macro-level, there are likely to be losers in particular localities in the short term. The problem of finding a fair balance between competing claims becomes even more difficult when the costs and benefits straddle different generations, and short-term jobs protection is at stake.

Fourth is the importance of political borders. Problems like acidification, ozone depletion and climate change cannot be solved by independent state action. Difficulties arise, however, when that action is required of states who can legitimately claim to be blameless. Over the years, developing countries have fiercely resisted legal curbs on the use of fossil fuels and CFCs on just these grounds, and insist their participation must involve some form of compensation or special treatment – what has come to be known as 'greenmail'. The response of the international community has been to develop institutional mechanisms to transfer finance and technology to the industrialized world to help it leapfrog the more polluting stages of economic development, though they are of questionable significance given the wider flows of finance and trade which nourish unsustainable development (*see* Fig. 14.3).

Environmental politics also involves extreme complexity and uncertainty. Policy-makers always face myriad constraints and conflicting demands, but scientific uncertainty is arguably more acute in the environmental sector than other policy domains (O'Riordan, 1999). Natural systems are multi-dimensional and extremely complicated, as are societies. Therefore, it is reasonable to expect environmental problems, which stand at the intersection of the two, to be doubly complex (Dryzek, 1997: 8). The key question is: how should decisions be made when the entire sequence of consequences cannot be predicted? Should politicians err on the side of caution or develop environmental resources in the hope that

Figure 14.3 Many environmental problems now involve complicated discussions between states, international organizations and international NGOs. Credit: Popperfoto

unfavourable consequences will not arise? This is one reason for widening the basis of science in policy-making, for incorporating the precautionary principle and for embarking on a more participatory democracy.

Sixth, is the problem of irreversibility. Once development exceeds the Earth's capacity for self-repair, environmental assets collapse and disappear, never to re-appear. Irreversibility complicates environmental politics, differentiating it from other policy areas like tax or social affairs where problems can be more easily re-visited. Irreversibility demands careful anticipatory planning not incremental, *ad hoc* policy responses. The problem is that whenever the state of the environment cannot precisely be determined before or after a decision is made, those with vested interests can twist scientific ignorance to suit their own cause. The US fossil fuel lobby's recent attempt to scupper the climate change agreement at Kyoto was a good case in point (O'Riordan and Jordan, 1998). The trouble here is that political institutions are not terribly good at coping with genuine uncertainty where intergenerational inequity is concerned. The political power game makes it difficult for policy actors to act re-distributively in advance of consensus. In any case, politicians find it difficult to convince their constituencies to make sacrifices when the 'payoff' may be generations away.

Finally, while the environment functions as an integrated whole, the bureaucracies set up to manage it are typically fragmented into different departments and sections. The environment is probably unique among policy problems in the extent to which it cuts across traditional sectors of government. When Chris Patten, as Environment Secretary of the UK Government, famously remarked that 'the most important parts of environmental policy are handled elsewhere – the levers aren't in my office', he revealed a considerable barrier to 'greening' the policy process. Protecting the environment often involves deep conflicts between environmental ministries and the older and often more powerful parts of government representing finance, agriculture, transport, and so on. One of the great unmet challenges of the 1990s has been to find mechanisms to integrate environmental concerns into all areas of decision-making. Somewhat paradoxically, if and when it is achieved, there will be no need for a separate environmental department.

Summary

- Environmental policy is concerned with the interaction between humans and nature. Technocentrists believe that the environment is there to be exploited by humans for their own utility. Economic growth is regarded as an inherently 'good' thing which, if managed, will generate the resources and material stability needed to conserve environmental assets. Others are suspicious of growth, believing that it has to be limited to safeguard natural resources.

- Politics is the process through which decisions are taken about the environment. Environmental politics has a number of characteristic features which distinguish it from other forms of politics.

PUTTING SUSTAINABLE DEVELOPMENT INTO PRACTICE

The 'sustainability transition' is the process of moving to a society which lives within environmental limits and is just and fair. In this important sense, environmentalism has passed its phase of being a high profile and clearly delineated agenda. As environmentalism tackles sustainability, however, the political stakes become much more resistant to the demands of tax reform, economic and political redistribution, and spatial reorganization that the new sustainability agenda is making. This is why the 'post-Rio' process is so lamentably failing. The nerve centres of power have recognized that sustainability needs to be tamed, and this process is now the centre of the struggle.

The implications for the various institutions whose origins and power bases, including their supporting client groupings are deeply rooted in unsustainable processes of development, are profound. To remain effective, environmental departments at all levels have to form coalitions and partnerships with colleague departments that heretofore have not been sympathetic – finance, industry, social services, public order, and so on. Furthermore, these structures of policy development and regulation can no longer rely on the tools of analysis and justification that have served them for over a quarter of a century, namely cost-benefit analysis, environmental impact assessment, risk management, life-cycle analysis and eco-auditing. These tools are not only insufficient to meet the new dawn of sustainability, they are also too tied to the apron strings of the political elite that accommodate the environmental agenda only on terms which are acceptable and tolerated. These tools substantially reinforce existing patterns of power and political disposition, by giving the appearance of legitimacy.

In a similar vein, the metamorphosis of environmentalism into sustainability has enormous implications for pressure groups and eco-activism generally. Arguably, the leaders of the mainline organizations are in a state of uncertain transition. Some remain time-warped in the old ways and cannot adjust to the fluidities of coalition building and opportunism that the new global–local sustainability agenda is offering. Others are realizing that globalization is an unavoidable outcome of globalizing economies, communications and actions. So they must match global with local and battle it out on the planetary stage at big meetings like the 1992 Earth Summit in Rio and the various global environmental convention conferences with the corporate and political giants. Others still, seeing that this is dangerously alienating for those at the grass-roots, recognize the power of local Agenda 21 – the United Nations blueprint for sustainable development signed in 1992 – as an organizing force for civic localism and seek to capture these energies that are created in 'do it yourself' activities. So once again, the environmental agenda is dispersing its energies, building partnerships, gathering strength in coalitions, and using fresh approaches to science and evaluative analysis.

Summary

- In the 1990s environmentalism has been transformed from its politically separate identity into a much more diffuse social and political arena that is loosely termed 'sustainability'.

- Sustainable development involves connecting technological inventiveness with social justice and ecological stewardship. A key challenge is integrating environmental protection requirements into and across 'non-environmental' policy domains such as transport, energy production, trade and industry.

CONCLUSION

These are exciting times for the decay of environmentalism and the phoenix reincarnation of sustainability. For the geographer, the sustainability transition offers a golden opportunity. Environmental science has evolved to the point where the rediscovery of space and place is back in vogue. Globalism is a process of change that seems to hold most social meaning through local impacts. Neither globalism nor localism is spatially or culturally separate from one another. In a similar vein, it is arguable that human interference on natural systems is now so ubiquitous and comprehensive that there is no longer a purely 'natural' biogeochemical process. The job of the geographer is to examine the coupled relationship between environmental criticality, where discontinuous thresholds of fundamental phase change are on offer, and societal vulnerability, when the established and locally fair social process of adaptation and response are also discontinuously disrupted. Because such interdependencies perforce occur through globalizing and localizing processes, so the outcome will be space and place determined.

Further reading

There are surprisingly few text books on environmental policy. The best are:
- Young, S. (1993) *The politics of the environment*. Manchester: Baseline Books.
- O'Riordan, T. (ed.) (1999) *Environmental science for environmental management* (2nd edition). Harlow: Longman.
- Carley, M. and Christie, I. (1997) *Managing sustainable develop*ment. (2nd edition). London: Earthscan.
- Porter, G. and Welsh-Brown, J. (1997) *Global environmental politics*. (2nd edition). Boulder, Col: Westview Press.

There is, however, a huge literature on politics and policy. The best introductory texts are:
- Ham, C. and Hill, M. (1993) *The policy process in the modern capitalist state*. London: Harvester Wheatsheaf.
- Greenaway, J., Street, J. and Smith, S. (1992) *Deciding factors in British politics*. London: Routledge.
- Parsons, W. (1995) *Public policy*. London: Edward Elgar.

CHAPTER 15

Environmental knowledges and environmentalism

Jacquie Burgess

The woman with the clipboard stops you in the street. It's an environmental survey. Could she ask you a few questions? 'OK, so long as it's quick.' So she starts by asking you how serious are each of the following problems.

'Global warming?'

And you think: *the IPCC statement – when was it? A couple of years ago? Global warming est arrivé – don't think it's chance variation any more. Um, Kyoto conference last December (reminder to self: must look up an atlas and find out where Kyoto actually is). Lots of front page coverage in the press. Didn't the Americans force everyone to stay up all night to try and reach some sort of agreement? Typical. Fossil fuel lobbies have been trying to rubbish the scientists, haven't they? Hey, I remember reading that piece in the New Scientist where some bloke from, oh hell, where was it? Said something about IPCC being dominated by 'guys from the bottom of the heap, such as geographers.' That really annoyed me. Um, can't think of anything else. But this is a really weird winter. Dawn chorus in the middle of January; the bulbs are up in the parks, already.*

So you say: 'I think global warming is quite serious.'

'Local air quality?', she asks.

And you think: *We're standing on the pavement and I can hardly hear what she's saying because of the noise of the traffic. The smell is absolutely disgusting! Look at the fumes belching out of the back of that bus! I gave up riding my bike 'cos it's so bad. Can't afford the bus fares, though. Read somewhere that the cost of motoring has stayed constant since 1974 while public transport costs have increased by 20 per cent. Where was that? My little brother's got asthma. They say it's not traffic related, but they would say that, wouldn't*

they? Something's got to be done. The road protesters did a good job, got a lot of publicity. I wouldn't mind immobilizing a bulldozer – probably more interesting than the tutorial I'm late for . . .

So you say: 'I think local air quality is a very serious issue.'

Then the interviewer says, 'Could you tell me whether you do any of the following, and if so, how often? Recycle your domestic waste?'

And you think: *What does she mean? Bottles? – well, mum does that at home, so I suppose that counts. Er, we had a compost heap but the smell was so bad we dug it in. Can't you get worms or something? Wonder what would happen if some escaped – or got 'liberated' by the animal rights people:* **TERROR IN THE TURF** *. . . Scientists warned today of a potential catastrophe as alien worms munch through their mild-mannered English cousins. Come on, concentrate – this is serious. I always leave the newspaper on the bus so someone else can read it – that's a sort of recycling. But then again, what's the point? There was a story on the 6 o'clock News the other day that recycling actually uses more energy and it would be better to burn paper in one of those, um, what are they called, combined heat and light stations? Er, don't do that much really. But I ought to be doing more; over-consumption of resources; think about the Third World and all that.*

So you say: 'I always recycle papers and bottles.'

You have just read a creative construction of how someone, not unlike you, might think as they respond to questions about environmental topics. We know very little about the internal deliberations, the arguments that go on inside people's heads, especially when they answer questionnaires. All researchers are able to capture on their survey forms are the bald responses. Contrast what my imaginary student was thinking with what was actually communicated and recorded. One of the major changes in geographical research over the last ten years has been a shift towards qualitative, discursive forms of inquiry where the aim is to engage people in extended conversation so as to better understand the nature of argument and evidence they draw upon to make sense of the world.

My student is imaginary but what she is thinking and how she is framing answers to questions is truthful of the ways people understand environmental questions. Complex knowledges of different status and different levels of certainty come together as half-remembered items in the news and fuse with personal memories. The intense, sensory impacts of immediate experience take precedence over the discursive, abstract knowledges of experts. The imaginary student accords different kinds of authority to these different sources of knowledge. Her stream of consciousness is an example of what Michael Billig, a social psychologist interested in rhetoric and social dilemmas, calls 'witcraft' – the skill of argumentation which includes thinking as much as speaking and writing. Witcraft is about asking, what are the crucial claims in this issue? What is a persuasively structured argument? What style of presentation is effective? How did we get to this point or, in other words, what is the history of this topic? A good place to start.

CONTEXTS

Geography claims much of its disciplinary identity from study of the relations between people and environment, although research which straddled the apparent divide between 'physical' and 'human' geography became rather unfashionable in the late 1970s and 1980s. This was a pity for it was at just this time that environmentalism emerged as a powerful new social movement, with the activities of campaigning groups such as Friends of the Earth and Greenpeace forcing environmental issues onto the front pages of the newspapers and the agendas of national governments. Sociologists, political scientists, economists and philosophers have joined the debate, each discipline offering its own slant on the characteristics of modern environmentalism.

In this chapter, I want to focus on one element of modern environmentalism – the production of environmental knowledges and the extent to which the communication of environmental problems is effective in persuading people to adopt more sustainable lifestyles. The movement to sustainable development involves major changes both to the structures of mod-

ern society and to individual identities. Over-consumption of resources, excessive levels of waste, and the release of pollutants into water and air are problems of the rich countries of the North. It is very hard to persuade people voluntarily to cut back on those things which, for so long, have been seen as central to 'the good life', and changes will only be achieved through programmes which combine education and persuasion with the development of new policies and regulations. At the turn of the millennium, like it or not, we are all participating in a massive social experiment as environmental and social scientists, campaign groups, governments and international agencies such as the UNEP seek to change the world.

Why now? There is growing appreciation that the meanings of 'nature' and 'environment' are socially constructed in both scientific and popular discourses (see Chapter 1). The late 1980s was the period when concerns about the possible consequences of global environmental change rose to the forefront of political and popular concern. It was a time when, to take the sociologist Anthony Giddens' (1991) apt phrase 'apocalypse became banal'. Through the development and widespread dissemination in the media of scientific theories about the impact of carbon dioxide and other gases on the atmosphere, the meaning of wind, rain and sun changed. No longer could extreme events be thought of as 'Acts of God', or 'elemental nature', rather, they became frightening consequences of apparently irrevocable, human-induced changes to nature (McKibben, 1990).

The mass media are playing a fundamental, if too often neglected part, in the social construction of environmental knowledges (Burgess, 1993; Anderson, 1997). Public awareness of global environmental problems such as global warming, ozone depletion, and the loss of rain forest comes primarily from the pages of the newspapers and the television news. But the media never provide a 'window on the world' – an unmediated view of events. Environmental news, like any other sort of news, is actively constructed, shaped by the professional, technical and economic demands of each medium. Because the environment has not traditionally been a mainstay of newsgathering practices, coverage tends to vary quite dramatically over time. Moreover, environmental problems do not fit easily within journalistic news values: the events are long term rather than immediate; they are complex which makes stories harder to explain simply, and scientists tend to use long words and difficult concepts. In addition, competition for space in the paper or time on the television news is intense.

One environmental correspondent put it this way: 'Everyday, I am working within the limits of what I can get in the paper' (Mike McCarthy, *The Times*, pers. comm. 14 February 1991). One outcome of these different pressures is that, when events are running fast, environmental stories are radically reshaped to fit media judgements of newsworthiness, or they are simply ignored at other times.

Summary

- Over the last 20 years or so, geographical studies of environmentalism have been joined by many other social sciences, each bringing their distinctive disciplinary perspectives to bear. A particularly important contribution is being made by academics interested in 'witcraft' – the deliberations which go on inside people's heads – as well as in public forms of communication.

- The production, circulation and consumption of environmental knowledges are of fundamental importance in understanding social constructions of 'the environmental crisis'; and the extent to which institutions and lay publics can be persuaded to change towards more environmentally-friendly practices.

- The mass media are playing a central role in helping to frame environmental issues for the public and in shaping political agendas. However, the media are not neutral in these processes. As active mediators, environmental science is translated in accordance with news values.

GLOBALIZATION: A CENTRAL ISSUE IN THE PRODUCTION AND CONSUMPTION OF ENVIRONMENTAL COMMUNICATIONS

There is general agreement that the age we live in differs fundamentally from what has gone before. We are living in 'a runaway world', Anthony Giddens suggests (1991: 16), a world characterized by the speed, scope and depth of social changes. There are many reasons for the sense of uncontrollable dynamism and endless change which characterizes modern life which have profound implications for environmental knowledge. Probably the most important factor is the ever-increasing capacity to produce, process and move information around the world. Information travels very fast across space and through time. It comes from everywhere and goes to everywhere else almost instantaneously.

As a result, human consciousness of the relations between time and space is changing. Harvey (1989) calls it 'time–space compression' – the sense that the world is 'shrinking' and time is 'speeding up' through ever-faster forms of electronic communications. Events in distant places now penetrate private, domestic spaces in particularly dramatic ways through tele-

vision images, resulting in unpredictable fusions of global and local knowledges (Meyrowitz, 1985; Morley, 1992). For some people, the increasing penetration of the global into the local encourages a greater consciousness of the interdependencies of societies around the world and a corresponding global environmental sensibility. For other audiences, the outcome is a sense of hopelessness and 'compassion fatigue' (Philo, 1993).

The incredible power of the new information technologies lies at the heart of globalization, proving scientific evidence for processes of global environmental change as well as the means of communicating them to the world. Figure 15.1 provides an illustration of how the process works. In this article, an arbitrary event (New Year's Day; when few journalists are working and little 'hard' news is being produced) is used as a hook to bring together disparate sources of scientific and 'folk' evidence about the vagaries of the world's weather patterns. Through articles such as this, the mass media continuously bring distant worlds, events and individuals into people's living rooms, creating a world which is both unitary and present, fragmented and dispersed. The drama of the story in Figure 15.1 lies in the frisson of anxiety it creates – the rising thermometer; people being killed by extreme weather events in different parts of the world, the unpredictability of nature. The media, through the ways in which they frame environmental science stories, constantly provide their readers with interpretative scripts that guide ways of making sense of these events. But the challenge of reporting the weather in the late 1990s is the level of scientific uncertainty about the causes of such extreme events. Are they 'natural nature' (El Niño? sunspots?) or 'technological nature' (global warming)?

Dependence on experts: the downgrading of local knowledge

New technologies and practices mean that individuals and social institutions are also lifted out of their places of origin or 'disembedded' in Giddens' (1991) terms. Pre-modern societies were characterized by secure, local places and communities where social relations were clearly demarcated, the workings of local nature were known intimately through everyday practices, and a moral order defined people's obligations to one another and to nature. These certainties have been replaced by scientific rationality and technological expertise which, paradoxically, create radical insecurity both individually and collectively. Individuals are now highly dependent on expert forms of knowledge, and in the proliferation of expertise, all knowledge becomes hypothesis, subject to challenge and

A year in the weather

Number key for map

November 2
Super Typhoon Keith blows through the Northern Mariana Islands in the Pacific with a central pressure of 872mb, one of the deepest typhoons ever

April 2
Record-breaking snow storms over New England with depths of up to 61cm over some parts of Boston

October 17
Eilat in Israel has 21mm of rain in 18 hours. This is 7 times the October average of 3mm, and some reports of hail too

November 1
Perth in Western Australia equalled its November lowest temperature with 6°C (average is 14°C)

July 6
Rome records a record low temperature of 11.2°C. The previous record was 12.1°C, and the average minimum is 18.3°C

November 24
Western Australia bakes: Shay Gap at the edge of the Great Sandy Desert reached 45°C at 4am, 10°C above average

September 15
Satellites confirm that El Nino is back. There are droughts, fires and famine in Indonesia

June 13
Bangkok reached a record high temperature of 40°C beating the previous record by 2°C

February 22
Gusts of up to 153mph are measured on the Cairngorm summit, and the mean wind speed is 110mph. (Hurricane strength is classified as anything above 74mph)

October 27
Denver has the worst snow for 30 years, temperatures down to -14°C, and 35 inches of snow recorded in the city (50 inches in the mountains)

December
El Nino washes against the American coasts. There are floods in Ecuador, Paraguay and Peru

October 28
Germany freezes up. Hamburg recorded -6.8°C: the previous record was -5.9°C

March 9
Verhoyansk, in the Siberian Arctic, and known as the coldest place on earth, had a temperature of -46°C, but there is still 12°C more to drop to beat the March record minima!

November 7
Temperatures reached 20.2°C in the Austrian capital Vienna, beating the previous November record of 19.6°C

June 18
The afternoon temperature in Naples soared to 36°C, beating the previous June record of 35°C

May 21
Acapulco, Mexico had temperatures up to 36.3°C (the previous highest for May was 35.6°C)

May 1
Britain's sunniest day of the year so far with London reporting 23.2°C. August then became our second hottest since records began

WINDY

SNOW

WET

COLD

HOT

Britain bakes in record summer temperatures

Clouds of haze cover Indonesia as a result of the fires

Snow storms in New England

El Nino causes flooding in South America

GRAPHIC: STEVE VILLIERS

Worst snow in Denver for 30 years

Figure 15.1 Hottest year for Planet Earth. Credit: Guardian; Associated Press; Guardian; Reuters; Popperfoto/*The Guardian*, 1/1/98

change. As one environmental advisor commented in a discussion group I was moderating: 'But how can we tell the truth when we don't know how long the truth will be for?' (Burgess *et al.*, 1998).

One consequence of the rise of expert systems is that new kinds of knowledge, new forms of thinking through and responding to problems are proliferating which leave people who lack expertise in these areas in a vulnerable position of feeling, and being 'deskilled'. People often feel they no longer have sufficient control over their everyday lives. At the same time, the penetration of expert systems into people's everyday lives means individuals are constantly acquiring new kinds of knowledge but these knowledges are always partial and fragmentary. They are also subject to expert revision and change in the light of new evidence, for example, or dispute between different experts. Again, the media play a particularly significant role in these processes as they communicate the findings of environmental science to the public. Researchers need the media to get their message across to the general public, but more importantly, to politicians, government and those responsible for making environmental policies. The media, in their turn, need scientific stories that can be told simply, stories where there is conflict between different scientific claims, and/or where the science can be placed in a political framing.

The reality (or not) of climate change is a case in point. Figure 15.2 shows the lead story in *The*

So this is global warming

Snow conceals hard choices

Nicholas Schoon
Environmental correspondent

A week ago *The Independent* reported that 1996 had been one of the world's warmest years on record, adding to scientists' convictions that man-made climatic change was well underway.

Ever since then, across Europe, it has got colder and colder. In cosy pubs across the country the lounge-bar talk has been: "Whatever happened to global warming?"

It is a tough but fair question. In a world still prone to spells of extreme cold which kill scores of people and cost hundreds of millions of pounds, how can electorates and politicians be convinced that global warming matters?

It has been a front-runner among green issues for almost a decade. World leaders – Clinton, Kohl, Thatcher, Major – have all made solemn speeches declaring that something must be done. But very little has been done. The new politics and diplomacy of weather and climate amount to little more than hot air.

We are curious, awed by the notion that a single species – us – can now alter our planet's entire climate. But in a bitter week of frozen winds from Siberia, it is hard to think of this grand, looming threat as anything other than an apocalyptic fantasy or an irrelevance.

But global warming is happening, it does matter, and we should take action now to reduce the threat. There is a golden opportunity to do so later this year. In December in Japan, environment ministers from around the world will meet in order to strengthen the very weak global warming treaty which was signed at the Rio Earth Summit in 1992.

Under that treaty, developed countries undertook to stabilise their rising annual emissions of heat trapping "greenhouse gases" at the 1990 level by the year 2000. Most of them – but not Britain – seem set to break that promise.

What is now required at the Japan meeting is a treaty commitment from the developed countries – which have produced the great bulk of the atmospheric pollution to date – to reduce their emissions. That means using less coal, oil and gas. Burning these fossil fuels adds carbon dioxide and other greenhouse gases to the atmosphere.

And that, in turn, means increasing taxes on these fuels. But electorates and politicians hate the idea. The débâcle over raising VAT on domestic electricity and gas, the griping about increased petrol and diesel duty which inevitably follow each UK Budget, the gutting of the first Clinton administration's energy tax, shows just how much.

Reducing emissions also requires giving householders, commerce and industry other, more popular incentives to conserve energy, which could in turn be funded by those larger fuel and power taxes. The expansion of non-polluting energy sources needs to be intelligently subsidised.

But back to today's bitter cold. Why should voters and politicians even *believe* in global warming – let alone make changes to their habits, homes and economies – with fresh memories of freezing weather?

They need to get the message that the climate will retain its natural variability, with continued extremes of hot and cold, drought and flood, against the background of a gradual, world-wide warming trend.

As we reported last week, scientists at the Meteorological Office's Hadley Centre for Climate Prediction have found that average annual world temperatures are now 0.6°C higher than they were a century ago.

The science of global warming has made giant strides in the past 10 years. There is now a consensus among climatologists that the warming trend will continue and, in all probability, accelerate into the next century. It will be the fastest rate of change since the last Ice Age ended 10,000 years ago.

There will be cooler years, even cooler decades, which buck the trend, because natural variability will continue. By changing prevailing winds and ocean current, some countries could even end up colder.

Rich countries like Britain may find it fairly cheap and easy to adapt to whatever climate change we experience here. Some recent studies have suggested that the US, Canada and Russia, which have done more in total or in *per capita* terms to raise the concentrations of greenhouse gases than any other nation, may actually benefit from climate change in the next century.

But poor, populous ones, such as Bangladesh, will find the shifts in temperature and rainfall, and the rising sea levels, much harder to cope with and possibly catastrophic.

Common sense and justice demand that the developed countries do act. Short-termism and selfishness make it quite likely that they will not. But they have been warned.

PRIESTLEY

Figure 15.2 So this is global warming. Credit: Chris Priestley/*The Independent*, 4/1/97

Independent newspaper, published on 4 January 1997. Another New Year story but this time linked to 'freak' cold weather conditions at home. The banner headline catches a tone of ironical scepticism about scientific claims but the body of the text berates governments for not doing more. Global environmental problems have been discovered – indeed, some would argue they have been *constructed* – through the instruments, models and measurements of experts. This means that lay people, including politicians, are utterly dependent on the risk assessments of expert and counter-expert. It is becoming much harder to know what to do. Not so for the newspaper, however, which takes a worthy moral stance with its 'one world' rhetoric in the last paragraph.

Environmental risks

Societies have always faced many kinds of natural hazards and disasters but the contemporary environmental crisis is fundamentally different. Over the last decade, one of the most dynamic 'knowledge environments' has been that of **environmental risk**. The discovery of the unintended consequences of science, technology and industrialization on the natural and physical systems which support life on earth has prompted Ulrich Beck (1992) to describe contemporary society as Risk Society. The 'mega-hazards' associated with nuclear power, chemical and bio-technological programmes, and ecological destruction embody threats of environmental catastrophe.

Further, and this is one of Beck's central arguments, it is simply not possible to insure against the risks of mega-hazards. Citizens can only place their trust in the claims of experts that the 'fail-safe' mechanisms of nuclear power plants are indeed safe; that there are in fact 'safe limits' of pollutants in air or water; and that the release of genetically modified organisms will not harm ecosystems. The public are thus forced increasingly to put aside their doubts and place their trust in the safety claims made by different kinds of expert systems, institutions and governments responsible for drawing up and implementing environmental regulations.

Expert claims and assurances are articulated in the public sphere through the mass media but, once again, issues are represented within their own communicative rules and strategies. The media are engaged in creative processes of risk translation as they make the news. Some commentators who take a rather pessimistic view of the power of the media in relation to the capacity of audiences to make sense for themselves, argue that public opinion is easily swayed by the amount of environmental coverage in the

media and the style of reporting. Evidence to support this position can be found. Figure 15.3 shows a graph tracking people's responses to MORI opinion polls on the relative importance of environmental and other key economic and social problems over the period 1989–94. The very high levels of public concern recorded in 1989–90 fell away rapidly; thereafter, environmental issues bump along the bottom of the graph, having been superseded by economic and social welfare issues. Now compare the shape of the graph with those in Figure 15.4, which record the number of stories with the key words 'greenhouse effect' and 'global warming' published between 1987–91. Comparing these graphs, it is clear that as media coverage of environmental issues rises, so too does public opinion of the importance of those issues; when media coverage falls, public awareness of environmental problems similarly declines. This relationship suggests that public commitment to environmentalism is very shallow and fickle, being driven largely by media agendas (see Lowe and Rudig, 1986; Worcester, 1993).

Others would disagree, seeing the media as instruments through which more positive and emancipatory public responses to risk may be achieved. The media, at least potentially, are able to create a public space where science and the judgements of expert systems can be subject to much wider and more democratic debate. After all, environmental reports do question science's achievements and thus raise doubts about the industrial structures which science legitimates (Wynne, 1993) In this vein, Beck (1992) comments that reading the daily newspaper 'becomes an exercise in technology critique' as he aspires to a new form of environmental politics where citizens can 'win back the competence' to make their own judgements by shaking off their dependency on experts.

From this perspective, people's responses to public opinion surveys about environmental issues are not

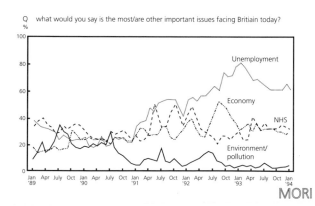

Figure 15.3 Public concern about the environment 1989–94.
Credit: MORI

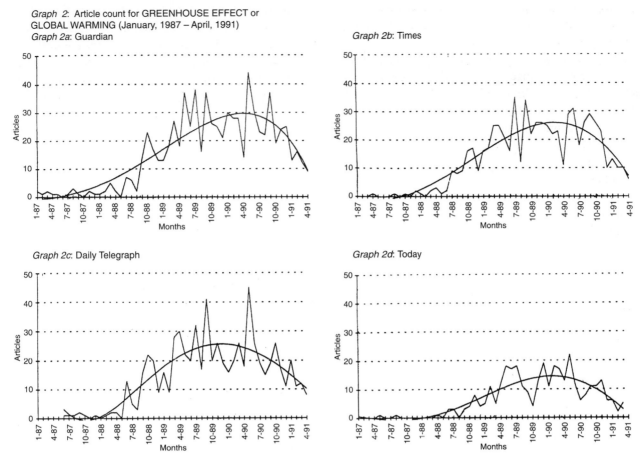

Graph 2: Article count for GREENHOUSE EFFECT or GLOBAL WARMING (January, 1987 – April, 1991)

Graph 2a: Guardian

Graph 2b: Times

Graph 2c: Daily Telegraph

Graph 2d: Today

Figure 15.4   Global warming stories in the press. Credit: Langman and Lacey, 1993

especially helpful because, as we saw at the start of the chapter, if researchers want to understand how people think about environmental issues, and to find out more about why individuals answer questionnaires in the way they do, then the best strategy is to listen to them discussing the issues. The 'cultural' or **'linguistic turn'**, commented on in several chapters in this book, has also penetrated environmental research. Most effort is being directed towards interpretation of ordinary people's knowledge, meanings, values and understanding of environmental issues. For example, several studies show that people's concerns about environmental issues encompass feelings of dread and fear, moral correctness and responsibility. But there is also an intensely political edge, as individuals express doubt and mistrust of those institutions responsible for creating and managing risk (Irwin, 1995).

Summary

- The ever-increasing capacity to produce, process and move information lies at the heart of globalization. Without the globalization of information there would be no global environmental crisis.

- In the last 150 years, a major feature of modernity has been the domination of local 'traditional' knowledge by a scientific rationality and technological expertise. We are now highly dependent on experts and their forms of knowledge; and in the proliferation of expertise, all knowledge becomes hypothesis, subject to challenge and change.

- Alarmed discovery of the global reach of mega-hazards has led to the concept of the Risk Society. There is no agreement about the extent to which members of the public understand the causes and consequences of contemporary mega-hazards. Either public opinion is shallow – merely reflecting the rise and fall in media coverage of these issues, or there is a subtle, rich understanding as individuals interpret environmental stories in the light of their personal and political experiences.

THINK GLOBAL, ACT LOCAL: THE CHALLENGE OF CHANGING LIFESTYLES

Reflexivity is the concept used to define the ability to produce knowledges of the future which form the basis for changes in current practices. Reflexivity is central to an understanding of how individuals and institutions are changing under conditions of late modernity. In other words, individuals and institutions are both organized and able to transform themselves through knowledge and information. Reflexivity is now a fundamental element of individual and institutional identity: both individuals and organizations constantly monitor their behaviour and experiences, and thereby are caught up in a process of continual change.

Reflexivity lies at the root of arguments for sustainable development, for example. By becoming aware of the risky consequences of currently unsustainable practices, political, economic and social policies may be put in place to turn development to another path. '*Think globally* – consider the mounting evidence about the consequences of human activity on the natural and physical systems of the world; of the effects of your actions on peoples distant from you in space; '*Act locally*' – because what you can do at home, in your immediate environment will have a beneficial impact right through to the global level and thereby *change* the future. The old slogan of Friends of the Earth is an excellent example of reflexivity.

Environmentalists believe that the only logical response to mounting evidence of ecological catastrophe is to change, fundamentally, the ways in which societies use and abuse the natural and physical systems which support human life on earth (*see* Figure 15.5). If people are given sufficient information about the consequences of continuing to live beyond the capacity of the environment to support them, then individuals and institutions can be reflexive in changing their values, beliefs and practices. The UK government's 'Going for Green' programme which was established in 1994 to raise public awareness of green issues, and what individuals can do to help, is a good example of the kind of mass advertising campaign which seeks to change hearts and minds (*see* Figure 15.6). The strategy is based on an assumption that if people are given the 'right' information in the 'right' way', they will change their behaviour accordingly.

Unfortunately, research evidence suggests that people are rather more resistant to these rhetorical appeals than might be hoped; there is real and active resistance to exhortations for people to change their lifestyles. Our own research with members of the public in the United Kingdom and the Netherlands (Harrison *et al.*, 1996) shows that two strands of argument are particularly important in people's rationalizations of why they are reluctant to change their lifestyles. The first relates

IF YOU'RE MAD ABOUT WASTE, THERE ARE THINGS THAT YOU CAN DO.

Remember the 3'R's
Reduce waste - don't buy over-packaged goods. Lots of goods come wrapped in more packaging than they need. The picture shows one example. Can you think of any others?

Reuse things when you can instead of throwing them away. You can do this by mending things, sending old toys and clothes to jumble sales and by using jars, bottles, bags and envelopes again.

Recycle bottles, paper, cans and other goods if they can't be reused as this saves energy, resources, and means less waste is produced.

Why not read the other leaflets in the **Mad About...** series?

Air Pollution
Water Pollution
Tropical Rainforests
The Ozone Layer
Natural Habitats
Climate Change
Energy

FRIENDS *of the*
earth

June 1994 © Friends of the Earth 1994
Published by Friends of the Earth Trust Limited
Charity number 281681
Illustrated by John Watson
Friends of the Earth, 26-28 Underwood Street,
London N1 7JQ. Telephone (071) 490 1555
Partially funded by the Department of the
Environment.

Figure 15.5 Recycling campaign leaflet. Credit: Friends of the Earth

TRAVELLING SENSIBLY by

Making fewer car journeys and sharing cars

Walking and cycling more

Using public transport more

Keeping cars properly tuned and maintained

Going for Green
Making a world of difference – together

Figure 15.6   Travelling sensibly. Credit: Going for Green

to people's sense of agency – the belief that what they do can make a difference. In both countries, there is a need to be convinced that actions would be effective in achieving the desired aims. Global environmental problems often seem intractable and people often speak of compassion fatigue. 'You've got enough problems of your own, without taking on all of the world's all of the time,' as one woman said.

Furthermore, people are unsure what to do for the best, and have a sceptical relationship to environmental expertise. For example, is it better to save newspapers and bottles and drive to the waste tip once a fortnight, or does the increased pollution from using the car outweigh the benefits of not creating more landfill? At the same time, stories of domestic waste that has been assiduously sorted at home, only to be tipped into the same hole in the ground is a powerful demotivator.

The second strand of argument concerns questions of governance; in particular, the extent to which citizens trust local and national government, and feel there is a strong social contract between them. If such a political culture exists, then individuals are more likely to accept shared responsibility for environmental action.

Environmental appeals fall on deaf ears when people can see no visible evidence of change in institutional lifestyles. Our research suggests that public and private institutions of all kinds must demonstrate their own commitment to the need to change by underpinning their environmental rhetoric with visible, tangible actions of their own (Burgess *et al.*, 1998). Co-ordinated programmes of waste reduction, reduced water and energy demand, and increased public transport use are necessary at work, in government offices, and at home, if public scepticism about the urgency of the environmental crisis, and the sincerity of those calling for changes in lifestyle is to be overcome.

Summary

- Reflexivity is a particularly important concept in sociology and social geography. It suggests that individuals and organizations constantly monitor their behaviour and experiences, and make adjustments in the light of new information.

- 'Think global, act local' is a reflexive injunction which, if obeyed, would lead to greater sustainability. However, there is dispute about the most effective means of achieving the changes in individual identity and practices required. Some, notably government organizations, believe top-down mass advertising campaigns will work. Others argue for a bottom-up strategy of working with members of the public and recognizing the psychological, social, economic and cultural barriers to change.

CONCLUSION: THE GEOGRAPHY OF ENVIRONMENTAL KNOWLEDGES

Public awareness and understanding of the causes and possible consequences of environmental degradation have increased dramatically over the last 30 years. The media have played important roles in this process, acting as translators of the environmental message as well as channels for communication – for both environmental pressure groups and national institutions. Academic research and environmental policy-making are being enriched by the multi-disciplinary interests in society–environment relations. Of particular note are social theories to explain why the environmental crisis is an inevitable consequence of modernity and the move towards greater discussion and dialogue between elites and lay publics in deciding what should be done to enable societies to become more sustainable. And what should be Geography's role in these developments? Geography is unique in having an internal dialogue between natural and social sciences – our own form of disciplinary 'witcraft'. At the same time, through our central concern with the distinctiveness of places and regions, with the specifics of how relations between people and environments vary across space and through time, we ensure that what the sociologists call the environmental *problematique* remains grounded in the real world of landfill sites, polluted drinking water, eroding soils and El Niño.

Further reading

One of the best introductions to the historical growth in environmental knowledge, and the dimensions of contemporary environmental politics is:

- Pepper, D. (1996) *Modern environmentalism: an introduction.* London: Routledge.

An introduction to mass media and environmental issues can be found in:

- Anderson, A. (1997) *Culture, media and environmental issues.* London: UCL Press.

Interesting case studies of public understanding of environmental change include:

- Hinchliffe, S. (1996) Helping the earth begins at home: the social construction of socio-environmental responsibilities. *Global Environmental Change* **6**, 53–62.
- Harrison, C.M., Burgess, J. and Filius, P. (1996). Rationalising environmental responsibilities: a comparison of lay publics in the UK and the Netherlands. *Global Environmental Change* **6**(3), 215–34.
- Macnaghten, P. and Jacobs, M (1997) Public identification with sustainable development: investigating cultural barriers to participation. *Global Environmental Change* **7**, 5–24.
- Kempton, W., Boster, J.S. and Hartley, J.A. (1995) *Environmental Values in American Culture.* Cambridge, MA: MIT Press.

Historical geographies

INTRODUCTION

Many Geography students begin their university degrees with little experience of Historical Geography. Some may even be surprised by its existence, especially given the promotional emphasis placed by schools and universities alike on Geography's ability to speak to contemporary events and up-to-date issues. In this brief introduction we therefore want to say a little about why Historical Geography is actually a very central and important part of Human Geography.

A simple, but overly simplistic, definition of Historical Geography is that it is concerned with the geographies of the past. If it's dead, old or a little bit musty, it's Historical Geography. To the extent that this is true it means that Historical Geographers study a vast range of thematic issues, time periods and geographical areas from a variety of perspectives. Importantly, and happily, this also means that Historical Geography is not confined within some sort of intellectual ghetto. In this book, for example, you will find historical research being discussed in many chapters outside of this particular section. But Historical Geography is distinctive. It has a special role to play in Human Geography more widely. Increasingly, this is seen as stemming less from its attention to the past *per se* than from its concern with '**historicity**', or 'historical specificity and historical transformation' (Gregory, 1986: 197). To elaborate, Historical Geography is important because first, it allows us to recognize the potential difference of other times and places from our own, both widening our horizons and preventing us from making uncritical generalizations about how things can and might be; second, it also discovers parallels in the past, qualifying a frequent sense that we live in unheralded times, and

helping to sharpen our sense of what may really be 'new' and different about our present circumstances; third, it ensures the geographical preoccupation with spatiality (see Chapter 2) is complemented by an attention to **temporality**; and perhaps most importantly, it explains how geographical worlds actually come into being. The landscapes, ideas, languages, economies and political systems we inhabit are not natural or timeless, but historical creations. Human geographies are made; Historical Geography can tell us about their making.

Historical Geography is therefore far from a dry, antiquarian pursuit devoted solely to the archiving and organization of relics of the past. In fact, thinking historically is something people do all the time, although often not in terribly rigorous ways. Historical notions infuse our ways of speaking about the world and its geographies: we talk of some things being traditional, others modern; some places developing, others not; we implicitly divide time (and space) up into past, present and future. And the past fascinates us; think of the popularity of costume dramas, westerns, or heritage tourism sites. In exploring the importance of historicity, the three chapters here deliberately pick up on these 'popular' historical geographies, developing a more critical interrogation of their themes of temporality, modernity and heritage. The chapters by Miles Ogborn and Peter Taylor both deal with the dominant framing of world history in terms of a progression towards a modern present. In Chapter 16, Miles emphasizes how the notion of 'creative destruction', with its combination of both loss and invention, may better capture the experience of this 'modernization' than ideas of progress or development, illustrating his argument with reference to debates over the national modernization of Turkey and the urban development of nineteenth-century

Paris. He also stresses how modernization is a profoundly geographical as well as historical process, centrally involving the re-making of places and their landscapes so that they can embody **modernity** (in modern housing, roads, countryside, and so on).

This point is taken up by Peter Taylor in Chapter 17, as he makes the case for bringing together questions of temporality and space in a genuinely 'geohistorical' understanding of modernity. In particular, Peter argues against what are known as 'diffusionist' accounts, in which aspects of modernity are seen as spreading outwards from some places (the most modern) to others (the less modern). This is the kind of logic we are using if we talk about the need for 'developing' countries to modernize their economies, political systems and cultural beliefs in order to become more like the modern, 'developed' world. Instead, he suggests that we need to understand the long-running existence of modern '**world-systems**', world orders in which all societies are caught up, both those cast as more modern and as more primitive. Historically, this involves a long-term perspective, in which contemporary developments are seen in the light of several hundred years of **globalization**. Geographically, it suggests both a large-scale global focus to grasp the character of the whole world system, and also a local focus to scrutinize the way particular places differently experience these global modernities.

In Chapter 18, Nuala Johnson demonstrates that whilst we may live in a modern world, the past is still of great importance to us, not least as evidenced by the creation of particular places devoted to it, such as museums and heritage sites. Nuala's discussion is framed through two case studies of the latter, namely an Irish country house and the Auschwitz death camp in Poland. In both cases, her analysis is particularly powerful in drawing out the constructed and contested nature of landscapes of the past, an issue that faces academic Historical Geographers just as much as it does producers of heritage tourism sites. Crucially, her cases also illustrate the contemporary importance of landscapes of remembrance. The past should never be a subject of interest to us only for what it can tell us about the present, but it is worth emphasizing that history matters today. In turn, as all these chapters demonstrate, Historical Geography matters, and not only because it shows this importance of history to our human geographies, but also because it demonstrates the importance of geography to those histories.

Further reading

- *Journal of Historical Geography.*
It might be useful to have a brief look through some of the papers in a recent volume of this journal to get a feel for the variety of work done by historical geographers, as well as some of the key contemporary debates they are engaged in.

- Overton, M. (1994) Historical geography. In Johnston, R.J., Gregory, D. and Smith, D.M. (eds) *The dictionary of human geography*, third edition. Oxford: Blackwell, pp 246–50.
A schematic history of the approaches adopted by post-war Anglo-American historical geographers, including references to exemplary studies.

- Philo, C. (1995) History, geography and the still greater mystery of historical geography. In Martin, R., Gregory, D. and Smith, G.E. (eds) *Human geography: society, space and social science*. London: Macmillan, pp. 252–81.
An exploration of the contribution Historical Geography can make to both Human Geography and History.

Modernity and modernization

Miles Ogborn

INTRODUCTION: THIS IS THE MODERN WORLD

What does it mean to call something '**modern**'? In part it simply means that it is new, up to date, or of the moment. This might relate to (modern) technology, art or life as a whole, and these descriptions are about understanding how the world is changing. Sometimes this idea of 'modernity' – the condition of being modern – is used to celebrate newness, perhaps to encourage people to adopt an innovation or, in the case of the British Labour Party, to attract a wider range of voters to a new political programme. Sometimes it is part of complaints that the modern world is moving on and leaving much that is valuable behind, as in some discussions of the difficulties found in understanding modern art. Whichever point of view is taken, the idea of 'modernity' situates people in time. It suggests that time is divided up into past, present and future. It gives a certain value or significance to the past (positive or negative), and it makes the present important as a time of change and of decisions about what the future should be. Calling something 'modern' makes people think about historical change (Baldwin *et al.*, 1999).

Understanding modernity as 'newness' allows an appreciation of the modernity of any point in time. It helps to explain the excitement which accompanied the coming of the railway, electricity, the cinema, or buildings made of concrete, steel and glass. It also helps in understanding the sense of danger they brought too – the fear that the world was changing too fast. In each case the relationship between past, present and future established by modernity is experienced by people and understood in terms of how it might change their lives for better or worse. For example, Raphael Samuel has examined the ideas of modernity in Britain in the 1950s to argue that '[t]he ruling ideology of the day was forward-looking and progressive, the ruling aesthetic one of light and space. Newness was regarded as a good in itself, a guarantee of things that were practical and worked' (1994: 51). He explores this by showing that post-war 'home improvement' meant – for those unlucky enough to live in 'ugly', 'old-fashioned' Victorian houses – tearing out fireplaces, removing draughty sash windows, knocking down partition walls, and covering over 'dust-collecting' plasterwork. In their place came central heating, fluorescent strip lighting, fixtures and fittings of easy-clean, smooth plastic, Formica and fibreglass, and kitchens and 'broom cupboards' full of labour-saving devices – washing machines, electric cookers and Hoovers (Figure 16.1). Houses and people's lives were to be transformed through this 'appetite for modernization' (ibid.: 56). This example shows both the intensity with which people experience modernity (in this case as something desirable) and its role in making new geographies. Here it is new domestic geographies. These houses are made into different sorts of *places*; they look and feel different. Indeed, the same impulse also transformed the geography of many cities and their inhabitants' lives and experiences. The motorways, housing estates and civic and shopping centres so characteristic of the urban planning of the 1950s and 1960s can also be seen as an attempt to make spaces which were new, clean and easy to use. In both cases modernization

This wash-off mark distinguishes all FORMICA surfaces

FORMICA
LAMINATED PLASTIC
DE LA RUE

FORMICA *for me!*
PLASTIC
LAMINATED

When you discover how easy it is to cover up old-fashioned wooden surfaces with FORMICA laminated plastic, you'll get ambitious. It brings such a wonderful feeling of light and cleanliness into your home. FORMICA laminates won't stain, chip or crack ; they resist heat up to 310°F and wipe clean in a flash ; stay new and fresh for years.

Why not start on the kitchen table ? In less than an hour, with the help of De La Rue adhesive, you'll have a spic-and-span FORMICA-topped table, a joy to look at, to eat off and work on. A 3 ft. x 2 ft. table need not cost you more than 35/-. You can buy 1/16" panels in any of the thirteen popular patterns at 5/- a sq. ft. or, cut to size, at 5/9 a sq. ft.

Write for the free 'Do it Yourself' leaflets to Thomas De La Rue & Co Ltd (Plastics Division) Dept 39C, 84 86 Regent Street, London W1

FORMICA is the registered name for the laminated plastic made by Thomas De La Rue & Co Ltd

Figure 16.1 Home improvement as modernization. Source: R. Samuel, 1994: *Theatres of memory.* p.53

involves geographical change – transforming places, spaces and landscapes – as a part of historical change.

These examples begin to suggest that modernity and modernization are about more than just 'newness'. There are particular sorts of historical and geographical change involved. Without setting this out too rigidly (there are many opinions about what 'modernity' is; see Ogborn, 1998), important processes can be identified: the application of scientific principles to human and natural worlds; the development of industrial economies (capitalist and non-capitalist); and the formation of states which govern many aspects of life through their bureaucracies (Giddens, 1990). These are all both historical and geographical processes. They create particular forms of 'modern life' through specific (and often very rapid) types of historical change, and a crucial part of this is the transformation of spaces, places and landscapes. For example, Figure 16.2 shows a dramatic modern landscape: the Hoover Dam built across Black Canyon on the Arizona–Nevada border between 1931 and 1936. It is 725 feet high, incorpo-

rating 2.5 million cubic metres of concrete and creating a lake of 210 square miles. Understanding this dam in terms of modernity means seeing it as a product of the technological and scientific control of nature by geologists and engineers, funded largely from public money, and planned by one of the world's most powerful states to provide irrigation and hydroelectric power for the development of capitalist industry and agriculture. The Hoover Dam has not only transformed the physical geography of Black Canyon, but also the economic geography and environment of the western United States. It should also be understood in terms of the responses it provokes:

Confronting this spectacle in the midst of emptiness and desolation first provokes fear, then wonderment, and finally a sense of awe and pride in man's skill in bending the forces of nature to his purpose. In the shadow of the Hoover Dam one feels that the future is limitless, that no obstacle is insurmountable, that we have in our grasp the power to achieve anything if we can but summon the will.

(Stevens, 1988: 266–7)

Figure 16.2 The Hoover Dam: a landscape of modernity.
Source: US Department of the Interior pamphlet, 1993

The idea of 'man' against nature is raised again later, but for now it is important to note the sense of wonder, excitement and fear in how the Dam seems to open up the future. The Hoover Dam is a landscape that is the product of the scientific, industrial and political processes of modernization and is experienced in terms of the challenges and dangers of modernity. It combines, therefore, the ideas of modernity as both 'newness' and as particular forms of historical and geographical change.

The historical geography of modernity and modernization (and the difficult question of '**postmodernity**') is a huge area. In many ways this whole book is about the geographies of the modern relationships between people, science and nature; industry, **capitalism** and space; and politics, power and **territory**. Instead I want to concentrate in this chapter on some of the different ways of understanding the historical and geographical changes involved in processes of modernization. I want to start with an

idea that is there in both the examples of the Hoover in the 1950s' house and the Hoover Dam: modernization as 'progress'.

Summary

- Modernity is a matter of the experience of 'newness' and the specific understanding of historical time that involves.

- Modernity is also a matter of a specific set of interconnected economic, political, social and cultural changes.

- Modernization involves changes in people's lives and in geographies at all levels. These changes can be understood in a range of different ways.

MODERNITY AND MODERNIZATION AS PROGRESS

Understanding modernization in terms of 'progress' suggests that a society makes a clean break with a problematic past and does what is necessary to move forward into a better future. An influential version of this was the 'modernization theories' which were applied to Latin America, Asia and Africa after 1945. In his book *The Stages of Economic Growth* (1960) W.W. Rostow suggested that the history of each society could be understood through five stages: traditional society; the preconditions for take-off; take-off; the drive to maturity; and the age of high mass consumption. Britain, with its 'industrial revolution', had been through this process first and could be followed by the other countries of the world (Figure 16.3). Rostow argued that 'traditional society' prevented regular growth through its non-scientific attitude to nature, lack of social mobility, failure to see the potential for change, and non-centralized political power. This situation was to be changed in the 'preconditions' period by removing the technological, social and political constraints on economic growth through, for example, newly centralized states (controlled by new, often nationalist, elites keen on modernization) building roads and railways and encouraging science, technology and key industries. These geographical changes in agriculture, industry, transportation and urbanization, combined with a positive attitude to 'modernization', would prompt 'take-off' – 'the great watershed in the life of modern societies . . . when the old blocks and resistances to steady growth are finally overcome' (ibid.: 7) – into a stage of continual growth and less state intervention. This economic growth would utterly transform the society into a complex modern industrial economy during the 'drive to maturity'. Eventually it would

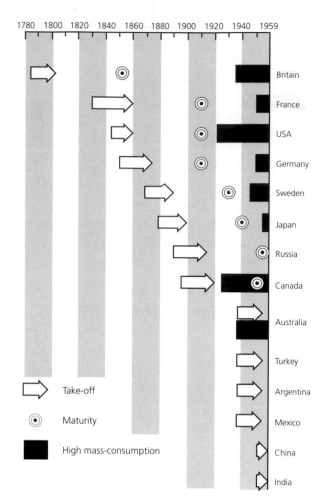

1780 1800 1820 1840 1860 1880 1900 1920 1940 1959

Britain
France
USA
Germany
Sweden
Japan
Russia
Canada
Australia
Turkey
Argentina
Mexico
China
India

Take-off

Maturity

High mass-consumption

Figure 16.3 Rostow's stages of economic growth. Source: W.W. Rostow, 1960: *The stages of economic growth.* p.xii

reach a final stage of 'high mass consumption' (modelled on the affluent parts of the USA in the 1950s) where production and consumption were based on consumer durables: a land of automobiles and suburban homes equipped with refrigerators and televisions.

Rostow presented modernization as progress. He argued that it was absolutely necessary for countries to make the decisions which would promote modernization or they would lose out to others. He also presented it – through the idea of 'take-off' – as a dramatic transformation. However, by understanding it as an orderly progression of stages that others had successfully gone through he was able to claim that there was only one path to follow for successful modernization, and that this break with the past and the move into a brighter future would be relatively painless and full of benefits in the long run. All of this meant that this form of modernization could be offered as the way forward for the countries of the 'Third World'. They could become like Western Europe and the USA.

One country that modernized in this way was

Turkey (although Rostow argued that the Turkish state pushed for 'take-off' too early). It is a useful example of some of the changes in people's lives and in geographies that 'modernization as progress' involves. It also suggests some of the problems with this version of modernity. Turkey's transformation involved opening itself up to the forces of Western modernization. This began during the nineteenth-century Ottoman Empire, but was dramatically accelerated with the foundation of Turkey as a **nation-state** in 1923 under Mustapha Kemal (he later called himself Atatürk, 'Father of the Turks'). The modernization process was to be a total transformation of Turkish economy, society, politics and culture, and a total break with the past. Mustapha Kemal declared that 'the new Turkey has no relationship to the old. The Ottoman government has passed into history. A new Turkey is now born' (Robins, 1996: 68). Moving into this promising new future involved economic policies which transformed parts of the country. During the 1930s the state intervened to modernize the economy and to make it serve national ends rather than being oriented to the export of raw materials (Figure 16.4). Import controls were coupled with agricultural policies aiming at national self-sufficiency in food through scientific farming and irrigation schemes. A modern capital city was built at Ankara. Extensive railway construction produced a national network, and a Soviet-inspired 5-year plan (1934–38) provided large textile, sugar, paper, cement and steel plants spread across the country in order to substitute Turkish goods for imports from the West. These were combined with earlier (1920s) policies designed to remove what were seen as cultural barriers to modernization. The influence of Islam on people's lives was to be reduced by separating state and religion, abolishing religious courts and centres of religious learning, and closing down shrines and religious brotherhoods. Turks were to be reoriented towards the West and its ideas by prohibiting the fez in favour of the Western hat for men; the adoption of Latin rather than Arabic script, the metric system, family surnames and the Gregorian calendar; and, along with the Westernization of new architecture, the playing of American jazz in public places. Finally, any identifications with causes other than that of modern Turkey were prevented by the prohibition of internationalist organizations and programmes of repression and assimilation aimed at ethnic minorities (Kurds, Georgians and Armenians). Turkey's modernizers understood the process of modernization as one of forcing people to make a clean break with the past and encouraging them to progress into a bright (Western) future (Parker and Smith, 1940; Landau, 1984; Schick and Tonak, 1987).

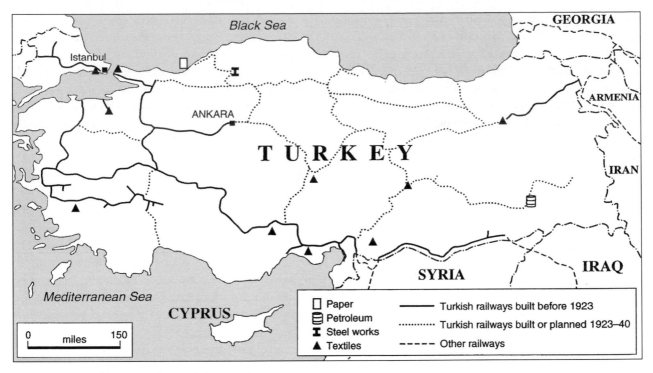

Figure 16.4 Turkish modernization: industry and transport

There are clearly some very real problems with this single, progressive and Western version of modernization. First, it can never be a matter of simply following a Western model. These countries are not separate entities ranged along a developmental path, they are connected together so that the development of one has consequences for others (Taylor, 1989). Also, within Turkey, the adoption of a model from outside has meant that 'modernization has been an arid and empty affair' (Robins, 1996: 67) unconnected to the cultures of the place and lacking dynamism. Second, this version of modernization underplays the disruptiveness of the transformation for people and the landscapes that they live in. Robins shows how Kemalist policies to suppress the past, the specific nature of Turkish society and the cultural differences within the country to achieve a Western modernity were very traumatic and are only just being dealt with now.

Why modernization theorists put forward such ideas is best explained by their political context. In an era of decolonization, the Cold War and a post-war boom in Western Europe and USA the idea of modernization as Westernization (and seeing that as orderly progress) was part of a political move to combat the appeal of Communism to 'Third World' countries. Rostow certainly saw himself as offering an alternative and subtitled his book 'A non-communist manifesto'. In the next section I want to look at some of the ideas that Rostow was reacting against to see

how modernization is understood in a different way: as 'creative destruction'.

Summary

- Rostow's modernization theory presented modernization as progress: a single (Western) path to modernity which was positive, beneficial and relatively painless.

- Turkey's experience of 'modernization as progress' shows the changes in lives and geographies that it involves, and that these ideas of progress simplify the processes and experiences of change.

MODERNITY AND MODERNIZATION AS CREATIVE DESTRUCTION

Understanding modernization as 'creative destruction' suggests that the changes involved are dramatic and unsettling ones, and that making a new future always means destroying many of the geographies and ways of life of the past and present. Marshall Berman (1982) has suggested that Karl Marx and Friedrich Engel's (1848) *Communist Manifesto* puts forward this view of historical and geographical change in its discussion of capitalism. He sums it up using a phrase (borrowed from Shakespeare) from the *Manifesto* – in the modern world 'All that is solid melts into air'. This poetic image suggests the sense of constant change and uncertainty which Marx and Engels argued is necessary for **capitalism** to be successful. This arises from the continual need to develop the 'productive forces' – labour power, raw materials, machinery, science, communications and transportation – needed to produce **commodities** and make a profit. This generates the constant search for new materials and new markets which drove global exploration and settlement and radically transformed the lives and landscapes of many in Africa, Asia and the Americas, as well as in Europe. It also meant a dramatic transformation in forms of work, most vividly seen in the coming of the factory system, where all sorts of labour – no matter how menial or prestigious – became activity that was done for wages and was often controlled by the workings of a machine. Finally, it involved radical geographical changes, the building of huge industrial cities, the making of modern nation–states, and what Marx and Engels described as the 'Subjection of Nature's forces to Man, machinery, application of chemistry to industry and agriculture, steam-navigation, railways, electric telegraphs, clearing of whole continents for cultivation, canalization of rivers, whole populations conjured out of the ground' (1848: 85). So, even though they opposed it, Marx and Engels were in awe of capitalism's capacity to make 'all that is solid melt into air', destroying entire ways of life and whole landscapes as it developed the forces of production. More importantly, they saw capitalism as destroying what it had itself created – factories, docks, whole cities – as they became unprofitable. For them the process of modernization is both dramatic and traumatic. It offers all sorts of possibilities for changing the world, but also destroys ways of life and places that had become known, accepted and familiar. This making and breaking of social relations and their geographies is modernization understood as 'creative destruction'.

An example can help here. Between 1850 and 1870 Paris was transformed by this sort of 'capitalist modernization'. Many of that city's most recognizable features – tree-lined boulevards, monumental architecture and pavement cafés – were constructed during that period. However, to create this new urban geography much of Paris had to be destroyed and many lives were seriously disrupted. How and why did this happen? David Harvey (1985) argues that the impetus was the economic crisis of the 1840s: the worst that France had experienced. There were many workers who could not find work and, at the same time, many investors who could find nowhere profitable to invest. In addition, Paris's eighteenth-century infrastructure made it more of a hindrance than a help in escaping from this crisis. The solution orchestrated by Emperor Louis Napoleon and his minister Baron Haussmann

was a massive reorganization of the geography of France and its capital through a huge public works programme – building new railways, roads and telegraph systems across France and transforming Paris with new streets, water supply, sewers, monuments, public buildings, parks, schools and churches. The rebuilding programme would put people and capital to work in the short term and, in the longer term, would provide the sort of modern city within which profitable investments in land, manufacturing or commerce could be made.

The most significant part of this new urban geography was Haussmann's boulevards. Figure 16.5 shows the streets built between 1850 and 1870. In all they were 85 miles long and, on average, three times wider than the ones they replaced (Pinckney, 1958). The extent of the remodelling of traffic flows is clear. The aim was to make Paris into a single, functioning unit rather than a series of separate neighbourhoods. Particularly significant is the cross at the heart of the city made by Rue de Rivoli, Boulevard de Strasbourg, Boulevard de Sébastopol and Boulevard Saint Michel which linked both banks of the River Seine to an axis that ran right across Paris. Other boulevards connected new railway stations to the city centre. These new streets brought other changes too. Building them combined modern science and engineering, as well as a lot of hard work (in the mid-1860s 20 per cent of the working population of Paris was employed in construction). Their planning necessitated the first accurate topographic and land ownership maps for the city. They were also politically important. Haussmann ensured that the boulevards provided vistas of monuments or buildings which were symbols of France, religion and empire – like the Arc de Triomphe – so that the new city combined both tradition and modernity. It was also rumoured that these wide, straight roads made it harder for revolutionaries to put up barricades as they had done in 1830 and 1848, and easier for the army to ride their horses or fire their artillery down them. These huge streets certainly made it easier to travel faster through the city by horse and carriage, and pedestrians were now confronted with huge numbers of speeding vehicles (Berman, 1982). Finally, they provided new opportunities for private investment in the plush apartment buildings, exclusive hotels, fashionable pavement cafés and dazzling shops (including vast department stores) which lined the gas-lit boulevards. As Harvey argues, they 'became corridors of homage to the power of money and commodities, [and] play spaces for the bourgeoisie' (1985: 204). Paris was transformed – a new geography was created – and some people benefited from the opportunities for profitable investment and pleasure.

However, to build the boulevards the old city of

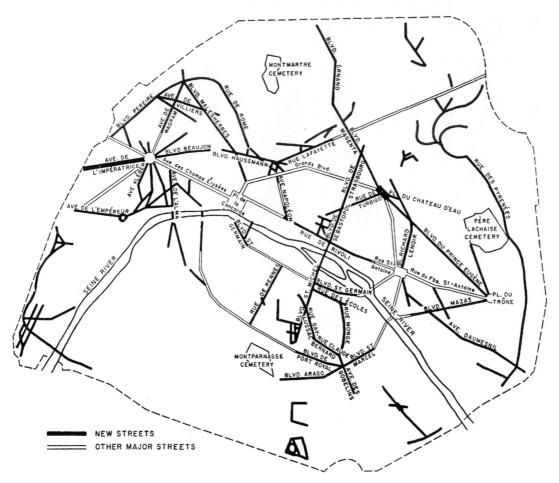

Figure 16.5   Principal new streets in Paris, 1850–70. Source: D.H. Pinkney, 1958: *Napoleon III and the rebuilding of Paris*. p. 73

Paris had to be torn apart. The new streets carved their way straight through ancient districts of winding streets, hidden corners and higgledy-piggledy buildings housing vast numbers of Parisian workers. In 1850 the Ile de la Cité (the island in the Seine where Notre Dame cathedral stands) housed 14 000 people. By 1870 only a few hundred were left to defy Haussmann's plans to devote the island to institutions of law, religion and medicine. For twenty years the centre of Paris was a building site and thousands of people were displaced. They were either forced out to the suburbs, and had to walk for several hours to get to work, or they crowded into the remaining central areas and paid exorbitant rents for appalling accommodation. Figure 16.6 shows a contemporary cartoon which, while making fun of the situation, gives a sense of the disruption felt by many ordinary Parisians as their city was destroyed. As well as the elimination of neighbourhoods it also hints at the disruption of family relationships. This man's wife may have not been in the shell of what had been their home because, as Harvey shows, the economic restructuring which accompanied Paris's transformation meant that both men and women were forced to work, often very long hours, if they and their families were to sur-

vive. In Paris, therefore, we can see an example of the wider process of 'creative destruction' whereby the forces of modernization benefit some and disadvantage others as they dramatically transform lives and landscapes (Pred, 1990).

Summary

- Understanding modernization as 'creative destruction' shows that it involves the destruction of lives and geographies as well as the construction of new geographies and new ways of life.

- The example of nineteenth-century Paris shows that some people benefit and some people lose out through the processes of modernization.

CONCLUSION: MANY MODERNITIES

'Modernization as progress' and as 'creative destruction' offer very different ways of understanding these historical and geographical changes. They give different versions of the relationship between past, present and future and, as presented by Marx and Rostow,

Figure 16.6 'But here is where I live - and I don't even find my wife'. Source: Honoré Daumier cartoon, 1852. In D.H. Pinkney, 1958: *Napoleon III and the rebuilding of Paris*

suggest very different political responses. However, there are three points that can be made about both versions which help to develop the idea of modernity. First, they both present modernity as a radical break with the past. This may be true for some parts of Turkey in the 1920s and 1930s or for many Parisians in the 1850s and 1860s. However, it is worth thinking about more subtle ways of understanding these historical geographies of modernization which recognize that change happens at different rates in different places, and that instead of being eradicated by modernity past lives and landscapes are often remade, reinvented or reincorporated with new ways of doing things and new geographies (see Chapter 18 by Nuala Johnson). Second, they are both primarily concerned with economic transformations, and particularly with capitalism. While this is clearly important we need to remember that historical geographies of modernity also need to be about political transformations (e.g. modern bureaucratic and territorial nation–states), technological transformations (e.g. the application of science to both nature and society), and social and cultural transformations (e.g. modernist art's attempts to deal with processes of modernization). All these processes shape the modern world and they have done so in non-capitalist as well as capitalist economies.

These two issues lead towards the third point: that there are many different modernities. 'Modernity' is different at different times, and it has many different strands. It is also different in different places. Both versions discussed here see Western Europe and North America as the most important places in these historical geographies of modernity. Instead, we might also ask more about other experiences of modernity which were connected to what went on in London, Paris and New York but were not simply the same (Miller, 1994). Robins' (1996) discussion of Turkey offers one possibility, another is Paul Gilroy's (1993) argument that black Africans transported to the Americas and to Europe under slavery can be seen as the first modern people because of their experience of (and resistance to) **globalization**, industrial work on the plantations and attempts to use science to justify racism. There are, therefore, also different modernities for different social groups defined in terms of **class**, '**race**' and **gender**. For example, this means that processes of modernization have different effects on and are experienced differently by men and women: modernity is often talked about in terms of dramatic confrontations between 'Man' and 'Nature' which present the whole process in gendered terms; Turkey's modernization programme meant that women were given the vote (as well as equal rights in many other areas); and, in Paris, the new pleasures and hardships of the modernized city were experienced differently by men and women who had different access to the new spaces of the city and very different expectations on them when they were on the streets or in the cafés or department stores (Pollock, 1988).

What we end up with, therefore, are many different modernities. There are various processes of modernization which have been transforming lives and landscapes across the globe in different ways for hundreds of years and which are experienced differently by different sorts of people. Each one makes for a different historical geography of modernity and raises different questions about understanding the past, present and future.

Further reading

Given that is a good idea to read original works and make up your own mind, the two books that I use for this discussion are Rostow and Marx and Engels:

- Rostow, W.W. (1960) *The stages of economic growth: a non-communist manifesto*, Cambridge: Cambridge University Press.

Especially Chapters 1–4 which cover the stages up to 'take-off'.

- Marx, K. and Engel, F. (1967, originally published in 1848) *The Communist manifesto*, Penguin Books, Harmondsworth.

Section 1: 'Bourgeois and Proletarians' includes the material discussed here.

Other, more general treatments of the issues raised are in Berman and Ogborn:

- Berman, M. (1982) *All that is solid melts into air: the experience of modernity*, London: Verso.

Offers a discussion based on the idea of 'creative destruction' that covers Marx (Chapter 2) and Paris (Chapter 3). Quite a difficult book to read but has many interesting insights and examples.

- Ogborn, M. (1998) *Spaces of modernity: London's geographies 1680–1780*. New York: Guilford Press.

Chapter 1 contains an overview of theories of modernity which relates them to questions of historical geography.

CHAPTER 17

A geohistorical interpretation of the modern world

Peter J. Taylor

The modern world has seen many modernities. In this chapter I will argue that among the many there have been three outstanding constructions of modernities which should particularly command our attention. These I shall term prime modernities because of their pervasive impact – Dutch mercantile modernity in the seventeenth century; British industrial modernity in the late eighteenth and nineteenth centuries; and American consumer modernity in the twentieth century.

There are two things to notice immediately about this short list. First, each **modernity** is specifically linked to a particular country: modernities do not just 'emerge', they are constructed through the practices and activities of people in real places. Every time a person buys a pair of blue jeans, no matter where they are in the world, they are reproducing habits which most clearly originated in the USA. Indeed, that has been a key part of the attraction of the blue cowboy trousers. Second, the three examples cover four centuries and constitute a sequence of prime modernities. In fact I will argue that they represent cumulative, wide-impact modernities so that they each have contributed importantly to our contemporary experience of being modern.

This concern for place and sequence I call a geohistorical interpretation. This way of putting geography and history together aspires to be more integrative than other **Historical Geography**. Obviously all human practices and events occur simultaneously in both time and place. We separate them in our studies for pedagogic reasons: it is very difficult to do justice to both geography and history in most

social science so that typically one or other, sometimes both, are removed from the analysis. For instance, economic analysis is often criticized for ignoring both space and time variations in market practices. In this chapter I use the prime modernities to illustrate a geohistorical interpretation of modernities.

Putting modernity under the scrutiny of a geohistorical interpretation is particularly necessary because of the relative nature of the concept. Typically, to be modern is understood as the opposite of traditional. But the truth is that the future of all modernities is that they become 'traditional' in their turn. For instance, in regional geographies of Britain in the twentieth century, traditional manufacturing is associated with the heavy industrial regions of northern Britain which developed in the nineteenth century. In contrast, light industry is viewed as modern manufacturing which became concentrated in the south of England in the first half of the twentieth century. But this neat space–time separation of tradition and modern is not as straightforward as it first seems. Traditional industry did not begin as traditional; quite the opposite in fact. In the nineteenth century northern Britain was the wonder of the modern world, a new industrial civilization, widely interpreted as the most advanced place on Earth, with its leading city, Manchester, described as 'the essence of modernity' (Arblaster, 1984: 260). So what was spectacularly advanced and modern became relegated to backward and traditional. Quite simply, modernity never stops changing; hence it 'has no fixed objective referent' (Osborne, 1996: 349), industrial or otherwise.

The basic lesson of all this is that whereas most

important discussions of modernity have a philosophical, sociological or cultural focus, it is equally important to develop a geohistorical understanding of being modern. After reading this chapter you should have gained two additions to your knowledge, one empirical, the other theoretical. First, you will have a skeleton framework about how, where and when the prime dimensions of modernity have developed. Second, you will appreciate that the places which are currently at the forefront of modern practices, the world cities such as New York, London and Tokyo through which trillions of financial transactions take place daily, will in their turn become 'traditional'.

STORY-AND-MAP

What is a geohistorical interpretation of modernity? Studies of modernity may be considered as lying along a continuum with ahistorical/non-geographical analyses at one end and geohistorical approaches at the other. The pole position opposite to my approach consists of carefully delineating a meaning of modernity and then applying that definition as if it were separate from, and autonomous of, any history and geography. For instance, Matthew Arnold˙ in the 1860s outlined a timeless and placeless view of modernity by emphasizing particular intellectual and civic virtues focusing on a rationality he thought existed in Victorian England. Since he identified these same virtues in classical Athens, he designated the latter to be a modern society. Similarly, Baudelaire in the Paris of the 1850s, by focusing on the aesthetic, was able to argue that every age has its modernity represented by the painter capturing the fleeting and transitory in his work. In complete contrast, here I take a much more grounded and concrete view of being and becoming modern. For many millions of people in the twentieth century, and for some time before, being modern is, and has been, taken for granted: for instance, readers of this book will probably consider themselves to be modern without giving the matter much thought. This implies it is embedded in everyday thinking and behaviour. Such a condition I shall term modernity. A geohistorical approach respects this embeddedness, never neglecting the contexts in which modern behaviour and thinking take place. Quite simply, embedding occurs in real time and space locations which are constitutive of the modernity under study. Hence a geohistorical interpretation of modernity is concerned to understand the specific periods and places where ideas and practices of being modern are created, challenged and changed.

Most studies of being modern fall between these two opposite poles. Periods and places of creation and reproduction of modernity remain part of the analysis but they are less central, often, for instance, being relegated to an illustrative role. At its worst this can lead to a relatively random ransacking of history and geography examples to find suitable cases to prove a point. Even in more sensitive studies, in using a case study to expedite parts of the argument, period and place may be brought back into the analysis as context but the wider time–space structure will still be missing. In a geohistorical approach periods and places are not simply 'used', they are interpreted as being the concrete face of modernity as a single inter-connected story and map. It is this combination – story-and-map – at the centre of the analysis which defines a geohistorical methodology.

Summary

- A geohistorical approach focuses on the embeddedness of social practices within specific space–time locations.

- A geohistorical approach identifies the concrete face of modernity, an inter-connected 'story-and-map'.

MODERNITY: BEING MODERN AND THE MODERN WORLD

Consider the following comparative sociological study by Inkeles and Smith (1974). Using a standard sociological methodology, a questionnaire was constructed to elicit the ability of individuals to carry out what were thought to be 'modern' roles. The answers of respondents were coded, aggregated and converted into a composite score ranging from 1 to 100 measuring the degree to which an individual was 'modern'. In this way the study was able to assess the modernization prospects of six Third World countries from which samples were drawn. Countries were ranked in terms of the average 'modern' score of their samples to suggest how rapid they were likely to become 'thoroughly modern' like First World countries. Studies such as this have been properly criticized on many counts but they do raise an interesting question, even if their approach is seen now as crude and simplistic.

The work quoted above takes what I call a liberal approach to defining modern. 'Being modern' is something individuals possess and therefore can be measured person by person. In the process every individual is abstracted out of her or his social context and tested by a universal measure. There is no sense of modernity as a geohistorical phenomena, a network of opportunities and constraints which vary by time and place. Rather, an individual attribute is measured without reference to the necessary social support which makes being modern possible or indeed meaningful. If being modern is thought of geohistorically, there is a query as to whether or not the idea of 'percentage modern' has any credence at all. Surely being modern is one of those all-or-nothing conditions: you either are or you are not. Modernity is a condition experienced by people who live in a modern society; I would deem all such people to be modern by definition. Of course this then begs the further question: what is a modern society?

Identifying modern societies seems to be an easy task. American society, British society, Australian society and Japanese society, it would generally be agreed, are four of many examples we could list as modern societies. One commonly overlooked feature of such a list is the fact that these societies are politically bounded; each society corresponds geographically to a single state. There seems to be little theoretical justification for this society/state congruence, it is simply part of our contemporary taken for granted interpretation of the modern world. There is a simple geohistorical critique of such common thinking. If indeed societies are defined by states, then they become as unstable as state existence. Was there, for instance, an Estonian society in the 1920s and 1930s, the time of the first Estonian state, which disappeared into Soviet society for half a century only to emerge again in the 1990s with the recreation of an Estonian state? Where there was one Yugoslav society just a decade ago, are there now five societies to match the political division of the original state? Of course, such definitional conundrums are not merely a feature of the rise and fall of communist states. What about former polities which are not on today's world political map, do their societies automatically vanish with them? For instance, did Newfoundland society disappear when the country become a province of Canada? For the counter case, consider Quebec; will a new society emerge only when the province wins independence? All of these examples suggest equating state and society is not a sensible thing to do. Society, as a much deeper social construction than politically created states, should be able to predate and outlast the political fact of a particular state's existence.

Equating society with state is a basically modern view of the world: the cartographic image of countries across the world forming a mosaic of bounded spaces is the most familiar of all maps. But this is very much a modern self-image we do not have to subscribe to. **Globalization** is an alternative vision of the world which is becoming increasingly popular. This is an argument which posits a new scale of economic organization transcending individual states. The global financial market is the classic case where private transactions in New York, London and Tokyo totally dominate the public currency reserves of any one country. If societies are built upon their economies, then it would seem that crucial aspects of what we think of as societal relations should now be viewed beyond the state, perhaps even globally. This contemporary argument is a controversial one but here I want to extend its implications historically. Although unusual in its scale and intensity, the current financial market is by no means unique in its geographical scale of operation (Arrighi, 1994). There were many important economic transactions beyond states which existed in previous centuries. Transcending states has been an integral part of economic development, and in the argument of this chapter contemporary globalization is a culmination of these processes and not a new departure. From this long-term perspective, the modern world is viewed as a system of inter-connected economic links. It is a 'world', not in the sense of global, but as the broad geographical area through which people and their social relations, loosely society, are able to be reproduced – that is, to have a history. Immanuel Wallerstein (1979) calls this the modern **world-system** which he dates from the expansion of Europe in the 'long sixteenth century' (*c.* 1450–1650) and which eliminated all other social systems to become global in the twentieth century; hence today this modern social world is global in scale.

The corollary of this position is that since all members of this system are part of the same story-and-map, from a geohistorical perspective they are all judged to be modern. Whatever their skills and aptitudes, people making a life in the modern world-system have specifically modern roles within the larger whole. Modern is ultimately societal, not individual.

Summary

- Modernity is a social condition experienced by people living in 'modern society'.

- Equating modern societies with societies defined by states is a typically modern state-centric view which we need to transcend.

- The modern world-system defines the scope of modernity in both time and space.

PLACING MODERNITY

Counting all people within the modern world-system to be modern does not make them equal, of course. The modern world is a very differentiated world in both space and time. One of the problems of calling it modern is that it tends to suggest a homogenous, or at least a homogenizing, world when all that a geohistorical interpretation argues is connectivity. Hence the question of power relativities must be kept to the fore.

Part of the difficulty arises from the word modern itself which has direct temporal but not spatial referens. Modern society implies contemporary society plus the antecedents that created it which between them define the modern period. Given that there is no equivalent spatial connotation, once in this period, modern can be treated as a universal process. But, of course, the development of the modern world was anything but non-geographical. Certain people and places are intimately implicated in its creation and growth and this is expressed in alternative geographically referenced descriptions of the process of modernization as 'Europeanization' or 'Americanization' or, more generally, 'Westernization'. Let us briefly consider each of these.

The modern world-system was originally created and developed through the spread of Europeans across the world as conquerors, settlers and traders. Like all migrants with the requisite power, they tried to recreate their new worlds in the image of their old so that the modern world grew in large part as Europe writ large. By the twentieth century many non-European elites in Africa and Asia were Europeanized, in the sense that they were educated in the schools and universities of the metropolitan centre or in new local educational establishments which purposively copied the latter. Thus by the time of world-wide **decolonization** after World War II almost without exception the leaders of the new independent states were modernizing elites entering politics with European qualifications (Davidson, 1992). This meant Europeanization did not end with decolonization but the new states, usually with constitutions modelled on that of their 'mother country', continued to have strong links with their erstwhile masters. The British Commonwealth and French Community are the political expressions of this fact.

However, while this Europeanization was continuing in the 'South', in the 'North' a different social transference was happening. The twentieth century has been an era of Americanization, a projection of US power through both coercion and by consensus. One of the key periods of such projection was in Western Europe after World War II. This was a time of adopting and adapting US economic practices so that in the 1950s Western Europe began to experience the 'affluent society' pioneered by the USA (*see* Fig. 17.1). In contrast, American power was experienced in the South largely as a coercive force; this was where the North's 'Cold War' was allowed to go hot in numer-

Figure 17.1 McDonald's in Prague. An example of the Americanization of a European city. Credit: Colorific/Telegraph Colour Library

ous wars, with the USA sometimes fighting them, as in Vietnam, at other times using local surrogates, as in Afghanistan. It was during this post-World War II period that American social scientists created their models of modernization (see Chapter 16). However, popular descriptions of these processes, such as coca-colarization, Disnification, McWorld and the Levi generation, left no doubt as to the generally understood place origins of this 'universal' process (*see* Fig. 17.2).

The idea of the world geographical division of North–South was devised in the 1970s as a means of diverting attention away from the **Cold War** and towards world poverty by placing the USA and USSR (plus Europe and Japan) in the same category. Although the Cold War was specified geographically as East versus West, this division is actually much older than the rise of either the USA or USSR. Thus East–West tensions have not ended with the demise of its particular manifestation as Cold War. Older civilization conflicts most notably relating to Islam have come to the fore as resistance to westernization. In Moslem states, the modernizing elites such as Kemel Atatürk's group in Turkey, attempted to westernize their countries, for instance by adopting the western alphabet, but with only partial success. Here, and in Egypt and Algeria – all three countries have genuine modern revolutionary credentials – westernization is being fiercely resisted by those who want to live in a non-western Islamic society. In this case the term westernization is particularly appropriate since it combines both the USA and Europe as a single threat albeit with the former as leader – Iran's 'Great Satan' no less.

These various geographical locatings of modernization are important but they do not necessarily lead us to a critical geohistorical understanding. Very often behind these ideas, and behind modernization itself, there is a taken for granted geohistorical perspective which combines a Whig history with a diffusionist

Figure 17.2   A painter works on a new Coca-Cola advertisement at a Colombo marketplace, Sri Lanka. An example of a Third World country with obvious Western influences. Credit: Popperfoto/Reuter

geography. The former is defined as history whose story celebrates the present (Carr, 1961). This takes the form of defining the important features of contemporary society and tracing their lineage back in time so that the story told is one which culminates in the success of today's society. Nineteenth century English history is the *cause célèbre* of this school of history in which the torch of progress was passed on, starting in ancient Mesopotamia, Egypt and Greece, until finding its final resting place in Victorian English hands. Diffusionist geography is the spatial equivalent of the Whig history (Blaut, 1993). In this case the important features of contemporary society are traced geographically in order to show that all progress emanates from the centre of the modern world-system, first Europe and then the USA. Twentieth-century American development theory, which introduced the concept of 'modernization', is the *cause célèbre* here with its assumption that modern American values could diffuse, even 'trickle down', to the rest of the world. Cutting out the ancient roots (except perhaps Europe's representative in the story,

classical 'democratic' Greece) from the history roster and focusing on modern examples allows for the construction of a combined Whig-diffusionist geohistorical world view: England passing on the torch of liberty and progress to her American 'cousins' across the sea to spread around the world.

The alternative to this smooth and virtuous modern morality story-and-map is to emphasize discontinuities in both time and space. Modernity does not just appear as a result of 'natural' evolution; there are discontinuities with both the rise and the development of the modern world creating quite different forms of what it is to be modern. Similarly, the modern does not simply exist as a continuous geographical gradient from high to low: there are discontinuities between core and periphery zones of the system again creating quite different forms of what it is to be modern. In short, there are different modern times and different modern spaces in a world of multiple modernities.

Summary

- Modernity implies a universal condition but in reality has a definite geographical lineage as expressed by the terms 'Europeanization', 'Americanization' and 'Westernization'.

- The Whig-diffusionist geohistorical interpretation of modernity assumes a smooth development of modernity in the centre of the modern world-system and a simple diffusion out from this centre.

- The alternative geohistorical interpretation emphasizes discontinuities in both time and space creating many different forms of what it means to be modern.

PRIME MODERNITIES

And so we return to prime modernities. These are important to our argument here because they define crucial stages in the developments of modernity which have had an inordinate influence on the modern world-system. Quite simply, these are the particular modern worlds emanating from world **hegemonic** powers. The concepts of modernity and **hegemony** come from two rather different theoretical traditions and, therefore, have not usually been considered within the same analysis. But they are not so far apart as may at first appear. If the watchword of modernity is change, the equivalent for hegemony is stability. The stability of hegemonic social relations can be interpreted as the creation of new social arrangements which, temporarily at least, 'tame' the inherent chaos in the system. In this simple logic, hegemony can be a producer of modernity, despatching old social relations to history while constructing a new modern world (Taylor, 1996).

The Netherlands in the seventeenth century, Britain in the nineteenth century and the USA in this century, have been identified as the three hegemonic states in the history of the modern world-system (Wallerstein, 1984). That is to say, they were not just great powers, or even *primus inter pares* in their respective heydays, they had something extra that marked them out as qualitatively different from their rivals (Arrighi, 1990; 1994). World **hegemons** have provided, to use Gramsci's terminology, 'moral and intellectual leadership' in the modern world-system. The result has been the creation of three modernities: Dutch-led mercantile modernity, British-led industrial modernity and American-led consumer modernity. Of all the many modernities which may be identified, I call these three prime because of their direct association with world hegemony. Their hegemonic nature means that they penetrate societal relations throughout the system within their era. In fact, they invade and pervade so much that they quickly transcend their hegemonic state origins to provide the common labels for their respective times: the age of mercantilism, the industrial age, and the consumer age, respectively.

Each hegemon has been responsible for creating its particular version of what is modern about the modern world-system. Unlike overt state modernizers, such as Revolutionary/Napoleonic France and the USSR, who have attempted to tame modernity through large-scale social planning, hegemonic states and their civil societies have been rather covert modernizers. They have operated within a particularly successful trajectory of social change, reinforcing it by their activities and without the need to attempt to 'control' change. Thus instead of simplistic planning, a more complex mix of political and economic elite behaviour is rewarded in the world market creating the success story which is hegemony. For a short period it even seems that the world economy is working for the hegemon and the modern world-system is the hegemon writ large as, for instance, when the twentieth century is called the 'American Century'.

The rest of the world responds by trying to emulate the success of the hegemon with many profound geographical consequences (Taylor, 1996). For instance, Czar Peter the Great worked incognito in a Dutch shipyard to learn the secrets of modern production – he finally created St Petersburg as a 'new Amsterdam'. In the early nineteenth century entrepreneurs came from Europe and the USA to learn how to industrialize and build 'new Manchesters' in their own countries. In this century, the growth of suburbia represents 'Los Angeles writ large' in cities across the world. But what exactly have these other countries been copying for four centuries?

Mercantile modernity was created through the everyday life of commerce and the massive coterie of activities that it generated. There had, of course, been many influential networks of merchants in the past but in the seventeenth century a new rationality came to dominate success in the world-system, in which the calculating behaviour of the merchant was the archetypal practical form. The Dutch, more than anyone else, made making money respectable. Navigation became the great enabling applied science and new market institutions (e.g. the Bourse) were invented as Amsterdam became the first modern information centre in the world (Smith, 1984) (*see* Fig. 17.3). It is in the second half of the seventeenth century that the idea of a self-consciously modern world appears. These were attempts to understand the changes occurring which seemed to make Western Europe appear to be more 'advanced' than its past 'golden age', the classical civilization of Greece and Rome. The intellectual victory of the 'moderns' over the

De Roowaensche Kaey.

Figure 17.3 Seventeenth century print of Amsterdam harbour. Credit: Mary Evans Picture Library

'ancients' culminated in the eighteenth-century **Enlightenment** with its abstract theories of progress and where the seventeenth-century Dutch were viewed, in Voltaire's words, as a 'terrestrial paradise' (Schama, 1987: 56).

In the nineteenth century industrial modernity was created through a new scale of production: new energy (steam engine), new machines, new factories. This made new levels of **capital accumulation** possible in concentrated zones of production: great commercial ports gave way to massive industrial towns as the archetypal modern. There is no doubt that Britain is the country most implicated in bringing this second modern condition about. In this industrial world, modern society became mass society – the modern became, in Marx's terms, an alienated way of life. Mechanical engineering became the enabling knowledge as new machines of iron and steel became the cutting edge of change. Hence, with industrialization, the British gave the Enlightenment idea of progress a concrete manifestation making it the dominant organizing principle for understanding social change. This 'Victorian cult of progress' was based upon the successes of technology and created an inordinate faith in the future (*see* Fig. 17.4).

In the twentieth century the social effects of **alienation** have been countered in selected countries by a spread of affluence to ordinary people. The USA has been the major creator of the new consumer modernity as suburbia and its ubiquitous shopping mall have become the focal modern place This modernity is defined by a uniquely close relation between mass production and mass consumption: new production practices coinciding with the birth of modern advertising. As more and more is produced, the focus moves clearly on to consumption in order to realize the capital. Management science becomes the new enabling knowledge culminating in the merging of computers and communication. It is at this time that progress is repackaged as development but loses none of its social optimism, promising 'high mass consumption' for all as modernization.

One final point needs to be emphasized about these modernities. They represent three distinct episodes within a single system, not three new systems. In terms of their political economies, for instance, at all times commodities had to be made (production), distributed (commerce) and sold (consumption). What each hegemon achieved was a particular structuring of these relations in which one of these was particularly emphasized: the Dutch cutting edge was in exchange, Britain's cutting edge was in production, and the US cutting edge has been in consumption (*see* Fig. 17.5). As far as socio-cultural relations are concerned, there is much overlap between the respective eras but there is no ignoring their distinctive *mentalités* with different ways of life dominating both practice and aspiration.

Summary

- The Netherlands, Britain and the USA have been the three hegemonic states of the modern world-system providing 'moral and intellectual leadership'.

- As hegemons they created very successful new social arrangements, modernities, which others emulated.

- Mercantile modernity derived from Dutch pre-eminence in commerce in the seventeenth century.

- Industrial modernity derived from British pre-eminence in production – the 'Industrial Revolution' – in the late eighteenth and nineteenth centuries.

- Consumer modernity derived from the US practice of creating a society of mass consumers as a necessary complement to mass production.

Figure 17.4 The smokey skyline of Glasgow c. 1880. Credit: Mary Evans Picture Library

Figure 17.5 US suburbia 1950s – the consumer age. Credit: Popperfoto

CONCLUSION

The USA is no longer the hegemonic state it was in the 1950s and 1960s. Other countries have emulated American practices and added some of their own to create a much more complex world. Nevertheless, the consumer modernity which the USA pioneered continues to dominate as the prime modernity. And American corporations continue to be prominent in this ever-consuming world. With China attempting to add another billion consumers to the great world shopping mall, the question arises as to how our geohistorical interpretation can feed into a geofuturistic prediction. The omens are not good: is the Earth big enough for more consumer modernity? But that is another story–and–map . . .

Further reading

Additional historical background on the hegemony–modernity link can be found in:

- Taylor, P.J. (1996) *The way the modern world works: world hegemony to world impasse*. Chichester and New York: Wiley.

The geohistorical interpretation is drawn at more length in:
- Taylor, P. J. (1998) *Modernities: a geohistorical interpretation*. Cambridge: Polity.

For a reasoned evaluation of 'modernization' influences in the Third World (specifically Africa) see:
- Davidson, B. (1992) *The black man's burden*. New York: Times Books.

The modern world-system as a framework for analysis is laid out in:
- Wallerstein, I. (1979) *The capitalist world-economy*. Cambridge: Cambridge University Press.

Finally, to investigate further the future growth of consumer modernity a good place to start is:
- Zhao, B. (1997) Consumerism, Confucianism, Communism: making sense of China today. *New Left Review* **222**, 43–59.

Memory and heritage

Nuala C. Johnson

INTRODUCTION

The wreaths have been removed. The ceremonies are over. As Remembrance Sunday passes once again the statue to fallen soldiers who were members of Queen's University in Belfast returns to its slightly invisible status as a memorial icon. During the days surrounding the 11 November each year this monument, like many others around the country, assumes renewed significance in the maintenance and cultivation of a public memory dedicated to Ulster soldiers killed in the First World War and subsequent conflicts. This spectacle of remembrance amplifies Nijinsky's response to the Great War: 'Now I will dance you the war, with its suffering, with its destruction, with its death' (cited in Eksteins, 1990: 273). This dance takes the form of a seasonal ritual of wreath-laying, memorial services held in the churches and the wearing of the poppy. All play a central role in the memory-work rehearsed annually in cities and towns across the United Kingdom. This memory-work, however, is not confined to the spaces of commemoration associated with the First World War but forms part of a larger network of sites which connect history, heritage and the geographies of identity in contemporary life.

There has been a huge proliferation in the number of such heritage spaces over the past 20 years coinciding with the expansion of tourist activity world-wide. Experiencing a growth rate of 5 to 6 per cent per annum tourism is expected to become the largest source of employment by the twenty-first century (Williams and Shaw, 1988). But tourism involves not only commercial transactions, it is as McCannell (1992: 1) notes 'an ideological framing of history, nature and tradition; a framing that has the power to

reshape culture and nature to its own needs'. Museums, stately homes, heritage centres, folk parks, memorials and the myriad of other sites designed to convey historical and geographical knowledge emphasize the ways in which our efforts to represent and remember the past are mediated through complex and sometimes contradictory lenses. The following discussion will underline the contention that the relationships between heritage and history, between tradition and **modernity** continue to play a significant role in contemporary society. The heritage site itself frequently forms the epicentre upon which these issues are scrutinized. If histories are constructed and memories are mapped on to the past, the manner in which these stories and recollections of the past are related is constantly open to contestation, to alternative renderings of history and to the spaces in which histories are mediated and interpreted. This chapter has two main objectives: first, I will briefly outline some of the key themes anchoring discussions about the relationship between history and heritage, second, I will examine in detail two particular sites of memory – a stately home in Ireland and the death camp at Auschwitz, Poland – as exemplars of the contested spaces in which history is materially and metaphorically translated to the public.

WHAT SHOULD WE MAKE OF HERITAGE?

The relationship between history, heritage and memory has been subject to much debate recently among geographers, historians, cultural critics and others. Conventionally a rigid line of demarcation ran between the past as narrated by professional historians, on the one hand, and by the heritage industry, on the other. Heritage, as a concept, begins with a highly individualized notion of what we either personally inherit or bequeath (e.g. through family wills and legacies). We are more concerned, however, with collective notions of heritage which link us as a group to a shared inheritance. The basis of that group identification varies in time and in space. It can, for instance, be based on allegiance derived from a communal religious tradition or a class formation or a 'nation'. Indeed, it is with respect to cultivating the 'imagined community' of nationhood that heritage is often most frequently linked. Three different, albeit interrelated, approaches to understanding heritage have gained currency in recent years. Briefly these comprise the view that (a) heritage is a form of inauthentic history; (b) heritage is primarily part of a process of tourism expansion and **post-modern** patterns of consumption; and finally (c) heritage is a contemporary manifestation of a longer historical process whereby human societies actively cultivate a social memory. The next few paragraphs will deal with each approach in turn.

While the **nation–state's** origins may be relatively recent, the national state is based on the assumption that this group identity derives from a collective cultural inheritance that spans centuries. As Benedict Anderson (1983: 15) has put it, nations are collectively imagined because 'members of even the smallest nation will never know most of their fellow-members . . . yet in the minds of each lives the image of their communion'. And that communion is traditionally conceived as historical. National states therefore attempt to maintain this identity through highlighting the historical trajectory of the cultural group through preservation of elements of the built environment, through spectacle and parade, through art and craft, through museum and monument. The heritage industry, then, is viewed as a mechanism for reinscribing nationalist narratives in the popular imagination (Wright, 1985). Lowenthal (1994: 43) claims that 'heritage distils the past into icons of identity, bonding us with precursors and progenitors, with our own earlier selves, and with promised successors'. As such, then, the historical narratives expressed through heritage are seen to be partial, selective and distorting. They offer us a 'bogus' history which sanitizes the past and ignores complex historical processes. This contrasts with the work of professional historians where 'testable truth is [the] chief hallmark' and where 'historians' credibility depends on their sources being open to general scrutiny' (Lowenthal 1996: 120).

The expansion in the number of heritage sites over recent years has also been examined as exemplary of the **post-modern** cultural forms associated with **post-industrialism** and late twentieth-century tourism. According to Urry (1990: 82) 'postmodernism involves a dissolving of boundaries, not only between high and low cultures, but also between different cultural forms, such as tourism, art, music, sport, shopping and architecture'. Consequently, the distinction between representations and reality, between genuine history and false heritage is made problematic. Baudrillard (1988) suggests that signs are all that we consume and that we do so knowingly. These signs representing the past can be found in history textbooks as much as at heritage sites. Heritage tourism, however, is seen as 'prefiguratively' postmodern because it has long privileged the visual, the performative and the spectacular for popular consumption. In this portrait, therefore, the past that is mediated through heritage is just one element, albeit an increasingly important one, in a whole suite of historical representations.

The links between memory and heritage are also important. While the early history of memory is relatively undocumented, the historian Nora (1989) suggests that before the nineteenth century memory was such a part of the practices of everyday life through storytelling that people were hardly aware of its existence. While elite classes (e.g. the church or aristocracy) had an institutionalized memory preserved through archives and biographies, ordinary people neither recorded nor objectified their past. This latter type of memory had taken 'refuge in gestures and habits, in skills passed down by unspoken traditions, in unstudied reflexes and ingrained memories' (Nora 1989: 13). In contrast, from the nineteenth century onwards modern memory became more democratized and it became self-consciously preserved and archival. In this light, rather than viewing heritage as a false, distorted history imposed on the masses, we can view heritage sites as forming one link in a chain of popular memory. Bearing these points in mind I will now turn to analysing the first of my examples – a stately home – which is often treated as a representation of an exclusively elite past.

Summary

- There has been a long-standing debate about the authenticity of the historical narratives offered at heritage sites. It has frequently been suggested that heritage is merely a form of bogus history.

- Recent writing about heritage suggests that it is part of a much larger cultural transformation associated with **postmodernism**.

- Recognizing the evolution of a modern collective memory in the nineteenth century contributes to blurring the distinction between heritage as a site of history or as a site of memory.

ELITE LANDSCAPES: FROM COUNTRY HOUSE TO OUR HOUSE

Tours of stately homes conventionally focus on the architecture and design of the house, accompanied by a potted history of the owners, details of the interior design, furnishings and art work contained in the house and a short commentary on the design of the **demesne**. Frequently presented from the viewpoint of the ruling class, preservation of country homes has been encouraged as they are seen to represent the prestige of a community's past or of its most successful landowners. In a British context, the country house and its estate have been linked to the evolution of a particularly English landscape tradition and English historical identity and thus have been preserved especially by the National Trust. Lowenthal (1991: 220) claims, however, that often 'the country house door – along with the countryside itself – is kept firmly shut'. The case I want to use here from Ireland, however, is rather different.

After the foundation of the Irish Free State in 1922, many houses of the gentry were either destroyed, abandoned or neglected. Not regarded as either an architectural or cultural icon to be preserved 'the destruction of the Big House was an ideal means through which the Free State could symbolically be seen to break with the past' (Dodd, 1992: 10). In the early years of Irish independence the stately home was viewed as a representation of the colonizer's **cultural landscape** and thus unworthy of state aid or public memory.

In recent decades some 'Big Houses' have been preserved by the state and independent trusts. The example of Strokestown Park House, County Roscommon, opened to the public in 1987 illustrates how the past can be represented multivocally, that is, from a range of positions. Located 90 miles from Dublin, the house is presented in its local geographical context but is also connected to regional, national and international historical geographies. In their analysis of museum culture Sherman and Rogoff (1994) suggest four conceptual keystones in the arch of museum politics and practices which will help to throw light on how representation works at Strokestown. First, museums are comprised of a series of objects, which are ordered and classified in a specific sequence to offer a coherent meaning to the display. Second, these sequences of objects are woven into an external narrative which may relate, for instance, to local history, class relations or the nation. Third, museums are designed to serve a specified public and exhibits are structured to disclose the story to that public. Finally, the audience's response to a display becomes an integral part of the design process.

The house at Strokestown was built on lands granted to a certain Nicholas Mahon in the 1650s. The house itself was built in the 1740s in the Palladian style, with the wings adjoining the central block being added later (*see* Fig. 18.1). At the beginning of the

Figure 18.1 Strokestown Park House, Co. Roscommon. Source: Strokestown Park House

nineteenth century the Mahons were the most significant gentry family in the area with an estate exceeding 30,000 acres. The owners of Strokestown established a planned estate village adjacent to the demesne. Its exceptionally wide main street underlines a principle of linearity popular among the gentry of the time, with an Anglican church at the western end of the town and the Georgian Gothic triple arch at the eastern end, forming the entrance to Strokestown Park House. While the house serves as the centre for the estate's public display, the morphology of the demesne and town constitutes the backdrop to the discussion of Strokestown's past (*see* Fig. 18.2).

The house at Strokestown is presented to the public through a guided tour lasting approximately 45 minutes. John Urry suggests (1990: 112) that

> heritage history is distorted because of the predominant emphasis on visualisation, on presenting visitors with an array of artefacts, including buildings (either 'real' or 'manufactured'), and then trying to visualise the patterns of life that would have emerged in them.

At Strokestown, however, the visual and the verbal are united into a coherent narrative where the 'tour is structured to use the house as a vehicle to explain social history' (Dodd, 1993). This is made possible by the availability of detailed records on the house's management and also by the internal geography of the house itself.

The tour visits the main reception rooms of the house, the first-floor living quarters and the kitchen. While the architecture of the house dictates, to some extent, the sequence of the tour, it is also arranged according to the type of history it seeks to tell. A typical tour is arranged in four parts: (a) economic and architectural history of the early estate; (b) the house during the years of the Great Famine; (c) gender relations and family history; (d) social relations between the gentry and the servant classes.

The tour begins with a discussion of the early acquisition and architectural evolution of the house. In the main reception room and ballroom the early economic history of the estate, the injection of new money into the estate in the 1800s and the pastimes of

Figure 18.2 An 1837 six inch to one mile Ordnance Survey map of Strokestown and the estate. The town has a cruciform design with an expansive main street. The house is located at the eastern side of the main street and the parkland surrounding the house conveys a sense of the naturalistic style of landscape planning popular among the Irish gentry

the owners are emphasized. The spatial division of labour between the landlord and servant classes is highlighted through a discussion of the invisible underground passageways built to disguise the routes taken by the servants in the administration of the house and demesne. In the study of the house the guides offer an extensive discussion of the role of Strokestown House during the time of the Famine. Thanks to the large volume of archival papers dealing with this period, the tour reconstructs the role of Denis Mahon (the landlord) in the administration of his estate during these years and tallies up the effects of the Famine which shrunk the estate's population of around 11,000 people by about 88 per cent. While much of the literature on heritage tourism focuses on the authenticity of past narratives and the tendency for popular histories to sanitize the past, in Strokestown the tour narrative presents the Famine as a critical moment in nineteenth-century Irish history, but the story is contextualized in the local geographical setting, an area severely affected by the potato blight of the 1840s. At this juncture in the narrative the local, national and international are interwoven. Equally impressive is the tour's handling of the *contested* nature of historical interpretation. In the case of

the assassination of the landlord in 1847 the guides offer several different documented versions of his death, which amplifies for the audience the equivocal nature of historical evidence and it illustrates how interpretation can be coloured by systems of belief.

In the upper floors of the house, comprising the family's quarters (bedrooms, children's playroom and school room), the narrative shifts to the themes of gender relations among the gentry, child–parent relationships and the spaces occupied by children. The tone of the tour is lighter here also, where objects move in to the foreground. For instance, the role of the governess/tutor in the social relations of a mansion of this type is discussed.

The final section of the tour is set in the dining room and galleried kitchen. The spatial and social distances maintained between the servants and landlords are reinforced through a discussion of the architectural practices and spaces occupied by each group. The galleried kitchen (the only remaining one in Irish country houses) serves as a poignant social metaphor for the hierarchical social relations cultivated through a system of domestic management (*see* Fig 18.3). The adjoining subterranean passageways ensured the invisibility of the servant class as they car-

Figure 18.3 The galleried kitchen at Strokestown Park House. The balustraded gallery allowed the lady of the house to deliver the daily menu without having to directly enter the space occupied by the domestic kitchen staff. Source: Strokestown Park House

ried out their daily duties. Ironically it is the very existence of these passageways today which enables visitors to visualize those people whom the landlords sought to hide from public view. The kitchen was the servant class's central demesne, linked geographically to the house but socially separated from it (Johnson, 1996).

So what is the relationship between heritage, memory and history as it is represented at Strokestown Park House? The presentation differs from other sites in several important ways. First, the house is currently occupied and thus it is not presented as solely a window to the past. Second, the house does not adopt a 'museumification' approach to preservation. There are no barriers or warning signings in the house. Visitors can freely touch objects and even though there are many items on display of high monetary value, these are not presented in a way that distances them from the viewer. Similarly, unlike many other country houses which are approached through the back door or some discreet entrance, the visitors to Strokestown enter the house through the main front door. This avoids 'perpetuating the class division that [the house] was made to represent' (Dodd, 1993). Visitors are encouraged to see the house as part of their own past, one in which their ancestors may have played an active role. Moreover the vocabulary used in the tour eschews an elite perspective by minimizing reference to the minutiae of Palladian architecture and portraiture.

Summary

- The representation of the past at stately homes has often been concerned with providing an elite perspective on the role and function of landed estates. It is not, however, inevitable that country houses open to the public offer an elite view of history.

- At Strokestown Park House the history of the estate is located in its geographical setting and the meanings are derived from an interplay of object, narrative and audience.

LANDSCAPES OF DEATH: REMEMBERING THE HOLOCAUST

Memory is blind to all but the groups it binds. History, on the other hand belongs to everyone and to no one, whence its claim to universal authority.

(Pierre Nora, 1989)

The silent landscapes of genocide perhaps present the most difficult task for European society in general and German society in particular to remember and to record. This second example focuses our attention on the representational, political and moral questions raised by attempting to preserve sites of mass human destruction for popular consumption. The term genocide was first coined in 1942 and after the 1960s the term 'holocaust' was adopted to describe the mass murder of European Jewry. It is worth noting, however, that attempts at premeditated extermination of a particular ethnic group did not originate with the Second World War. In the twentieth century alone the Great War had witnessed the mass murder of Armenians by Turks in Ottoman territory. The fact that this genocide receives less attention both in the canon of scholarly work and in public memory reminds us of the fragile processes involved in the making of history and memory. Indeed, the distinction between memory and history becomes intensely blurred when confronted with the challenge of representing large-scale murder. While history texts can describe, document, diagnose Nazi policy and practice towards Jews and these insights can be incorporated into school curricula and university programmes, there is a danger that the holocaust is then consigned solely to the past to be accessed through the work of professional historians, and to be represented as an aberration of the human spirit or an iniquity bred from geopolitical and eugenicist imperatives. But of course, genocide does not literally take place in the pages of a textbook or in the files of Nazi Chiefs of Staff, but in concentration camps scattered throughout Germany and Poland. Memory, then, is not just a recollection of times past but it is an intense recollection of spaces past – spaces of mass death and destruction – where history is writ large on the landscape. These sites of memory have been subject to vociferous debate over the last decades and have been central to discussions about how to represent genocide. While the landmarks of destruction – the death camps – seem to reveal a timeless property, their meaning is continuously bound up with new cultural and political demands on memory.

Between 1939 and 1945, 3.2 million of Poland's 3.5 million Jews were killed by Nazis and their sympathizers. This extermination took place primarily in the six Polish death camps established under the National Socialists (see Fig. 18.4). The surviving remnants of Jews largely evacuated Poland in 1946 due to post-war Polish pogroms. Three million non-Jewish Poles were also killed during the war and half of Poland's population was either killed, wounded or imprisoned during the conflict (Young, 1993). Consequently, for both Jews and Poles, the war represents an episode of unprecedented suffering, a fact which has both a unifying and divisive effect on the

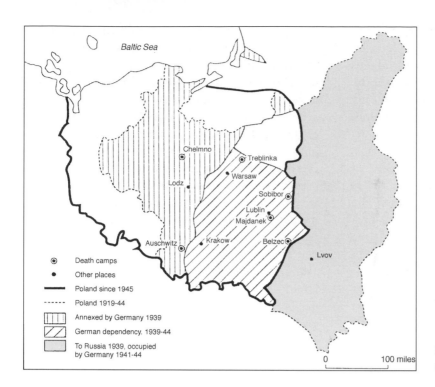

Figure 18.4 The location of the main Nazi death camps. Credit: Pion

conjugation of a collective memory. Although the ruins of some of the death camps survived after the war it is their conversion from 'death' camps *per se* to 'memorial' camps which highlights the contested nature of public memory and popular interpretations of the past. Of the six death camps in Poland, only two – Majdanek and Auschwitz – had large portions of the complexes left intact after the war. Both were camps where Jews, Poles, Romani, Sinti and Soviet prisoners of war had suffered. Both were turned into memorials and museums, although Auschwitz is commonly regarded as the central symbolic site of the Holocaust. In each case 'Guard towers, barbed wire, barracks, and crematoria – mythologized elsewhere – here stand palpably intact' (Young, 1993: 120).

In 1947 the Polish Parliament, dominated by a Soviet-style Communist Party chose Auschwitz as the site which best represented the 'Nazi occupation from which the Soviet Union had liberated Poles' (Charlesworth, 1994: 581). In an attempt to combine memorialization of Jews and Poles, and to stress the role of the Soviet Union in defeating fascism, the site was seen as significant in nurturing Polish–Soviet relationships. Located closer to the German border than other death camps it underscored the Soviets wish to situate Polish suffering solely in the context of German aggression. Second, although it is thought that 1.6 million people were killed at the camp and that 87 per cent of them were Jews, it turns out that less than one-third were Polish Jews. The significance of this fact is that the camp came to be represented as a site for international genocide which would link

Poland to other Warsaw Pact countries (Charlesworth, 1994). The Polish Parliament declared that the camp would be 'forever preserved as a memorial to the martyrdom of the Polish nation and other peoples' (cited in Young, 1993: 130). Cold War geopolitical considerations were important in the maintenance of specific sites of memory.

Auschwitz comprises three main camps: Auschwitz I which housed the administrative headquarters and a concentration camp (with one gas chamber/crematorium); Auschwitz II, also known as Birkenau, built as a death camp for the annihilation of Jews but also containing a concentration camp; and Auschwitz III, primarily a camp for prisoners working in the nearby industrial plant. The camp has been memorialized through a variety of strategies: the conversion of barracks into national pavilions documenting the deaths of different groups (e.g. Belgians); museum spaces; sculptures of remembrance; and exhibits of artefacts of the deceased which include piles of suitcases, human hair (shorn from women), spectacles, and prosthetic limbs belonging to Jews. It is this accumulation of artefacts that many tourists remember most clearly about their visit to the camp (*see* Fig. 18.5). It is also the **representation** of mass extermination through the paraphenalia of those on the eve of their death that raises questions about the types of history that the camp seeks to elucidate. The lives of those killed are literally represented as the clothes on their backs. Young (1993: 132) suggests that 'these remnants rise in a macabre dance of memorial ghosts ... victims are known by their absence, by the

Figure 18.5 Tourists visiting a site of memory at Auschwitz, Poland. Credit: Popperfoto/Reuters

moment of their destruction'. The difficulty lies in the fact that the varied and complex histories of those brought to Auschwitz are reduced to the collective geography of their death in the camp. Little attention, therefore, is devoted to the lives they had led – culturally, politically, spiritually – before the war. The complexity of their lives is masked by a representation which emphasizes the clinical manner in which their executions were orchestrated. While it is important that the manner of their murder is visible, it is equally important that the roots and scale of European anti-Semitism be exposed.

There is also a blurring of the distinctions between those that were killed at Auschwitz. While one may legitimately argue that each life taken at the camp is of equal value and worthy of equal inclusion in the narrative, there is a suggestion that an emphasis on Polish suffering under Nazi occupation undermines the specificity of the Jewish experience at Auschwitz. Charlesworth (1994) suggests that the site allows for a particularly Polish reading of the holocaust which connects with a nationalism that focuses on Polish victimization by its neighbours over the previous centuries. The remit for the establishment of the museum in 1947 reads that 'a monument of the martyrdom of the Polish nation and of other nations is to be erected' (cited in Charlesworth, 1994: 584). The absence of any direct reference to Jews is revealing. More recently there have been efforts to Catholicize Auschwitz by the activities of Cardinal Karol Wojtyla (now Pope John Paul II). Through the beatification of Father Kolbe, a rabidly anti-Semitic priest who had exchanged his life for a Polish prisoner at Auschwitz, Cardinal Wojtyla suggested that the martyrdom of this priest should be memorialized at Auschwitz through

the erection of a church (Young, 1993). The establishment of a Carmelite convent inside the camp in 1984 complicates the issue further and illustrates the *contested* nature of the spaces of memory at Auschwitz. That Jews, the church, the state, Polish nationalists, professional historians and the international community all have had influence on how the holocaust is represented at Auschwitz underlines the contention that memory-work is rarely an act of mimesis, that is, imitation, but is woven through complex cultural, political and symbolic processes.

Summary

- Genocide highlights the complex issues involved in creating, maintaining and representing sites of memory.

- The reading of history through a geographical space such as a death camp can simplify the complex lives of those exterminated by focusing primarily on the manner of their death.

- The case of the death camp at Auschwitz underscores how heritage spaces can be sites of contested memory. The distinctions between different groups of people killed at Auschwitz can become blurred and this raises the question of whose past is being recorded and remembered.

CONCLUSION: SPACES OF HISTORY OR SPACES OF MEMORY?

Geographers are increasingly concerned with the representation of landscapes and how the past can be read through landscape interpretation. Apart from a few exceptions, however, geographers have paid scant attention to the historical geography of heritage landscapes and how they represent the past. Criticisms of the heritage industry's attempts to narrate the past as little more than bogus history are often overdrawn and in the case of Strokestown Park House I have attempted to highlight how the past can be explored provocatively through a heritage landscape without diminishing its popular appeal. Rather than focusing on whether heritage conveys inaccurate history, the more interesting questions for geographers relate to examining the ways in which the spaces of heritage translate complex cultural, political and symbolic processes into the popular imagination. Through the example of Auschwitz, a provocative case certainly, an analysis of the site reveals how these processes are intermixed in complex ways and it exemplifies the moral dilemmas that are thrown up by the seemingly contradictory desires to remember and forget the past.

Further reading

For the most comprehensive overview of the distinction between history and heritage by a geographer read:

- Lowenthal, D. (1996) *The heritage crusade and the spoils of history*. London: Viking.

For a general introduction to tourism, postmodernism and heritage in Britain and North America respectively, see:

- Urry, J. (1990) *The tourist gaze*. London: Sage, Chapters 1, 2.
- McCannell, D. (1992) *Empty meeting grounds: the tourist papers*. London: Routledge, Chapters 1, 4.

The issue of remembering the Holocaust is dealt with in:

- Young, J. (1993) *The texture of memory: holocaust memorials and meaning*. London: Yale University Press.

Political geographies

INTRODUCTION

The statement that the 'personal is political' has achieved quite widespread currency since it was first used as part of the feminist movement in the 1960s. It neatly sums up the idea that politics is to be found in each and every aspect of our daily lives, and is not restricted to the more formal machinery of parliament and government. The latter concern with formal politics dominated political geography for many years and led to an emphasis on a seemingly distant and specialized sphere of activity to do with political parties, elections, governments and public policy. Although there was the notion that everyday life was affected by such processes, there was little conception that politics was actually part of our day-to-day lives. It was something that was carried out by other people (politicians and civil servants) and which went on elsewhere (in government institutions).

More recently, however, this view has changed in two key ways. First, there has been a realization that the formal politics of government and the state have much more impact on our daily lives than hitherto thought and, second there has been a far broader examination of a whole host of informal politics – taking place in the home, in the workplace in the street and in the community. The three chapters in this section all examine these twin developments, but do so at different levels of enquiry. Initially, in Chapter 19, Joanne Sharp takes a global perspective by concentrating on the issue of geopolitics or the relations between states. She shows how these relations are not fixed, but instead are fluid and always specific to particular historical and cultural circumstances. She then looks at how the geography of international relations is constructed through particular sets of discourses, and at how even these very dis-

tant formal politics are connected to our everyday lives via elements of popular culture such as Hollywood movies. These elements in turn help to shape the 'geographical imaginations' that we all hold of the world and its political interrelationships at a global level.

In Chapter 20, Mark Goodwin examines the twin concepts of governance and citizenship, and in doing so moves the scale of analysis down to the more local level. He shows how the two themes are related and how together they cover the issues of what it means to be a citizen, of what rights and obligations one has as a citizen and of the ways in which we as citizens are governed. The chapter points out how both citizenship and governance are experienced differently by different social groups, and examines how these divisions create differential spaces where different degrees of citizenship and governance are experienced and played out. Crucially, these differential geographies have become critical to the building and sustaining of new forms of participatory citizenship around localized community issues.

In Chapter 21 Pyrs Gruffudd helps to show that the scale of analysis is also fluid by looking at the contested notions of the 'nation' and nationalism. He shows that what for some people is only a region within a bigger entity, is to others a fully fledged 'nation'. Often the tension between these two views boils over into war and turmoil, as witnessed by recent events in the former Yugoslavia. Pyrs looks at how the symbolic role of geography is hugely implicated in the very construction of national identity and nationalist movements, which are often centred on a struggle for land and territory.

Taken together, these chapters provide excellent examples of the movement of political geography away from a concentration on formal politics to an exploration of the myriad spaces of informal politics. Yet they also show the continued importance of the formal political sphere. The challenge for the future, perhaps, is to examine these two spheres together, by looking at how each affects the shaping of the other, rather than continuing to see them as separate. Each chapter gives a pointer to the very exciting areas of enquiry which emerge if this is done.

Further reading

Two excellent and very readable textbooks on political geography are:

- Painter, J. (1995) *Politics, geography and political geography.* London: Arnold.
- Taylor, P. (1993, 3rd edition) *Political geography: world-economy, nation-state and locality.* London: Longman.

A good collection of readings which span a wide range of subject material can be found in:

- Agnew, J. (ed.) (1997) *Political geography: a reader.* London: Arnold.

For the most up to date writings on economic geography you should scan the recent editions of the journal *Political Geography*. This will give you a sense of the current topics being pursued in this field.

Critical geopolitics

Joanne P. Sharp

INTRODUCTION

For many people, the downfall of the Soviet Union, symbolised so dramatically by the destruction of the Berlin Wall and the toppling of statues of communist leaders, heralded a new world order of peace and stability after the fear and anxiety of the **Cold War**. The moral global battle of the previous half century – the ultimate battle between freedom and the 'Evil Empire' as it was often put – was over. Images of world leaders shaking hands at global summits seemed to offer an image of a unified world (*see* Fig. 19.1). Rather than seeing images of stability, however, many commentators quickly started to sense danger lurking in this New World Order. The multipolarity of the post-Cold War order seemed to offer the potential for greater dangers to world peace. One commentator suggested that we would soon miss the stability of the superpower opposition of the Cold War (Mearsheimer, 1990), while another suggested that the emerging world 'is likely to lack the clarity and stability of the Cold War and to be a more jungle-like world of multiple dangers, hidden traps, unpleasant surprises and moral ambiguities' (Huntingdon, quoted in Ó Tuathail, 1996: 242). Such challenges were to come from the emergent nationalist loyalties to which people would turn in the vacuum left by the fall of the Cold War (in Rwanda or the Balkans, for example), from heightened inter-regional economic competition (especially the threat to US economic hegemony from Japan), and from 'rogue states' refusing to accept the rules of the world order (such as Iraq during the Gulf Crisis). Rather than seeing each of these phenomena as resulting from specific historical or cultural contexts, they were seen to emerge from changes to the global political system.

Figure 19.1 A world united? Gorbachev and Reagan shaking hands at world summit. Credit: Associated Press

Each of these models of the contemporary world order is structured by a geographical imagination of some sort. During the Cold War the Free World was pitted against the Communist World (if you happened to live in the USA) with maps geared towards proving the superiority of the Soviet Union and her allies, and demonstrating the inevitability of countries falling under Communism like dominoes unless the Soviet Union was contained behind the Iron Curtain. This political geography has moved into history and now seems somewhat quaint and rather absurd in its crude moral geography of American freedom versus the 'Evil Empire'. This does not mean that we are now without political geographies. The theorists mentioned above have offered new visions of a globe of chaos and fragmentation offering threats and dangers from all around. The use of geographical imaginaries in global political models is called 'geopolitics'.

The *Dictionary of human geography* defines geopolitics as an element of the practice and analysis of statecraft which considers geography and spatial relations to play a significant role in the constitution of international politics. Certain 'laws' of geography, such as distance, proximity and location are understood to influence the development of political situations. In geopolitical arguments, the effect of geography on politics is based upon 'common sense', rather than ideology: the 'facts' of geography are seen to have predictable influences upon political processes.

However, recently certain authors have challenged such arguments about the political innocence of geography to suggest that rather than being a timeless concept, geographical relationships and entities are specific to historical and cultural circumstances: the nature of the influence of geography on political events can change. Given this, the meaning of geography can be *made to* change: there is a politics to the use of geographical concepts in arguments about international relations. After a brief introduction to traditional geopolitical concepts, this chapter will explore the alternative political arguments of 'critical geopolitics'. This chapter will use the example of the Cold War to explore the uses and critiques of geopolitical concepts.

THE GEOPOLITICAL TRADITION: GEOGRAPHY AS AN AID TO STATECRAFT

The term 'geopolitics' was first used by the Swedish political scientist Rudolf Kjellen in 1899, but did not become popular until used in the early twentieth century by British geographer and strategist Halford Mackinder. Mackinder wanted to promote the study of geography as an 'aid to statecraft', and he believed

that geopolitics offered one such way in which geographers could inform the practices of international relations. The study of geopolitics focused upon the ways in which geographical factors shaped the character of international politics. These geographical factors included the spatial layout of continental masses and the distribution of physical and human resources. As a result of geography, certain spaces are seen as either easier or harder to defend, distance has effects on politics (proximity leading to susceptibility to political influence), and certain topographical features promote security or lead to vulnerability.

The concept of security is fundamental to the study of geopolitics. This refers to the maintenance of the state in the face of threats, usually from external powers. Geopoliticians argue that they can aid national security by explaining the effects of a country's geography, and that of potential conquerors on future power–political relations. A student of geopolitics claims to be able to predict which areas could strengthen a state, helping it to rise to prominence, and which might leave it vulnerable. An oft-quoted line from Nicholas Spykman illustrates the necessity of a geopolitical vision by insisting that 'geography is the most important factor in international relations because it is the most permanent' (quoted in Nijman, 1994: 222). As a result, geopolitics has traditionally been considered to be a very practical and objective study: the actual *practice* of international relations has been seen to be quite separate from political theory.

One central feature of geopolitical reasoning is that it presents the world as one closed and interdependent system. It is perhaps not accidental that the rise of geopolitics as a way of understanding the world occurred at a time when global space was 'closing', the entire world was now fully explored by Western colonists and imperialists so that it was now all available for state territorial and economic expansion (see Agnew, 1998). European **colonialism** had reached its height. Geopolitics offered a way states could protect territorial holdings at a time when the 'blank spaces' on the world map were finally all filled in by European powers.

Mackinder's best-known geopolitical argument is presented in his 'Heartland Thesis' which insisted upon the importance of the Asian Heartland to the unfolding history of great powers (*see* Fig. 19.2). Mackinder believed that controlling the territory of the Heartland provided a more or less impenetrable position and could thus lead to world domination. For Mackinder, unless checked by power in the 'outer rim' of territory proximate to the Heartland, the occupying power could quite easily come to control first Europe and then the world. In 1919 Mackinder famously stated that

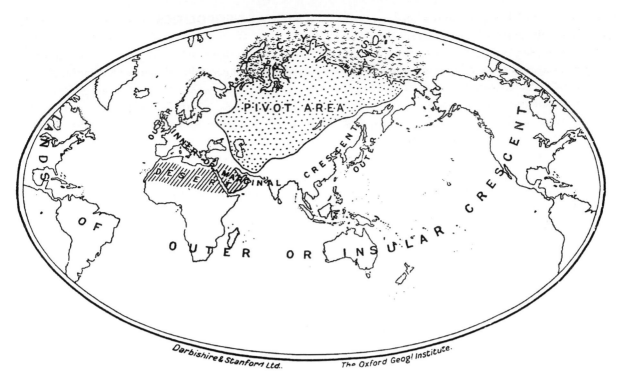

Figure 19.2 Mackinder's Heartland map

Who rules East Europe commands the Heartland;
Who rules the Heartland commands the World Island;
Who controls the World Island commands the World.

(quoted in Glassner, 1993: 226–7)

His conclusion was that British statesmen would need to be wary of powers occupying the Heartland, and should create a 'buffer zone' around the Heartland to prevent the further accumulation of power that might challenge the hegemony of the British Empire.

Such geopolitical reasoning was heady stuff indeed and there is evidence that it has both influenced foreign policy and the popular imagination. However, despite geopoliticians' insistence upon their geographical laws, their conclusions to the location of power differed. For example, whereas Mackinder promoted the power of territory, American strategist Mahan viewed control of the sea as paramount, and later others highlighted the importance of air power. Each came up with different core areas from which political dominance could be exercised.

Associations with Nazi expansionist *Geopolitik* policies (also inspired by Kjellen's work), meant that geopolitics, expressed formally in formal spatial models, fell out of use. Models still remained in textbooks, however, and were periodically updated to keep up to date with changing technology (especially the dominance of airpower and the introduction of inter-continental ballistic missiles). More significantly, a form of implicit geopolitical reasoning persisted in international relations theory and state practice throughout the Cold War.

COLD WAR GEOPOLITICS

One of the formative documents of the Cold War was sent as a telegram from Moscow, by George Kennan – 'Mr X' – a US official in the Soviet Union at the end of the Second World War. Kennan argued that the Soviet Union was so different from the USA that there could not be compromise between the two. This image of two distinct and incompatible territorial blocks was reinforced by the political rhetoric of various political figures: in Stalin's pronouncements of the threat of capitalist expansion, Churchill's image of an Iron Curtain dividing Europe, and more recently, Reagan's depiction of the Soviet Union as an Evil Empire. A number of inter-related geopolitical concepts reiterated this binary geography in political discourse. These were, most importantly, containment, domino effects, and disease metaphors.

Containment, first outlined by Kennan, referred to the military and economic sequestration of the Soviet Union. Russia's historical geography, and not simply its political and cultural difference, was invoked to give this argument scientific respectability: the USSR was seen as an inherently expansionary force which had to be kept in check. Pietz suggests that in Cold War rhetoric, the USSR was presented as nothing more than traditional Oriental despotism plus modern police state technology (Pietz, 1988: 70).

The inevitability of Soviet expansion was also

expressed in metaphors of dominoes or disease. Such metaphors saw the spread of Communism or socialism not as a complex political process of adaptation and conflict but instead merely as a result of proximity to territory ruled by Soviets. The Domino Theory assumed that Soviets, Communists and socialists everywhere 'were, and are, unqualifiedly evil, that they were fiendishly clever, and that any small victory by them would automatically lead to many more' (Glassner, 1993: 239). For US Admiral Arthur Redford speaking in 1953, for example, an American nuclear strike on Vietnam was essential in order to halt a Viet Minh victory which would set off a chain reaction of countries falling to the Communists, 'like a row of falling dominoes' (in Glassner, 1993: 239). The Domino Effect can actually be seen to underlie the Vietnam war more generally. As Glassner (ibid.: 241) put it: 'The argument went that the United States had to fight and win in Vietnam, for if South Vietnam "went communist," then automatically, like falling dominoes, Cambodia, Laos, Thailand, Burma, and perhaps India would as well.' This process would not stop until it reached the last standing domino, the USA, and made future political action appear inevitable, unless proactive action – such as containment or pre-emptive strike – were enacted here and now.

The Domino metaphor simultaneously embodied a power political system where only two forces existed (the USSR and the USA), where only force could oppose force and where the unfolding of the process was inevitable – once started, the continuing fall of states was as unavoidable as stopping a line of dominoes from toppling once the first had been pushed. Disease metaphors were structurally very similar, relying upon notions of contagion or the malign spread of infection, again depending upon a simple notion of geographical proximity as the basis for social and political change. Even more so than with dominoes, disease metaphors illustrated the necessity for immediate action in order to prevent the further spread of the malady.

Summary

- Traditional geopolitical models explain the effects of geography on international relations.

- Especially important was the question of territorial security and the danger of proximity to territory ruled by an opposing power.

- Traditional geopolitics were seen to represent a very practical application of geographical 'laws' to understanding international politics.

CRITICAL GEOPOLITICS

Some recent approaches to geopolitics, sometimes called 'critical geopolitics', refuse to accept the objectivity and timelessness of the effects of geography on political process. Critical geopolitics encompasses a range of engagements with more traditional forms of geopolitics. Some have highlighted geopoliticians' over-emphasis on the state as the main, or only, actor in international politics. Clearly other powers are involved both at the sub-state level such as ethnic, regional and place-based groups, and at the supra-state level, such as transnational corporations and international organizations including the UN and NATO.

Critical geopolitics has been especially interested in questioning the language of geopolitics, or 'geopolitical **discourse**'. Language is not unproblematic, somehow *simply* describing what is there. Language is metaphorical, explaining through reference to other, known, concepts. Thus, there is always a choice of words and metaphors. The type of terms used – the conceptual links made – affects the meaning of what is being described. There is, as a consequence, a politics of language.

Geopolitical discourse

Critical geopolitics is influenced by postmodern concerns with the politics of representation, with 'the use of particular modes of discourse in political situations in ways that shape political practices' (Dalby, 1990: 5). To Gearoid Ó Tuathail (1996: 1), geography is not a collection of incontrovertible facts but is instead about power. What he means is that geography is not an order or facts and relationships 'out there' in the world awaiting description. Instead, geographical orders are created by key individuals and institutions and then imposed upon the world. Geography is thus the product of cultural context and political motivation.

Critical geopolitical approaches seek to examine how it is that international politics are imagined spatially or geographically and in so doing to uncover the politics involved in writing the geography of global space. Ó Tuathail (1996) calls this process 'geo-graphing' – earth-writing – to emphasize the creativity inherent in the process of using geographical reasoning in the practical service of power. Those adopting critical geopolitical approaches grant a range of power to language, from Agnew and Corbridge (1989) who see language becoming out of synch with the geopolitical reality it seeks to describe and so causing inappropriate state practice, to a figure like French philosopher Jean Baudrillard for whom language and representation are everything (he famously suggested that the Gulf War only occurred on television [see Norris, 1992]).

Rather than arguing over the true effects of geography on international relations – whether land or sea powers are strongest, as Mackinder and Mahan might have debated – critical geopolitics asks whose models of international geography are used, and whose interests these models serve. This approach owes much to the work of Michel Foucault (1980) who argued that power and knowledge are inseparable. For geopoliticians, there is great power available to those whose maps and explanations of world politics are accepted as accurate because of the influence that these have on the way the world and its workings are understood, and therefore the effects that this has on future political practice.

Critical geopolitics aims to challenge the objectivity of the geopolitician. For example, the privileging of sight (especially with the use of maps and diagrams) over other senses in geopolitical reasoning allows the geopoliticians to write as if from afar, as if somehow unconnected to the world being surveyed. This reinforces the idea of an objective account rather than one written from a position grounded within the events being discussed. It hides the fact that the geopolitician has his or her own point of view and loyalties. Although it is generally accepted that Nazi *Geopolitik* had a political agenda, this is considered to be an aberration of the 'science' of geopolitics. Yet other geopoliticians have not been innocent of interest. For example, Mackinder wanted to help maintain the British Empire and its **hegemony** over world affairs and Mahan, a naval historian, was interested in building up the US Navy at a time when other technologies seemed to make naval power less important.

Critical geopolitics looks to analyse the geography in any political description of the world. As Ó Tuathail and Agnew (1992: 194) have suggested, 'geopolitics is not a discrete and relatively contained activity confined only to a small group of "wise men" who speak in the language of classical geopolitics'. Simply to describe a foreign policy is to engage in geopolitics and so normalize particular world views. Any statement concerning international relations involves an implicit understanding of geographical relationships or a world view.

Similarly any geographical description can influence political perception. Descriptions of other places and the character of the people who inhabit them can be as significant as measurements of distance and calculations of location in constructing people's geographical imaginations. For example, the constant use of terms such as 'Evil Empire' to describe the USSR in America reinforced a binary geography of super-power stand-off that legitimated US military build up and intervention.

Critical geopolitics and identity

Perhaps the most important claim of critical geopolitics is that traditional geopolitical arguments are in fact profoundly a-geographical. Rather than being concerned with understanding geographical process, geopolitics reduce spaces and places to concepts or **ideology**. Space is reduced to units which singularly display evidence of the characteristics which are used to define the spaces in the first place (Asia *is* exoticism, the USSR *is* communism, Iran *is* fundamentalism, the USA *is* freedom and democracy, and so on).

In the contemporary political system, dominated by the territorial state, the geography of geopolitics tends to reduce the complex workings of politics into two spheres, the domestic sphere under the control of the modern territorial state and the international realm facing anarchy without higher power to control it. Thus the state invokes discourses of security through which any different characteristics are excluded through practices of territorial control, such as patrol of borders. In security discourse, it is difference that threatens the states so that for critic Simon Dalby (1990: 185):

> the essential moment of geopolitical discourse is the division of space into 'our' place and 'their' place; its political function being to incorporate and regulate 'us' or 'the same' by distinguishing 'us' from 'them,' 'the same' from 'the other.'

In arguing this, critical geopolitics suggests that geopolitics is not something simply linked to describing or predicting the shape of international politics, but is central to the ways in which identity is formed and maintained in modern societies. National identity is not simply defined by what binds the members of the nation together but also – perhaps even more importantly – by defining those who exist outside as different from members of the nation. Drawing borders around territory to produce 'us' and 'them' of the nation and those who are different, does not simply reflect the divisions inherent in the world but helps to create differences. Again, geopolitics does not simply reflect the facts of geography but in dividing the world into a state and the international realm helps to form geographical orders and geographical relationships.

The construction of 'otherness', and particularly the sense of danger that this presents, have implications for the practice of domestic affairs in addition to foreign policy. Thus Dalby (1990: 172) suggests that geopolitics is 'about stifling domestic dissent; the presence of external threats provides the justification for limiting political activity within the bounds of the state' (*see* Fig. 19.3). The construction of otherness simultaneously presents a normative image of identity. So, for example, when the USSR is constructed as being completely unlike the USA, any description of the USSR as evil, aggressive and unreasonable, implies goodness, tolerance and reason on the part of Americans.

Critical geopolitics and popular culture

One of the effects of broadening the scope for analysis in critical geopolitics is a consideration of a wide range of sources for analysis. More traditional approaches to geopolitics have concentrated upon the writings and pronouncements of political leaders and their academic advisers. More recently some theorists have considered popular culture to be an important source of information.

Although not given much attention until recently, it is possible to see the influence of popular culture on state practice, whether directly in the central role of CNN as a source of information during the Gulf War for American leaders, or indirectly in the role of popular culture in the construction of hegemonic cultural values that shape both the actions of politicians and the expectation of societies. Popular culture is entwined with more formal geopolitical visions as James Der Derian (1992: n1) explains:

> We are witnessing changes in our international, intertextual, inter*human* relations, in which objective reality is displaced by textuality (Dan Quayle cites Tom Clancy to defend anti-satellite weapons) . . . representation blurs into simulation . . . imperialism gives way to the Empire of Signs (the spectacle of Grenada, the fantasy of Star Wars serve to deny imperial decline).

For example, it is accepted that the film *The Manchurian Candidate* (1962) encouraged the CIA to pursue the possibility of controlling secret agents without their knowledge, and to envisage methods of combating the prospect of the Soviets having already developed such technology. The film starts from the plausible premises of both the Cold War exchange of spies, and the 'brainwashing' of soldiers in the Korean War. It then develops the possibility of the programming of people who are unknowing of their condition and therefore undetectable, the perfect spy. Ultimately the film influenced the political situation that it initially sought to reflect (*see* Fig. 19.4). More recently Reagan suggested enthusiastically after seeing *Rambo: First Blood Part II* (1985) that the next time American hostages were taken in the Middle East he would know what to do about it.

This can be seen in other US media too. Whether in films, TV programmes or in print culture, the opposition between a heroic and moralistic America and the Communist Evil Empire is taken as a given. Although never using the term, the American magazine the *Reader's Digest* frequently invoked geopolitical reasoning in its explanation of world affairs to its readership (see Sharp, 1993; 1996). The *Reader's Digest* believed in the limitless expansive potential of Communism, and that only power could effectively oppose power, so that the USSR 'will not stop at international frontiers unless it is opposed' (Chennault, 1948: 121). The magazine used geopolitical arguments to warn its readers about their country's vulnerability: 'The United States is naked – incredibly naked – against a Russia atom-bomb attack' (Taylor, 1951: 85).

As a result of the influence of cultural context, different country's geopolitical traditions draw upon specific metaphors to create images of international geography. Political elites must use stories and images

Figure 19.3 Nice to have an enemy again.
Credit: © Joel Pett, *Lexington Herald*

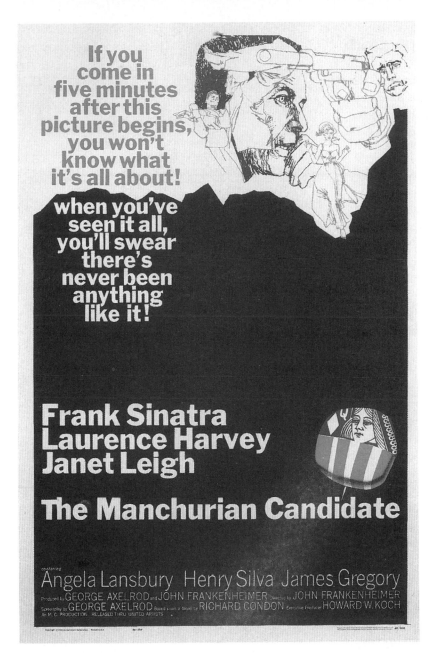

Figure 19.4 (Re)creating the popular geopolitical imagination: The Manchurian Candidate. Credit: Cinema Bookshop

that are central to their citizens' daily lives and experiences. By reducing complex processes to simple images with which their audiences would be familiar, geopoliticians could render political decisions quite natural, or could make the result of the process appear predetermined (as the domino example has demonstrated). For example, sport metaphors have been particularly prominent in the USA. Such language points to the 'essential' differences between national potentials for world-class performance and naturalizes a global arena in which the rules of the game are understood, and within which there are clear (unequivocal) winners and losers. Agnew (1998) argues that in so doing, the ambiguities of conflict are reduced to technicalities in game play. Michael Shapiro (1989: 70) points out that comparing world politics to sporting contests serves the geopolitical purpose of emptying world space of any particular content: places lose their uniqueness and world politics becomes a strategy played out on a familiar sports field.

American Presidents (particularly Presidents Nixon and Bush) have been particularly fond of sports metaphors as applied to world politics. In a Vietnam bombing campaign, Nixon adopted the code-name 'quarterback' (Shapiro, 1989: 87) and used terms such as 'end-run' and 'play selection' in foreign policy. As Agnew suggests, such metaphors 'allowed a notoriously socially awkward man to appear as "one of the boys," engaging in dialogue with other sports-loving men'. This not only explained the condition of conflict, but also why those hearing about it should accept the interpretation that Nixon offered: it provided taken-for-granted cultural referent that the majority of his American audience would accept.

Feminism and critical geopolitics

Critical geopolitics has also begun to address feminist concerns. Some feminist commentators have remarked on the lack of women in the history of geopolitics and contemporary theory and practice of international relations. Cynthia Enloe suggests that women have been written out of international politics. The story of international politics has traditionally been one of the spectacular confrontation of mighty states led by powerful statesmen, of the speeches and heroic acts of the elite, and the specialist knowledge of 'intellectuals of statecraft'. Enloe (1989) refuses to accept this story as covering the full extent of the workings of international relations, and instead focuses on those elements the traditional story excludes and silences: the role of international labour migration, the availability of cheap female labour for transnational corporation investment, the availability of sex workers for the tourist industry in southeast Asia, and so on. Enloe's is a very different account of international politics than the traditional story and certainly one that lacks its glamour. She links international geopolitics to everyday geographies of **gender** relations. Her account links the personal and the political, arguing that these alternative political geographies need to be uncovered because, 'if we employ only the conventional, ungendered compass to chart international politics, we are likely to end up mapping a landscape peopled only by men, most elite men' (Enloe 1989: 1).

Summary

- Critical geopolitics understands geography to be a 'discourse', created by powerful individuals and institutions, and used as a map or script with which to make sense of the world.

- Critical geopolitics seeks to denaturalize geographical statements in international relations which appear to be so self-evident as to be 'common sense'.

- Critical geopolitics recognizes the power of geopolitical arguments not only in the context of elite debates but also in popular accounts which help to form the 'geographical imaginations' that all people hold of the world and its political interrelationships.

> **Spykman on geopolitics**:
> 'Geography does not argue; it just is.'
>
> **Ó Tuathail and Agnew on critical geopolitics**:
> 'Geography is a social and historical discourse which is always intimately bound up with questions of politics and ideology.'

Further reading

The best examination of traditional forms of geopolitics and recent critical approaches can be found in:

- Ó Tuathail, G. (1996) *Critical geopolitics*. Minneapolis: Minnesota University Press.

To see the range of work going on under the title of 'critical geopolitics', see the special issue of the journal *Political Geography* in 1996 volume **15**(6/7).

Works that explore the interdependencies between geopolitics and American identity are:

- Campbell, D. (1992) *Writing security: United States foreign policy and the politics of identity*. Minneapolis: University of Minnesota Press.
- Dalby, S. (1990) American security discourse: the persistence of geopolitics. *Political Geography Quarterly* **9**(2): 171–88.

For a different map of international relations that includes everyday processes and gendered images rather than just high profile events and confrontations see:

- Enloe, C. (1989) *Bananas, beaches and bases: making feminist sense of international relations*. Berkeley: University of California Press.

Citizenship and governance

Mark Goodwin

INTRODUCTION

Issues of citizenship and governance have recently assumed a prominent position in political and academic debates. Indeed, during the summer of 1997 when this chapter was being written, fierce arguments were raging, for example, over the questions of closer European integration, the peace process in Northern Ireland, devolved governments in Wales and Scotland, the age of sexual consent for gay couples, the rights of single parents to work if they wish to, the closure of underperforming schools, and the charging of tuition fees to students. Running through these and other similar debates are the twin themes of citizenship and governance – covering the issues of what it means to be a citizen, of what rights and obligations one has as a citizen, and of the ways in which we, as citizens, are governed. A good example of how these twin themes come together has recently been provided by the debates over Scottish and Welsh **devolution**. At the heart of these debates are concerns about how different parts of the United Kingdom are governed, and about the different identities felt by those who are citizens of this supposedly 'united' nation. As part of these, *The Observer* newspaper (20 July 1997) carried a three-page article in which twelve media and political personalities were asked about their national identity, and where they would place such identity in a list of their own defining features. Despite the fact that technically all are British citizens, only two mentioned that they felt British, and in both cases this was placed after, or alongside being Scottish. In fact most linked their

national identity not to Britain, or the United Kingdom, but to England, Wales and Scotland, which are only constituent parts of the nation–state. Others identified with even smaller territorial units, such as London and Yorkshire, while one person said they felt European.

You might like to try this simple exercise for yourself – which area, or country, do you identify with? To which place do you feel you belong? Of where do you consider yourself a citizen? Asking these relatively simple questions raises a host of issues about the complex mixtures of states and territories, and areas of government, within which we all live, and about how these relate to our shared understandings of what it means to be a citizen of one or other of these spaces. These questions promise to become more complex as we enter the new millennium. As the debates over devolution indicate, the material spaces of government and the immaterial spaces of identity both seem to be moving away from the established order of the nation–state in two directions. They are moving downwards, to a regional or sub-regional space, and upwards to a supra-national space. Government by the United Kingdom Parliament, and a sense of British identity, is being replaced by government from Edinburgh, Cardiff, or Brussels and an identity which draws on notions of being Welsh, Scottish, or European. Strategic government for the English regions and for London will only serve to make matters more complex. This chapter explores such complexity by investigating the twin notions of citizenship and governance. It looks at the ways in which geography is bound up with both concepts and with how they operate in practice, and at how an appreciation of these issues helps to broaden the way we look at human geography.

THE CONTESTED SPACES OF CITIZENSHIP

Issues of citizenship are usually seen as the province of political science, and not geography, and the usual definition of citizenship is provided in political terms as referring to 'the terms of membership of a political unit (usually the nation–state) which secure certain rights and privileges to those who fulfil particular obligations. Citizenship is a concept, rather than a theory, which formalises the conditions for full participation in a community' (Smith, 1994: 67). A moment's thought, however, will indicate that even this narrowly political definition is bound up with geography. The political unit that one is a member of has a certain territory – thus, by definition, one is a citizen of a particular place. Also, the community, or communities, that one participates in, whether fully or not, are also bounded, and geographically situated. Indeed, the whole concept hinges around notions of inclusion and exclusion – to be a citizen is to be included in both a social and spatial sense, and this has offered fertile ground for geographical work.

In particular, geographers have questioned the basis on which such inclusion and exclusion takes place. The political definition of citizenship stresses the inclusive nature of the term – it implies that anyone within a certain territory who meets certain obligations will be included as a citizen, with corresponding rights and privileges. Yet matters are not this simple and the act of residence within a definable and bounded space does not necessarily secure citizenship. Somewhat paradoxically, the **globalization** of both capital and labour which has led to increased flows of people around the world, (see Chapters 7 and 30) has resulted in attempts by many governments to legislate for tighter immigration controls. This has two main implications – some people are excluded from residence altogether, and others are denied full citizenship rights. In Britain, for example, there is an explicit link between immigration and those means-tested benefits which as part of the social security system should be available to all citizens. As Oppenheim points out, (1990: 89) claiming income support, family credit, or even access to housing under the homeless legislation can endanger the chances of bringing the rest of one's family to the UK, or create problems for the claimant themselves or their sponsor. The 1971 Immigration Act meant that the wives and children of Commonwealth citizens could only enter the country if a sponsor could support and accommodate them 'without recourse to public funds', defined clearly for the first time in 1985 to include the three major means-tested benefits referred to above. Thus people who may have worked and paid taxes here for decades, are only allowed to be joined by their families on condition that they do not claim benefits for them or turn to the welfare state for accommodation. More surreptitiously perhaps, the increasing frequency of passport checks on black claimants at benefit offices – regardless of whether they were born in the United Kingdom – helps to create a climate of opinion which views welfare as the entitlement of white Britons rather than of black 'outsiders' (Oppenheim, 1990).

What we might term the bounded spaces of citizenship are therefore not 'straightforwardly inclusionary' (Painter and Philo, 1995: 112). Divisions along racial or ethnic lines as referred to above are fairly widespread and long established – in Germany, for instance, migrants who arrived to fill labour shortages in the 1950s and 1960s were labelled as 'guest workers' and denied social rights and freedom of movement. This labelling denied them even the status of ethnic minority migrants – they were expected to return 'home' when no longer needed. Figure 20.1 shows the abject living conditions which many such workers endured, often in spatially segregated 'ethnic enclaves'. The denial of full citizenship rights to 'guest workers', and their geographical and social marginality in special hostels or dormitories have played no small role in the rise of racism and facism in the newly unified Germany. This is one example of the ways in which inclusionary citizenship is actually riddled with divisions which are at once spatial and social.

Other examples of these divisions have emerged as recent debates about the term have broadened the scope for new avenues of geographical enquiry. It quickly became apparent that many groups who seem to enjoy full citizenship are actually limited in terms of the places and spaces in which this can be exercised. A highly visible example has been provided by the recent debates about **sexuality**. One of these debates has concerned the campaign to equalize the age of consent for heterosexual and homosexual sex – the former is currently 16, whilst the latter is 18.[1] Some citizens clearly enjoy more rights in this sphere than others. But aside from this general issue of equality, geographers have been increasingly investigating the links between sexuality and citizenship by exploring what Bell (1995: 139) has called 'the spaces of sexual citizenship'. His starting point for this exploration is to ask the question, quoting Diana Fuss, 'What does it mean to be a citizen in a state which programmatically denies citizenship on the basis of sexual preference?' (ibid.). His answer begins from the

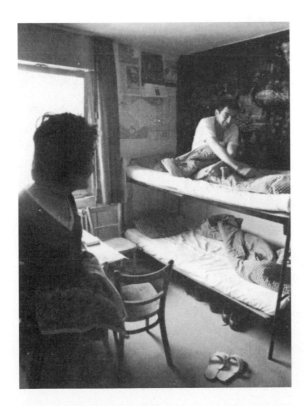

Figure 20.1 The social and spatial marginality of migrant workers denied full citizenship. Source: J Berger & J Mohr (1975) *A Seventh Man.* Harmondsworth: Penguin Books

fact that such preferences can be played out in some locations and not in others. What it does mean to be a citizen in this case is partly dependent on where you are. As Bristow observes, 'In Britain, it is possible to be gay [only] in specific places and spaces: notably, the club scene and social networks often organized around campaigning organizations' (quoted in Bell, 1995: 141). It is only in these spaces that those who are gay can feel comfortable, and as Painter and Philo (1995: 115) put it, 'if citizenship is to mean anything in an everyday sense, it should mean the ability of individuals to occupy public spaces in a manner that does not compromise their self-identity, let alone obstruct, threaten or even harm them more materially'.

Valentine (1993), in her exploration of the geographies of gay friendships, has confirmed that gay people are often forced to inhabit marginal spaces. The result is the growth of 'dense and heterogeneous networks formed around a limited geographical base [which] foster a sense of community' (Valentine, 1993: 113). As this suggests, the more marginal spaces of sexual citizenship can be positive as well as negative. Valentine goes on to explain that 'because lesbians find it difficult to make friends and express their lifestyles outside these gay contexts . . . their identities can become embedded in the networks formed in and around these places'. In some senses then, we can see how these restricted geographies can offer the chance for a reconstituted citizenship to emerge around a series of 'alternative' or 'underground' spaces. In the case of gay people, whole neighbourhoods have now grown up as places of deliberate congregation, such as Castro in San Francisco (see Castells, 1983), West Hollywood in Los Angeles (Jackson, 1994), Soho in London or the City of Amsterdam (Binnie, 1995). In these areas we can find a clustering of bars, restaurants, bookshops, theatres, clothes shops and other retail outlets all catering for a gay clientele (*see* Fig 20.2). In these spaces we can see the flowering of an alternative culture, which can act as the basis for an alternative kind of citizenship, in which members of certain groups can establish rights and obligations to each other.

Such alternative forms of what we may call 'participatory citizenship' have recently emerged over many issues, not just that of sexuality. A recent collection of writings from the USA documents how different marginal groups are refashioning spaces of social control into sites of resistance, and how in this process they are contesting dominant views and assumptions about their 'place' in society (Smith, 1995). The groups include squatters resisting urban development in Michigan; the homeless campaigning for decent housing in Chicago; an anti-gentrification coalition seeking to preserve low-cost housing in New York; low income African-American women living in a public housing project in New Orleans involved in community development; and immigrant Mexican agricultural labourers in California campaigning for employment rights. All the groups are using the appropriation of space to claim what they see as legitimate citizenship rights for their section of society.

One chapter describes the setting up of 'Tranquility City', a homeless encampment of 22 plywood 'huts' in a run-down industrial district of West Chicago. The author concludes that:

> for a brief period Tranquility City became a *mini-movement area* in which a different way of living poor was experimented with: a possibility was created for the formation of a homeless community free of institutional shelter restraints. Within this mini-movement area, residents of Tranquility City were able to construct a collective identity centred around issues of social justice for other homeless individuals and collective action in helping each other acquire housing and needed services.
>
> (Wright, 1995: 39, original emphasis)

As this suggests, the spaces that such groups inhabit can be crucial in helping to form a group identity, which in turn builds a localized community that can experiment with different ways of living, which in turn is able to campaign around issues of social justice and citizens rights. In this way geography becomes critical to the establishment of new forms of participatory citizenship.

Summary

- The concept of citizenship is inherently geographical, hinging around social and spatial inclusion and exclusion.

- Such inclusion is never straightforward and is riddled with divisions.

- These divisions create differential spaces where different degrees of citizenship are experienced and played out by different groups in society.

- These spaces can be positive as well as negative, and can be used for resistance as well as control.

THE CHANGING GEOGRAPHIES OF GOVERNANCE

The renewed interest in issues of citizenship within both academic and policy circles can be viewed as evidence of a rethinking of the relationship between individuals, the communities they are part of, and government. Citizenship lies at one pole of this rela-

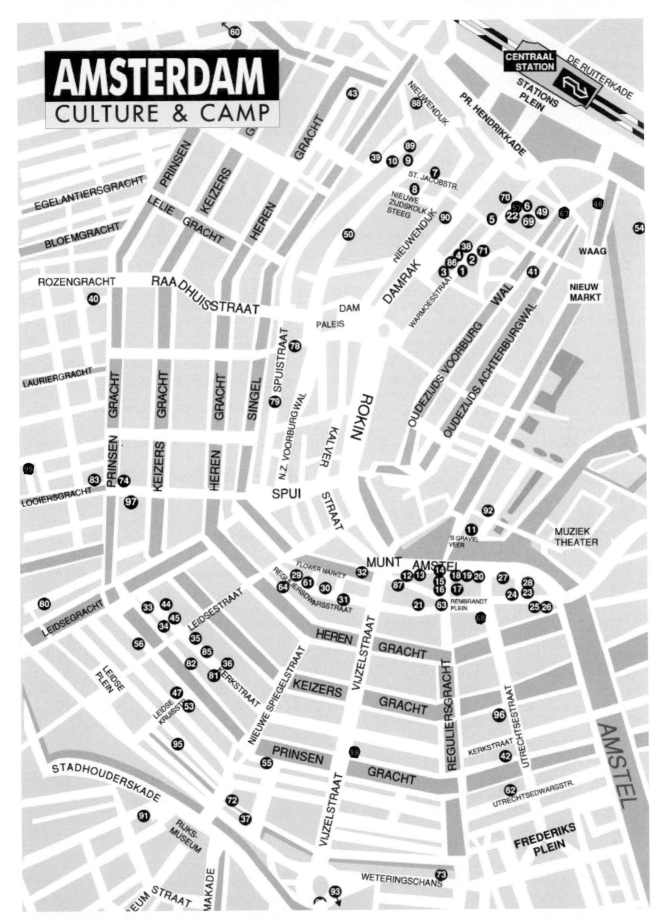

Figure 20.2 The alternative spaces of sexual citizenship – the clustering of gay bars, clubs, hotels, restaurants, shops, cinemas and fitness centres in Amsterdam. Source: www.matchoman.com/gay/community/map

tionship, concerning as it does the rights and obligations of those being governed. At the other pole are the processes and institutions of government, and the actions of those who do the governing. And just as new forms of citizenship have emerged as part of this rethinking, so too have new forms of government. Indeed, so prevalent are these new forms that the term governance, and not government, is now used to describe these new structures. The term governance is now widely used and accepted across a variety of academic and practitioner circles. Put simply, it 'refers to the development of governing styles in which boundaries between and within public and private sectors have become blurred' (Stoker, 1996: 2). Thus the term governance is not simply an academic synonym for government. Its increasing use signifies a concern with a change in both the meaning and the content of government. As Rhodes puts it (1996: 652–3), the term is now used to refer 'to a new process of governing, or a changed condition of ordered rule, or the new method by which society is governed'. Where government signals a concern with the formal institutions and structures of the state, the concept of governance is broader and draws attention to the ways in which governmental and non-governmental organizations work together, and to the ways in which political power is distributed, both internal and external to the state.

The way in which we are governed touches all aspects of our lives. It determines the type of education we receive, from nursery school to university. It dictates the level of health care we are provided with, and it concerns the provision of housing and the provision of jobs and training. It concerns planning and environmental issues, as well as transport and social services. Previously, these services were mainly provided by central and local government, through a combination of policy decisions and service delivery. The new structures of governance have transformed this system into one which now involves a wide range of agencies and institutions drawn from the public, private and voluntary sectors. They will still include the institutions of elected government, at central and local level, but will also involve a range of non-elected organizations of the state, as well as institutional and individual actors from outside the formal political arena, such as voluntary organizations, private businesses and corporations and supra-national institutions such as the European Union (see Goodwin and Painter, 1996: 636). The concept of governance focuses attention on the relations between these various actors, and crucially from our perspective draws attention to the complex geographies now involved in the act of governing any particular area. The governing of localities is no longer exclusively, or even

mainly, a local matter, but instead is a complex, differentiated and multi-scale process.

Geographers have investigated these issues at a variety of levels. Some have studied the international scale and looked at the emergence of multinational or global forms of governance. Others have looked at the way sub-national or regional governance is developing, both in Britain and elsewhere, and many have looked at the changing nature of local governance, especially at the urban level. Some common threads can be discerned running through many of these studies. In particular the new emphasis on governance raises questions about:

- the purpose of the new governing mechanisms – how and why were the particular agencies involved brought together, and what are their interests and rationales?

- the effectiveness of the various agencies involved at working together – how does each blend its particular capacities with the others?

- the links between different forms of governance and different rates of economic and social development – how do the new mechanisms of governance cope with uneven development and with the specific problems that might emerge in one place rather than another?

- the nature of democracy and accountability in non-elected agencies – just how does the public influence what are in the main unelected and appointed institutions, and how is the declining scope of elected state activity squared with the fashionable political notions of inclusion and empowerment?

We can illustrate these concerns by reference to some of the new work on urban governance, which has been both theoretical and empirical. Figure 20.3 charts the growing complexity of those institutions involved in urban policy, and shows the various initiatives which have been tried by successive governments in an effort to successfully manage urban change. As we can see, since the late 1970s there has been a proliferation of these initiatives, and our cities are now governed by a somewhat bewildering array of institutions. What was once the job of local and central government is now performed by a system which has become increasingly differentiated. There has been a huge growth in the number of centrally appointed bodies or quangos involved in urban governance at the local level (See Duncan and Goodwin, 1988; Atkinson and Moon, 1994, for details). Urban Development Corporations took over the powers of development control and economic regeneration

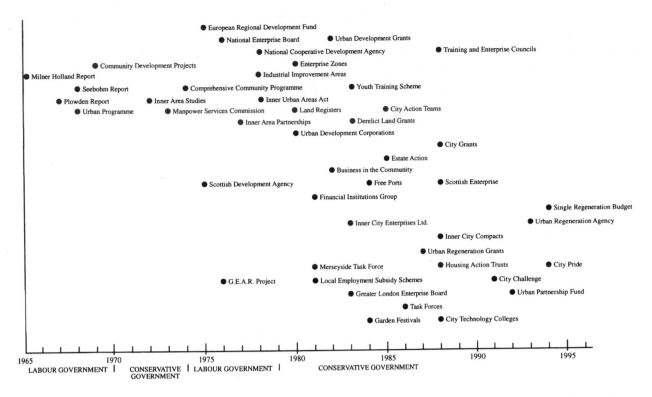

Figure 20.3 The complexity of urban governance – note especially the proliferation of agencies involved in managing the city after the Conservatives came to power in 1979. Source: M Pacione (ed.) (1997) *Britain's Cities: Geographies of Division in Urban Britain.* London: Routledge.

from local authorities in certain parts of our biggest cities; Training and Enterprise Councils are now responsible for training and business development; Housing Action Trusts, Estate Action and Housing Associations now provide and manage social housing; and the private sector has been heavily involved in many initiatives to promote economic development, through institutions such as Business in the Community, City Pride, Inner City Enterprise and the Financial Institutions Group.

There has also been a flowering of locally based networks and **partnerships**. Some of these have emerged from local initiatives, as in Manchester where the City Council has also been involved in promoting partnerships with an extensive network of local business organizations. These included Manchester 2000 (which co-ordinated the unsuccessful bid for the 2000 Olympics, and the successful bid for the 2002 Commonwealth Games); the North West Business Leadership Team; the City of Drama Committee; the East Manchester Partnership; and the Phoenix Committee. As Tickell and Peck conclude 'the partnership fever which has gripped . . . Manchester . . . has over recent years fostered a dense web of interconnected agencies, lobby groups and committees' (1996: 597). This 'dense web' typifies the new structures of urban governance. In other

cases, however, the impetus for the formation of the partnership has come from central government. Central funds for urban regeneration are now delivered through the Single Regeneration Budget, a policy introduced in 1994 which forces places to bid against one another for government funding. It also encourages the creation of local partnerships, as bids are expected to involve the public, private and voluntary sectors working together. Local authorities cannot bid on their own, and thus by definition governance replaces government as the guiding force behind urban policy. Figure 20.4 shows the diversity of those agencies involved in just one partnership in Leeds, as well as the range of projects that they are involved with.

In order to explain and understand the emergence of these new forms of urban governance, geographers have increasingly turned to two particular concepts – the notions of urban growth coalitions and urban regimes. These were originally employed in the US context, where such partnerships have a longer and more established history, but they have recently been used to examine the changing nature of urban politics in the UK. They show how geographers are turning to new theoretical frameworks as they seek to understand the changing world around them. The idea of growth coalitions was first developed in the late 1970s

Leeds Initiative objectives
- Promote the city as a major European centre
- Ensure the economic vitality of the city
- Create an integrated transport system for the city
- Enhance the environment of the entire city
- Improve the quality and visual appeal of the city
- Develop the city as an attractive centre for visitors

Partner organizations:
- Leeds City Council
- Leeds Chamber of Commerce and Industry
- Leeds Training and Enterprise Council
- Leeds Chamber of Trade
- University of Leeds
- Leeds Metropolitan University
- West Yorkshire Playhouse
- Yorkshire Post Newspapers
- Regional Trade Union Congress
- Government Office for Yorkshire and Humberside
- Leeds Civic Trust
- West Yorkshire Police
- Railtrack
- West Yorkshire Passenger Transport Executive

Initiative projects:
- The Leeds Flower Initiative
  – the North's 'City of Flowers'
- Education Business Partnership
  – targets for local education and training
- The Leeds European Initiative
  – promotion of Leeds as 'a major European city'
- The Leeds Retail Initiative
  – marketing and promotion of shopping
- The Leeds Financial Services Initiative
  – 'the UK's leading financial centre outside London and Edinburgh by the year 2000'
- The Leeds Engineering Initiative
  – awareness raising, training and marketing
- Gateways and Corridors
  – transport infrastructure rebuilding, lighting and landscaping
- The Leeds Lighting Initiative
  – floodlighting schemes and lighting improvements
- Leeds City Station
  – redevelopment
- The Leeds Printing Initiative
  – training and promotion
- Leeds Environment City Initiative
  – a range of environmental action schemes and a Business Forum
- Opp2k – Working for Women Working in Leeds
  – initiatives to advance the position of women in the labour market
- Leeds Media Initiative
  – development of the city as a media centre
- Leeds Architecture and Design Initiative
  – promotion of the city as a 'centre of design excellence'
- Leeds Initiative Regeneration Board
  – £17m 5-year programme to increase employment in inner city areas

Source: The Leeds Initiative *Annual Report* 1994–5, Leeds City Council.

Figure 20.4 The range of institutions and projects from just one urban partnership in Leeds. Source: M Pacione (ed) (1997) *Britain's Cities: Geographies of Division in Urban Britain*. London: Routledge.

and early 1980s to account for the pro-growth nature of much urban politics in the USA. Logan and Molotch (1987) argued that a key feature of US urban government was the formation of coalitions of local business leaders seeking to promote growth. The key players in such coalitions (also known as growth machines) were local property owners and property developers who stood to gain the most from a rise in property values, due to economic growth which would fuel intensification of land use and demand. They were joined by other pro-growth interests – financiers, regional banks, construction companies, estate agents, retailers, etc. – who wished to mobilize the powers of local government in order to create a business climate conducive to economic growth. They were acting out of self-interest, but the key to success was to present this to local politicians and the local electorate as value-free development which was self-evidently good for the locality. This model of local politics seeks to explain how economic agents in the city realize their vested interests through influencing local politics – the key point here is that the new mechanisms of governance are giving much more scope for this influence to be exercised.

The theory has proved difficult to apply in a British context, and critics have claimed that it concedes too much power to the capacities of individuals and neglects wider constraints, and that it concentrates too heavily on the interests of one particular section of the business community. In an effort to find a more suitable theoretical framework, researchers have recently turned to regime theory, also developed in the USA during the 1980s. This seeks to overcome some of the difficulties with the growth coalition concept. In particular, it recognizes that growth machines are only one form of urban political activity, and that some governing coalitions may well favour policies which are not exclusively pro-growth and pro-property interests. In making this claim the literature focuses much more on the management of a range of local interests (rather than on their origin) through the formation of different types of governing coalitions, or regimes. A regime is formed when a variety of local interests mesh together to form a relatively stable governing coalition, and involves a number of groups co-operating behind a certain set of policies to achieve their own ends (which may well vary from one another). Attention is paid to the more informal arrangements that surround and support the official workings of local governance, and to the ways in which the interests of different members of the regime are realized.

Viewing the new arrangements of urban governance in these terms allows us to ask questions about who is holding the levers of power in our contemporary cities, and about how they are using that power. By definition we can also ask questions about who is not represented in the ruling regime, and about why they have been excluded. The networks and partnerships which are emerging in British cities can be investigated in these terms. We can ask questions about the interests each member of the regime brings to the coalition, or partnership, and about the goals which they intend to achieve from their membership. Thus if we use the governance perspective, we can open up a whole host of research questions about the exercise of power at the urban level.

This promises to be crucial, because urban-based governance will become more rather than less complex in the years to come, as places will increasingly have to sell themselves and their partnerships – whether to central government or the private sector – if they wish to remain competitive in an era of global competition. However, in an era of reducing state support and increasing capital mobility this competition is a zero-sum game. This means that existing resources are simply shifted around rather than new resources being added. The places that win, can only do so at the expense of those who lose. Territorial competition and difference can only serve to deepen rather than reduce uneven development, which means that a geographical perspective will continue to be invaluable in analysing the trajectory of the new urban governance.

Summary

- The concept of governance refers to new ways of governing society involving a range of participants drawn from the public, private and voluntary sectors.

- The new mechanisms of governance are operating at a variety of scales, from the local to the global.

- There is no guarantee of governance success, and in many instances there are concerns over the co-ordination, accountability and legitimacy of governance structures.

- The concept of governance opens up new avenues for the study of urban politics, focused on the exercise of power via a collective capacity to act.

CONCLUSION: SPACES OF CITIZENSHIP AND GOVERNANCE

There are very real concerns over the nature of contemporary citizenship and governance which promise to keep these issues at the forefront of geographical enquiry. Official policy statements, concerning many areas of economic and social development at all levels from the local to the European, emphasize the important role which is envisaged for partnerships and networks operating beyond the formal structures of government. Yet once these new mechanisms of governance are in place they raise a number of critical questions about participation and citizenship. In many instances the new structures of governance, especially those which involve the private sector, are blurring the distinction between the public as citizen and the public as customer. This blurring is given official recognition through the various Citizen's Charters which now regulate the operation of many areas of governance. These may be useful in guaranteeing procedural rights of proper treatment, whether by a health authority or privatized rail network, but the user (or rather customer) has no powers to hold the agency to account, and no substantive rights to receive services which they need. Yet the citizen has rights which differ from the customer, and in the long term the legitimacy of the new structures of governance will rest on the granting of consent and support from the public.

A crucial issue, however, is the territory to which the citizen feels they belong. Increasingly, as we saw at the very beginning of the chapter, people are identifying not with the established nation–state but with entities which are either supra-national (such as the EU) or sub-national (such as the region, or the locality). The debates over devolved government in Scotland and Wales, over elected mayors and regional assemblies in England, and over an increasingly integrated Europe are evidence of this. As this uncertainty indicates, the very complexity of the new structures of governance may well ensure that all kinds of new spaces of citizenship emerge to replace those previously linked only to the nation–state. As we saw, many of these spaces will be created and defined by those who presently feel that they can only exercise partial citizenship. Already there are signs that, as a response to new management structures, self-organization by user groups in the fields of health and social services is redefining the relationship between previously excluded citizens, their communities and processes of governance (Barnes, 1997). These kinds of redefinitions point to the ways in which the processes of citizenship and governance will continue to impact upon one another, and they indicate that these issues will be of continuing importance for the political geographer.

NOTE

1 This difference is inspite of ongoing parliamentary attempts at reform, which at the time of writing have been unsuccessful.

Further reading

• Smith, S. (1989) Society, space and citizenship; a human geography for the 'new times'? *Transactions of the Institute of British Geographers*, **14**, 2, 144–56.
This is the paper which set the agenda for the study of geography and citizenship, and it is still worth reading for the suggestive links it draws between geography and social justice.

• Painter, J. and Philo, C. (1995) Spaces of citizenship. *Political Geography*, **14**, 2.
This is a special edition of the journal which contains nine separate articles on various aspects of geography and citizenship. It is still the best collection of articles on this theme.

For excellent case studies of the new urban partnerships in the UK see any one of a series of papers by Peck and Tickell on Manchester. Articles by them on this theme appear in the *International Journal of Urban and Regional Research*, **19**(1) (1995); in *Urban Studies*, **33**(8) (1996), with Alan Cochrane; and in *Transactions of the Institute of British Geographers*, **21**(4) (1997).

Nationalism

Pyrs Gruffudd

INTRODUCTION

According to the sociologist Anthony Smith (1991: viii) 'Nationalism provides perhaps the most compelling identity myth in the modern world.' Feelings of identification with, even loyalty to, a particular 'nation' remain powerful despite the emergence of 'global culture' and appeals to other forms of identity like class, gender, and ethnicity. If anything, the appeal of national identity is growing. On the one hand, we have seen political turmoil in the United Kingdom recently concerning the relationship with the rest of Europe, with much of this defined in terms of threats to national sovereignty. On the other, we have witnessed the relatively peaceful re-emergence of small nations in Eastern Europe – Estonia, Latvia, and the Ukraine – whose identities had been suppressed for generations by the Soviet Union, but also the bloody re-emergence of Serbs, Croats and Muslims in Bosnia following the break-up of the former Yugoslavia. And simmering away for decades, and occasionally bursting into violent life, are conflicts in places like Northern Ireland, the Basque region of Spain, and the Middle East (*see* Fig. 21.1).

This chapter tries to highlight some of the complexities of nationalism. The first concerns precisely which situations are commonly considered 'nationalist'. Billig (1995: 5) argues that 'In both popular and academic writing, nationalism is associated with those who struggle to create new states or with extreme right-wing politics. According to customary usage, George Bush is not a nationalist; but separatists in Quebec or Brittany are; so are the leaders of extreme right-wing parties such as the *Front National* in France.' He claims that this narrow understanding of nationalism always locates it on the periphery while overlooking the nationalism of established Western states. Members of those states often suggest that they are 'patriotic' whereas others are 'nationalistic'. But that distinction might usefully be collapsed. In its broadest sense, nationalism is simply an **ideological** movement that draws upon national identity in order to achieve certain political goals. It therefore covers a far wider range of contexts and narratives than we

Figure 21.1 Half a million Spaniards march during a massive protest against the kidnapping and killing of Miguel Angel Blanco by Basque terrorists, 1997. Credit: Popperfoto/Reuter

might, as a gut reaction, suppose. The usual tendency is to imagine that nationalism is a negative phenomenon characterized by bigotry or racism, violence (be it terrorism or warfare), and aggressive social exclusion. For many others, however, nationalism can refer to a positive celebration of identity in the face of oppression or marginalization and the eventual attainment of social and political liberation. The conditions against which nationalist groups struggle can be extreme – racial or religious intolerance, for instance – but they can also be simple dissatisfaction with the structure of government. Nationalist struggle can, therefore, be carried through by force of arms, but also through democratic process.

NATIONALISMS

Nationalism is a complex phenomenon. In this section I will try and highlight some key concepts that can help us recognize and make sense of nationalisms. The first is the distinction between civic nationalism and ethnic nationalism. Civic nationalism (sometimes called 'territorial nationalism') is a modern, liberal phenomenon geared towards the creation and regulation of an efficient social, economic and political unit. Its origins lie in eighteenth- and nineteenth-century Europe when states such as France and Britain modernized, industrialized and sought to create homogeneous societies based on the capitalist system. This form of nationalism, according to Ignatieff (1993: 3–4)

> maintains that the nation should be composed of all those – regardless of race, colour, creed, gender, language or ethnicity – who subscribe to the nation's political creed. This nationalism is called civic because it envisages the nation as a community of equal, right-bearing citizens, united in patriotic attachment to a shared set of political practices and virtues.

Civic nationalism is often seen as a set of state-building practices, concerned with the political, economic and cultural systems which serve to bind a people together. Sometimes, however, its lack of conceptual regard to **race**, creed, colour and so forth can be problematic when the nation assumes that cultural homogeneity is desirable and refuses to acknowledge difference. Ethnic nationalism, on the other hand, replaces this formal, rationalistic language of 'rights' and 'systems' with the language of 'belonging'. What makes the nation a place to which people feel they belong and to which they pledge allegiance is 'not the cold contrivance of shared rights, but the people's pre-existing ethnic characteristics: their language, religion, customs and traditions' (Ignatieff, 1993: 4).

Modern nations are often built on ancient ethnic origins, and the myths and legends of that **ethnicity** are resilient and powerful. This form of nationalism can be a positive celebration of cultural heritage and diversity, often in the face of an oppressive regime, but it can also be exclusionary and can promote intolerance. An ethnic group that is denied autonomy by an existing state of which it is a part will often build its nationalism on ethnic identity and, as 'national separatists', will seek political autonomy as a vehicle for protecting and expressing its ethnicity. In truth, however, most nationalisms combine both 'civic' and 'ethnic' elements – either at different times, in different proportions, or according to the different audiences to whom messages are being addressed. Smith (1991) suggests that we think, rather, in terms of the functions that these various types of nationalism perform for a people. He draws a distinction between 'external' functions, which govern the territorial, economic and political relationships with the rest of the world, and 'internal' functions that socialize members into a national community.

Some of the most interesting recent work has addressed these 'internal' processes – the way we are made to feel we 'belong' to a particular community. Benedict Anderson (1991) argues that the nation is not 'real' as such, but that subtle mechanisms work on us and blend us into what he calls an 'imagined community'. The nation is 'imagined' because you will only ever meet, or even be aware of, a miniscule proportion of your fellow 'nationals' but you have a strong image in your mind of the nation as existing simultaneously none the less. And that image is hardly neutral 'because, regardless of the actual inequality and exploitation that may prevail in each, the nation is always conceived as a deep, horizontal comradeship' – a form of community (ibid.: 7). These processes include overt forms of national bonding like the singing of national anthems (*see* Fig. 21.2), which provide an experience of simultaneity:

> At precisely the same moments, people wholly unknown to each other utter the same verses to the same melody. The image: unisonance. Singing the Marsellaise, Waltzing Matilda, and Indonesian Raya provide occasions for unisonality, for the echoed physical realization of the imagined community.
>
> *(1991: 145)*

Famously, Anderson also suggests that 'national newspapers' provide a daily, visible and shared reminder of the imagined nation. We might extend this to the 'national news' on television or radio, particularly at times of 'national crisis' or 'national rejoicing'. Billig goes even further than Anderson in highlighting the

Figure 21.2 Wales' daffodil in the crowd, Cardiff Arms Park, illustrating the role that sport plays in national identity. Credit: Popperfoto

very mundaneness or banality of this process of national socialization. He argues (1995: 8) that 'because the concept of nationalism has been restricted to exotic and passionate exemplars, the routine and familiar forms of nationalism have been overlooked. In this case, 'our' daily nationalism slips from attention.' For Billig, 'banal nationalism' is not the flag being passionately waved, it is the flag hanging unnoticed on public buildings like schools and post offices in the United States. He extends his analysis to language, to politicians' clichés, even to newspapers' sport and weather coverage. Together, these banalities make the homeland look 'homely, beyond question and, should the occasion arise, worth the price of sacrifice' (Billig, 1995: 175).

Summary

- National identity is, arguably, becoming more – rather than less – significant in the modern world, despite appeals to 'global culture', etc.

- Nationalism is a phenomenon whose complexity has often been overlooked. It can, for instance, be both a positive and a negative social force, and one found both in the core and in the periphery.

- There is a major distinction between 'ethnic' and 'civic' nationalism. The former is concerned with citizenship, laws, rights, etc. whilst the latter is concerned with ethnicity, culture and belonging. In most cases, nationalism is a blend of both types, perhaps at different times.

- Some of the most interesting studies of nationalism examine its day-to-day presence in our lives – its 'banality' or its role in creating 'imagined communities'.

GEOGRAPHIES OF NATIONALISM

There are several reasons why nationalism is of critical interest to geographers. Patterns of electoral support for nationalist parties, for instance, are open to geographical analysis (e.g. Levy, 1995). But I will concentrate here on the symbolic role of geography for national identity and nationalist movements. This is a major role, for nationalism is primarily a territorial **ideology** that derives its logic and inspiration from the relationship between a particular group of people and a particular parcel of land. This leads on to a whole raft of cultural relationships though which a people make *a* land *their* land.

One of the best summaries of this relationship is still Williams and Smith's (1983) article in which they identify eight major dimensions of national territory – habitat, folk culture, scale, location, boundaries, autarchy (self-sufficiency), the idea of 'homeland', and processes of nation-building. Some of these dimensions are 'external' in that they refer to attempts to distinguish one nation from another. Attempts may be made to 'solidify' or 'confirm' national space by delimiting and maintaining borders. The changing of place names from an earlier phase of occupation serves to harden the identity of that piece of land and to deny the claims of others upon it. 'Internally', however, territory and landscape can become symbolic of national identity and powerful agents of social cohesion. This process frequently draws on an awareness of history:

> The nation's unique history is embodied in the nation's unique piece of territory – its 'homeland', the primeval land of its ancestors, older than any state, the same land which saw its greatest moments, perhaps its mythical origins. The time has passed but the space is still there.
>
> *(Anderson, 1988: 24)*

Legends are therefore placed within the nation's space and national heroes are located through their birthplaces, graves, or the sites of their greatest acts, thus confirming the link between a particular people and that place. These sites – houses, battlegrounds, etc. – are often conserved or 'museumized' by the State and can often become places of pilgrimage at which overtly political ceremonies might be staged.

Landscape and specific physical features can also become emblematic of national identity. As Stephen Daniels (1993: 5) puts it: 'Landscapes, whether focusing on single monuments or framing stretches of scenery, provide visible shape; they picture the nation. As exemplars of moral order and aesthetic harmony, particular landscapes achieve the status of national icons.' In many cases, rural landscapes are imagined as the 'real', 'authentic' essence of the nation. Daniels (1993) argues that the gentle, pastoral lowlands of southern England – and paintings of them by artists like John Constable – have come to symbolize 'Englishness' and have been used at times of social tension (the two World Wars, for instance) as emblems of national identity. This idealization also extends to the people – or folk – living within them in idealized, organic communities. Many nationalist movements had close ties with groups protecting the legacy of the 'folk', be it their customs and way of life or simply their buildings, costume or music. Some nationalist groups went so far as to advocate moves 'back to the land' – that is, to resettle the population away from the cities and in the rural 'heart' of the country – in order to regain some essential, and lost, form of national identity (e.g. Gruffudd, 1994).

There are few better examples of this mythologizing and politicization of territory than Israel. Hooson (1994: 10) argues that 'The endowment of religious symbolism upon a piece of land . . . precipitated by the establishment of Israel after the Second World War . . . alongside the Moslem religious significance of the area, will make that tiny piece of land a tortured example of multiple overlapping national identities for a long time.' The maintenance of Israel's external and internal borders has assumed immense significance since the creation of the new state in 1947. Newman (1989) argues that this maintenance has been achieved not only through military action but also through civilian settlement in farms, *kibbutzim* and industrial villages in border regions. More recently, the controversial and frequently violent building and defence of Jewish settlements in areas hitherto exclusively or overwhelmingly Arab-Palestinian transfers this spatial form of nation-building to the internal space of the Israeli state (see Falah, 1989). Elements within the Israeli state have also, at various times, used history and myth as a form of nation-building, and geography has been a potent factor in that process. Zerubavel (1995) notes how Zionists sought to legitimate Jewish nationalist ideology through the recovery and reinvention of a settler history. The 'science' of archaeology played a prominent part in this recovery as it did in other aspects of Israeli life – uncovering traces of Jewish settlement in Jerusalem, for instance, thus adding to claims about the legitimacy of present-day settlement and control. One of the most dramatic elements of this historical reconstruction is the legend of Masada, a mountain-top fortress established above the Dead Sea by King Herod and whose Jewish occupants were besieged by the Romans at the end of the first century AD. Legend has it that, as the Romans were about to break through the defences, the occupants of the city committed mass suicide rather than be enslaved by the Romans. The production of modern translations of the legend 'led to its reconstruction as a major turning point in Jewish history, a locus of modern pilgrimage, a famous archaeological site, and a contemporary political metaphor' (Zerubavel, 1995: 63). Despite its grisly and ultimately hopeless end, the legend became symbolic of Jewish resistance in the face of overwhelming odds. Hebrew teachers – socializing their charges into a new language and, thus, a new nation – organized youth trips to Masada from the 1930s onwards and a pilgrimage to Masada and a pledge that it will not fall again is part of Israeli military training to this day (*see* Fig. 21.3).

Summary

- Nationalist ideology is frequently 'geographical' in that it is based on a territorial claim and it proclaims a clear sense of place.

- This geographical aspect can be manifested equally in material acts of nation-building (e.g. road networks) and in cultural forms like landscape painting, etc.

- History and folk culture provide a nation with a longstanding bond to the land. Apparently 'neutral' conservation policies and 'sciences' like archaeology can thus play a role in nationalist movements by defining the authorized 'national past'.

INFLECTING NATIONALISM

Smith argues that the key to understanding nationalism is an appreciation of its multidimensionality. It is this 'that has made national identity such a flexible and persistent force in modern life and politics, and allowed it to combine effectively with other powerful ideologies and movements, without losing its character' (1991: 15). To understand this I will briefly look at inflections of nationalism in different contexts and countries. This will also help us grasp the distinctions

Figure 21.3 Israeli soldiers taking their oath of allegiance in a ceremony on top of Masada. The Hebrew inscription reads 'Never again shall Masada fall'. Source: Israel Defence Forces' Spokesman's Office

between civic and ethnic, external and internal, even positive and negative forms of nationalism. I will consider, briefly, issues of economics, 'race', and language.

Some of the most influential early writings on nationalism sought an explanation for its persistence or re-emergence in relative socio-economic conditions. Hechter (1975), for instance, saw nationalism as a phenomenon born out of poverty and oppression. His historical analysis of Britain identified a developed economic core and an underdeveloped periphery, with the latter concentrated in the 'Celtic fringes' of Wales and Scotland. Because capitalism had flowed along ethnic divides to the disadvantage of the minority Celtic groups, it heightened ethnic consciousness due to a growing awareness of disadvantage. Similarly, Tom Nairn (1977) claimed that nationalist cultural resurgence in Scotland and Wales was a romantic and populist response to uneven development. There are parallels in David Harvey's (1989: 306) argument that the insecurity caused by capitalist globalization in part explains the resurgence of nationalism: 'there are abundant signs that localism and nationalism have become stronger precisely because of the quest for the security that place always offers in the midst of all the shifting that flexible accumulation implies'.

But there is no inevitable correlation between poverty, ethnic identity and nationalism. In Spain, the regions with the most overtly developed senses of identity are the traditionally more prosperous ones. The Basque Country and Catalonia have been for centuries the core regions of the Spanish economy, but modernization actually served to heighten – rather than diminish – their senses of regional identity. The Spanish state came to be seen as a parasite. The Basque country was, however, hit by an economic recession in the 1980s and this *did* give Basque

nationalism a new surge of energy, including a dramatic upturn in the activities of the terrorist group ETA (*Euskadi Ta Azkatasuna* – Basque Homeland and Liberty), established in 1959. Catalonia, on the other hand, has established itself as one of the core regions of the European economy, and its own parliament projects it as a self-confident, dynamic and creative 'nation' within Europe, rather than a region within Spain. There is a strong – though less attractive – echo of this in Italy and its long-standing disparity between the affluent north and the poorer south. There the *Lega Nord* (the Northern League), formed in 1991 – and its charismatic leader Umberto Bossi – have harnessed northern grudges against the south and its supposedly 'corrupt' politicians that are preying on the industrious north. In a ceremony on the banks of the River Po in 1996 Bossi declared an independent north Italian state called Padania, with its capital in Venice, and demanded a separate currency (*see* Fig. 21.4). The League's electoral success has lagged way behind its historically-inspired rhetoric though (see Agnew, 1995).

The *Lega Nord*, however, is also notable for its racist rhetoric and opposition to immigration. As I have already noted, nationalist politics are frequently articulated around the issue of ethnic 'belonging', with 'ethnic cleansing' in Bosnia – the forcible construction of ethnically homogeneous areas – perhaps the most painful recent example in Europe. While the origins of Basque nationalism were, in part, based on racial distinctiveness and superiority (Conversi, 1990), very few of the 'mainstream' nationalist movements in Europe (i.e. those organizing as political parties) are now overtly racist. However, that does not mean that issues of 'racial belonging' are not prescient in any understanding of the nation. At the level of the state, immigration policy serves to exclude and include pri-

Figure 21.4 Northern League supporters waving flags of their self-styled state of Padania, on the Grand Canal in Venice. Credit: Popperfoto/Reuters

marily on ethnic lines, and below that level a whole range of popular discourses about race are also woven together with nationality (see Jackson and Penrose, 1993). In Australia, a party opposed to immigration and **multiculturalism** significantly calls itself One Nation, and in Britain possibly the most blatant example of this alliance between racialism and nationalism is the use of the Union Jack by right-wing racist groups.

Ethnicity more generally is, however, central to any understanding of the vast majority of nationalist struggles, but that ethnicity may be expressed as the politics of language or of religion, rather than some crude notion of 'racial belonging'. Often, we need to think of these nationalisms in terms of the celebration of ethnic diversity in the face of homogenizing or oppressive forces, rather than in terms of aggressive exclusion. Recent Spanish nationalisms can be partly understood as a response to the ethnic suppression that characterized the Fascist dictatorship of General Franco from 1935 until 1975. Franco abolished the historic parliaments and legal rights of both Catalonia and the Basque Country and ruthlessly repressed their regional identities and languages. Books in Catalan were destroyed and the names of villages and towns changed to Castilian (what we call 'Spanish'). Despite

this, Catalan nationalism was almost entirely peaceful and based on the maintenance of folk culture. Since the restoration of Catalan autonomy in 1980 the language has regained its status in all aspects of life. According to the first leader of the new regional government in the 1980s 'If some issue is crucial to Catalonia, it is language and culture, because they are core elements of our identity as a people … Catalonia did not want autonomy for political or administrative reasons, but for reasons of identity' (quoted in Conversi, 1990: 56).

Welsh nationalism too has been a predominantly cultural movement. *Plaid Cymru* (The Party of Wales) was formed in 1925 in response to the perceived decline of traditional Welsh rural life and, crucially, the Welsh language. The language had been marginalized and outlawed by the British state since Wales's incorporation under the Act of Union of 1536, but the modern world (radio, tourism, etc.) further threatened the language. *Plaid Cymru* believed Welshness to be primarily rural and they idealized the *gwerin* (folk). Many of their campaigns therefore opposed the British state's incursions (for water, military training, even tourism) into those parts of rural Wales imagined as the cultural heartlands of the nation. Since the 1970s, campaigns by the more radical nationalist groups against holiday homes and in-migration to rural Wales attempted the same kind of defence, though using direct actions such as arson attacks. For most of the century, then, Welsh nationalism has operated within a linguistic definition of identity. More recently, however, this ethnic stress on cultural maintenance has been assisted by government support for the language through schools and through the establishment of S4C (the Welsh language TV channel). There has also emerged a growing civic nationalism in Wales – of which *Plaid Cymru* is a part – newly confident about issues of identity and language and critical of the unaccountable systems of government that emerged in the 1980s and early 1990s. Remarkably, the Welsh referendum on devolution in September 1997 which produced a narrow vote in favour of the establishment of a Welsh Assembly was, on the surface at least, less to do with national identity than it was with governance. Though this does not mean that tensions over the issue of identity have gone (and the closeness of the result proves that) they have, at least, been placed within a wider spectrum of national concerns.

Summary

- Nationalism is a very flexible or multidimensional ideology that can be manifested in a number of contexts.

- It has frequently been associated with groups highlighting poverty and economic and social injustice along ethnic lines, although several nationalist movements represent 'rich' areas.

- More commonly, nationalism interacts with ideas of ethnicity, race, language, religion, and so forth. In these forms it can be either culturally repressive or a reaction against cultural repression.

CONCLUSION

Nationalism, then, is an extremely complex – and even bewildering (*see* Fig. 21.5) – phenomenon, and one that is not always (as the popular use of the term implies) reducible to simple measures of 'good' and 'bad'. It can be liberating as well as oppressive, peaceful as well as violent, progressive as well as reactionary, traditional as well as modern, even rural as well as urban. Little surprise then that it has often been characterized as 'Janus-headed', after the two-headed Roman God. Also wrong is the tendency within established Western states to think of nationalism as something 'out there' on the margins. As Billig and Anderson have shown, nationalism (though maybe in the apparently more neutral and benign form of 'national identity') is as much a feature of those estab-lished states as it is of those nationalist groups strug-gling for expression and for sovereignty. A thorough and sensitive analysis of nationalism, then, should consider the context within which it is being expressed, the issues that it identifies as central, the balance between ethnic and civic strategies, and – importantly – the way in which these factors evolve over time. It should also look at how national identity – or competing identities – come to be politicized. And, of course, for geographers, considerations of space, territory and landscape are not only crucial but also open avenues of study that reveal in fascinating detail how powerful, creative, and often destructive, nationalism can be.

Further reading

- Samuel, R. (1989) (ed.) *Patriotism: the making and unmaking of British national identity*. London: Routledge.
A 3-volume series that, despite the title, was overwhelmingly focused on England! See in particular Volume 3 on 'National Fictions'.

- Lowenthal, D. (1998) *The heritage crusade and the spoils of history*. Cambridge: Cambridge University Press.
A provocative study of how history – commodified or packaged as 'heritage' – has been used to legitimate or gloss over certain political points of view.

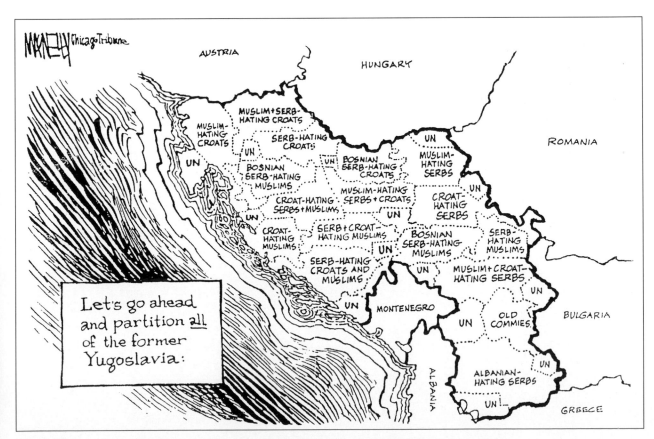

Figure 21.5 Let's go ahead and partition all of the former Yugoslavia. Credit: MacNelly/*Chicago Tribune*

- Short, J. (1991) *Imagined country: society, culture and environment*. London: Routledge.

A very readable introduction to the environmental myths of England, America and Australia, covering everything from city planning to westerns.

- Paasi, A. (1996) *Territories, boundaries and consciousness: the changing geographies of the Finnish-Russian border*. Chichester: Wiley.

An historical study of the role of geography in Finnish national identity – at the level of formal politics, in the school curriculum, and in popular memory.

- Leonard, M. (1997) *Britain™: Renewing our identity*. London: DEMOS.

Some marketing suggestions on what Britain's identity is and might be, from a think-tank favoured by the British government.

Social and cultural geographies

INTRODUCTION

It is understandable that people tend to resent being reduced to crude social determinants of class, gender, race, age, sexuality or nationality. Whilst we may often be guilty of stereotyping others, we want to be seen as individuals. On the other hand, with just a little reflection, it is also fairly obvious that we only possess and express individuality through these wider social and cultural dimensions: through the social relations we have to others (the three of us writing this could not be lecturers without students, fathers without our children); or through the social and cultural materials we use to construct our individuality (the languages we speak, the music we love, the places we cherish, the clothes we wear, and so on). Taking that even further, we might say that we only exist as individuals through social labels and categories, however much we may chafe against them at times. Think about our gendering. It is quite right to question simplistic judgements about all men or all women having certain kinds of characteristics. But being women or men is something that we all have to deal with. It is a part of what makes us who we are, both to ourselves and others. In conjunction with our **class**, **ethnicity**, **sexuality**, and so on, it structures our lives, too, impacting on how we are treated, the life chances we have, the expectations placed upon us. Generally, then, we cannot be ourselves outside of social and cultural contexts; and those contexts also structure how differing selves are differently valued.

Social and Cultural Geographers are interested in these social and cultural dimensions that shape our life circumstances and make us what we are. They address the ways they are constructed and the ways people, both individually and collectively, contest them. Whilst increasingly the two sub-disciplines have come

together, hence our combined treatment of them here, Social Geography pays particular attention to social relations, groups and inequalities, while Cultural Geography emphasises the values, beliefs, languages and meanings that are bound up with them. In both cases, a major concern is with exploring the two-way relations between the social–cultural and the geographical: first, then, with investigating how key geographical phenomena (such as landscapes, places and peoples) are socially and culturally constituted; and second, with understanding how social and cultural processes operate through these geographies of space, place and environment. The latter is worth emphasizing (see also Chapter 2 for a sustained analysis of it). In recent years Social and Cultural Geography has gone far beyond merely detailing the spatial distributions or environmental impacts of social groups and their cultural beliefs, while leaving direct investigation of these groups and beliefs to others such as sociologists and anthropologists. Instead, it has emphasized how geography is central to these social and cultural processes themselves.

The three chapters here take interconnected cuts through these relations between the social, the cultural and the geographical. In Chapter 22, Felix Driver discusses the enormously influential notion of 'imaginative geographies'. Crucial here, and initially this may seem paradoxical, is an emphasis on the imaginative and the fantastic as an important part of geographical reality (see also Chapter 6). More specifically, Felix illustrates this through thinking about the divisions made of the world into different cultures and nations. We may take these kinds of geographies for granted, as almost natural, but what this chapter shows is how in fact they are products of our imaginations, feeding our senses of who we are. In the process, this chapter also demonstrates how particular places and landscapes do not just have a material existence, but symbolic meanings, for example as national icons. This point is developed by Catherine Nash in Chapter 23 as she considers how Cultural Geographers study landscapes. Her argument is that landscapes are not just entities that we see in the world but 'ways of seeing' that world (Berger, 1972). To get a feel for this, think for a moment about how the word landscape is used in landscape painting or photography. Landscape here is not just something 'out there' that the artist or photographer looks at. It is also a kind of picturing, with its own particular, if changing, conventions for what counts as a 'good' landscape, for what is beautiful and what is not, and of where the viewer should be positioned. Catherine takes the example of so-called 'picturesque' landscapes to illustrate this, and in the process draws out how ways of seeing the world are far from being narrow, artistic concerns. Instead, and initially this may surprise you, we see how they are shot through with social relations of class, **gender**, **race** and **colonialism**.

In Chapter 24 Tim Cresswell shows how our understandings of place are also far more loaded than we may realize. Using as examples our distinctions of public and private and rural and urban places, Tim focuses on how we come to see people, things, or actions as being 'in' or 'out of' place. This, he argues, is an important component of social and geographical processes of exclusion, as well as opening up possibilities for social protest through acts of 'transgression' (being or doing something 'out of place'). You may feel this example is itself 'out of place' in a Geography textbook, which is telling, but think for a moment about a gay or lesbian couple holding hands or kissing in public. Heterosexual couples do this all the time. For gay or lesbian couples, however, such displays of affection are likely to attract voyeuristic attention from straight people, if not outright hostility (a classic homophobic refrain is 'I don't care what they get up to in private, I just don't want it done in public'). These actions are out of place, despite their legality, because dominant senses of public space are 'heterosexist'. Gays and lesbians are only included in the public arena on restricted and second-class terms, though this can be challenged through acts of geographical 'transgression', as when a Gay and Lesbian Pride march temporarily reclaims urban public space as not exclusively straight. Our understandings of places can therefore play a central role in processes of social and cultural exclusion and the contesting of these.

Further reading

- Cosgrove, D. (1994) Cultural geography. In Johnston, R.J., Gregory, D. and Smith, D.M. (eds) *The dictionary of human geography*, third edition. Oxford: Blackwell, pp 111–13.
A brief history of the approaches adopted by post-war Anglo-American cultural geographers.

- Crang, M. (1998) *Cultural geography*. London: Routledge.
A textbook review of current research topics and approaches within the so-called 'new' Cultural Geography.

- Hamnett, C. (ed.) (1996) *Social geography: a reader*. London: Arnold.
A collection including some classic papers in social and cultural geography, set in context through editorial commentaries on the scope and history of social geography. Particular emphasis is placed on the relations between gender, class, race and space; the construction and contestation of places and their meanings; and the contribution geographers can make to progressive social politics.

- Jackson, P. (1989) *Maps of meaning*. London: Routledge.
A highly influential book that signalled the coming together of social and cultural geographic concerns. Introduces some key theoretical concepts, and has examples ranging across issues of class, gender and sexuality, and race and racism.

Imaginative geographies

Felix Driver

INTRODUCTION

Geography is a subject which has always had a reputation for being down to earth: its focus is after all on the real world, the shape of its landscapes and the pattern of its use by human beings. This sense of the rootedness of the discipline in the material world is often associated with an image of the geographer as an active inquirer, engaged with the world rather than distanced from it. In the eighteenth and nineteenth centuries, this image was embodied in the figure of the intrepid explorer, determined to seek out the truth with his own eyes rather than rely on the speculations of 'armchair' geographers. Today, relatively few geographers consider themselves explorers in quite this sense; indeed, one British geographer has gone as far as describing himself as an 'extrepid implorer'! (Lowenthal, 1997). None the less, a sense of engagement with the world, and in particular a commitment to solving real-world problems – whether those of environmental degradation, poverty or injustice – remains a strong feature of a modern geographical education. This is one of the reasons why we continue to say that geography matters.

The purpose of this chapter is to consider how people imagine the geography of places, and why this matters. The inclusion of such a topic in a geography book might seem fanciful, given the emphasis on practical relevance which has had such an influence on the writing and teaching of modern geography. Why focus on 'imaginative' geographies when there are so many practical problems to deal with? Can't we leave that to other disciplines, like Literature or Art, concerned with fictions rather than facts, with subjective impressions rather than objective realities? Before going any further, we must first address these fundamental questions. The argument of this chapter – indeed, much of the work which has been done in cultural geography over the last forty years and more – depends on the answer we shall give.

IMAGINATIVE GEOGRAPHIES AND WHY THEY MATTER

In this chapter, the term 'imaginative geographies' refers to more than individual perceptions. While every human being is unique, in the sense that each of us experiences the world in a particular way, the images we construct are at the same time inherently social. Think of the words we use or the pictures we draw: these depend on shared systems of communication, which we call languages, available to a wider community. While writers in the West have long understood 'imagination' in subjective terms, associating it with creative licence or individual genius, there is no reason why we cannot think of imaginations in other ways. Ever since the origins of modern anthropology in the eighteenth century, cultures have been understood in terms of shared beliefs or values, common ways of thinking about the world; and at least since the birth of modern psychology, we have come to recognize that these beliefs or values may lie in the unconscious realm as well as in the world of thoughts and actions. Of course, there are many different ways of conceiving these ways of thinking or patterns of belief: the fundamental point here is that they are more than the offspring of individual minds. In other words, imaginations are social as well as individual.

Let us now consider the content of these imaginative geographies. Perhaps the easiest way to do so is to reflect on how we think about ourselves: how do I conceive my own identity, or yours? I have a birth certificate which tells me where I came from, and a passport which tells me I am a British subject; I may identify myself as belonging to a certain generation, **class**, **gender** and **ethnic** group. But do these alone define my identity? It very much depends how I imagine them to be related to each other, and how I imagine myself to belong, or not to belong, to a number of different communities; and, not least, on how others identify me. These senses of identification – both subjective and imposed – may well change as I grow older. Moreover, my answer to the above question – and I suggest yours too – will vary depending on the circumstances in which it is asked: at a border post, in a bar, on a train, at home. In other words, identities are very complicated things: they are shaped not just by our biological characteristics or our social positions, but also by images – those we ourselves compose to make sense of the world, those of others and those in the culture which surround us (Dyer, 1997). In this chapter we shall consider only a few of the ways in which cultures, nations and societies are imagined: the more general point is that these imaginative geographies help to shape our sense not only of places, but of our selves.

So imaginative geographies make a difference: that is to say, they are real. Think again about passports: these are documents made up of images – words, stamps and photographs – which together compose one kind of identity. That identity shapes our lives by making certain things – residence, nationality, mobility – available to us on certain terms. It simultaneously restricts our access to these things elsewhere in the world, though this will vary according to the nature of the images involved. This is one example of the real effects that images may have. Obviously it is a particular kind of example: the images which constitute a passport are enormously powerful, and mark the extent to which our lives are bound up with the power of states. Let us then take another, less obvious, example: that of images of childhood. How do these images shape the geography of the world we inhabit? The answer to such a question is inevitably complex, especially given the wide variety of ways in which childhood has been imagined across time and space. In Europe, for example, images of childhood have undergone considerable change, notably in the nineteenth century, the era in which mass schooling began. Many of our ideas about what it means to be, and look like, a child date from this era: think for example of the gradual exclusion of young children from the world of paid work, the emergence of child protection movements and the development of ideas about juvenile delinquency, all of them associated

with this period (Cunningham, 1992). The point here is that these changes were in some measure imaginative – they required new 'ways of seeing' childhood – and all of them had practical consequences for the geography of children's lives. The same might be said for other aspects of identity – such as our conceptions of masculinity and femininity, or able-bodied and disabled bodies, or madness and sanity. Images, too, have real effects (see also Chapter 6 by Mike Crang).

Summary

- The study of imaginative geographies takes images seriously: it treats words and pictures as both objects of study in their own right and as clues towards an understanding of the ways in which identities are constructed.

- Human geographers are concerned with the realms of the imagination, not as a contrast to, or an escape from, the real world 'out there', but because they help to make sense of, and indeed shape, that world.

IMAGINING CULTURES

How are cultures 'imagined'? While the word 'image' refers to more than pictures – for words are images too – it is with a graphic image that we will begin. Consider the picture in Figure 22.1. It forms the frontispiece of a *Concise History of the World*, published in 1935 by Associated Newspapers, which you might think of as the inter-war equivalent of the multimedia encyclopaedias widely available today. These reference works aim to provide within a single volume – today within the space of a compact disk – the complete story of humanity, from the very beginnings to the present day. As Figure 22.1 indicates, this story is sometimes illustrated by a kind of family tree, quite literally in this case, as the birth of 'man' is represented by a primitive figure emerging from a rather uncomfortable-looking tree of life. The subsequent history of humanity is portrayed through a series of emblematic figures, each representing a distinct civilization or culture, culminating at the foot of the page with an image of the modern family, complete with domesticated pet. The figure as a whole condenses a larger theme in pictorial form: history as a procession of figures through age after age, culminating in the modern era.

Let us examine this image a little more closely. The first point to make is that it represents the evolution of humankind as a sequence of stages, through historical time: indirectly, this owes something to the impact of

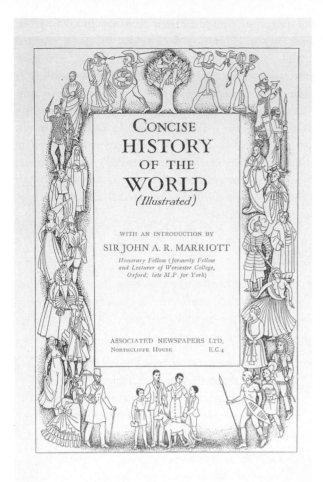

Figure 22.1 Frontispiece of *Concise History of the World (illustrated)*. Source: Associated Newspapers 1935

Darwin's writings on human descent. H. G. Wells' famous book, *The Outline of History*, first published in 1920, set the pattern for a large number of similar works, intended for both children and a wider popular audience (it is very likely that your parents or grandparents will remember similar volumes). In this case, it is not Wells' name but that of an eminent historian and man of affairs that graces the title page. Yet while this view promises nothing less than a total history of the world, an epic staged within the pages of a single volume, it presents this spectacle in a very particular way: as a stately procession or pageant, in which colour is more important than detail. Moreover, a fundamental parting of the ways appears early in the story. The figures on the left-hand side of the image clearly correspond to a sequence of civilizations: through the Persians, Greeks and Romans, to the medieval, the renaissance, the Victorian, and so on. Each figure is represented in subtly different ways, their posture and dress signifying the characteristics of their time: and there is a distinct sense of progression, from the era of martial prowess, through that of courtly ritual, to the sedate and modest world of the

nuclear family. This is a story of progress by domestication: an evolutionary tale in which, ultimately, the values of the modern win out not only over the primitive, but perhaps over evolution itself. But on the other side of the image, the same logic does not seem to apply. Here we find a mix of cultures, including the ancient Egyptian, Chinese, Japanese, Indian and native American. There is no sense of progression: as if to emphasise this, a naked African couple have been placed at the foot of the chain, complete with spear and shield. The message is simple: human history has been divided into two, the West portrayed in evolutionary terms as the domain of progress, the destiny of the human race; the rest pictured more as a spectacle than a pageant, located firmly in geography rather than history.

Looked at in this way, the image in Figure 22.1 can be related to much wider traditions of thought about world history: it may, for example, be interpreted as a popular version of the more scholarly traditions of 'Orientalism' analysed by Edward Said in a very influential work published in 1978, in which the term 'imaginative geographies' was first coined. Two aspects of Said's argument are relevant here. First, he suggests that non-Western cultures in general (and those of the 'Orient' in particular) have often been represented by Western commentators as being static and backward. Second, he argues that these images have played an important part in the historical construction of a contrasting image of Europe – and the West in general – as dynamic and progressive. For Said, such imaginative geographies – based on this binary opposition between the West and the Rest – have played an important role in the history of global power relationships during the nineteenth and twentieth centuries. Whatever the merits of this argument, which has been hotly debated (Gregory, 1995; MacKenzie, 1995), it does seem that something like this pattern is pictured in Figure 22.1. This image raises questions not only about our images of global history, but also about the ways in which geography is imagined (Hall, 1992). The question then becomes: how might we imagine a different picture for a world history, one which was not structured by a binary opposition between the West and the Rest; one, for example, which paid attention to the *interactions* between Europe and the rest of the world? We might begin to answer such a question by looking again at the parting of the ways represented at the top of the image; or perhaps by considering the point at which the two streams come together, in what appears to be an unresolved point of tension between the African warrior and the nucleated family group at the bottom.

Edward Said's work emphasizes the role of the

Other in the construction of imaginative geographies of culture; that is to say, he is concerned with the ways in which European history has often been conceived on the basis of an essential opposition between the civilized European and the barbarous native. While it would be too simple to suggest that this was the only way in which Europeans have imagined other cultures, it is remarkable how common such stereotypes have been, even where the intention has been to disrupt them. Look, for example, at Figure 22.2: this image was designed to represent the philosophy of a mid-Victorian pressure-group, the Aborigines Protection Society, which was a forerunner of modern organizations such as Anti-Slavery International and the Minority Rights Group. The image contains a group of figures, each representing a different 'branch' of the human family (see also Barthes, 1972). Beneath is a Latin motto – *Ab Uno Sanguine* ('From One Blood') – which summarized the Society's faith in the unity of the human race, in opposition to those who maintained that the world's different ethnic groups had entirely separate origins. The image portrays a single group of men (evidently women were not thought to be suitable models for

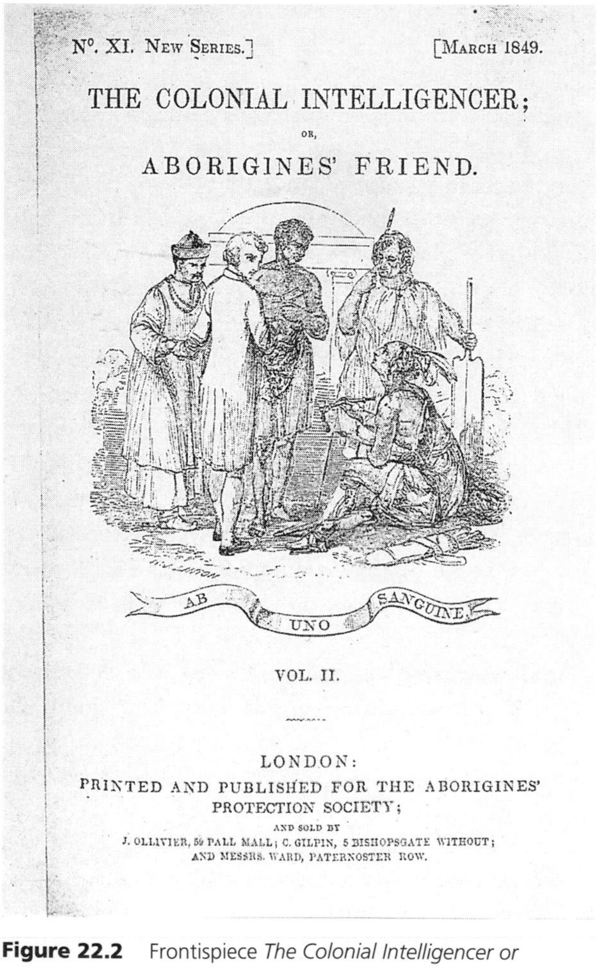

this purpose), united by their common humanity. Yet, on closer inspection, it is clear that this is an unequal family: the white man, fully clothed, is essentially portrayed as an enlightened philanthropist, standing in relation to the others as a father does to his children. On the one hand, then, the image affirmed the unity of humanity, in marked contrast to those Victorians who maintained that there were essential and innate differences between different '**races**'; on the other, however, a sense of hierarchy was maintained, though this time it was cultural (i.e. learned) rather than biological (i.e. innate).

Summary

- Images of culture (for example, ideas about the 'family of man') draw upon more extensive collective imaginations about cultural difference. They are, in other words, rarely the product of mere fancy.

- Such images transmit messages about the global geography of cultures to a wider audience, in school texts, popular books, film or television, for example. This assumption provides the starting point for works such as Edward Said's influential book, *Orientalism*.

- These images can be regarded as 'real', not because they reproduce the world accurately, but because they reflected and sustained people's imagination of that world; and in turn, helped to influence the worlds we still inhabit.

- Consider the image in Figure 22.3, which is taken from a missionary school-book published at the end of the nineteenth century. In what ways do you think this image might reflect 'imaginative geographies' of Africa then prevalent in Europe? You might consult Brantlinger (1985) on this issue.

Figure 22.2 Frontispiece *The Colonial Intelligencer or Aborigines' Friend*. March 1849

IMAGES OF THE NATION

Both the images we have discussed so far were produced in England; indeed, their Englishness is readily apparent in their composition. As this suggests, it is difficult to separate out ideas about cultural difference from ideas about nationality. Indeed, some people regard the two as identical: the English are thus said to have a different 'culture' from the French, the Japanese or the Brazilians. It is not uncommon to find such views expressed in debates about immigration or education: in England, for example, more than one politician has claimed that English culture needs to be defended against the 'threat' of outside influences, whether these are American, Jamaican or European. But what is a national culture? Is it something natural

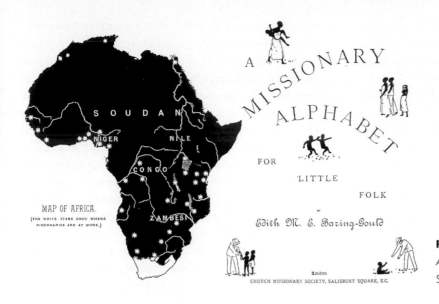

Figure 22.3 Frontispiece to *A Missionary Alphabet for Little Folk*. Church Missionary Society, London, 1894

or created? Does it grow from within the nation, or does it depend on relations with other cultures? Is it homogenous or heterogeneous? Such questions are addressed elsewhere in this volume (see Chapter 21). Here, we shall focus on the ways in which nations, too, are imagined. They do not simply exist by virtue of location: they are constructed, not least through the mobilization of images.

Take the example of Englishness. How do you imagine England? Of course, different people may have different answers to that question, not least depending on whether they are English; and, moreover, even those of us who are English might well imagine England differently depending on where we are located, in both space and time. The England of Ealing is different from the England of Exeter; the England of the 1930s would be pictured very differently (one imagines) from the England of today; and the England remembered when we are in other parts of the world is something else still! In other words, to speak of a single way of imagining the nation is clearly misleading: the word 'England' means different things to different people in different times and spaces. Nevertheless, it would be wrong to conclude from this that images of national identity are simply a matter of personal preference. As geographers and historians have shown, there are some images of national identity which become dominant at particular times; and frequently, though not always, these images take geographical forms. England has often been pictured through landscape imagery supposed to represent the 'essence' of the nation: sometimes these have been rural landscapes (as in the image of Constable country popularized from the late nineteenth century), at others they have been urban (as in the image of St Paul's during the Blitz). The task of the cultural geographer working on these images is to show how they have become such potent icons of Englishness, in dif-

ferent ways at different times (Daniels, 1993; Rose, 1995). Two conclusions may be drawn from such work. First, the imagery of national identity is liable to constant change: while images may endure, their meanings may not. Second, the imagination of national identity is not a process which works exclusively from within, as if emerging from the heart of the nation: it is always constructed relationally, by contrast with an 'other', wherever that other is located (Hall, 1995).

These two points may be illustrated with reference to two quite different examples. First, consider the role of a building such as Tower Bridge in the making of the images of British national identity. Images of this Bridge continue to circulate world-wide, thanks to the tourist industry, in the form of postcards, jigsaws, tea towels, posters and models of all kinds, and the Bridge has been used in many films to provide an authentically English backdrop. If the building appears to represent the perfect tourist **icon**, it is because it simultaneously represents so many different aspects of the imagined nation: indeed, Englishness is only one aspect of its fame, and perhaps a rather minor one at that. In its neo-gothic grandeur, the Bridge literally towers over the Thames, dramatizing the role that the river has played in the history of London, and by extension the course of the British empire; and its location, adjacent to the Tower of London, evokes the ancient history of royal London, complete with Beefeaters, jewels and princes. And yet, the Bridge itself was a very late addition to this landscape: it was completed little over a hundred years ago, in 1894, just in time for Queen Victoria's Diamond Jubilee. This was in fact a thoroughly modern construction, whose mock-medieval exterior hid from view the huge steel mechanism which actually operated the bridge. The contemporary response to its erection was mixed, to say the least: dismissing it as 'a pretentious piece of bad medievalism', one critic complained that 'a more wretched piece of architectural falsehood and vulgarity was never perpetrated' (Driver and Gilbert, 1998: 16).

To the late twentieth-century reader, these contemporary reactions may come as something of a surprise: we find, after all, that the meaning of this landscape was contested at the moment of its creation. Yet, for all its fakery, Tower Bridge soon established itself as a symbol of imperial London *par excellence*; indeed, given its design and location, it is difficult to imagine a more appropriate monument to an age of 'invented traditions' (Hobsbawm, 1983). A thoroughly modern building, made to look old – the last word in imperial kitsch? As if this were not enough, we soon find the Bridge appropriated for different sorts of meanings – as a prop for commodity advertising, for example (*see* Fig. 22.4). Here the Bridge has become a vehicle for a different kind of imagining: one that associates national greatness with mass consumption. This example illustrates a larger process, in which images of the nation are exploited for a range of different ends: once again, it is clear that such landscapes have no single, eternal meaning. They are how they are imagined.

Because of their location, buildings like St Paul's or Tower Bridge have often been represented – sometimes seriously, at other times ironically – as the heart of the nation (Daniels, 1993). But nations are imagined in other ways too, often by contrast with other peoples (see Ryan, 1997). For a second example, consider the images in Figure 22.5, which appeared in a book entitled *Life in the Southern Isles*, published by a missionary (and incidentally, a Fellow of the Royal Geographical Society) in 1876. They show two contrasting engravings of a Pacific Island village scene, before and after conversion to Christianity. The first scene depicts the 'natives' engaged in a ceremonial dance, wearing little but grass-skirts and shells, abandoned to a hedonistic life; in the second, they have been transformed into a sedate community, domesticated and cultivated, acquiring the virtues of labour, exchange and civilization. Such 'before and after' images were a staple part of missionary writings during this period, and were clearly designed to promote the missionary cause in Europe: they often had little to do with the realities on the ground. Yet these images mattered, not just because they reflected the imagination of missionaries, but also because they helped to shape the actions of their supporters 'at home'. Indeed, looking more closely at the second image, you will find that the 'heathen' natives have not simply been converted to Christianity and commerce: they and their landscape have evidently been Anglicized. The domestic scene in the second image portrays a decorous pastoral scene, reminiscent of some English landscape, in which people know their place; and the landscape is as ordered as the society (see also Chapter 23 by Catherine Nash). Together, the two scenes combine an historical sense of transformation (the impact of Europeans on the South Seas) with a geographical restructuring: the island has, imaginatively at least, become England.

Summary

- Nations may be understood as imagined communities, constructed not merely on the basis of a shared location but through the mobilization of images.

- Images of the nation (of England, for example) vary considerably over time and space; they do not have a single, eternal meaning. The processes by which some images come to be regarded as dominant is a major theme for Cultural Geographers to study.

- Images of national identity do not work exclusively from within; they are frequently constructed relationally, by

Figure 22.4   Gateway to Empire – or the Commodity Triumphant? (from R. Opie, *Trading on the British Image*, 1985)

A VILLAGE IN PUKAPUKA, UNDER HEATHENISM

THE SAME VILLAGE, UNDER CHRISTIANITY.

Figure 22.5 An imaginative geography of missionary work: a village in Pukapuka (a) under Heathenism and (b) under Christianity. Source: W. Gill, *Life in the Southern Isles*, 1876

contrast with an imagined 'other'. This process is particularly evident in the cultural geography of imperialism.

CONCLUSION

Images of the kind we have considered in this chapter help to compose 'imaginative geographies': they structure people's understandings of the world, and in turn help to shape their actions. They have a real existence and real effects; in other words, they matter. In some circumstances, it may be argued that particular instances of an imaginative geography are false or partial – the example in Figure 22.5 clearly reflected a particular point of view, which was evidently not that of the 'natives' themselves. Yet to make this case is not to say that such images are inconsequential. Whether or not we regard them as essentially true or false, such

imaginative geographies have significant implications for the way in which people behave.

While most of the examples in this chapter have been drawn from British history, you might want to consider the ways in which 'imaginative geographies' are produced and consumed in today's world, through television, film, popular magazines, advertising or tourism, for example (Anderson and Gale, 1992; Lutz and Collins, 1993). In this context, the impact of '**globalization**', discussed in many other chapters in this book, is of particular concern. Some geographers maintain that, culturally speaking, this process is reducing the differences between places over the globe; others argue that in many respects it is amplifying them (Massey, 1994). In this context, we might take the historic theme park as an exemplary site for the study of imaginative geographies. What, for example, are we to make of the Japanese fascination for reproductions of 'exotic' cultural landscapes in their own country? In Maruyama, a major transnational corporation was commissioned by the city authorities to construct a 'Shakespeare Country Park' (completed in 1997), effectively a concrete abstraction of Elizabethan culture, complete with townhouse, inn, theatre, garden and village green. We might perhaps expect this to be no more than an exercise in consumerism, the Disneyfication of Merrie England. Yet it has been argued that the attention to detail manifested in the architecture and design of the site, together with the broader tradition of recreational pilgrimage within Japanese culture, may be interpreted more positively (Chaplin, 1998). Whatever the merits of this particular project, it highlights the intensification of the role of imaginative geographies within contemporary forms of consumption (see Chapter 12).

The case of the contemporary theme park raises a more general methodological issue: how should we go about interpreting the meaning of imaginative geographies? This is no easy task: indeed, if it were, there would be no need for you to read any further! To interpret a theme park, a text or an image, it is clearly necessary to enter the world of those who created it; to understand their ideas, values and relationships. Once we accept that the process of creation, or production, involves much more than a single intention, and usually requires the participation of many different people, the matter begins to look much more complex than at first sight. But that is not all: a theme park or a museum, for example, is likely to convey quite different meanings for its visitors, and they too will carry with them a wide variety of experiences, preconceptions and desires (Karp and Lavine, 1991). Interpreting an imaginative geography, then, turns out to require us to think not only about their content, but about their form; and not only about what they say, but how they have said it and to whom.

While we may find it useful to think of imaginative geographies as 'maps of meaning', we must not forget that they are also social processes, moments in the making of the worlds that we inhabit.

Further reading

- Daniels, S. (1993) *Fields of vision: landscape imagery and national identity in England and the United States*. Oxford: Polity Press.

A series of thoughtful studies of landscape imagery and national identity by a cultural geographer, making useful connections beween the visual arts, popular culture and imaginative geographies.

- Hall, S. (1992) The West and the rest. In Hall, S. and Gieben, B. (eds) *Formations of modernity*. Oxford: Polity Press.

An accessible account of the significance of 'imaginative geographies' for contemporary understandings of global history and culture, paying particular attention to the history of European exploration of the 'new world'.

- Lutz, C. and Collins, J. (1993) *Reading National Geographic*. Chicago: University of Chicago Press.

A challenging study of the world's most popular geographical magazine, focusing on the role of photographic imagery in the representation of place, race and gender.

- Rose, G. (1995) Place and identity: a sense of place. In Massey, D. and Jess, P. (eds) *A place in the world?* Oxford: Oxford University Press, 87–132.

An excellent account of the idea of a 'sense of place', exploring the role of visual and other images in the construction of national and other identities.

- Said, E. (1995; orig. 1978) *Orientalism*. Harmondsworth: Penguin, 2nd edition with afterword.

The classic work on the 'imaginative geography' of the Orient as produced in political, scholarly and literary writings in the West. I suggest that you read this after having introduced yourself to the themes discussed in Rose (1995) and Hall (1992).

CHAPTER 23

Landscapes

Catherine Nash

UNDERSTANDING LANDSCAPE

The study of landscapes has been one of the longest traditions in human geography. Human geographers have mapped distinctive landscapes, tried to reconstruct landscapes in the past and traced the social processes which have produced and continue to shape the landscapes of today. Landscapes are also understood not just as physical environments, but as outcomes of particular ways of thinking about places, depicting them and giving them meaning. Geographers have used different **metaphors** to help explain how landscape has worked in different contexts. Landscape has been talked about as a way of seeing (Cosgrove, 1985), a text that can be read (Duncan, 1995) or a theatre where society acts out its dominant values (Cosgrove and Daniels, 1993). Whichever metaphor is used, geographers today understand ideas of landscape as ways of shaping the world and its meanings that have real social, economic and political effects. Cultural Geographers are especially interested in the connections between ideas of landscape and other ideas about the world and how it is organized, especially relationships between people, that are expressed in cultural forms like writing or painting. This chapter explores the idea of landscape from the approach of Cultural Geography.

Guides to better photography for amateurs and enthusiasts often follow technical instructions with advice on how to take different kinds of pictures: action shots, portraits, still life and landscapes. Like many other systems of classification, these different categories of things have been devised by people to try to understand and organize the world. Landscape is one of these important categories through which people try to make sense of the world and their place within it. Though it might seem like a natural or obvious way of looking at and representing the countryside, the familiar format of rural landscape imagery is a cultural convention established in particular times and places and by particular people. Even with a simple camera it is obvious that choosing to take the photo when the view finder contains a scene focused on the mid-ground with a distant horizon and trees framing the view is just one option. In other times and in other places there have been very different ways of thinking about representing the natural world. Western ways of organizing, understanding and representing natural environments as landscape are only one set of traditions of relating to and thinking about the world. There are many other cultural ways of considering and managing the environment. This chapter explores Western landscape traditions, as one example of the ways in which how landscapes look and how they are represented are caught up in the shaping of ideas – about **race**, morality, beauty, political authority, national identity, geographical knowledges – and social relations – between women and men, between different classes, between the colonial powers and the colonized. Since the category of landscape is most often associated with the countryside, it is also linked to ideas of nature, the environment and **ecology**. Landscapes reflect the values of a society, especially those of its most powerful members, but they can also be used to try to reinforce or challenge dominant and taken for granted ways of thinking about and organizing society. As we will see in this chapter, the questions of who owns what (the distribution of property or wealth) and who does what (the social division of labour) are key examples of issues which have been connected to the control, organization, representations and meanings of landscapes.

This chapter first of all explores the origins and implications of one Western landscape tradition in order to show that what may seem the only, 'natural', or best way of representing landscape is a particular cultural and social tradition tied up in particular ways with other culturally and historically specific ideas of **gender**, **class** and **race**. Second, it traces the role of this influential visual tradition in broader processes of Western **imperialism** and **colonialism** (see also Chapter 29). The third and final section considers the way in which this tradition is resisted or reworked in contemporary contexts, particularly in countries that were formerly colonies, and affected by ideas of British cultural superiority, as well as white settler policies towards indigenous people.

Summary

- Landscapes on the ground and images of landscapes often convey messages about the societies in which they were produced.

- In turn, landscapes are shaped by ideas about society.

- Landscapes are always caught up in social relations.

THE PICTURESQUE AT HOME: CULTIVATING CIVILIZED VISION

The late eighteenth and early nineteenth century was a particularly important period in the development of landscape traditions in Britain, especially for the style called 'picturesque'. Though this term is used nowadays to describe pretty or picture postcard scenes, at this time the term had very specific meanings and connotations. The development of this style was linked to the new importance of images of the countryside in art. As the traditional landed classes became more involved in the active development of their estates at home and abroad and as the middle class emerged from the growing commercial and industrial sector, new frameworks for understanding their place within the world and depicting their land developed. Not only did landowners commission paintings of the land they owned but there was a new and growing metropolitan middle-class market for landscape painting. Many guides to travel and taste were produced for this market with advice on where to visit, what to see and how to see and sketch 'properly'. Being able to share an appreciation for picturesque scenes became a measure of good taste and civilized status. Picturesque landscapes seen on canvas or in travel were panoramic but usually framed by large trees or craggy outcrops. Organized into foreground, midground and background they depended on the viewer looking down

and out over a vista along a zig-zag path to a distant horizon and back to foreground details (*see* Fig. 23.1). This way of seeing emphasized a distant and elevated viewpoint, unobstructed by the branches or hills that would limit what could be seen. This approach to landscape was not confined to visual representations. Written descriptions in poetry or topographical writing could also convey ideas of an ordered, distanced and linear vision instead of a sense of immersion in the landscape and its sounds and fragrances and sights. Being able to write about, sketch or look at landscapes in this way was thought to be proof of good taste, of being cultivated, refined and of high social standing. Images and ideas of landscape were deeply tied to attempts to make sense of what was felt to be a modern world and to define the modern individual.

But this idea of landscape was not just a matter of taste. The idea that there was a certain 'right' way of seeing landscape was not neutral but was used to legitimate political authority. The picturesque was based on the idea that by eliminating all the messy and confusing detail in the scene and through the objective detachment of the viewer, the true and abstract qualities of the landscape could be seen. Upper-class thinkers argued that this ability to think with rational detachment, in general terms and about abstract ideas was at the same time a requirement for political authority (Barrell, 1990). These men who defined their good taste and social standing through their ability to recognize and enjoy picturesque landscapes did not extend this privilege or pleasure to others. A private income was needed, they suggested, to both govern impartially and see landscape with a proper distance. The lower classes and especially those who worked the land, they suggested, were not able to see landscape in this way, since they supposedly lacked the distance and objectivity which defined both the ability to govern and to see the world in appropriate ways. They were thought to be too close to the land to see it 'properly', and too caught up in making their own living to consider the good of society as a whole. This role of governing for the good of all, claimed by the upper classes, usually also meant securing and reinforcing their own privileged position. This tradition of landscape, then, was developed by a particular section of society and used to make claims about the characteristics of different classes and their place in the social order, especially the place of the rural poor.

The people who worked the land were seen as part of the landscape, like the trees and crops, not as viewers of it in their own right. But they were also depicted in particular ways. As the conditions of labour for the poor worsened in the early nineteenth century, the middle and upper class grew increasingly anxious about the possibility of rural unrest. At the same time

Figure 23.1 Picturesque landscapes.
Source: W. Gilpin, 1792: *Three Essays: on Picturesque Beauty, on Picturesque Travel and on Sketching*

there was a shift away from images in poetry and painting of a harmonious and idyllic rural world in which 'nature's' abundance required little labour, to images which emphasized the morality and necessity of hard work (a tradition sometimes known as Georgic). With this new model of 'nature' and human relationships to the natural world, landowners could justify both their schemes of agricultural improvement to increase the profitability of the estates and their philosophy of economic individualism – the freedom to pursue profit unhindered by the constraints of tradition – often at the cost of the rural poor. Hard work was seen as natural and a productive landscape was seen as beautiful. So, if the rural poor were depicted within the landscape they had to be seen as emblems of hard and sober labour rather than as individuals with all their potential to resist a social order that oppressed them (*see* Fig 23.2). Like the light and shade of landscape, society, it was suggested, was ideally and 'naturally' stratified into the rich and poor. The rural poor were located firmly 'on the dark side of the landscape' (Barrell, 1980). But as the estates of landed gentry and the new middle class were landscaped to make them more picturesque, landscape taste was increasingly caught up in questions of money, morality and refinement (Daniels and Seymour, 1990). Landscape was deployed not only to

make distinctions between classes but to define the civilized and polite amongst those with money to buy landscape images and landscape their estates. Ideas of landscape were thus deeply connected to the development of a specifically upper-class model of freedom and individualism upon which commercial **capitalism** depended. When English aristocratic men made connections between landscape taste and political authority in the eighteenth century, they were attempting to define and differentiate themselves from the lower classes. But they also used ideas of landscape to claim that they were different from women and from other 'races'. As well as lower-class men, women in general and other races were not thought to be civilized enough to be able to appreciate landscape or to govern themselves.

Summary

- The development of picturesque landscape imagery in Britain in the eighteenth century was tied to questions of money, subjectivity and political authority.

- Ideas of landscape involved questions of how to see landscape and how to represent the poor in the countryside.

Figure 23.2   John Constable, Landscape, Ploughing Scene in Suffolk (A Summerland) c. 1824. Credit: The Yale Center for British Art, Paul Mellon Collection

THE PICTURESQUE ABROAD

These picturesque landscapes do not only matter in Britain, but have much wider global significance, especially through histories of British colonialism and settlement. Ideas of landscape improvement and beauty in Britain were linked to places overseas, as the English gentry and growing middle class were increasingly making fortunes within early global networks of trade. The development of colonial plantation systems in the Caribbean or the settler colonies of Australia, Canada, New Zealand and South Africa involved, not only the movement of goods and people but also of ideas. The criteria of landscape beauty and the models of describing and representing landscape developed in Europe were exported to very different places with the assumption that they could be universally applied. Artists and writers frequently celebrated in paint or in print those places that seemed most similar to picturesque places in Britain and criticized other colonial landscapes for their difference from English ones. For some arriving in Australia in the early nineteenth century, the land was most unpicturesque; others reordered the landscape in drawings and engravings in order to make them conform to English tastes in landscape painting (Smith, 1985). This sense of the superiority of English landscape aesthetics was linked to a broader certainty that English ways were the best ways of doing things and of their natural superiority and authority over other people and places. Ideas of landscape were involved in the multiple ways in which European expansion within imperial trade and **colonization** was **naturalized** and legitimated (Mitchell, 1994). This again involved both the way in which landscape itself was supposed to be seen and depicted, and the **representation** of indigenous people within landscape imagery (Bunn, 1994). These ways of seeing had important implications for how places and relationships between people were organized.

We can use recent work on the West Indies as an example here. Recently geographers and cultural critics have begun to explore the colonial dimensions of key texts of English literature and key icons of Englishness. One such icon is the stately home with its allusions to the glamour and splendour of a different kind of England. In the eighteenth century, attempts to make them more productive and picturesque were frequently funded by wealth being made through colonial trade and overseas plantations. So as places such as Herefordshire were being depicted in painting, poetry and agricultural texts as picturesque, Georgical and national landscapes central to British economic prosperity, places such as the West Indies were also being understood and organized through English discourses of landscape productivity and beauty (Seymour, Daniels and Watkins, 1994). English travel guides and topographical accounts frequently concentrated on the description of places that could be most easily accommodated to the conventions of the picturesque and likened the scenery to the famous examples of European landscape painting. Like the detailed descriptions of cider making or the importance of oak trees in England within Georgical poetry, English poets encountering the West Indies wrote in similar ways of sugar cane, increasingly the staple of English plantation agriculture in the Caribbean. Thomas Hearne who painted estate portraits for landowners in Britain also depicted colonial estates through the framework of the picturesque and the Georgic (*see* Fig. 23.3). His image of Antigua both depicts the landscape through the conventions of the picturesque – here the palm trees frame the scene – and paints a productive landscape of sugar cane cultivation and harvesting. Slaves are depicted as happy workers. These ways of understanding landscape defined the West Indies within English cultural frameworks and as British colonies. English estates and their owners were often deeply involved in imperial systems of exploitation. Timber from estates in Herefordshire, for example, was transported to Bristol down the Wye and Severn Rivers, where ships designed for the slave trade were built and sailed between Bristol, Africa and the West Indies (Seymour, Daniels and Watkins, 1994: 42). In these ways, the landscapes of estates in Britain and the West Indies were connected through concerns about labour, productivity, land improvement, landscape aesthetics and imperialism.

In this colonial context, images of landscape could be used to legitimate and support colonial rule. This involved not only establishing authority but in securing a sense of belonging for settlers in these new places. British settlers in overseas colonies were concerned not only with the potential of land to support them but also with their feelings of belonging in these places. The African landscape, for example, was often represented as an arena through which the settler was free to roam (Bunn, 1994). In the poem 'Evening Rambles' by Thomas Pringle, who in the 1820s led a group of setters to Britain's Cape Colony, the poet wanders at ease through the landscape unimpeded by its native inhabitants. This image of a landscape empty of indigenous people was a significant feature of much European colonial travel writing and topographical description and was used to legitimize its colonization. These panoramic descriptions met European landscape tastes but they also allowed a commanding and prospecting gaze to travel across the land, scanning 'possibilities for the future, resources to be developed, landscapes to be peopled or repeopled by Europeans' (Pratt, 1986: 144). This idea of ordered picturesque vision and free movement was linked to a concern for the order imposed by colonial settlement and the opportunities afforded by free market capitalism. But freedom and mobility were dependent also on the idea of permanence represented in the image of the settler homestead and on reproducing at least some of the familiar features of the environment of Europe. Both visual and literary representations of the landscape used the conventions of the picturesque to make the landscape seem familiar (*see* Fig. 23.4). But the environment was also altered in material ways. In the nineteenth century, for example, species of plants, animals and birds from Britain were introduced in New Zealand to make settlers feel more at home. Again, ideas of landscape were used to make sense of the place and make the presence of British settlers seem natural and legitimate. These ideas of belonging were based on highly gendered roles for European men and women and on very specifically subordinate roles for native people. By depicting the landscape as an arena for *men* to work and play, images of landscape were also used to naturalize particular roles for men and women within European settler societies. White women, instead, were firmly located within the private domestic world of the home (*see* Fig. 23.5). In colonial landscape imagery women and men, natives and settlers were assigned different locations within both landscape representations and the landscape

Figure 23.3 T. Hearne, A Scene in the West Indies.
Credit: © The British Museum

Figure 23.4 Encampment in the Great Namaqua Country. Source: Francois Le Valliant, 1790: *Travels into the Interior Parts of Africa*

Figure 23.5 The Bechuana boy. Source: T. Pringle, 1834: *African Sketches*

itself. Indigenous peoples frequently figured as the potential or actual work force for European capitalist projects, as exotic fauna merged in nature or were simply missing from the landscape. These colonial landscape traditions continue to play their part in conflicts over the cultural value, meaning and use of places today.

Summary

- Western traditions of landscape were part of broader processes by which cultural and political authority were asserted in colonial contexts.

- Images of empty colonial landscapes were part of a discourse which supported colonial control and capitalist development overseas.

- Landscape imagery was used in colonial contexts to shape and reinforce ideas of racial and gender difference.

POSTCOLONIAL LANDSCAPES

This work on picturesque landscape conventions in colonial contexts draws on '**postcolonial**' theoretical approaches. Though sometimes used to describe countries which were formerly colonies, the term is most usefully used to describe perspectives which are critical of the past and ongoing effects of European colonialism. The visual traditions of Western landscape **representation** defined the meaning of places through how they looked. But there are other ways in which land can be meaningful. The metropolitan painters and travel writers who visited the countryside in Europe or the colonial world in the late nineteenth century had little access to the ways in which local and indigenous people understood their environments through stories, family histories and genealogies, myths, rituals and memories. Though metropolitan visitors often romanticized a nostalgic view of the deep local knowledges of those they encountered, these alternative traditions are again not a 'natural' way of knowing landscape for people instead often imagined as close to nature, but offer different sets of cultural meanings. These conflicting ways of knowing the land or environment have been and often still are linked to social inequalities. These postcolonial politics of landscape can be explored through Jane Jacobs' work on Aboriginal land claims in the 1980s (Jacobs, 1988) and the more recent production of Aboriginal heritage landscapes (Jacobs, 1996).

Both early settlers and their later descendants in Australia used ideas of landscape to justify their presence and later to find a sense of nationhood or Australian identity through claiming a meaningful relationship with the landscape. They also had to negotiate the challenge of indigenous ways of living in and knowing the land. For economic, political and cultural reasons Aboriginal claims to the land had to be denied. In Australia today attempts are being made to redress the injustice of Aboriginal dispossession. In 1993 the Australian Federal Government passed the Native Title Act which acknowledges the pre-colonial authority over the land of Aboriginal people and provides for contemporary Aboriginal claims to land and compensation. This undid one of the founding myths of White Australia; the idea of *terra nulis* or land belonging to no-one before colonization. This process of 'reconciliation' through funding Aboriginal community projects, recording indigenous sacred sites and responding to land claims has a complex politics, caught up as it is with ideas of nature, identity, nationhood and landscape.

The most successful Aboriginal groups in claims to land rights are often those whose can prove to mainstream Australian society and most especially to the government that their claims are legitimate. This may seem a simple process but it often involves Aboriginal groups negotiating external sets of criteria as to what counts as a legitimate claim and 'authentic' tradition. As Jane Jacobs shows, white Australians frequently perceive some Aborigines to be more authentically Aboriginal than others especially if they

live on the land over which they hold customary rights and interests, they speak an Aboriginal language as a first language, they still know the details of their traditional mythology and lore and can demonstrate this by their continued participation in the various rituals which among other things, sustain the spirituality of the land.

(Jacobs, 1988: 250)

Thus when urban Aboriginal groups appeal for land rights they not only have to compete with powerful mining or agricultural investors with economic interests in the land but must emphasize the kinds of cultural attachments which meet the expectations of external groups and adapt their ways of knowing the land to the official methods of recording the Aboriginal **cultural landscape**: making lists of sacred sites and mapping territorial tribal boundaries. This means emphasizing particular sites over the mythological tracks linking these sites, and emphasizing a clearly delimited territory over the more fluid and flexible boundaries of traditional tribal territories. What is understood as the Aboriginal landscape is constructed in this process as these groups negotiate often very romantic and **ethnocentric** notions of indigenous Australian society.

Ideas of authenticity are also deeply embedded in the production of Aboriginal heritage sites and parks such as rock art trails. Early this century the development of key sites for White Australians to visit was part of a process of affirming a secure sense of belonging in Australia. It also entailed displacing native knowledges and people themselves as Aboriginal groups were forced to live in state reserves (Huggins, Huggins and Jacobs, 1995). Recently tourism is drawing not so much on the European aesthetic of an empty land of grandeur, but on often romantic notions of a better relationship to the land and to nature supposedly found in pre-contact Aboriginal cultures. Again as Jane Jacobs shows, Aboriginal artists and communities involved in state-sponsored heritage projects, have to deal with the ways this notion of a true, timeless and pure Aboriginal relationship to the land seems to afford value to Aboriginal culture as at the same time it devalues existing urban and de-tribalized Aboriginal lives, cultures and identities.

In 1993 Brisbane City Council commissioned artists Laurie Nilsen and Marshall Bell of an Aboriginal visual arts company to create a walking trail at J. C. Slaughter Falls, Mt Coot-tha, on the outskirts of the city (Jacobs, 1996). In the earliest days of white settlement Mt Coot-tha attracted sight-seers who could survey the surrounding landscape and its prospects for settlement. Like the gardens of English estates, here the vegetation was also cut to afford the best vistas. At one level the new Aboriginal rock art trail which guides the viewer through an alternative set of cultural meanings can be seen as a form of postcolonial appropriation: reclaiming place and re-asserting a different identification with the landscape. But this site can undermine simple ideas of a timeless and fixed native culture. The rock art is not original but re-created by the artists working with native elders' permission and the public funds (*see* Fig. 23.6).

Though the rock art looks like an original site, its art is deliberately made to need periodic re-painting by members of Aboriginal communities. Instead of creating a static image of Aboriginal landscape **iconography**, it is a site where culture is shaped through landscape imagery as an ongoing project. The artists undermine the notion of a fixed set of landscape meanings and upset the simple oppositions between authenticity and inauthenticity. Instead of neat ideas of native and Western landscape meanings, identities and cultures, the landscape can be part of an oppositional politics as it is marked through an 'unruly' mix of inherited traditions and **hybrid** cultural forms. In turn, in Britain today Black artists are exploring how images of English landscape have been based on a very white version of Britishness and are creating alternative and more inclusive images (for a fuller discussion, see Kinsman, 1995).

Summary

- Colonial traditions of depicting empty landscapes or of romanticizing indigenous groups' relationship to the land often continue to work against the interests of native peoples in postcolonial contexts.

- Postcolonial approaches to landscape can involve reasserting alternative cultural meanings that have been overlooked or devalued but they can also involve using landscape to shape more complex and hybrid versions of cultural identity.

CONCLUSION

Geographers are interested in landscape in part because of Geography's traditional concern with how

Figure 23.6 Artists Laurie Nilsen and Mark Garlett (standing) of Campfire Consultancy, at the Main Gallery at J.C. Slaughter Falls. Credit: Laurie Nilsen and Marshall Bell

places look, what makes them distinctive, how they are organized and how they have come to be this way. Landscape is clearly also a focus because of Geographers' interest in social relations. The concept of landscape brings the two areas of the material and social together. But as this chapter has shown, how people have understood landscape, shaped the world and treated each other has been also about the ideas, attitudes, symbols, institutions and practices of culture. Landscape, then, involves the social, material and symbolic. This is what is meant by saying that 'a landscape is a cultural image, a pictorial way of representing, structuring or symbolising surroundings' (Daniels and Cosgrove, 1989: 1). The cultural meanings of landscape are not universal but tied to particular societies and groups within them. These meanings are not simply a extra dimension that is less important than its concrete forms – they are inseparable from it (Matless, 1992). Landscape imagery, as Stephen Daniels has written 'is not merely a reflection of, or distraction from, more pressing social, economic or political issues; it is often a powerful mode of knowledge and social engagement' (Daniels, 1993: 8). Conflicts over what landscapes mean or how they should be organized are part of complex processes through which individuals and groups define themselves, claim and challenge political authority.

Further reading

A useful introduction to the different traditions of analysing landscape as either a portion of natural and cultural environment or as a pictorial convention is:

- Duncan, J. (1995) Landscape geography, 1993–94. *Progress in Human Geography* **19**: 414–22.

For an accessible introduction to how landscape has been linked to problematic ideas of race, class, gender and sexuality see:

- Kinnaird, V., Morris, M., Nash, C., and Rose, G. (1997) Feminist geographies of environment, nature and landscape. In Women and Geography Research Group *Feminism and geography: diversity and difference.* London: Longman, 146–89.

The specific feminist debates about landscape, power, vision and gender are explored in depth in:

- Rose, G. (1993) *Feminism and geography: the limits to geographical knowledge.* Cambridge: Polity, 86–112.
- Nash, C. (1996) Reclaiming vision: looking at landscape and the body. *Gender, Place and Culture: A Journal of Feminist Geography* **3**: 149–69.

An influential article in the development of approaches to landscape in contemporary human geography was Denis Cosgrove's exploration of the origins of the term landscape in the culture and economy of the ruling class in fifteenth- and early sixteenth-century Italy.

- Cosgrove, D. (1985) Prospect, perspective and the evolution of the landscape idea. *Transactions of the Institute of British Geographers* **10**: 45–62.

To explore in greater depth current theoretical approaches to landscape interpretation see:

- Matless, D. (1992) An occasion for geography: landscape, representation and Foucault's corpus. *Environment and Planning D: Society and Space* **10**: 41–56.

Examples of the ways in which landscape and national identity have been connected can be explored in:

- Daniels, S. (1993) *Fields of vision: landscape imagery and national identity in England and the United States.* Cambridge: Polity Press.

CHAPTER 24

Place

Tim Cresswell

PLACE, ORDER AND CATEGORIZATION

Place is one of the central terms in Human Geography. It is a term which eludes easy definition and has been used in a number of disparate ways throughout Geography's history (see Entrikin, 1991; Massey, 1993; Sack, 1997; Tuan, 1977). Place has been used as an alternative to 'location'. While location refers to position within a framework of abstract space, often indicated by 'objective' markers such as degrees of longitude and latitude, or distance from another location, place has come to refer to a mixture of 'objective' and 'subjective' facets including location but adding other, more subtle, attributes of the world we inhabit. John Agnew (1987) has argued, for instance, that place consists of:

- location – a point in space with specific relations to other points in space;

- locale – the broader context (both built and social) for social relations;

- sense of place – subjective feelings associated with a place.

Sense of place refers to the subjective feelings evoked by a place for both insiders (people who live there) and outsiders (people who visit). We can see, then, that place is a much richer idea than its precursor, location. It is not surprising, therefore, that Social and Cultural Geographers have studied place and places in a number of ways. While **Humanistic Geographers** have looked at places as both personal and universal centres of meaning and care, 'new' Cultural Geographers have tended to look at places as expressions of power or as landscape. The approach I take below, then, is only one possible way of thinking about place.

Place is more than an academic term – it is a word we frequently use in our everyday lives. Some of the ways we use it point to the richness of place as an idea. Here are some examples:

- He knew his place.

- She was put in her place.

- Everything in its place.

Terms such as these point towards the social and cultural significance of place. In each of these phrases the

Figure 24.1 A woman's place is in the home?/A stereotypical image of a post-war U.S. housewife. Credit: Popperfoto

word place suggests simultaneously a geographical location and a position on a social hierarchy. Think, for example, of a dinner table, either at home or in a more formal setting such as an annual dinner of an organization. Everything from the flowers, to the position of the cutlery to the seating arrangement is in some way related to place in the social sense. Old-fashioned notions of the patriarch sitting at the head of the table live on in households today and are formalized in the formal business dinner with its 'head table' and peripheral space for secretaries and janitors. A place for everything and everything in its place.

The human mind makes sense of the world by dividing it up into categories. As the examples above reveal, place and space are fundamental forms of categorization. Philosophers (most famously Kant) have insisted that the two basic dimensions of life are space and time which form the basis for all other forms of categorization. Indeed, our conceptions of space and time are so fundamental they appear to pre-exist our conception and **representation** of them – that is to say they appear as nature. When we say that something is natural we are saying that it is not social – it 'just is' and is therefore unchangeable. This makes categories of space and time potent **ideological** weapons. They are ideological because they are laden with meanings that tend to create and reinforce relations of domination and subordination. As the French theorist Pierre Bourdieu (1984; 1990) has claimed, categorization schemes that remain unarticulated (seemingly as nature) inculcate adherence to the established order of things. This is the case because categorizations in space and time, for the most part, are not recognized discursively (we do not speak about them, write about them or even think about them) but practically (we act upon them). As we cannot possibly think about everything we do throughout our lives, the vast majority of our actions are fairly uncritical acts which conform to the expectations of those around us. In the remainder of this chapter we will examine the way in which place acts as a category which serves to reproduce the existing 'order-of-things'. In addition, we will see how challenges to the taken-for-granted relations between place and actions provide profound challenges to the 'order of things'.

PRIVATE PLACES AND PUBLIC PLACES

Mothers of the Disappeared

Public space is often simultaneously the site of the assertion of power and **ideology** by the **nation–state**, corporations and local governments and counter-ideological practices. Just as power is spatialized as place is

given meaning, resistance can take the form of spatialized disobedience. A case in point is the actions of the Plaza de Mayo Madres (sometimes called the Mothers of the Disappeared) in Buenos Aires, Argentina during the late 1970s and early 1980s. The application of meaning to space (the creation of place) is a supremely political process which tends to inscribe a particular idea of order on the lives of the people who inhabit (but do not build) that space (Duncan, 1990). The creation of public places such as streets, parks and public squares are often acts of ordering of the first magnitude. These spaces are constructed in order to convey the legitimate order to citizens.

One of the master codes by which such places are constructed is the division of public and private. The public spaces are supposed to be non-political. Proper politics is supposed to occur in designated political spaces – the chambers and corridors of government. Similarly, public space is supposed to be masculine space. Public spaces are thus material manifestations of masculine ideas of order and authority. A 'feminine' presence on the street, in public space, is often seen as threatening (Wilson, 1991).

Jennifer Schirmer recounts some of the history of this attitude in Western spaces. She tells the story of the French eighteenth-century assertion of freedom for men to speak, move and think which arose at the same time as the banishment of women's public speech and political life. Women in public were 'out of place'. The best kind of women, the most virtuous woman, was one who knew her place and did not speak out of turn: 'Boundaries between the public and private, the political and social, the productive and reproductive, and justice and family were established, and justified by women's absence in the first and presence in the second (Schirmer, 1994: 188).

Schirmer goes on to tell the story of women who have taken it upon themselves to transgress the boundary of public and private space in order to make political points (in places supposedly non-political). Thus the Plaza de Mayo Madres inserted their bodies into the public space of Plaza de Mayo in order to protest against the 'disappearance' of family members (*see* Fig. 24.2). In the period 1976–82 the military junta of General Videla fought against a so-called International Conspiracy of Subversion which was said to be against all the (Western) values that Argentina stood for. This included getting rid of all those who were out of place in Argentina – the 'alien bodies' who sought to subvert the state. Over 30,000 people became targets for official and unofficial state security forces to abduct, torture and disappear. The victims were erased from public consciousness. In protest, the Plaza de Mayo Madres began to circle the main square in Buenos Aires in 1977.

Figure 24.2   A protest by the Plaza de Mayo Madres (the mothers of the disappeared) in Buenos Aires 1985. More than 10,000 attended. Credit: Popperfoto

The square itself, like many grand squares around the world, had been built to symbolize elements of official history and ideology. In this case it was a symbol of the Inquisition in addition to contemporary ideas of governance. Every Thursday at 3.30 p.m. they would walk arm in arm, their heads covered, demanding the return of their disappeared relatives and the punishment of the people responsible. The presence of these women in public space is an immediate transgression of two place-based categorizations – the association of public space with masculinity and the association of such places with an absence of politics. In addition to these transgressions, however, the women used many of the symbols of motherhood and domesticity to make their case.

> The Plaza de Mayo is flanked by monumental buildings that are incongruous with the private lives of domesticity: the presidential palace (the Casa Rosada), which was used by the juntas; the cathedral; and the Ministry of Social Welfare. This site of masculinist power is demystified by older women, humbly circling the plaza wearing on their heads diapers first, and later white headscarves, embroidered with the names of their disappeared son or daughter or husband, together with worn photographs of their loved ones pinned on their breasts or placed on large placards at marches and demonstrations.
>
> *(Schirmer, 1994: 203–4)*

Here the Madres are confusing the relationship between place and meaning in complicated ways, both transgressing the expectations of **gender** and politics; public and private space, and reaffirming (strategically) the 'normal' and 'proper' association between femininity and the nexus of the family and the domestic. A similar process has been identified at Greenham Common where peace protesters in the 1980s pinned symbols of domesticity and reproduction such as nappies and sanitary pads to the fence at the monumental and masculine airbase which housed Cruise missiles (Cresswell, 1994; Schirmer, 1994). Relations between genders are not the only social relations that are maintained through the division of places into public and private. Other geographers have shown how the same divisions are used to construct the relations between adults and children.

The place of children

Childhood as a social category varies over time and place. In the Western world childhood is a lifestage

THE CONVOY KIDS

TRAVELLING ON Peace convoy people wearing a selection of headgear.

Long road ahead for the hippy youngsters

By SYDNEY YOUNG and RONALD RICKETTS

FACES streaked with mud, wearing tattered denims and even bovver boots, the children of the peace convoy played happily in the road yesterday as their hippy parents dreamed up another plan to outwit the police.

Their mobile—often immobilised—homes are a bizarre collection of clapped out buses and vans. Nearly 100 strong, the so-called peace convoy snaked along a country lane as farmers and locals looked on fearfully.

The hippies, rarely popular in areas where the convoy grinds along, have come in for loathing since their occupation last week of a field at Leslie Attwell's farm near Yeovil, Somerset.

Barricades

When they moved out on Friday, under threat of eviction, Mr Attwell was left with a mountain of rubbish and widespread damage.

Farmers in Dorset, where the hippies camped overnight yesterday near Corfe Castle, were determined not to suffer the same fate.

They barricaded gates with heavy machinery and dug trenches to stop the convoy invading meadows.

One farmer had a tanker of liquid manure standing by to spray invaders.

Meanwhile, as the hippies considered their position and the police stood by, the children scampered cheerfully among the parked vehicles.

They roamed along the road with their dogs, cats, and even goats.

The barefoot convoy kids rode their bikes, weaved in and out of the vehicles on skateboards, and good naturedly hurled the occasional insult at a policeman.

After police ordered the hippies to disperse and stop blocking the road, parents rounded up the youngsters.

The bikes were piled back into the vehicles — ranging from an ancient hearse to a vintage military police van.

Roadside fires were doused and the convoy was on the move again.

The police didn't know its destination. Maybe the hippies weren't sure either.

But the farmers round Corfe Castle relaxed.

The travellers' children grinned farewell from the windows of the cavalcade.

And cheerfully stuck a tongue out at the society they reject — and which rejects them.

MUDLARK Cheeky pose from a happy hippy.

BOOTBOY Convoy kid.

Figure 24.3a Press coverage of traveller children in the UK during the 1980s. Source: *Daily Mirror* 2/6/86, p.4

associated with the home which is constructed as a safe space which children leave when they become adults. Gill Valentine has noted how parents associated public space with potential danger for their children ranging from abduction to traffic accidents. She suggests that the equation of stranger = danger, in particular, helps to reproduce the idea of the street and public space in general as a space 'populated by "deviant" others, a space in which the male body (particularly the black male body) is saturated with threat and danger' (Valentine, 1996: 210).

The restrictions placed on the activities of children by their parents equate types of behaviour with categories of place. Because children are seen to be dependent, incompetent and under threat, they are protected through segregation and the restriction of access to 'dangerous places' in the public realm. Public places are thus constructed as adult spaces where children are 'naturally' absent. It is not just adults that produce public places in this way, however. Children internalize expectations about danger in public places and withdraw from interaction with adults in places such as playgrounds and parks. Thus 'children contribute through their own performative acts, towards producing public space as an adult space where they are not able to participate freely' (ibid.: 211). The inverse, Valentine suggests, is that children no longer produce their own street space in the way they once did.

It is not just the streets of cities where children have been labelled out of place. Some of the labels of

Figure 24.3b Press coverage of traveller children in the UK during the 1980s. Source: *Daily Mirror* 30/6/83, p.23

deviance in 1980s Britain were partly built on images of misplaced children. The convoys of travellers (New Age and otherwise) that have become a part of rural life in many parts of the United Kingdom have been met with considerable hostility by the media, some local residents and government at both the local and national level. One story used to indicate their manifest deviance concerned 'Emma', a young child who travelled with one of the groups attempting to reach Stonehenge in 1986. She was shown to be dirty, 'not toilet trained at four' (*Sun*, 7 June 1986, pp. 4–5) and having no place to go. She was called 'Emma, the Kid from Nowhere' (*Daily Mirror*, 30 June 1986, p. 23) (*see* Fig. 24.3). Emma was made out to be indicative of the 'dirtiness' and 'lack of discipline' among traveller children. In an article entitled 'The Convoy Kids' (*Daily Mirror*, 2 June 1986, p. 4) we are informed at some length of children 'their faces streaked with mud, wearing tattered denims and even bovver boots'. The

travellers were accused of depriving their children of a 'normal family environment'. The theme developed by these stories was the out-of-placeness of the children and the nomadic travellers as a group. Children are particularly associated with a so-called 'normal family environment' and thus the existence of children such as Emma among the travellers pointed towards the travellers' transgression of the place-based norms of home, family, work and privacy.

Summary

- Place is a complicated term which refers to both objective location and subjective meanings attached to it.

- Place is simultaneously geographical and social.

- The division of the world into private and public places denotes particular relationships between geography, social group and behaviour.

- Examples of the way in which public and private places are implicated in the construction of acceptable behaviour can be seen through investigations of the social divisions between, for instance, men and women and adults and children.

RURAL AND URBAN

So far we have focused on the relationship between place and categories surrounding the important dualism of public and private. As we have seen, relationships between men and women and adults and children are constructed through this dualism at a number of scales. It is certainly not the only dualism that gets mapped onto particular places though. Another dominant spatialized dualism is that of urban and rural.

In Chapter 27 Paul Cloke discusses the recent march on London by British people claiming to represent the 'rural way of life'. A significant part of their complaint is that present attitudes to rural pursuits are driven by urban perceptions of the rural. Thus the claim is made that fox-hunting is an age-old rural tradition under threat from urban do-gooders who do not understand the intricacies of country life. Indeed, there is a long tradition of people and actions labelled 'urban' being labelled out of place in the country just as there are things associated with the rural which are labelled out of place in the city.

Protest in the country

As we have seen in the case of Emma and the convoy children, New Age Travellers are frequently targeted as being out of place in the British countryside. Another way they are so labelled is by saying that they are essentially urban people who do not belong in a rural environment. Keith Halfacree (1996) has looked at reactions to travellers in the countryside. He points out that they were consistently differentiated from 'real' Gypsies who (in an idealized form) did fit into a vision of the countryside as a bucolic rural space. Unlike these mythical 'real Gypsies', New Age Travellers moved around in large groups indicating their inappropriateness to rural life. Comments in the House of Commons pointed towards the invasion of a sleepy, rural realm by noisy, disrespectful travellers: 'The new age travellers displayed some dreadful antics: they invaded peaceful countryside, decimated peaceful villages, went on the rampage and had raves lasting two or three days, showing a total disregard for the area' (Nigel Evans MP, cited in Halfacree, 1996: 62). Talk of 'invasions' of peaceful rural life indicates the belief that the travellers are not themselves from the countryside as invasion involves arrival from elsewhere. This is supported by consistent reference in the media to the travellers as criminals which, Halfacree claims, is based on a belief that crime is an urban phenomenon and has no place in the countryside.

Similar strategies of representation were prevalent in my own investigations of reactions to the Stonehenge 'Convoy' in the early 1980s (Cresswell, 1996). The convoy was often compared unfavourably to the countryside that surrounded it. 'The hamlet of Lytes Cary has never seen anything like it. A scatter of houses along a lane of head-high cow parsley, it is normally a quiet place near Somerton, home to about 30 quiet people. Somerton is pretty quiet too' (*The Guardian*, 29 May 1986, p. 1). In addition to the issue of noise versus quiet, the travellers were spoiling the natural beauty of the countryside. In a telegram to the Prime Minister, Margaret Thatcher, local landowners wrote that 'The fair face of this unique area is being disfigured and fouled. The New Forest is recognized internationally as of prime ecological importance and as a place of quiet recreation for our people' (quoted in the *Daily Telegraph*, 4 June 1986, p. 1 and 36). A day earlier the *Daily Telegraph* had reported that: 'The situation at the moment is that these anarchists are soiling this beauty spot and harassing both residents and holiday makers' (3 June 1986, p. 1). By placing the so-called 'hippies' in direct contrast to a landscape of nature, quiet, and beauty the 'hippies' were clearly being portrayed as deviant and 'out-of-place'. The countryside, as a place, is one that has been constructed as the site of a rural idyll in which nature, innocence, beauty, quiet and order prevail. Residents and holiday makers belong there while travellers, marked by noise, dirt, and anarchy do not.

Animals in the city

The division between urban and rural mirrors the more fundamental division between culture and nature (see Chapter 1). Cities are supposed to be places of culture and society while rural landscapes have often been thought of as natural landscapes. Particular problems arise, therefore, when 'nature' appears in the city. Animals, for instance, are subjected to many of the place-based forms of control that marginalized social groups experience. Animals, like people, have their place. Chris Philo (1995) has examined the 'possible responses of a social geographer when confronted by the intrusive reality of (say) cows, sheep, and pigs mingling with people in the spaces of a large urban area' (Philo, 1995: 657). Philo discusses how various aspects of domesticated animal behaviour intrude upon ideas about the 'proper' place of animals. During the nineteenth century, just as the urban poor and other marginal groups became the object of sanitary and environmental discussion, animals and the places associated with them (abattoirs, meat markets, etc.) began to be removed from the city due to, amongst other things the 'odours, flies and unseemly sights associated with animal husbandry' (Fielding, cited in Philo, 1995: 666). As the city became progressively segmented into functional containers for people and activities, animals became matter-out-of-place. While some animals became acceptable urbanites, such as cats and dogs, others were expelled to the country where they apparently 'naturally' belonged. Of particular concern was the meat industry and the spaces associated with it. Most cities had large areas devoted to the slaughtering and processing of meat products by the early nineteenth century. In Britain Smithfield Market in London was the most prominent of these (*see* Fig. 24.4). Philo reveals how as the century progressed people surrounding Smithfield became disturbed by the mixing of people and animals in the city. Take, for instance, the account of a shop assistant:

> On Monday last we had one beast put its head through the window; we are obliged to have a person at the door to keep them off; and last Monday week we had a sheep got into the shop and fell down the cellar steps into the cellar amongst the workmen: I think that fewer customers come to the shop on Monday; the ladies would not come to the shop if there was a crowd of bullocks . . .

> *(Padmore, cited in Philo, 1995: 667)*

Or, alternatively there is the case of a Mr J.T. Norris, the owner of a printing establishment located near Smithfield. When asked by a committee looking into the presence of animals in the city if he had seen any immorality in and around the market he replied:

> I . . . know that I have seen that which to me in a refined city is very unbecoming: I have seen bullocks driven from the market, which have been imperfectly operated upon, jumping on the backs of the cows in public streets, in the presence of passengers of both sexes. I think that it is an offence to decency: it is unbecoming in a great city, and forms a reason in my mind why such scenes should be at a distance, and out of the way of the observation of females and children.

> *(Norris, cited in Philo, 1995: 669)*

Here we have the obvious displeasure of people observing the country in the city, nature in culture, animals out of place. Philo charts the way in which the natural behaviour of animals in the city was seen to be a threat to human morality. Smithfield Market was not just associated with people out of place but with people engaged in immoral acts, spurred on by the sight of animals copulating. Nature and animals needed to be put in their place as they increasingly transgressed the links between space and behaviour.

So while the New Age travellers have been seen as the urban in the rural, the animals of Smithfield Market were seen as the rural in the urban. Both provoked responses that sought to define the 'proper' activities for such places and expel the intruding transgressors.

Summary

- As with the conceptual division between private and public, the categories of urban and rural are signified by particular places and particular actions.

- Rural places have been portrayed as quiet, natural, beautiful and harmonious while travellers have been

Figure 24.4 Smithfield Market during the nineteenth century, London. Credit: Mary Evans Picture Library

portrayed as out of place in the countryside with reference to noise, dirt, and disorder.

- Certain kinds of animals were progressively seen as out of place in the City during the eighteenth and nineteenth centuries as their behaviour was too close to nature for an ordered, civil urban life.

CONCLUSION: PLACE, CATEGORIZATION AND TRANSGRESSION

In the introduction to this chapter I discussed the way in which the association between place as meaningful location and categories of people and actions is often invisible because it is so deeply engrained. Knowing one's place thus seems 'natural' or 'inevitable'. In the examples which followed, the relationship between place and categories became most apparent when the relationships were transgressed – when people (or animals) were said to be 'out-of-place'. Thus the construction of public space as adult space becomes obvious when children appear in it. Likewise beliefs about what happens and belongs in the countryside are underlined when something (such as New Age travellers) is described as 'out of place' there. The order which is constructed by and through place is not inevitable and is often transgressed. It is in these moments of transgression that a great deal is said (in the media, by politicians and figures of authority) about what and who belongs where. Transgression, then, is a key concept for social and cultural geographers who want to describe and explain the construction of 'normality' through the creation and maintenance of particular types of place.

Further reading

- Cresswell, T. (1996) *In place/out of place: geography, ideology and transgression*. Minneapolis: University of Minnesota Press. In this book I expand upon the connections between place, meaning and power through an examination of events and people labelled out of place. It includes discussions of graffiti, travellers and Greenham Common Peace Protesters.

- Sibley, D. (1995) *Geographies of exclusion: society and difference in the West*. London: Routledge.
David Sibley develops a similar theme in this excellent book. His focus includes a consideration of psychoanalytic theory and he provides numerous examples of how space and place are implicated in the exclusion of people and actions defined as 'other'.

Some of the case studies discussed in this chapter are discussed in more depth in:
- Halfacree, K. (1996) Out of place in the country: travellers and the 'rural idyll'. *Antipode* **28**(1), 42–72.
- Philo, C. (1995) Animals, geography, and the city: notes on inclusions and exclusions. *Environment and Planning D: Society and Space* **13**(6), 655–81.
- Schirmer, J. (1994) The claiming of space and the body politic within national-security states. In Boyarin, J. (ed.) *Remapping memory: the politics of timespace*. Minneapolis: University of Minnesota Press, 185–220.
- Valentine, G. (1996) Children should be seen and not heard: the production and transgression of adult's public space. *Urban Geography* **17**(3), 205–20.

Contexts

INTRODUCTION

Part I of this book considered some of the foundational concerns that inspire and preoccupy Human Geographers. Part II offered illustrated reviews of how these concerns are being translated into more specific geographical research agendas, organized in terms of thematic sub-disciplines and topics. This third part now goes on to show how the questions and approaches developed within the different sub-disciplines of the subject are brought together in the study of particular kinds of places and spaces, or as we term them here, contexts. In this brief introduction we will say a little about Human Geography's concern with studying human life in context, and explain the selection of contexts presented.

Human Geography has perhaps always had two apparently contradictory ambitions. On the one hand, it wants to be able to contribute to our understanding of the human condition, to generalize about the nature of life on our planet. This has often been reflected both in the scale of analysis adopted – the global reach of the Geographer's eye – and in the kind of geographical theory developed – the search for geographical processes, and sometimes even laws, that are widely applicable. On the other hand, Human Geography has also always been interested in the specific. This interest has perhaps been most obviously and tellingly expressed in the long tradition of regional geography, where emphasis is laid on really getting under the skin of a particular place or region. You may already have encountered this. Most university Geography degree programmes have regionally focused courses, either at the continental level, the national or regional level.

At various points in Geography's history these two ambitions, to generalize and to understand the specific, have been set in opposition to each other, and picked up as standards by warring factions debating the direction that the subject should take. However, over the last 20 or so years, as part of the development of a so-called 'new regional geography', they are increasingly being recognized as two complementary and interconnected parts of the **geographical imagination** (see also Chapter 3). After all, virtually no region exists in total isolation from the wider world, so getting under the skin of one place always involves having some sense of its position in that wider world. One could not, for example, think about the human geographies of Canada without some attention to its past and present links to Europe, other parts of North America and the Pacific Rim. Moreover, in terms of the kinds of geographical knowledge produced, it is too simplistic to suggest a stark alternative between universally applicable theories, laws or models (sometimes termed a 'nomothetic' emphasis) and entirely specific, non-generalizable regional studies (what is called an 'idiographic' emphasis). Studies of particular places and regions are both informed by and informative of more general theories. They are shaped by, and speak to, the kinds of sub-disciplinary research themes set out in Part II.

A key geographical concept in overcoming the sterile oppositions between the general and the particular has been that of 'context' (for a classic formulation see Thrift, 1983). To somewhat oversimplify, contextual thinking emphasizes how any one thing we are interested in can only be understood through a consideration of the context or contexts it exists in. We apply this argument quite often in our everyday lives; for example when we claim that something we have said has been 'taken out of context' and in consequence misunderstood. A couple of examples may help illustrate its application in Human Geography. Let's start with television (see also Chapter 33). If we want to study the meanings of a television programme an emphasis on context suggests that we cannot just look at the programme itself, out of context. We need to know about what different viewers make of that programme in the different contexts in which they watch it. For instance, an Israeli study of the globally distributed 1980s' American soap opera, *Dallas*, found not only that people watched it in particular sorts of domestic settings, and hence in the context of family and friendship relations, but also that different cultural groups in Israel picked out very different elements from it (Liebes and Katz, 1990). Recent immigrants from the then Soviet Union denied watching it (even when they did) or dismissed it as American, capitalist propaganda. They saw it, then, in ideological terms. Middle-class kibbutzniks, in contrast, had a more psychological take on the series, concentrating on the personalities of the protagonists and gaining enjoyment from trying to understand their motivations. And the interviewees from Arabic and North African backgrounds emphasized the convolutions of the plot, seeing it as an epic family saga. What *Dallas* meant was context-dependent. Its meanings had geographies.

Now what is the case for a wonderfully trashy piece of American TV (which by the way is itself a particularly British way of enjoying *Dallas*) is equally true for more weighty matters. **Capitalism**, for example. Chapters 10, 11 and 12 emphasized how the economic geographies of the world are, following the collapse or reform of state socialism, nearly universally capitalist in character. However, what capitalism is and how it is done is 'subject to regional and localized difference' (Miller, 1997: 13). Studies from Geographers, Anthropologists and Sociologists have, for example, identified different 'styles of commerce' or ways of doing business around the world – one of the most cited is so-called Eastern or Chinese capitalism, which it is claimed places particular emphasis on networks of trust as cemented through family ties or gift giving – and different national forms of economic **regulation** reflective of the differing social and political matrices in which economic relations are embedded (see for example Gertler, 1997). Human Geographers would emphasize, then, that one cannot properly study capitalism solely in the abstract, out of context(s), despite the highly influential attempts of much economic theory to do precisely that. One needs to adopt a contextual and comparative approach.

However, there is a complication. Contexts are themselves not static or innate. They have to be produced and are subject to change. In consequence, when the Anthropologist Danny Miller wrote his recent book on capitalism in Trinidad, he was at pains to emphasize that the relations between capitalism and this Trinidadian context were two-way. Capitalism here was distinctively Trinidadian. But, and apologies for our clumsy phrasing, this 'Trinidadianness' was not fixed and itself was partly a product of capitalist processes. As Miller puts it, 'Trinidad continues to evolve in tandem with capitalism, and is not separable as a "context" to capitalism' (1997: 5).

So, Human Geography's fascination with the contextuality of human life involves two moves: first, seeing how the kinds of processes divided up into separate sub-disciplinary themes in Part II of this book come together and interact in particular times and places; and second, investigating how these contextual times and places are themselves constantly evolving products of those processes and their interactions. To illustrate these two sides of the coin of contextuality, our approach in this third part of *Introducing Human Geographies* is to present a series of chapters on the generic kinds of contexts that Human Geographers study. This generic framing in part reflects the impossibility of doing justice to a full range of more specific geographical contexts in the space available, but it also has intrinsic benefits. One of these is that it allows us to signal that while concerns with contextuality emerged from geographical studies focused on the regional scale, they are now being developed at scales ranging from the human body right up to those of international and transnational relations. In this progression some of the contextual scales addressed here may already be familiar to you; almost certainly the city and the countryside (discussed by Chris Hamnett and Paul Cloke, respectively in Chapters 26 and 27), and perhaps also the continental scale (explored here through the example of Europe by Kevin Robins in Chapter 28) and the global scale of the world-wide colonial empire (analysed by Richard Phillips in Chapter 29). On the other hand, the human body (as discussed by Ruth Butler in Chapter 25) is quite likely to be a new geographical topic. This is despite its absolute centrality to all our lives, and reflects still significant if unjustifiable conventions about the scales of human life geographers study as well as a **masculinist** disinterest in embodiment.

In addition, though, what the chapters collected here also illustrate is how geographical contexts do not always conform to the model of a bounded region, whatever its scale. In that vein Chapters 30 through to 34 focus not on places but on what Manuel Castells has termed **'spaces of flows'** (see for example Castells, 1989). Chapters 30 and 31 deal with *flows of people*: Claire Dwyer discussing processes of migration and the 'transnational diasporas' they produce; Luke Desforges focusing on the temporary but huge movements of people that are the lifeblood of the world's largest industry, tourism. In Chapter 32 Michael Watts investigates *flows of things*, through an analysis of commodities as 'organized space-time systems', that is circuits connecting together people and places (chickens and cocaine are the main examples). And Chapters 33 and 34 address *flows of ideas and imagery* through the modern media: James Kneale discussing the geographies of the book and audio-visual media; Ken Hillis offering a critical dissection of the cyberspace and cyberculture produced through new forms of electronically mediated communication such as the Internet.

It is worth emphasizing that in all these contributions no cut and dry division is set up between context as place and context as space of flows. The place of the city, for example, is in part given its character through the multiple flows of people, things and ideas that meet up within it; the flows of cyberspace are productive of new, simulated places. But the distinction does help to identify two rather different ways of approaching the contextuality of human geographies, and is suggestive of how that contextuality may be changing in the dying days of the twentieth century. Contexts matter to economic, political, developmental, environmental, social and cultural geographies, but those contexts themselves are subject to transformation. This final part of *Introducing Human Geographies* aims, then, to give you insights into some of the key geographical contexts that matter today and will matter tomorrow.

CHAPTER 25

The body

Ruth Butler

INTRODUCTION: NOT JUST FLESH AND BLOOD

The human body comes in an infinite array of shapes and sizes (*see* Fig. 25.1). Definitions of them in terms of their physical structure, social utility and symbolic value vary widely from place to place and time to time. They are many things to many people, a tomb of the soul, a temple, a machine, and the self. They are caressed and killed, loved and hated, thought beautiful and ugly, sacred and profane (Synnott, 1993). However, despite their heterogeneity, society's narrow expectations of any particular body are usually closely allied to the crude and broad social classifications of **gender**, age, **race**, ability, **class** and **sexuality** into which they are placed.

These complexities and contradictions are what make the body of great interest and value to Geographical enquiry. As this chapter will show, it is one site around which the different sets of sub-disciplinary themes discussed in earlier chapters of this book can be drawn together. The personal space of the body, as well as its interaction with the wider environment, is just one of the sites for geographical enquiry discussed in Chapter 2. In relation to Chapter 1, it will be shown how the roles of biology and society in constructing the embodied experiences of different individuals illustrate the continuing debate over the relative power of nature and culture in shaping people's lives, and in turn the societies in which they live. The complex, interacting internal and external forces of biology, society, culture, politics and economics at all levels will be clearly iden-

Figure 25.1 Pride against prejudice. The diversity of forms of the female body as depicted by the cartoonist Angela Martin. Source: Morris (1991)

tified in reflections upon bodily presentation and behaviour which make clear the interactions of local and global, micro- and macro-scale events and processes, as discussed in Chapter 3. Through their interaction with the environment and one another bodies are an essential aspect of human agency; people's attitudes, behaviour and achievements (see Chapter 4). Their physiological make-up and social image are of key significance in defining both the self and others (*see* Fig 25.2 and Chapter 5). Patterns of bodily presentation and performance at different times and in different places show how the body is used to adopt or resist others' images of an individual relative to what that individual believes their reality to be (Chapter 6).

In order to illustrate these issues further this chapter considers the strategies we all use in order to cope with the pressures that are placed on us to conform to socially accepted 'norms' of behaviour in public space, according to the physical structure and social representations of our bodies. Drawing on examples from the lived experiences of women, disabled people and lesbian and gay individuals, the chapter will fall into three sections. It first considers the nature of the stereotyped roles that people are expected to follow in society, how such social images have been constructed and how they continue to function. Second, it discusses the reactions people have to such expectations of them and their methods of coping with the demands society places upon them. It will finally question what implications these reactions have for our broader understandings of social interaction.

The human body is not just flesh and blood. An object for the mind to use at its will. The body is an active and reactive entity which is not just part of us, but is who we are. Its presentation and **performance** project an image of the person to both the individual themselves and the others who observe them. In Western, consumer society, obsessed with the 'body beautiful', there exists an unspecified, yet desirable bodily 'norm' to which we are all encouraged to conform. The costs of failing to do so can be high. Social beliefs mean that being labelled as 'different', 'deviant', 'marginal', as well as physical limitations, can make the experiences of public places and spaces uncomfortable, unpleasant, dangerous or even fatal. The next section considers why this is so.

BIOLOGICAL AND SOCIAL REGULATION OF THE BODY

In philosophical and academic discussion the body has been understood in two main ways; as a 'biological' phenomenon or as a 'social' phenomenon (Shilling, 1993). The first theory suggests that the biological

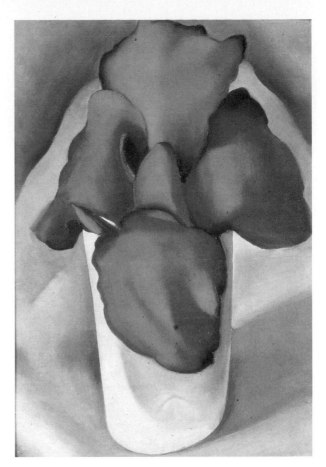

Figure 25.2   Red Gladiola in White vase, 1928 (oil on canvas) by Georgia O'Keeffe (1887–1986). Credit: © ARS, NY and DACS, London 1999/Private collection/Christie's Images Bridgeman Art Library, London/New York

structure of the body determines an individual's behaviour. In turn, it is assumed that the physiological limitations of its members determine the state and development of society and social relations. This theory has been most recently developed with reference to genetics. For example, it has been suggested that both feminine and masculine characteristics and an individual's sexuality may be the result of genetic make-up. The alternative social constructionist argument suggests that the body is developed and constrained by social processes and relations. For example, it suggests it is the different ways girls and boys are nurtured through childhood that gives them different characteristics. While both theories are **reductionist**, each ignoring the influence of the other and the interaction of the two phenomena, these two extreme views are a useful starting point from which to consider the nature of the images and resulting expectations of different people which operate in Western society.

Biology is destiny?

The revival of feminist politics in the 1960s and 1970s had an important focus on the rejection of the idea of biology as a reasonable explanation for the differences between men and women in society and in turn the justification for sexism. Female biology, it is argued, does not mean an individual is necessarily irrational, passive or maternal. Feminists argue that what are often seen as innately female characteristics arise from social conditioning (Oakley, 1981). Attention was drawn to the social pressures on women to conform to male expectations of them in terms of both their behaviour and bodily presentation. In order to underline their point a distinction has been drawn between the terms 'sex' and 'gender', the former relating to an individual's biology and the latter to the role of social, political and economic structures in the construction of male and female identity. In a similar manner the disability rights movement has recently drawn a distinction between 'impairment' as a biological fact and 'disability' as 'the disadvantage or restriction of activity caused by a contemporary social organisation which takes no or little account of people who have physical impairments and thus exclude than from the mainstream of social activities' (UPIAS, 1976). These two dichotomies both make a clear distinction between the results of social power relations and biological capacity (Butler and Bowlby, 1997).

Political awakening to the significance of social constructs does not, however, mean that the physicality of the body is of no importance in people's lived experiences. In relation to disability Morris (1991:10) writes:

> There is a tendency within the social model to deny the experience of our own bodies, insisting that our physical differences and restrictions are entirely socially created . . . to suggest that this is all there is to it is to deny the personal experience of physical and intellectual restrictions, of illness, of dying.

We all have our physical limitations. The Women's movement has come to recognize that there are differences between men and women, but have celebrated and valued them, rather than accepted them as the sole controls on women's destinies. The impossibility of performing certain physical actions when desired for some individuals, menstruation and other physiological facts of life mean that it is difficult to believe that an idyllic society where 'impairment' or 'sex' will be of no importance to a person or to those around them is attainable. However, less tangible social, economic and political structures remain, at present, the dominant controls on people's **embodied** experiences and it is to these issues which I now turn.

Presentation of self

A person's image of him/herself is strongly influenced by the opinion of others and by the representations in language and media images of desirable identities. In a self-defensive manner, we tend to see ourselves as 'normal', and hence socially acceptable, and those we view as 'other', as deviant (Shakespeare, 1994; Butler and Bowlby, 1997). This is one method we employ to distance ourselves from people we think to be on the undesirable margins of society. It produces a social structure where we are all measured against an unspecified, yet apparently desirable 'norm'. People's fear of marginalization encourages them to take on board ideas of social 'norms' of behaviour and appearance, and build them into their own evaluation of their identity (Goffman, 1963; Corbett, 1994).

It should be recognized that such 'norms' are not constant over time, space or between individuals. The social, economic and political categories that an individual places themself in, or is placed in by others will strongly affect the image others expect them to comply to. Some categories are themselves considered more 'normal' and hence desirable than others. Images of heterosexual couples, for example, are displayed in the media as having the desirable lifestyle to which everyone should try to conform. The stigmatism and marginalized status of the categories of lesbians, gay men and bisexual individuals can result in individuals attempting to hide their identity and 'pass' as heterosexual in spaces outside the security of the gay scene or homosexual household (Valentine, 1993).

This need to conceal personal identity is, however, not an experience of lesbians and gay men alone. It has also been recognized of disabled people, regardless of their sexual orientation. Corbett (1994: 344) notes that the contention that 'closeted gays pay too high a psychological price for passing', has been applied to disabled people. An able body is seemingly preferable to impairment.

Feminist writers have noted how women's bodies are often treated, both by others and ourselves, simply as objects to be viewed and evaluated (Young, 1990). Our efforts to shape our bodies into an image that both we and others are comfortable with in consumer culture have made the fashion and beauty industries highly profitable. While it is still open to debate to what extent we choose to decorate our bodies to our own tastes and to what extent we feel forced to do so (Baker, 1984), it is true that disabled men and women are encouraged by media representations of 'normal' bodies to obscure by dress and bodily decoration what are seen by others as bodily inadequacies.

Even for individuals who feel confident enough to take pride in their bodies and associated identities, problems can arise. A positive image, to be effective, must be mediated to those around us. If behaviour an individual is happy with is misunderstood by others it can be painful. For example, a woman may be at ease with her own identity as a mechanic, but wolf whistles and sexist remarks from passers-by may be unpleasant to deal with. Many disabled people may be considered strange and ridiculed due to their body language. Uncontrollable spasms, dribbling or a simple lack of awareness due to congenital blindness may mean their real intentions in an action are, frustratingly, not appreciated by a society disturbed by anything too out of the ordinary.

Summary

- In philosophical and academic discussion the body has been understood in two main ways; as a 'biological' phenomenon or as a 'social' phenomenon.

- Both theories are reductionist. Clear distinctions have been made between the results of social power relations and biological capacity. The interaction of these phenomena has often been ignored.

- Both phenomenon are of importance in people's lived experiences. However, social, economic and political structures remain, at present, the dominant controls on people's embodied experiences.

- Differences between people (negative and positive) can be celebrated and valued.

- There is a social structure where we are all measured against an unspecified, yet apparently desirable 'norm'. Hence there is a pressure to present our bodies and perform in a manner that both we and others are comfortable with.

- The social, economic and political categories that an individual places themselves in, or is placed in by others will strongly affect the image others expect them to comply to.

Social expectations

Understandings of and reactions to people are socially and culturally created and perpetuated. The ways that we react to people, what we expect or do not expect of them, where we expect to see them, what we expect to see them doing are all issues affected by images of people which are entwined in British culture. A wealth of literature on society's marginalized minorities has clearly illustrated the role of social expectations and economic marginality in controlling people's behaviour, presentation of self (as discussed above) and simple presence in public space (see for example Goffman, 1963; Winchester and White, 1988; Barnes, 1991; Valentine, 1993).

This is something to which the political fights of Gay Pride, Disability Rights and Women's Rights

organizations have also drawn attention. They have attempted to point out the ineffective nature of policies and legislation which do not address what they see as the crux of problems of discrimination, namely social opinion.

Public reactions to difference can be both negative and positive. At one extreme they can mean hostility, backed by legal expulsion from space. For example, the 'ugly laws' once upheld in many American cities, removed 'offensive' bodies from public space (Gilderbloom and Rosentraub, 1990).

> No person who is diseased, maimed, mutilated or in any way deformed so as to be an unsightly or disgusting object or improper person to be allowed in or on the public ways or other public places in this city shall therein or thereupon expose himself to public view.
>
> *(Ordinance from Chicago, cited in Burgdorf and Burgdorf, 1975: 863)*

In Nazi Germany disabled and gay individuals, amongst others, were not only removed from the streets, but removed from society via the gas chambers of asylums and concentration camps (Morris, 1991).

At the other extreme, being different can mean being met with pity, offers of medical 'help' for what are seen as biological 'faults', and unwanted acts of charity. Electric shock treatments for homosexuality, and painful 'corrective' surgery of no health benefit to impaired patients are extreme cases of efforts made in the name of 'caring' to 'normalize' different bodies.

The reasons for these various public reactions to difference may include fear, a desire for superiority and control, and distancing and othering processes. The human body is not immortal, and it has been suggested that disabled bodies are used as 'dustbins for disavowal' to relieve others' fears of their biological frailty and eventual death (Shakespeare, 1994: 283). The attention drawn by men to their physical strength relative to women may equally be linked to such fears, as may the connections made between homosexuality and AIDS.

Much of the behaviour which society sees as unfit for public spaces, such as spasmic movements and dribbling (as discussed above), menstruation, breast feeding, and homosexual behaviour has been considered animal-like. Characteristics which in the sixteenth and seventeenth centuries were considered to link human life to sin and the Devil – only by subduing the animal through reason and religion could people become fully human. Women and disabled people have been seen as closer to nature and our animal ancestors than men and able-bodied individuals, respectively. They have been seen to represent what Shakespeare (1994: 296) describes as 'the physicality and animality of human existence'. As conquerors and controllers of the natural environment, white, able-bodied men have equally seen themselves as superior to their more animalistic neighbours, including women, disabled people, lesbians and gay men, and members of ethnic minorities. Such individuals are seen as dangerous, though exciting '**others**'.

The visible appearance of any body is also important in society's expectations of people. Beauty has been taken as a sign of good morals in an individual, whilst ugliness has been seen as a sign of evil (Goffman, 1963; Synnott, 1993). The failure to make the most of your body and look after it can be seen as a moral failing (Synnott, 1993). It is an easy step from this to the view that the 'ugly', 'disabled', 'different' are morally lacking.

Appearance is also important as a means of communicating explanations for behaviour. Visible evidence of 'difference', deviation from the accepted 'norms' of social appearance and behaviour can result in individuals being the butt of jokes and hostile behaviour, but misunderstandings, resulting in an equal variety of reactions can arise through the invisible differences between people. For example, invisible impairments such as dyslexia or ME, if not disclosed, can lead to an individual being considered stupid, or lazy, respectively.

Physical and social structures of places

The social controls on disabled people are not necessarily obvious through the direct reactions of people in face-to-face encounters. The social, political, economic and physical structuring of the environment can be of great importance. The freedom of movement someone can enjoy in any given area and their acceptance in it by others can be highly influential in society's view of them, their view of themselves and, in turn, their acceptance in other spaces and places.

The physical structuring of the environment is an obvious and direct control on movement. It has been argued, for example, that the built environment in Western society has been designed by and constructed for white, able-bodied, heterosexual men. The fact that such accusations have been made make clear the inadequate nature of an environment structured to meet the physical needs of a mythical 'average' body. The very different bodies and hence different needs of individual people are not being met.

The most obvious and possibly most well-publicized failings relate to the needs of disabled people. The lack of induction loops for people with hearing impairments, large print and Braille information sources, adapted toilets for disabled people and ramped access ways in public spaces have all received a great deal of media attention (*Which?*, 1989).

However, it is not just disabled people who are paying the cost of an inaccessible environment. Ramped walkways are of value to wheelchair users, mothers with push chairs, elderly individuals and individuals with sporting injuries alike. Other issues affect other specific groups, such as debates over the need for greater numbers of women's toilets (Little *et al.*, 1988).

Social, economic and political structures equally play their parts in the structure of discrimination and different people's experiences. The effects of a consumer culture on social expectations of bodily appearance have already been made clear. **Capitalism** and the **commodification** of labour place expectations on individuals to work in a manner considered productive for an 'average' individual, but not all disabled individuals who may require specialized equipment will be able to work at the same pace as their work mates. Images of different bodies and their abilities, or inabilities, have become entwined in the economic, social, political and physical fabrics of daily life.

Summary

- The role of social expectations and economic marginality in controlling people's behaviour and presentation of self has been illustrated by a growing literature on society's minorities and has been central to the political fights of many social movements.

- Public reactions to difference can be both positive and negative, ranging from pity to physical expulsion from space.

- The reasons for these various reactions to difference may include fear, a desire for superiority and control, and distancing and othering processes.

- Appearance both affects others' expectations of an individual and can communicate explanations for behaviour.

- Social, economic and political structures all play their parts in the structure of discrimination. The nature of the built environment, consumer culture and the commodification of labour are just three influential factors in people's embodied experiences.

PERFORMANCE – COPING STRATEGIES

How different individuals choose to cope with the images and expectations others have of them varies enormously with the visibility and acceptability of the social, economic and political categories they find themselves in, their strength of character, self-confidence, experience, support networks and a multitude of other variables. What is common to us all is the need to reconcile the social constructions and lived realities of our bodies. Goffman (1963) has suggested that there are two ways of doing this. An individual can either choose to accept and play the role others expect of them, or they can fight against it and behave in a manner more suited to their personal desires. It must be stressed that these are two extreme positions in a fluid socialization process, but they are useful starting points from which to consider the reactions of people to the images and expectations others have of them.

Putting on an act

Self-image, as already discussed, is strongly influenced by social representations of desirable and undesirable social characteristics. Internalizing ideas of how we believe others see us or of how we want to be seen helps us to feel more comfortable with the performances we are expected to give in social interactions (Goffman, 1963; Scott, 1969). It also helps others to predetermine our expected performance in any given situation due to the social, economic and political categories in which they put us (Goffman, 1963; Scott, 1969). By performing in socially acceptable ways we also simultaneously create and re-create our social identities. Such performances are sometimes largely unreflective or part of a deliberate strategy of self-presentation.

This tactic of 'passing', projecting a necessary, socially accepted and expected, if not always desirable, image of self when negotiating the social environment is a common and well-documented one amongst the gay community as already mentioned (Valentine, 1993). Lesbians and gay men also draw on recognized, stereotypical appearances and patterns of behaviour in a positive manner to communicate amongst themselves, advertising their sexuality to like-minded individuals in possession of the relevant social capital through subtleties of dress and topics of conversation.

Disabled people and women will equally play to the advantages and disadvantages of the stereotypes they are expected to conform to as they see fit and find to be of use to them. A woman who is genuinely in need of assistance at the road side may well use her looks to gain the sympathies and attention of passing male, heterosexual motorists. On the other hand she may choose to power dress in a business meeting where she wishes her male colleagues to take her seriously on an equal footing.

In a similar way, disabled people may draw on images of them as vulnerable in order to gain assistance. A visually impaired person may wave a white stick for attention when seeking help in an alien environment. They may equally remove the white stick from public sight if worried about their apparent vulnerability to would-be pickpockets and other petty criminals (Butler, 1998).

Fighting back

If an individual feels uncomfortable and/or dissatisfied with the behaviour expected of them they may choose to fight others' expectations by making clear what they believe to be their true personality and desires. To this end, members of many minority groups are finding the courage to contest and resist the 'normality' with which they are supposed to comply.

Gay Pride festivals flaunt homosexual behaviour. Many women and disabled people have equally taken to the streets to display with pride their 'true' identities in political demonstrations. Disability politics has encouraged disabled people to take pride in their bodily impairments (Morris, 1991; Corbett, 1994). Charity groups such as SCOPE have made conscious efforts to use advertising campaigns to change social attitudes and raise awareness (see Fig. 25.3), rather than reinforcing stereotypical images of helpless, dependent individuals by using them to pull on people's heart strings for donations.

These political actions draw attention to the discrimination people face and the neglect of their human and civil rights. To this end, legal actions have been brought by many individuals in both the British and European courts. The exclusion of lesbians and gay men from the armed forces, the failure of many local authorities to provide adequate support for disabled youths in mainstream schools and the employment rights of women are among the issues which have been brought before the judiciary.

These extremely demonstrative actions increase the visibility of members of marginalized groups offering role models, and feelings of belonging and self-worth to previously isolated individuals. It also helps to raise the broader community's awareness and understanding of the experiences and needs of members of marginalized groups.

Extreme forms of confrontation between members of minority groups and social expectations of them have an undeniable value, but only involve a small number of people. It would be misleading and inaccurate to suggest that all members of minority groups choose to partake in such activities. The majority of individuals are not so assertive in their actions. Support groups and networks of like-minded individuals work at many levels offering support, information and friendship as well as political action. Many

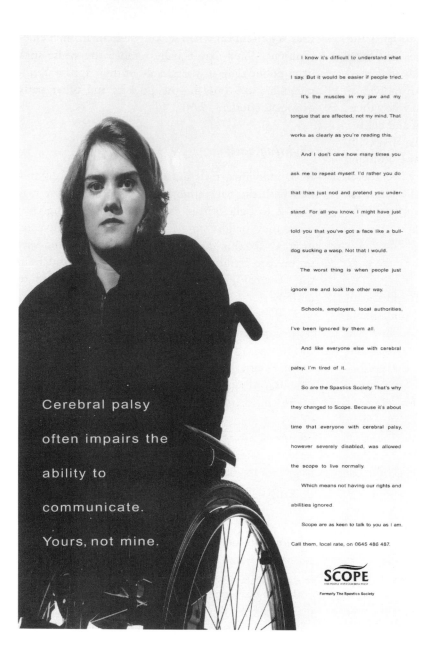

Cerebral palsy often impairs the ability to communicate. Yours, not mine.

I know it's difficult to understand what I say. But it would be easier if people tried.

It's the muscles in my jaw and my tongue that are affected, not my mind. That works as clearly as you're reading this.

And I don't care how many times you ask me to repeat myself. I'd rather you do that than just nod and pretend you understand. For all you know, I might have just told you that you've got a face like a bulldog sucking a wasp. Not that I would.

The worst thing is when people just ignore me and look the other way.

Schools, employers, local authorities, I've been ignored by them all.

And like everyone else with cerebral palsy, I'm tired of it.

So are the Spastics Society. That's why they changed to Scope. Because it's about time that everyone with cerebral palsy, however severely disabled, was allowed the scope to live normally.

Which means not having our rights and abilities ignored.

Scope are as keen to talk to you as I am. Call them, local rate, on 0645 486 487.

SCOPE
For people with cerebral palsy
Formerly The Spastics Society

Figure 25.3 An advertisement by SCOPE drawing attention to the social discrimination faced by people with cerebral palsy. Credit: Reproduced by kind permission of SCOPE

individuals have no recognized support other than that of close family and friends, but even these individuals, often unconsciously, challenge the images of others by simply having the courage to live their daily lives as they wish, in the mainstream of society. They break down the images they are expected to conform to through face-to-face encounters with others at the grass roots. They educate others through the same socialization processes which we all must partake in and must all learn to survive in the social jungle. It is to the broader understanding of social interaction that I now turn.

Summary

- We must all reconcile the social constructions and lived realities of our bodies in order to cope with the images and expectations we and others have of ourselves.

- At one extreme an individual may choose to accept and play the role others expect of them.

- At the other extreme an individual may choose to fight against others' images of them and behave in a manner more suited to their personal desires.

- It is most likely that an individual will use a combination of these tactics to varying extents as different circumstances in different places and at different times make it desirable to do so.

IMPLICATIONS FOR SOCIAL INTERACTION

The body is an always unfinished social and biological phenomenon which is transformed by its participation in society, and in turn, is the basis for, and affects social relations (Shilling, 1993). People's embodied experiences are simultaneously the result of both their physiology and more complex social reactions to it. Members of marginalized groups often suffer from unfavourable power relations in society, what Foucault observed as the social imprisonment caused by normalization processes (see McNay, 1994). The development of the dichotomies of sex/gender and impairment/disability have offered a new base on which to demand anti-discrimination legislation and civil rights. They have also acted as catalysts for the development of new theoretical explanations of inequality. An awareness of the body's inter-related social and biological properties in the recent growth of literature on the body in geography is 'playing a vital role in retheorising geography, that involves problematising the mind/body split, and making the body explicit in the production of geographical knowledge' (Longhurst, 1995: 102).

By developing a deeper understanding of the pressures society brings to bear on different people in public space and what coping strategies are used in resistance to them, a greater awareness of the complex social structures which affect all our lives can be constructed. Furthermore, what is socially constructed can by definition be deconstructed. The personal is most definitely political, as by recognizing and learning from micro-scale, individual experiences, macro-scale processes of marginalization and discrimination can be identified and ultimately eradicated.

Further reading

The impact of studies of the body in geography are considered at greater length in:
- Longhurst, R. (1995) The body and geography. *Gender, Place and Culture* **2**(1), 97–105.

For more detailed discussions of the nature of embodied experience and its social symbolism and values see:
- Shilling, C. (1993) *The body and social theory*. London: Sage.
- Featherstone, M., Hepworth, M. and Turner, B. (eds) (1991) *The body: social processes and cultural theory*. London: Sage.

The complexities of the biological state and social construction of the body are discussed further in relation to disability and gender, respectively, in:
- Butler, R. and Bowlby, S. (1997) Bodies and spaces: an exploration of disabled people's use of public space. *Environment and Planning D: Society and Space* **15**(4), 411–33.
- Young, I.M. (1990) Throwing like a girl: a phenomenology of feminine body comportment, motility and spatiality. In *Throwing like a girl and other essays in feminist philosophy and social theory*. Bloomington: University of Indiana Press.

Issues of identity management relative to sexuality are discussed in:
- Valentine, G. (1993) Negotiating and managing multiple sexual identities: lesbian time-space strategies. *Transactions of the Institute of British Geographers* **18**, 237–48.

CHAPTER 26

The city

Chris Hamnett

INTRODUCTION: LIVING IN AN URBAN WORLD

We are living in an increasingly urbanized world. In Britain over 80 per cent of the population live in urban areas, and in most developed Western countries the proportion is over 70 per cent. Although there has been rapid suburbanization and urban population decline in recent years, most people in Western countries live in urban environments. Nor is the process of urbanization confined to Western countries. The most recent UN figures (1996) suggest that 50 per cent of the world's population will be living in cities by the year 2000. Much of this growth will be in the Third World and much of it will be in large cities such as Mexico City, São Paulo, Lagos, Jakarta and Shanghai. Urban living will be a defining characteristic of life in the twenty-first century for the majority of the world's population. Consequently, the changing form, economic base and social structure of cities will continue to be of immense importance. We need to know how cities are changing, and what the implications for urban life are and will be in the future. These issues are frequently a subject of Hollywood movies ranging from *Escape from New York*, Woody Allen's *Manhattan*, Tom Wolfe's *Bonfire of the Vanities*, *Wall Street*, *Working Girl*, *Do the Right Thing* and *Desperately Seeking Susan*, to list just a few films which focus on life in New York.

Films and novels give us an insight into urban life, or representations of it, but as geographers we want to probe a little more deeply under the surface.

In this chapter, I want to concentrate on urban change and experience in the contemporary developed Western world, focusing on three very different aspects of urban living – first, counter-urbanization and the rise of ex-urban 'edge cities', second the experience of urban economic decline in the black ghettos of Chicago, and third the 'back to the central city' movement seen in some major western cities where it is associated with **gentrification** and the rise of what Sharon Zukin (1982) termed 'loft living'. Finally, I want to look briefly at the growth of inequality in **global cities**, focusing on Jonathan Raban's journalistic view of New York.

FROM URBANIZATION TO COUNTER-URBANIZATION

Urbanization in many Western countries peaked in the early decades of this century. In Britain, 70 per cent of the population lived in cities by 1900. Although urbanization has grown in many Eastern and Southern European countries in the post-war period, with rapid rural depopulation and urban growth in Spain, France, Portugal, Italy and other countries, much of North-West Europe has experienced a process of rapid suburbanization and counter-urbanization (Champion, 1989; Cheshire, 1995) and urban population decline. A growing share of population and employment have moved out of what were frequently seen as dirty, declining, derelict and crime-ridden central and inner cities, first to the growing suburbs but, more recently, aided by a rapid increase in car ownership levels to smaller towns and villages, far beyond the city boundary. This trend was accurately foreseen by H.G. Wells, the novelist and science fiction writer, almost 100 years ago in his book *Anticipations*, published in 1902. As Wells put it:

We are – as the Census returns for 1901 quite clearly show – in the early phase of a great development of centrifugal possibilities. And since it has been shown that a city of pedestrians is inexorably limited by a radius of about four miles, and that a horse using city may grow out to seven or eight, it follows that the available area of a city which can offer a cheap suburban journey of thirty miles an hour is a city with a radius of 30 miles . . . But 30 miles is only a very moderate estimate of speed and the available area for the social equivalent of the favoured season ticket holders of today will have a radius of over 100 miles . . . Indeed, it is not too much to say that the London citizen of the year 2000 may have a choice of nearly all England and Wales south of Nottingham and east of Exeter as his suburb . . . The country will take to itself many of the qualities of the city. The old antithesis will disappear . . . to receive the daily paper a few hours late, to wait a day or two for the goods one has ordered, will be the extreme measure of rusticity save in a few remote islands and inaccessible places.

What Wells anticipated has come to pass, and the trend towards population de-concentration was discussed by Berry (1970) in another perceptive paper which envisaged the urban population of the USA becoming increasingly dispersed by the year 2000 into low-density, ex-urban settlements, with their own shopping malls, factories, office parks, and entertainment facilities. In the USA Garreau (1991) terms them 'edge cities', and in Britain Herrington (1984) coined the term the 'outer city'.

There is no doubt that, in aggregate terms, population dispersal and counter urbanization have been the single most important trend in urban structure in many Western countries, particularly in North America, Australia and North-West Europe. There has been a major redistribution of the population in the USA from the older, industrial 'snow-belt' cities of the North East such as Detroit and Pittsburgh towards the 'sun-belt' cities of the South and West such as Pheonix, Tuscon, Las Vegas, Atlanta, Dallas and Houston. This has been aided by what Beauregard (1994) in his book *Voices of Decline* views as a long-standing and powerful **discourse** of urban decline concerning the nature, causes and consequences of urban change in older, industrial, cities. In Beauregard's view, this discourse has often functioned as a rationalization of, and a response to, racial change in American inner cities.

The magnitude of this change is indisputable. In the space of a few decades, the racial and ethnic composition of American cities has changed dramatically as a result of both high levels of immigration, particularly into key 'gateway', cities (Clark, 1995) and **white flight** to the suburbs (Massey and Denton,

1993). The changing ethnic composition of the city of Los Angeles has been clearly shown by Bill Clark of UCLA, who claims that the Los Angeles of the 1930s or the 1950s depicted in films such as *Chinatown*, and *Back to the Future* has been replaced by a Los Angeles more akin to that of *Blade Runner* in ethnic composition. Los Angeles is now a 'majority minority' city, the second-biggest Spanish-speaking city outside Mexico City, and what Ed Soja terms 'the capital of the Third World'. The Anglo population has fallen from 80 per cent to 41 per cent of the total, while the Hispanic population has risen from 10 per cent to 38 per cent and Asian-Americans make up another 11 per cent. Similar trends characterize Miami which has the highest rate of immigration in the USA. In 1960 Latin Americans acounted for just 5 per cent of the metropolitan population but 50 per cent by 1990, while the Anglo share of the population decreased from 80 per cent to 32 per cent (Nijman, 1996) as many of the Anglo population moved north out of the city in the 1980s. Amost half of the present population was born abroad and 60 per cent do not speak English at home. Nijman states that Miami has made the transformation from a 'southern US city' to a 'northern Latin American city'. In Detroit (Deskins, 1996) the white population fell by 71 per cent 1970–90 from 851,000 to 250,000, but rose by 240,000 in the suburbs. As a result, Detroit is now 75 per cent black against 43 per cent black in 1970. Similar trends have characterized Washington, DC and other cities (Knox, 1991). A black inner city is surrounded by a predominately white suburban ring.

One of the defining characteristics of edge cities in the USA is the emergence of what are termed 'gated communities': safe, socially selective, high security residential environments in which the predominantly white upper middle-class residents can turn their backs on the growing social and economic problems of the ethnically diverse central cities and retreat behind the walls, protected by security staff, electronic surveillance and 'rapid response' units. The rise of such gated communities is particularly marked in California and the South and West though white suburbanization is common in many large American cities. This phenonemenon has been documented by Mike Davis (1990) in his book *City of Quartz*, and by Ed Soja (1992) who discusses the growth of Mission Viejo and other gated communities in the southern half of Orange County, south of Los Angeles. Davis suggests that we are seeing the emergence of what he terms 'Fortress LA' characterized by the withdrawal of the affluent behind defensive walls and the erosion of public space. One of the best commentaries is by Christopher Parkes (1997) in a *Financial Times* article entitled The birth of Enclave Man (*see* Fig. 26.1). He

COMMENT & ANALYSIS

The birth of Enclave Man

Growth of US gated communities is imperilling the social contract, says **Christopher Parkes**

Even after dark the children of Avalon Gardens whirl and chatter in their new playground. Adults mingle over barbecues and in fitness facilities. All are relishing the unfamiliar sense of security and community that comes from the new 12ft iron fence, security floodlights and bullet-proof guard boxes ringing their homes.

Such simple freedoms are increasingly attainable throughout the US, as Americans, exhausted by urban violence and disintegration, retreat to the cosy sanctuary of so-called gated communities. There are now 30,000 such fortresses, home to some 8m people. In some parts of America a third of all new houses are being built behind walls.

So common have such communities become that they are now spreading even to ganglands like Avalon Gardens, a former free-fire zone in South Central district of Los Angeles.

The gang "tags" spray-painted on walls, and menacing clusters of young men loitering inside the fence betray this pilot scheme as a pale imitation of suburbia's orderly walled communities. But its ambitions are identical.

The fortress community has come to represent in the popular imagination a safe haven for people who want to live among their own kind in a compound shuttered against outsiders. The mainly black residents of Avalon Gardens – who include members of this local chapter of the Crips gang – now have an unprecedented measure of protection from the Bloods who dominate the surrounding districts. The shooting sprees have stopped.

Another housing project on Bloods turf in nearby Watts has since asked for similar privileges. Their request has awakened local politicians to the danger of sanctioning the establishment, at public expense, of unruly fiefdoms reminiscent of medieval Europe.

Others see even more danger in the cosy middle-class version of these communities. The prospect that the number of such developments may double within five years dismays social scientists. In the suburban separatism, defensive localism or plain Balkanisation that lies behind the trend, they see a rejection of the mutuality on which the country was founded.

Whites-only segregation, buried by federal legislation more than 30 years ago, is being restored by consumer demand for orderly, safe living conditions. Although anyone can buy a home in a gated community, today's inhabitants are predominantly white, earning between $60,000 and £200,000 a year.

In most aspects, they mirror the people who led the so-called "white flight" to the suburbs in the 1970s and 1980s. Most proceeded fight for, and win, municipal incorporation for their bedroom communities, thus escaping the obligations and taxes of the fast-rotting cities they left behind. To some observers, the rise of the gated community marks a critical further stage in the process of white flight.

According to Edward Blakely, dean of the University of Southern California urban planning school, the

retreat from the "basic ethic of mutuality which is the notion behind the US" is nowhere more apparent than in California, Texas, Arizona and Florida – the main points of arrival and concentration of Latino immigrants.

"You will see the walls go up wherever you see large numbers of immigrants," he says. "We will get to the point where [white] people will not feel safe unless they live in one of these places."

The mass retreat is well under way in parts of California, according to Dale Maharidge of Stanford University, who has logged what he calls "the browning" of the state and the withdrawal of monied whites to walled and gated developments in Orange County which keep "those people" out.

"No white society of the industrial world has ever evolved into a mixed culture," he writes in his recent apocalyptic book, *The Coming White Minority*.

According to state data, whites will account for less than half California's population in two years. The simultaneous rise of Latino political power – as 150,000 people of Hispanic origins reach voting age each year – foreshadows an unprecedented upheaval, Mr Maharidge argues.

Mr Blakely, co-author of *Fortress America: Gated Communities in the US**, concedes that the browning of California represents a "deep psychological threat" to some. But he is more concerned about the white withdrawal, which he believes could threaten the very structure of democracy.

This is a hefty charge to lay against what others, especially in the building and property business, see as simply a trendy lifestyle choice: one formerly restricted to the rich and famous and now affordable for the middle classes. Mr Blakely bases his fears on the argument that "fragmentation undermines the very concept of *civitas* – organised community life".

Most gated developments pay for their own private police patrols and security guards. Traditionally communal services such as schools, parks, entertainment facilities and even street cleaning and maintenance are often privatised within the enclaves. The inhabitants are ever more reluctant to pay higher taxes to maintain government and city services outside their walls.

Such reluctance has already capped Californian property taxes. In New Jersey some community-dwellers have won abatements on local levies after arguing that they already pay for their own private services.

According to Mr Blakely's theories, such tendencies threaten the social contract. The lack of social contact fostered by exclusive lifestyles serves only to accentuate the peril. Economists say one result of social protectionism is to cut ambitious minorities off from the contacts and opportunities they need to advance up the economic scale.

On the available evidence, these are not issues that impinge greatly on the thinking of the enclave-dwellers. Instead of a social contract – a bewildering concept in places such as California where more than 100 languages and cultures mingle – gated communities offer a straightforward option.

Membership of the homeowners' association and the payment of monthly dues to cover the cost of maintaining amenities – whether the newcomer chooses to use them or not – are mandatory. Association rules, typically drawn up by the builders and strictly enforced, establish a code of conduct.

In return for the security and other benefits offered by the community, people will commonly be required to sacrifice many of the freedoms available in the world outside. Rules may dictate the colour of their front door, curtains or other indoor furnishings visible from the street. Sticking political posters in front windows is usually forbidden. In some places homeowners may plant only shrubs from an approved list, and may not station gnomes in their gardens, make noise after 10pm, or own a dog weighing more than 20lbs.

"In effect," according to Jo Anne Stubblefield, a community governance specialist, "the property owner becomes a citizen of the association, subject to its governing and assessment powers."

This "contractual government" – in the words of a recent economic analysis of the gated community phenomenon – "appears to be the closest thing to a real-world social contract that can be found".

This is a curious development in a country where generations have brought up on the concept of limited government. As Evan McKenzie, professor of political science at Illinois University, pointed out in a recent radio broadcast, the Bill of Rights, which protects a citizen's right to hang out a flag or put a wishing well on the front lawn, is ineffective inside the confines of a private community.

"We teach [people] that they have rights that are protected against intrusion from government authorities," Mr Mckenzie says. "But what nobody told them was that local government was going to be privatised to this degree."

In spite of these disadvantages, the desire of ordinary Americans to live in walled communities is far from slackening. Some planners say that, in parts of Orange County, Montgomery County, Maryland and southern Florida, the concentration of gated communities is so dense that middle-class home buyers have little choice but to move into one of them.

One result is that the economic and social segregation characteristic of societies worldwide is, in the US, taking on a physical form. "Social barriers have always been there, but Americans climb over them," says Mr Blakely. "That's what America's always been about."

But walls, gates and armed guards – the essential elements of a prison – present more substantial impediments: both to those trying to get in and those who want to get out.

Figure 26.1 The birth of the Enclave Man. Credit: © Ingram Pinn. Source: *The Financial Times* 21/9/97

argues, like Davis, that the growth of gated communities is undermining any notion of the USA as a mixed society and is leading to the hardening of social divisions in space as social protectionism firmly excludes minorities from suburban homes and jobs.

Summary

- In many northern European and American cities counter-urbanization has proceeded apace in recent years, with population decentralizing from the old urban cores to new suburbs and, most recently, to ex-urban centres.

- This outmigration has been very socially and racially selective in the USA with large-scale ethnic immigration into the cities accompanied by a growing suburbanization of the white middle class.

- Many American cities, particularly in the southern and western states, have seen the growth of 'gated communities' which are highly socially selective by income, in an attempt to shut out the problems of crime and violence.

WHEN WORK DISAPPEARS: GHETTO POVERTY IN CHICAGO

Many large Western cities have experienced dramatic transformations in their social and residential structure in recent decades as a result of a combination of a massive decline of manufacturing employment, large-scale immigration, rapid suburbanization, white flight, inner city decay and social despair. This transformation has been particularly marked in many of the older American cities of the mid-West and North East such as Pittsburgh, Philadelphia, New York, Detroit and Chicago. The transformation of New York has been well studied by Mollenkopf and Castells in their book *Dual City*, but the city I want to concentrate on here is Chicago which has been the subject of major research by the black sociologist William Julius Wilson (1987, 1996) in his books *The Truly Disadvantaged*, and *When Work Disappears: The World of the New Urban Poor*.

Wilson's work is important for a number of reasons, but first and foremost it addresses a crucial urban social issue: that of the new black urban poor and the social dislocation and breakdown that follow from it. In this respect, his work is committed to real social issues, not to the intellectual games of **postmodernism**. This is the social world treated in the gritty and demanding films as Spike Lee's *Do the Right Thing*, *Grand Canyon*, and *Boyz 'n' the Hood*. The latter focuses on two talented young blacks who live in the ghetto of South Central Los Angeles and struggle against the forces of crime, violence and despair to get a college education. The French equivalent – *La Haine* – is the world of young immigrant males in one of the big social housing estates. They are depressing but worth seeing. *Trainspotting* offers a less bleak and humorous Scottish equivalent set in Edinburgh's social housing estates.

Wilson takes on two sets of intellectual opponents: the social conservatives who attribute the problems of the black ghetto to the attitudes and behaviour of its residents who are seen to be irresponsible, criminal and feckless welfare dependents, thinking only of the moment and happy to live by a combination of crime and welfare handouts. The second set of opponents are the liberals who ignore or deny the reality of anti-social behaviour in the black ghetto and have yielded the field to conservative popularizers such as Charles Murray in his book *Losing Ground*. Wilson argues that ghetto social problems are only too real and cannot be ignored or brushed aside, but he argues that they are a result of the changing employment, racial and demographic structure of American inner cities aided by systematic discrimination, and he argues that the attitudes and behaviours found in black poverty areas are a response to the problems faced by their residents rather than reflecting an innate culture of lawlessness, criminality and immorality as conservatives often believe.

As Wilson (1996) points out:

It is important to understand and communicate the overwhelming obstacles that many ghetto residents have to overcome just to live up to mainstream expectations involving work, the family and the law. Such expectations are taken for gran5ted in middle-class society. Americans in more affluent areas have jobs that offer fringe benefits; they are accustomed to health insurance that covers paid sick leave and medical care. They do not have to live in neighbourhoods where attempts at normal child rearing are constantly undermined . . . and their family's prospects for survival do not require at least some participation in the informal economy . . . I argue that the disappearance of work and the consequences of that disappearance for both social and cultural life are the central problems in the inner city ghetto.

(ibid.: xix)

Wilson's research is set in the neighbourhoods of Chicago's Black Belt (*see* Fig. 26.2) and incorporates a powerful mixture of quantitative and qualitative sources. He points out that less than one in three of the poor in the United States lived in metropolitan central cities in 1959 but by 1991 the figure had risen to almost half, and that much of the increase in concentrated poverty has occurred in African-American neighbourhoods. In the ten communities that constitute the historic core of the Black Belt, eight had poverty rates in 1990 that exceeded 45 per cent, three

had rates over 50 per cent and three of over 60 per cent. In 1970 only two neighbourhoods had poverty rates of over 40 per cent. Overall, the poverty rate in the Black Belt rose from one-third in 1970 to half in 1990. The increase in poverty has a simple explanation according to Wilson: the rapid growth of joblessness. In 1990 only one in three adults held a job in the Black Belt in 1990 compared to 60–70 per cent in the 1950s. The increase in joblessness is, in turn, is a result of **deindustrialization** and the replacement of manufacturing, transportation and construction jobs, traditionally held by males, by jobs in high technology and services which hire more women. These changes, says Wilson, are related to the decline of mass production in the USA or, perhaps more accurately, the automation of mass production, and the consequent job losses. He points out that in the 20-

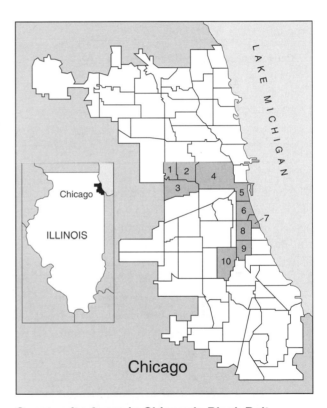

Community Areas in Chicago's Black Belt

1. West Garfield Park
2. East Garfield Park
3. North Lawndale
4. Near West Side
5. Near South Side
6. Douglas
7. Oakland
8. Grand Boulevard
9. Washington Park
10. Englewood

Figure 26.2 Chicago black belt. Source: Wilson 1996

year period from 1967 to 1987, Philadelphia, Chicago, New York City and Detroit each lost between 50 and 65 per cent of their manufacturing jobs. These manufacturing job losses particularly affected black males who disproportionately worked in the sector and had lower education levels which do not equip them for the new jobs in business services and high technology. In addition, many of the new jobs are either located in the city centre or in the expanding suburbs. He notes that in the last two decades, 60 per cent of the new jobs created in the Chicago metropolitan areas have been located in the northwest suburbs of Cook and Du Page counties in which African Americans comprise less than 2 per cent of the population. Consequently, blacks living in the inner city have less access to employment, and they are far less likely to own a car to enable them to get to the new jobs in a country where public transport is particularly poor. As one of his respondents, a 29-year-old unemployed South Side black male noted:

> You gotta get out in the suburbs, but I can't get out there. The bus go out there but you don't want to catch the bus out there, going two hours each ways. If you have to be at work at eight that mean you have to leave for work at six, that mean you have to get up at five to be at work at eight. Then when wintertime come you be in trouble.

(ibid.: 39)

Wilson notes that nearly half the housing stock in the black neighbourhood of North Lawndale has disappeared since 1960 and the remaining units are mostly run-down or dilapidated. And whereas in the past the Hawthorne plant of Western Electric employed 43,000 workers, International Harvester employed 14,000, and the world headquarters of Sears, Roebuck the mail-order firm employed 10,000, all have now closed. The departure of the big plants, says Wilson

> triggered the demise or exodus of the smaller stores, the banks, and other businesses that relied on the wages paid by large employers . . . In 1986, North Lawndale, with a population of over 66,000, had only one bank and one supermarket; but it was home to forty-eight state lottery agents, fifty currency exchanges, and ninety-nine licensed liquor stores and bars.

(ibid.: 35)

Another of his interviewees, a 29 year old unemployed black male, stated that:

> Jobs were plentiful in the past. You could walk out of the house and get a job. Maybe not what you want but you could get a job. Now you can't find anything. A lot of people in this neighborhood, they want to work but they can't

get work. A few, but a very few, they just don't want to work: but the majority they want to work but they can't find work.

(ibid.: 36)

The social consequences of mass joblessness are profound. Wilson argues that: 'Neighborhoods plagued by high levels of joblessness are more likely to experience low levels of social organization' and 'High rates of joblessness trigger other neighborhood problems that undermine social organization, ranging from crime, gang violence and drug trafficking to family breakups' (ibid.: 21).

The decline of job opportunities among inner city residents has increased the incentives to sell drugs, and addiction to crack cocaine has been paralleled by the rise of violent crime among young black males. Wilson points out that whereas the homicide rate for white males aged 14 to 17 increased from 8 to 14 per 100,000 between 1984 and 1991, the rate for black males more than tripled over the same period from 32 per 100,000 to 112, and he argues that neighbourhoods plagued by high levels of joblessness and disorganization are unable to control the volatile drug market and the violent crimes related to it. As the informal social controls weaken, so the social processes which regulate behaviour change and Wilson instances the spread of gun culture: 'Drug dealers cause the use and spread of guns in the neighbourhood to escalate, which in turn raises the likelihood that others, particularly youngsters will come to view the possession of weapons as necessary or desirable for self-protection, settling disputes, and gaining respect from peers' (ibid.: 21).

Wilson argues that many inner-city ghetto residents clearly see the social and cultural effects of living in high-jobless and impoverished neighbourhoods, particularly the effects on attitudes and behaviour. A 17-year-old black male living in a poor ghetto neighbourhood on the West Side stated that:

> Well, basically, I feel that if you are raised in a neighborhood and all you see is negative things, then you are going to be negative because you don't see anything positive . . . Guys and black males see drug dealers on the corner and they see fancy cars and flashy money and they figure: 'Hey, if I get into drugs I can be like him'.

(ibid.: 55)

The problem of the black ghettos, as Wilson sees it, is one of historic racial discrimination and segregation compounded by deindustrialization which has dramatically reduced the formal employment opportunity structure and led to destructive social behaviours which are pulling neighbourhoods down.

It is sometimes asserted that European cities are

being Americanized and that some areas are becoming ethnic ghettoes. But, as Loic Waquant (1993) and Peach (1996) have argued, this is a fundamental misconception. The overall proportion of ethnic minorities in European cities is far smaller than in the USA, and there is nothing approaching the levels of ethnic concentration in American cities. In answer to the question 'Does Britain have ghettoes', Peach unequivocally says 'No'. There are areas with high minority concentrations in some British cities but minorities comprise a majority of the population in only a very small number of enumeration districts. Ethnic minorities comprised 20 per cent of the population in Greater London in 1991 and 25 per cent in Inner London. The equivalent figure in city of Paris is about 17 per cent. This compares to 64 per cent in New York City in 1990 and 61 per cent in Los Angeles (Clark, 1997). Nowhere in European cities do we find the extensive concentrations of ethnic minorities that are found in American cities. This is not to deny that some groups fare badly in labour and housing markets (Madood, 1997), but this is not the same as the concentration of ghetto poverty found in the USA. The scale of the problem is quantitatively and qualitatively different.

Summary

- In the predominantly black inner cities of the North Eastern United States, large scale deindustrialization has been associated with a massive increase in unemployment and poverty. These problems are found to a lesser extent in some British and European inner city areas.

- The collapse of inner city manufacturing jobs, particularly for males, and the growth of predominantly low-wage service sector jobs, linked to the out-migration of jobs to the white suburbs have generated major social problems.

- The social and behavioural problems found in inner city black areas are very real, but they should be seen as the consequence of deindustrialization and discrimination rather than innate social characteristics: They represent a response to a changed set of economic and social conditions.

GENTRIFICATION AND LOFT LIVING IN THE CENTRAL CITY

The economic decline of older, industrial cities such as Detroit, Pittsburgh, Manchester, Liverpool, Glasgow, Lille and the Ruhr, has been paralleled by the rise of a small number of major world or global cities (Sassen, 1990) as the command and control cen-

tres of the world economy and finance system. These cities, which include London, New York, Paris, Tokyo, Toronto and others, have all experienced massive deindustrialization, but they have also seen the rapid expansion of business and financial services such as banking, legal services and management consultancy as well as the continued growth of a number of creative industries such as advertising, film and video, music, fashion and design.

These cities have been characterized by the transformation of their industrial, occupational, income and residential structure. As Ley (1996) shows in the context of Toronto and Vancouver, the rise of a service-based economy has been paralleled by the growth of a new professional, managerial, technical and creative middle class, generally highly educated and highly paid. The rise of this group, with their cultural interests and housing market demands has, in large part, been responsible for the growth of **gentrification** in post-industrial inner cities. Many of them work in business or creative industries in the central city or its environs and have long or irregular hours and want to live close to work and the cultural and entertainment facilities offered by the central city. But traditional central and inner city high status residential areas are expensive and in short supply. Consequently, the new middle class have sought out new living opportunities in the inner city, aided by developers and estate agents who have seen the prospects for profitable transformation of these areas.

There is a large literature on traditional forms of gentrification (Butler, 1997; Ley, 1996; Smith, 1996) which commonly involve conversion of nineteenth-century multi-occupied rental housing (much of it originally built for middle-class occupancy) back into single family houses or apartments. In New York and London, however, there has been a trend towards conversion of older industrial buildings into spacious if expensive city centre apartments. In New York, this was first concentrated in the SoHo area of downtown Manhattan, adjacent to the financial district and characterized by elegant late nineteenth-century multi-storey industrial lofts but it has since spread into Tribeca and other areas where industrial buildings are available (Zukin, 1982). Unlike the gentrification of single family housing, conversion of such buildings generally involves property developers who have the finance and expertise to carry out the work. An insight into loft living in SoHo is seen in the film *Desperately Seeking Susan* featuring Madonna and Rosanna Arquette.

In London, the process got underway in the 1980s in Docklands with conversion of some the old riverside warehouses along the Thames in Wapping and elsewhere, aided by the efforts of the London Docklands Development Corporation to socially transform the area (Goodwin, 1990). But in the last few years, there has been a dramatic expansion of loft conversions in the Clerkenwell area of London, just west of the City of London, and north of the Inns of Courts. This area was formerly one of industrial districts of London, as SoHo was for New York, but with the rapid decline of manufacturing in the 1960s onwards, it became increasingly derelict and empty. One or two pioneering developers such as the aptly named Manhattan Loft Corporation saw their potential and their proximity to the City and initiated the process of conversion. They have been an instant success, and many of the buildings sell out immediately, often straight from the plan, with prices ranging from £150,000 to £450,000 a unit.

Some of the conversions are very dramatic, such as the old headquarters of the New River Company (Thames Water) in Rosebery Avenue, near Sadlers Wells just south of the Angel, Islington. The conversion was undertaken by a consortium of Kennet Properties, a subsidiary of Thames Water, Berkeley Homes and the Manhattan Loft Corporation and it includes 129 flats as well as some spectacular private public space. Prices started at £160,000 for a one bedroom apartment to penthouse apartments at £400,000 each (see Fig. 26.3).

As the area has increased in desirability, aided by marketing and promotion as a fashionable place to live, prices have soared as they did in Soho in the 1970s. Clerkenwell lofts have become home to bankers, lawyers and the highly paid. The rise of Clerkenwell is illustrated in the *Hampstead and Highgate Express* property pages. In October 1997 they ran an article 'Clerkenwell seen as the new Belgravia' which focused on a review of central London property trends. This found that the average income of an owner of a central London property is £67,000, more than double the Greater London average of £26,800. In January 1998, in a clever piece of promotion, the *Hampstead and Highgate Express* article 'Keeping up with the Griff Rhys Joneses' said that they were selling their penthouse loft apartment in a building they bought and developed in 1989. The price, including a basement swim-

Figure 26.3 New River Head Development.
Source: Berkeley Manhattan

ming pool and lift, a mere £795,000 (*see* Fig. 26.4). This example is exceptional, but it illustrates the attraction of life in central London for a specific group of highly paid workers in London's financial, legal and creative industries.

Summary

- In addition to the rise of edge cities and ex-urban development, and inner city decline, there has a been a widespread growth in the middle classes in the central

Keeping up with the Griff Rhys Joneses

Miranda Norris

GRIFF AND JO RHYS JONES only discovered the true history of their home when a taxi driver dropped them off after an evening out. "He recognised the place and told us that he had done his apprenticeship here," Jo recalled. Rather glamorously, it turned out that he had serviced chicken-plucking machines.

Back in 1989, Jo and her husband, rising star of Not The Nine O'Clock News Griff Rhys Jones, had needed a home and somewhere in which to expand their young family. A semi-derelict former machine works in Sebastian Street in what was then unsalubrious Clerkenwell echoed with possibilities.

"We wanted a large flat and something lateral because the children were small and it was driving us mad going up and down stairs."

This was long before the loft/flat revival in the area had taken off and although the couple knew they wanted something a bit different they didn't start out thinking of a loft. It evolved that way because of the space they wanted and the building they found.

The Rhys Joneses bought and redeveloped Sebastian House, which they are now selling, into three two-bedroom flats, as well as their own penthouse-plus-basement home. Gaining planning permission was apparently the hardest part – completing the work took only a year but "thousands" of meetings with Clerkenwell architects Paxton Locher.

Both Jo, who is a graphic designer, and Griff had strong ideas about what they wanted to achieve. "We like fairly minimalist things but we also realised we had to live in it. And, having children, a certain amount of mess is generated."

The result was a penthouse, carved through the second and third floors of the building with fairly fluid living, dining and kitchen areas. The sleeping quarters – five bedrooms – are at the back.

Entering via the gallery, stairs lead down to the double-height sitting room where slouchy sofas sit around an enormous 17th century limestone fireplace, believed to have come from a chateau in the Loire.

Griff and Jo were going to create their own "Stonehenge" effect but went to Paris for the weekend and saw what they wanted in an expensive-looking antique shop opposite the Louvre.

"The shop was closed but the fireplace was so perfect that, even though we had to leave at about 11 on the Monday, we dashed back to get it. It was a bit of money but no more than you would pay for a really nice one in London. We had it brought over in five pieces."

The kitchen peninsula occupies one side of the gallery and on the other is a dark wood, monastic-style table with settles. A walkway leads from here to two interconnected studies with glazed walls, hanging over the sitting room.

"I suppose we do have his 'n' hers studies," Jo confessed, "but the wall is not central so mine is a bit smaller."

Jo works from home most of the time but Griff has an office at Talkback Productions,

● The sitting room (above) centres around an enormous 17th century French limestone fireplace (not seen); the basement swimming pool complex with den (left).

Figure 26.4 Clerkenwell seen as new Belgravia. Source: *Property Express/Ham & High*. Credit: Stirling Ackroyd

and inner areas of some major cities where economic change in the structure of employment has created new jobs in the creative industries and financial services.

- Many of the workers in these new growing industries have chosen to live in the central cities, leading to the growth of gentrification and 'loft living'. This latter trend has been associated with the conversion of industrial buildings to residential uses.

- Areas such as SoHo in New York and Clerkenwell in London have become fashionable residential areas for the new wealthy professional middle classes.

INEQUALITY IN THE GLOBAL CITY

It is clear from the discussion of the loft conversion market in London and New York, and ghetto poverty in Chicago that the modern city is marked by sharp inequalities. Recent analyses of earnings in London (Hamnett and Cross, 1998), New York (Mollenkopf, 1998) and Paris (Preteceille, 1998) reveal that inequalities have grown very sharply in recent years, aided by the rapid rise in earnings and bonuses in financial and legal services where earnings of hundreds of thousands of pounds are not uncommon. Friedmann and Wolff (1982) and Sassen (1991) argue that these trends are inscribed in the **global city** rather than being merely incidental. These cities, says Friedmann, are run for the benefit of the transnational elite, and social polarization is an integral and inevitable part of such cities: the prosperity of the elite rests on the exploitation of the poor.

In his book on America *Hunting Mr Heartbreak*, Jonathan Raban (1990) makes the brilliant distinction between the 'air people' and the 'street people' of Manhattan in terms of the total separation of their economic and social circumstances and their ways of life. Although his treatment is journalistic, and perhaps rather overdrawn, he puts his finger on the massive inequality which characterizes contemporary cities.

Raban's picture of New York is impressionistic jour-

Extract from Johnathan Raban's Hunting Mr Heartbreak

The beggars slept much of the day away on benches on the subway platform. By night, they scavenged. Returning home late after dinner, I would meet them on the cross streets around East 18th, where small knots of them went tipping over trashcans in search of a bit of half-eaten pizza, or the lees of someone's can of Coors. . . .

The current term for these misfortunates was 'street people', an expression that had taken over from bag ladies, winos and bums. The Street People were seen as a tribe, like the Beaker Folk or the Bone People, and this fairly reflected the fact that there were so many more of them now than there had been a few years before. In New York one saw *a people*, a poor nation living on the leftovers of a rich one. They were anthropologically distinct, with their skin eruptions, their wasted figures, poor hair and bony faces. They looked like the Indians in an old Western . . . (pp. 77–8)

There were the Street People and there were the Air People. Air People levitated like fakirs. Large portions of their day were spent waiting for, and travelling in, the elevators that were as fundamental to the middle class culture of New York as gondolas had been to Venice in the Renaissance. It was the big distinction – to be able to press a button and take wing to your apartment . . . access to the elevator was proof that your life had the buoyancy that was needed to stay afloat in a city where the ground was seen as the realm of failure and menace.

In blocks like Alice's, where doormen kept up a 24 hour guard against the Street People, the elevator was like the village green. The moment that people were safely inside the cage, they started talking to strangers with cosy expansiveness (p. 80) . . .

Everyone I knew lived like this. Their New York consisted of a series of high-altitude interiors, each one guarded, triple locked, electronically surveilled. They kept in touch by flying from one interior to the next, like sociable gulls swooping from cliff to cliff. For them, the old New York of streets, squares, neighbourhoods, was rapidly turning into a vague and distant memory. It was the place where TV thrillers were filmed. It was where the Street People lived . . . . (p. 81)

For Diane, places like Brooklyn and the Bronx were as remote as Beirut and Teheran. *Nobody* went there. The subway system was an ugly rumour – she had not set foot in it for years . . . I sometimes joined her on evenings when she was dining out uptown – evenings that had the atmosphere of a tense commando operation. At eight o'clock, the lobby of her building was full of Air people waiting for their transport. A guard would secure a cab, and we'd fly up through New York to the West 60's or the East 80's . . . . (p. 84)

It was a white knuckle ride. Diane sat bolt upright, wordless, clinging to the grab-rail in front of her, while the cab flew through the dismal 30s. At this level, at this hour, all of New York looked ugly, angular, fire-blackened, defaced – bad dream country. The sidewalks were empty now of everyone except Street People. This was the time when things began to happen that you'd see tomorrow on breakfast television, and read about, in tombstone headlines, in the *Post* and *Daily News* . . . . (p. 85)

Few of these journeys last more than ten or eleven minutes: they were just long enough to let you catch a glimpse of the world you feared. Then, suddenly, there was another guard, dressed in a new exotic livery, putting you through Customs and Immigration in another lobby. (p. 86)

nalism and travel writing, not social science, and for a more considered approach you should look at Mollenkopf and Castells (1991) book *Dual city? restructuring New York*, which provides an excellent overview of trends and inequalities. But Raban's piece is one of the most powerful and vivid insights into New York today as is Tom Wolfe's novel *Bonfire of the vanities*, since made into a film. Getting to grips with the culture of the modern city is as much about film, video and novels as it is about census data, interviews and questionnaire surveys. What matters is that we try to understand what is going on, and writers, journalists and film makers provide us with powerful visions and interpretations which can stimulate social scientific research.

CONCLUSION

Contemporary cities are changing in complex and often contradictory ways. Continuing suburbanization is paralleled, in some cities, by inner city urban decline (and the two are frequently causally linked) and by central city urban regeneration and gentrification. As a consequence, modern Western cities are frequently characterized by growing inequality – both between rich and poor and between different ethnic groups. In some cities this is accompanied by growing social segregation between those with greater resources and choice, and those with limited resources and limited choice. While some changes are clearly the result of a degree of choice and preference for different lifestyles and environments, others are often unwilling victims of economic and social processes largely outside their influence and control. While some people may be living in a post-modern urban lifestyle playground, others have to live in a post-industrial wasteland.

Further reading

To understand what is happening in the black inner areas of some American cities you can do no better than read one of W.J. Wilson's two major books. The second book is particularly interesting as it combines quantitative and qualitative material in an illuminating way. The urban poor are given voices and outline their situation in their own words.

- Wilson, W.J. (1987) *The truly disadvantaged*
- Wilson, W.J. (1996) *When work disappears: the world of the new urban poor.*

D. Deskin's chapter in O'Loughlin and Friedrichs (1996) gives a powerful picture of the changes which have affected Detroit in recent decades.

- O'Loughlin, J. and Friedrichs, J. (eds) (1996) *Social polarization in post-industrial metropolises.* Berlin and New York: De Gruyter.

For an insight into the impact and implications of large-scale immigration in the United States see:

- Clark, W. (1998a) *The California cauldron: immigration and the fortunes of local communities.* New York: Guilford Press.
- Clark, W. (1998b) Mass migration and local outcomes: is international migration to the United States creating a new urban underclass. *Urban Studies,* **35**(3): 371–84.

For an assessment of the scale of ethnic segregation in Britain see:

- Peach, C. (1996) Does Britain have Ghettoes? *Transactions of the Institute of British Geographers* **21**(1), 216–35.

For an analysis of gentrification, its causes and effects, see:

- Ley, D. (1996) *The new middle class and the remaking of the central city.* Oxford: Oxford University Press.
- Smith, N. (1996) *The new urban frontier: gentrification and the revanchist city.* London: Routledge.

For an understanding of the changes taking place in the edge cities of America, see:

- Sorkin, M. (ed.) (1992) *Variations on a theme park: the new American city and the end of public space.*
- Davis, M. (1990) *City of quartz.* London: Verso.

For a rigorous social scientific analysis of counter urbanization see:

- Champion, A.G. (1989) Counterurbanisation in Europe, *The Geographical Journal,* **155**, 32–59.
- Cheshire, P. (1995) A new phase of urban development in Western Europe: the evidence for the 1980s, *Urban Studies,* **32**(7) 1045–64.

For a good textual analysis of the representation of urban decline in the USA is:

- Beauregard, B. *Voices of decline: the post-war fate of US cities.* Oxford: Blackwell.

CHAPTER 27

The country

Paul Cloke

INTRODUCTION: THE COUNTRYSIDE COMES TO TOWN?

In July 1997, and again in March 1998, more than 100,000 people gathered in Hyde Park, London to protest about the gradual encroachment of urban-based bureaucracy into country life, as epitomized by the government's proposal to ban fox-hunting (*see* Fig. 27.1). In the words of the *Daily Telegraph*, 'in the annals of popular protest, there can seldom have been a noisier plea to a British government to do absolutely nothing than yesterday's Countryside Rally ... they had come from farms, moors and fells, emptying villages and leaving nature to its own devices for a day in order to let the urban majority know that the rural minority wishes to be left alone' (July 11 1997).

In many ways such a protest is indicative of a peculiarly British collection of landscapes, traditions and cultural practices associated with the countryside. Here we are offered the view of a somewhat timeless, highly valued and all-embracing country life which needs to be preserved at all costs from the ravages of urbanism. It is, however, the view of a small but powerful minority which can grab the imagination about what country life stands for. Our **geographical imaginations** of the country are often produced and reproduced from 'stuff' such as this. By contrast the distinguished travel writer, Jonathan Raban, records a visit to rural Alabama in his book *Hunting Mr Heartbreak*. Here he emphasizes the shock experienced by some Europeans when they encounter some of the countrysides of America. The scale, colour and 'savagery' of nature in the American outdoors do not easily accommodate direct comparisons with more familiar European landscapes.

Figure 27.1 The country comes to town. Source: *The Times,* 11/7/97 p.1

It was how Europeans had always seen American nature – as shockingly bigger, more colourful, more deadly, more exotic, than anything they'd seen at home. When the urban European thought of the countryside, he imagined a version of pastoral that was akin to, if a good deal less exaggerated than, that on offer in Ralph Lauren's Rhinelander Mansion on Madison Avenue. The 'country' was an artefact – hedged, ditched, planted, well patrolled . . . The European landscape was a mixture of park, farm and garden; the nearest we come to wilderness was the keepered grouse moor and the occasional picturesque crag. We were astonished by America, its irrepressible profusion and 'savagery'.

(Raban, 1990: 153)

In highlighting these differences, Raban also shows us that countryscapes represent a vivid and often specific facet of the geographical imagination. Not only do we carry around with us an **idyll-ized** sense of what our rural areas look like, and therefore of what they are like to live in and visit, but we are often shocked when encountering other stereotypes of our country-sides or other countrysides.

In this chapter, I want to discuss how rural areas have become exciting contexts for study in human geography. There has been, over recent years, something of a resurgence of interest in rural studies, partly as it has embraced the 'cultural turn' (see Chapters 22–24) which is evident in the broader social sciences, but also in part because the significance of 'nature' and 'rurality' have gone beyond rural geographical space. Rural areas themselves have offered fertile ground for the study of more mainstream cultural ideas, of which three have been of particular importance:

1. A focus on landscape, emphasizing the meanings, myths and ideologies which are represented therein. Geographical study of landscape can range from deep countryside to urban street and from deep history to the imaginative futuristic landscapes of science fiction. However, countryside landscapes demonstrate particular power relations as well as being objects of desire which many would wish to conserve.

2. A focus on *how nature relates to space*. Again, nature is by no means confined to rural areas, but countrysides are often represented as the 'obvious' spaces of nature. Here the relations between culture and nature (see Chapter 2) are often a visible element of country life – as in the 'hunting' debate mentioned above – and consequently the country provides fertile ground for the study of how humans and non-humans interact.

3. A focus on *'hidden **others**'*. Countrysides are rich in myth, and they represent territories where an overriding cultural gloss on life can mask very significant socially excluded groups. Issues of **gender**, **sexuality**, poverty, and alternative lifestyles are important in this context.

What links these themes together is the importance of an idyll-ized view of the rural. Countrysides are seen as places where people can live close to nature and in harmony with surrounding landscapes. Country living is characterized by a happy, healthy and close-knit community and a problem-free existence which differs markedly from urban life. Such an idyll reflects the power of those who can afford to buy into and enjoy rural life, and deflects any 'problems' which don't fit the image. And it is this idyllic cultural image which transfers itself into broader society such that the country is no longer confined to the spatial boundaries of recognizably rural areas. Through the *commodification* of nature and rurality within contemporary consumption (as indicated not only through media attention and in advertising but also in 'country' consumer goods ranging from 4-wheel drive vehicles to furnishings and clothing) the importance of the country, and the meanings attached to it, have spread throughout society.

It is important to note that these cultural themes of the country are not the only geographies to be told of rural areas. Indeed, these more recent cultural geographies are being overlaid onto existing accounts of behavioural and political studies of economic change, and demographic and relational studies of social change. What makes the country important in human geography, however, is a combination of the idyll-ized *imagined geographies* peddled by media and advertising and held by people as significant reference maps for spatial behaviour, and the specific material changes occurring in rural geographical areas. This mix of imagined and spatialized countrysides begs questions about how we recognize rurality when we see it, and it is important here to reflect briefly on these debates surrounding the nature of rurality.

THE BLURRING OF COUNTRY AND CITY

When we study rural georaphies we have to keep hold of two rather different kinds of change. First, there are the changes which are occurring in rural areas themselves. For example, over the past 25 years there has been a hugely significant reversal of the trend of the previous century whereby population had been concentrating into urban centres in most Western countries. In the United States, for example, the 1950s and 1960s saw a strong positive correlation between settlement size and population growth rate, but during the 1970s 'smallness' became associated with growth. In the UK, the 1981 census revealed that the only local authority districts to increase their rate of population growth over the previous decade were the remoter, largely rural districts (Table 27.1). By 1991 the more complex pattern of growth and decline still revealed that most rural areas were continuing to experience population growth (*see* Fig. 27.2). In-migrant populations were at one and the same time seeking out the perceived advantages of rural lifestyles, and bringing with them attributes of urban living and expectations which were likely to transform the very communities they had been attracted to. Traditional rural life had already been transformed by the near universal availability of urban-based media, and now this has been reinforced by the infusion of migrants, often from urban places, who were seeking to live out imagined geographies of rural life in particular geographical places and spaces, often leading to some turbulence with more 'indigenous' populations.

Table 27.1 Percentage rural population change, 1961–81

	1961–71	1971–81	Difference
Rural Wales and Scottish Islands	−0.2	7.0	7.2
Rural, mainly West	7.2	8.8	1.6
Rural, mainly East	15.0	12.7	−2.3
Rural, mainly Scotland	−1.9	9.3	11.2
Rural Growth Areas	22.0	8.6	−13.4

Demographic change has usually gone hand in hand with economic change. As the size of the agricultural workforce has diminished, the notion that rural areas are dominated by agriculture has in many places become more applicable to the dominance of agricultural landscapes than to the agricultural economy. Counterurbanization was often accompanied by an urban-to-rural shift in new manufacturing growth, and although the economic impact of this shift has sometimes been short-lived, new forms of service sector employment have often added to the eco-

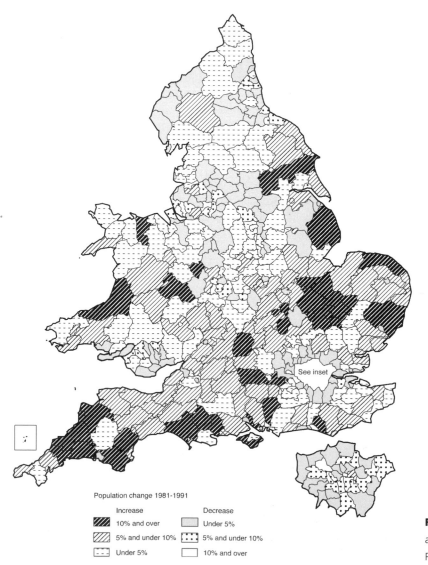

Population change 1981-1991

Increase		Decrease	
10% and over		Under 5%	
5% and under 10%		5% and under 10%	
Under 5%		10% and over	

Figure 27.2 Population change in England and Wales, 1981–91. Source: Office of Population Censuses and Surveys

nomic potential of these non-metropolitan areas. Indeed it is now commonly assumed that the growing importance of telecommunications and information technology will metaphorically 'shrink' the geographic distances between rural areas and major urban centres and thereby favour service-sector growth in rural areas. An on-line personal computer allows many contemporary work tasks to be performed from the rural home, although the degree to which such '*telecottaging*' will obviate the need for face-to-face work contact more generally is as yet arguable.

This general picture of change itself masks considerable variation both within and between nations. Indeed, there is evidence (for example in Scandinavia and parts of the United States) that the 'population turnaround' may have a limited duration. So, it is useful to talk of rural geographies rather than a rural geography. It is an obvious but often forgotten fact that what we regard as Western or 'Developed' nations vary enormously in scale. As Figure 27.3 demonstrates, the scale of influence exerted by major metropolitan areas differs widely, such that the urban pressure on the country in Britain will be far more intense than those on certain areas of the United States and Australia. Such variation means that particular places will be located rather differently in the mosaic of change described above. For some, it

would be no exaggeration to suggest that they reflect *suburban* characteristics, performing a dormitory role for metropolitan labour markets. Elsewhere, agriculture will remain as the dominant economic as well as landscape feature. Elsewhere again, extreme geographical marginality reflects characteristics of 'outback', 'wilderness', or even desert – each posing particular questions of nature–culture relations, and of the potential for commodification. So, the changes occurring in rural areas themselves are irregular, and particular attention has to be given to the geographies of particular places within the overall framework of change.

Accepting the importance of these differences in nature–culture relations, Marc Mormont (1990) has suggested that another key question about rural change concerns the changing relationship between space and society, and it is increasingly clear that this relationship is no longer only about the traditional divisions between rural and urban, or town and countryside. He argues that such dualisms have been completely overtaken by events, and outlines a series of changes relating to personal mobility, and new economic uses of the countryside which indicate the outmoded nature of any view which sees rural society and rural spaces as being welded together. Mormont's analysis of change relates to Belgium, but appears rel-

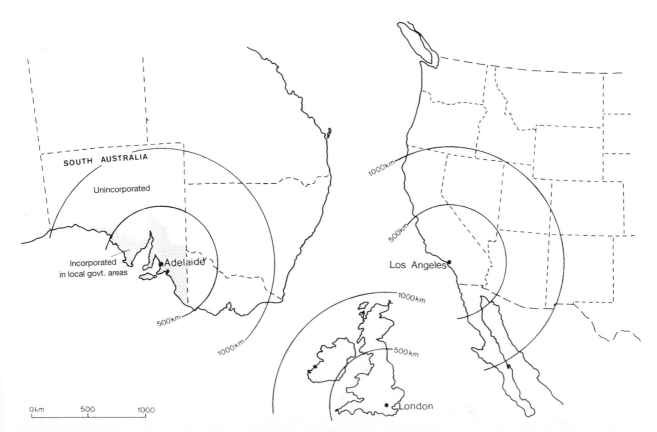

Figure 27.3 Different scales of urban influence. Source: G. Hugo and P. Smailes (1985) Urban rural migration in Australia. *Journal of Rural Studies*, 1, 19

evant to many countrysides as is his conclusion which is to suggest that there is no longer a single rural space, but rather a multiplicity of social spaces which overlap the same geographical area. The supposed opposition between the geographic spaces of city and countryside is being broken down, but oppositions between the *social* significances of city and countryside remain. For Mormont, then, rurality is a category of thought – a **social construction** – and in contemporary society the social and cultural views which are *thought* to be attached to rurality provide clearer grounds for differentiating between urban and rural than do the differences manifest in geographic space.

Other commentators (see for example, the seminal writings of Raymond Williams in the British context) have also noted the blurring of the country and the city. A very interesting contributor to this debate is Alexander Wilson (1992), who in his book *The Culture of Nature* suggests that recent land development in North America – suburbs, theme parks, shopping centres, executive estates, industrial parks, tourist developments and the like – has served to reproduce misleading ideas about city and country. He argues that the form of this development fragments geographies into those devoted to work and leisure, and production and recreation, which are oppositions which obscure more than they reveal about the nature of city and country. He cites the West Edmonton Mall (Edmonton, Canada) as an example of this jumble of country/city compromises. Its suburban location, 45-hectare size, and two and a half kilometre-long concourse suggest the monstrous urbanism of an indoor shopping centre. However, it includes a one-hectare lake with dolphins, sharks (and four full-size submarines which give rides in the lake). Moreover, the mall also houses an 18-hole golf course, a water park with six foot surfing waves, and hundreds of animals in aquariums and cages.

> The relentless mission of the West Edmonton Mall is to bring everything into its climatized, commodified space, especially objects and species from the natural world. There are hundreds of animals – 'hand-picked specimens,' the brochures say – in aquariums and cages. The species chosen are displayed in the ways that we've come to know them. First come the most glamorous and evil beasts, familiar from James Bond movies: piranhas, octopuses, alligators, and sharks. Then there are the performers: seals and penguins, peacocks, flamingoes, and dolphins. Then there are the cute animals: spider monkeys, emus, and angel fish. And the 'wild' ones we know from TV: black bears, mountain lions, jaguars, iguanas, ostriches. Some of the animals are available for photos. A sign at one stall says:
>
> Have your photo taken with a live cougar cub, $5.99. Extra persons in photo, $2.00 each. Small cub $5.99. Large cub $7.99. Special cuddly cougar prints, regular $30.00 value, now only $5.00 while they last. Lovable lynx prints only $5.00.
>
> There is little effort wasted on contextualization at the Mall. Why bother recreating the pre-industrial farm or a simulated jungle, conceits insisted on in Disney environments? Here the animals are just another commodity form, alongside Yves St Laurent and Shoppers Drug Mart. As if to emphasize the point, the mall maintains a 'retreat' for the animals at an undisclosed suburban site in Edmonton. There the animals can rest from their work at the mall, a fact that hasn't discouraged the local gossip that these non-human employees have a very short lifespan. See them while they last!
>
> *(Wilson, 1992: 198–9)*

Here, then, is nature, but in a controlled, climatized, commodified space. The 'landscape' is artificial, but the control over nature is very much part of the attraction. In yet another way, the binary opposition between city and country, and indeed culture and nature is blurred in such place-making.

These examples suggest that the assumed differences between the geographical spaces of city and country have been somewhat undermined by changes occurring in the social, economic and built environments concerned. It is again important to emphasize that such changes differ in scale and intensity in different places. Some countrysides will appear relatively untouched while others will have been visibly transformed. Nevertheless, the increasing importance of rurality as a social rather than geographical construct applies very widely. These factors constitute the first set of changes which rural geographers have to grasp.

The second kind of change relates to the way in which geographers have offered different ways of understanding rurality itself. Any given rural geographical space can *appear* different according to the theoretical perspectives adopted. For example, geographers have traditionally mapped rurality by equating it with particular functions: thus rural areas are dominated (currently or recently) by extensive land uses, such as agriculture and forestry, or large open spaces of undeveloped land; rural areas contain small, lower-order settlements which demonstrate a strong relationship between buildings and surrounding extensive landscapes, and which are thought of as rural by most of their residents; and rurality engenders a way of life which is characterized by a cohesive identity based on respect for the environment and behavioural qualities of living as part of an extensive landscape.

This type of analysis has been useful in generating indicators of rural territory, and remains useful, especially in those areas which are less transformed by the process of blurring described above. However, different theoretical epochs in social sciences have produced critiques of these definitions of rural space. From *political–economy* approaches came the insight that rural areas were increasingly linked into changing international economies, with the causes of 'rural' change usually stemming from outside of the rural areas concerned. From this viewpoint 'rural' places were not particularly distinct, and for some, this realization led to a call to do away with rural as an analytical category. Other windows on rurality have been opened by more **postmodern** and *post-structural* ways of thinking. It will be evident from the example of Marc Mormont's work, used earlier, that rurality can be regarded as a **social construct**, and that the importance of the 'rural' lies in the fascinating world of social, moral and cultural values that are thought to be significant there. Far from 'doing away with' the rural, then, the idea of rurality as a social construct invites researchers to study how behaviour and decision-making are influenced by the social and cultural meanings attached to rural places. In particular, there is considerable interest in how meanings of rurality are constructed, negotiated and experienced (Cloke and Milbourne, 1992). While such meanings may have much in common, there will be many different versions of rurality perceived by different individuals and organizations.

Drawing on the work of Baudrillard, Keith Halfacree (1993) discusses these multiple meanings of 'rural' in terms of three levels of divergence. The *sign* (= rurality) is being increasingly detached from the *signification* (= meanings of rurality) as social representations of rurality become more diverse. Equally, sign and signification are also becoming more divorced from their *referent* (= the rural geographical space). He points out that it is a characteristic of postmodern times that symbols are becoming more detached from their referential moorings, and therefore that socially constructed rural space is becoming increasingly detached from geographically functional rural space. Indeed, Jonathan Murdoch and Andy Pratt (1997) believe that we have reached the stage of 'post-rural' studies, reflecting that rurality can no longer be mapped out as any kind of totality.

These different approaches to 'mapping' or 'knowing' the country introduce a constructive tension to rural studies, especially when held together with the material changes occurring in what are commonly recognized as rural geographical spaces. For some, the country will be seemingly knowable, and apparently atheoretical. For others, the complexities of power, practice and process will render the category 'rural' unknowable as any kind of geographical or social entity. The more postmodern the country

seems, the more blurring seems to occur between country and city. Many residents and visitors do appear to act as if the countryside exists in some knowable form. Others, however appear to know their 'rural' places differently seeing them in other regional (e.g. 'The Borders) or local (e.g. as specific settlements) ways.

Summary

- Rural areas themselves are changing demographically, socially and economically. A geographically 'rural' space may now be overlapped by many different social spaces, thus transforming traditional countrysides.

- Many new land developments – theme parks, shopping centres, tourist developments etc – also blur the difference between country and city.

- Geographers have also changed the ways in which they have sought to understand rurality itself. Defining rural space by the functions that go on there has been challenged by those who view rurality as a social construct – a category of thought.

- The cultural meanings associated with the country have become increasingly detached from rural geographical space and are now important throughout society.

COMMODIFYING THE COUNTRYSIDE

There is evidence that the continuing importance of the country will in part lie in attempts to commodify the countryside as a particular type of attraction within postmodern consumption. Popular culture now serves us up with what Raymond Williams referred to as 'a continuing flood of sentimental and selectively nostalgic versions of country life'. Films such as *Remains of the Day*, *A River Runs Through It*, and *Four Weddings and a Funeral*; television series such as *Heartbeat*, *Peak Practice* and *Pride and Prejudice*; children's favourites such as *Postman Pat* and *Sylvanian Families*; magazines such as *Country Life* and *Countryman*; all merely add to classics in art, literature and media in their focus on cosy and nostalgic aspects of countryscape.

Moreover, the advertising industry repeatedly borrows from the treasure chest of positive meanings vested in the countryside: the 'goodness' of nature, to sell bread; the 'classiness' of the country house to sell cars; the 'pioneer spirit' of rural America to sell jeans or cigarettes – and in so doing reinforces these references to nature, heritage, nostalgia and so on in popular constructions of contemporary rural life. In this way, the meanings of country are attached to products which are themselves often aspatial. The country escapes from its geographical referent and inhabits the wider world of taste and consumption. An excellent example of this escape can be found in the way in which the Laura Ashley company purposefully commodified the appeal of country tradition to create a style which is applicable to many kinds of geographical space. In this example the past rustic traditions are sieved through the 'colourful mixture of prints and textures' and the Welsh farmhouse, Long Island house, and Swiss chalet are made available to anyone, anywhere. Rustic tradition becomes contemporary commodity, and nature's countryside is bought and sold as fabric and furnishings.

The country is thus being commodified within both the geographical spaces and social spaces which it inhabits. There has been a significant shift in the nature and pace of **commodification** in rural areas in many developed nations, giving rise to a series of new markets for countryside commodities: the countryside as an exclusive place to be lived in; rural communities as a context to be bought and sold; rural lifestyles which can be colonized; icons of rural culture which can be crafted, packed and marketed; rural landscapes with a new range of potential, from 'pay-as-you-enter' national parks, to sites for the theme park explosion; rural production ranging from newly commodified food to farming for the conservation of landscape, and so on.

As the country becomes commodified, particular meanings and characteristics are emphasized which come to represent its very essence. For example, in a study of how new and revamped rural tourist attractions were being advertised in parts of Britain, it was found that particular meanings, signs and symbols of countryside were clearly being represented (Cloke, 1993). These socially constructed ruralities reflected the perhaps predictable themes of nature, outdoor fun and history. They also reflected the slightly less predictable themes of family safety, 'hands-on' or 'up-close' experiences of nature; and the specific commodity links with souvenir craft and particular foods and drinks which form integral components of the packaged day out in the countryside. The study demonstrated that many of these new countrysides were based on the production of a *spectacle* for visitors. For example, Morwellham Quay on the border of Devon and Cornwall recreates a Victorian copper port (*see* Fig. 27.4) and in so doing offers an outdoor theatre of rural history. According to the brochure,

(a)

(b)

Figure 27.4 Rural attractions. (a) Morwellham Quay and (b) the National Shire Horse Centre

the quay workers, cooper, blacksmith, assayer and servant girls dressed in period costume, recreate the bustling boom years of the 1860s . . . Chat with the people of the past. Sample for yourself the life of the port where a bygone age is capture in the crafts and costumes of the 1860s . . . Try on costumes from our 1860s wardrobe.

The invitation is to spectate and participate in the history which is 'captured' by the attraction and presented to visitors in the form of spectacle.

Most of these kinds of attractions offer spectacle of varying degrees. Some spectacle makes claims to authenticity, such as happens at the National Shire Horse Centre in South Devon where traditional farming methods using heavy horses are displayed (Figure 27.4). Here, however, the attraction is augmented with 'special events' – western weekends with 'authentic' shoot-out displays; Teddy Bears' Picnics and Dollies' Tea Parties; steam and vintage rallies; and classic car shows. In cases such as this, the spectacle seems to emphasize signs and symbols which are only loosely related (if at all) to the specific rural geographical space, its landscape and its history. As Halfacree noted above, sign and signification are becoming increasingly divorced from their referent.

Examples of commodification of country spaces abound. Indeed, Howard Newby (1988) has suggested that rural Britain in general has become a theatre for visiting townsfolk, with rural people, and especially farmers, being the scene-changers and bit-part actors for that theatre. In other nations, this commodification of often reconstructed ruralities is also strongly represented in the changing rural scene. Alexander Wilson (1992) discusses examples in Southern Appalachia:

Scattered along the roads of Tennessee, Kentucky, and the western part of North Carolina are restored villages that recall and reconstruct ideas about the way things once were in those mountains. Some are within nature parks, some are part of theme parks, some are simply an assemblage of buildings that evolved out of someone's backyard, some promote religion, others consumerism

(Wilson, 1996: 206)

His examples range from gaudy theme parks such as 'Dollywood' (The Dolly Parton story) to 'authentic' museums, but each attempts to recreate a lost time and culture in this part of rural America. Here, too, then, the nostalgia of rural life – and in this case the pioneerism and specific culture of rural mountain folk – are commodified as attractions for visitors to the contemporary countryside. Once again, the character of the present is vested in the symbols and meanings of the past.

At this point, though, we need to appreciate that the process and practices of commodification do not always reinforce nostalgic countrysides notions of English rural idylls or the pioneerism of rural America. Although rural areas will usually be trading on their past, there are interesting examples now whereby the commercialism of place-making has begun to forge *new* identities for the country. An example of this can be drawn from rural New Zealand, where the growth and commodification of adventure tourist facilities, practices and subcultures have added new dimensions to the lives of many of its people, and to its landscape (Cloke and Perkins, 1998). Adventure tourism has influenced the production and reproduction of new imagined geographies of the country in South Island New Zealand. In particular, whatever the activity and place being advertised, there are repeated allusions and references to freshness:

1. *a fresh look at spectacular environments* – what was previously thought to be the spectacular scenery of New Zealand can be made more spectacular by participation in (or watching others participate in) adventurous pursuits in places of natural or historical significance.

2. *fresh, youthful thrills* – adventure tourists are provided with the white-knuckle excitement of contemporary theme parks, but is a 'natural' outdoor setting.

3. *the freshness of eager experimentation* – rural areas of New Zealand are being 'branded' by continual experimentation with bigger, better and more exciting thrills in the outdoors environment.

Invitations to 'crack the canyon with the Awesome Foursome' (*see* Fig. 27.5) reflect a different form of countryside commodification, where the relationship between tourist and landscape reflects a far more active, participatory and '**embodied**' experience of nature relations than is provoked by conventional countryside nostalgia.

This account of commodification in the reshaping (yet often reinforcing) of countrysides has tended to emphasize the natural and positive cultural attitudes of rural areas. Against these potentially idyllic representations of rurality, however, can be set rather more dystopian narratives of rural life and landscape (see, for example, David Bell's (1997) analysis of 'horror' films set in small-town America). Commodified rurality is not unambiguous. The glorious isolation of 'wilderness' or 'outback' can also be attractive to those wishing to dump hazardous waste which is unwanted in more populated regions (*see* Fig. 27.6).

Summary

- The country is being made into a commodity in many different ways, and this process contributes to the

Figure 27.5   The Awesome Foursome: helicopter, bungy, jet-boat and white water rafting experience near Queenstown, New Zealand

"REMEMBER WHEN THEY USED TO SEND US POVERTY PROGRAMS!..."

Figure 27.6   Issues in rural Appalachia.
Source: *The Charlotte Observer*, © Doug
Marlette

emphasis on particular meanings and symbols as being important indicators of rurality.

- Some meanings reflect a nostalgic representation of the country as being dominated by the virtuous values of the past, involving closeness to nature and close-knit community. These meanings are often presented in various kinds of spectacle.

- Other commodification has led to new 'country' meanings such as the fresh and adventurous encounters with nature sponsored by the rise of adventure tourism in New Zealand.

OUT OF SIGHT AND OUT OF MIND: RURAL OTHERS?

Figure 27.6 also emphasizes that beneath the innocent idyll of the country there lie other characteristics which hit the headlines less often. The underbelly of idyll-ized countrysides is rarely exposed. Indeed, the stories used to promote the country as idyll, usually serve to mask any contradictory social conditions. Unemployment or underemployment, the scarce availability of affordable housing, the rationalization of local services into larger centres, and the shrinking of public transport services have all served to disadvantage low-income households in rural areas. Moreover, two recent volumes (Cloke and Little, 1997; Milbourne, 1997) have emphasized different processes and practices by which certain rural people can be marginalized. Here the emphasis is on individuals and groups which are 'other' than the mainstream, with identities characterized by gender, race, sexuality, age, class, alternativeness and so on. I use the 'otherness' relating to rural poverty as an example here, but the story of how a rural problem is hidden discursively away in public discourses as well as geographically could just as easily refer to the rural homeless, or indeed any of the identity groupings listed above.

A recent study has shown that the percentages of households in or on the margins of poverty in twelve case study areas in England ranged between 39.2 per cent and 12.8 per cent, with ten of the twelve areas having more than 20 per cent levels of poverty (see Table 27.2). Yet, in the most recent government White Paper on the countryside – *Rural England: A Nation Committed To A Living Countryside* (1995) – the word poverty does not appear. Two reasons may be advanced for this. First, the political propaganda of 1980s and early 1990s Britain has effectively pronounced the end of poverty, claiming that *absolute* poverty has been eradicated by economic success, and that relative poverty is a figment of the academic

Table 27.2 Percentage of households in or on the margins of poverty* in 12 case study areas in rural England.

Nottinghamshire	39.2
Devon	34.4
Essex	29.5
Northumberland	26.4
Suffolk	25.5
Wiltshire	25.4
Warwickshire	22.6
North Yorkshire	22.0
Shropshire	21.6
Northamptonshire	14.8
Cheshire	12.8
West Sussex	6.4
Across 12 areas	23.4

Note: * measured as less than 140% income support entitlement.

Source: P. Cloke, P. Milbourne and C. Thomas (1994) *Lifestyles in Rural England*, Rural Development Commission, London.

imagination and should more properly be labelled as 'inequality'. Second, despite the censorship of the word 'poverty' from government pronouncements, there is a strong sense in which it has been easier to deny the existence of poverty in rural areas than in the cities. Here, we can link the imagined geographies of idyll-ized rural lifestyles with the idea that poverty in rural areas is being hidden or rejected in a cultural dimension both by decision-makers with power over rural policy, and by rural dwellers themselves (including those who appear, normatively, to be poor). Rural people can be recognized as 'deprived' of ready access to the advantages of urban life, but are not as such impoverished by such deprivation because rural living somehow offers perceived compensations for any such disadvantage. In this way, rurality appears to signify itself as a poverty-free zone, and constructs of rural idyll at the same time *exacerbate* and *hide* poverty in rural geographic space.

Poverty in other developed nations such as rural America is an equally important issue, and one which is also influenced by the impact of dominant cultural constructions of rurality. In contrast to Britain, the United States does have an official poverty line, and so by state-defined statistics, levels of rural poverty are currently reckoned to be around 20 per cent of households. Rather than attempting to deny poverty politically, the emphasis has been to differentiate between the deserving and the undeserving poor. In this way, urban underclasses are signalled as 'undeserving' while non-metropolitan low-income workers are cast as 'deserving'; the spotlight of publicity falls on the former not the latter. As a result, cultural construction of poverty in America has a distinct spatial outworking, with impoverished rural Americans being lost in the shadows. And even when poverty is recognized as an issue in rural areas, there is a tendency to assume that it is restricted to key problem areas, notably Appalachia and the South. But as McCormick (1988) graphically indicates: 'Today, the problem has no boundaries. A tour of America's Third World can move from a country seat in Kansas to seaside Delaware, from booming Florida to seemingly idyllic Wisconsin' (ibid. p. 22).

The tendency to regard social problems in the country as out of sight and out of mind is directly related to the dominance of prevailing social constructions of rurality. These in turn are essentially interconnected with the relations of power at work in and beyond rural areas. It is unsurprising, then, that the most significant demonstration of 'rural' opinion in Britain in recent years should be focused on the specific issue of fox-hunting, and the more general issues of being left alone to exercise personal freedom in the idyllic villages, farms, moors and fells. Neither is it surprising that the major demand of demonstrators is that nothing should change. Such is the conservatism of the powerful, not the powerless.

Summary

- Low income households in rural areas are disadvantaged by changes to job markets, housing markets and the provision of services.

- 'Other' individuals and groups are marginalized in rural society on the grounds of gender, race, sexuality, age and so on.

- The example of rural poverty demonstrates how idyllic representations of the country mask the occurrence of significant social problems and nullify the perceived need for policy responses to those problems.

CONCLUSION: POWERFUL GEOGRAPHIES OF COUNTRYSCAPES?

These complex and often ambiguous relations between nature and culture, and society and space, which underpin the mosaic of geographical space of the country make for an important and exciting territory of geographic enquiry. It should be made clear that there are many specific fascinations within these countryscapes that have hardly been touched on here, for example the restructuring and reregulation of agricultural landscapes in a *post-productivist* era; the cultural importance of food; the conservation of valued landscape and habitat; the evolution of community and social relations in countrysides; the power

relations and governances which pertain there; and so on. Those wishing to delve into particular issues such as these might like to use the Further Reading suggested at the end of this chapter. None of these concerns, however, is immune from the emerging core of significance in rural studies that has been outlined in this chapter – namely, the interconnections between the socio-cultural constructs of 'country', 'rurality' and 'nature' which seem to be so important in (re)producing our geographical imaginations of rural space (geographical and social), and the actual experiences of how lives are practised within these spaces. Such practices will need to be viewed both from the outside looking in (taking full account of 'structuring' influences) and from the inside looking out (taking full account of individual difference, embodiment and identity).

Chris Philo (1992) catches the mood of these significances in his review of what he regards as neglected rural geographies in Britain. His contention is that most accounts of rural life have viewed the mainstream interconnections between culture and rurality from the perspective of typically white, male, middle-class narratives. There is therefore an urgent need to look through other windows onto the rural world.

Myths of rural culture often marginalize a range of individuals and groups from a sense of belonging to, or in the rural. We need to make our geographies of the rural more open to the circumstances and voices of other people in order to overcome the neglect of 'other' geographies. Such a conclusion is exciting, but not unproblematic. As Jonathan Murdoch and Andy Pratt (1993) have suggested, simply by 'giving voice' to others we do not necessarily uncover the power relations which lead to marginalization or neglect. A range of important questions arise here, relating in particular to the power of the researcher, the potential reinforcement of marginalized identities by labelling them as 'other', and the potential for flippant rather than politically grounded engagement with marginalized people. However, the mosaic geographies of the country will be richer for the addressing of these questions than for their neglect. After all, when the huntsmen come to town, are they really *the* voice of the country?

Further reading

For a comprehensive account of the parallel importance of the changing nature of rural areas and the changing ways in which geographers and others have sought to interpret rurality, read:

- Marsden, T., Murdoch, J., Lowe, P., Munton, R. and Flynn, A. (1993) *Constructing the countryside*. London: University College London Press.

- Murdoch, J. and Marsden, T. (1994) *Reconstituting rurality.*London: University College London Press.

For a view of the idyll-ization of the country in North America and Britain, see:
- Bunce, M. (1993) *The countryside ideal*. London: Routledge.

The blurring of the country and the city is discussed in:
- Wilson, A. (1992) *The culture of nature*. Oxford: Blackwell.

The issue of marginalized 'others' in the country is dealt with in:
- Cloke, P. and Little, J. (eds) (1997) *Contested countryside cultures*. London: Routledge.

Several other interesting aspects of the rural geographies of community, landscape and agricultural change are discussed in:
- Bell, M. (1994) *Childerley: nature and morality in a country village*. Chicago: University of Chicago Press.
- Ilbery, B. (ed) (1988) *The geography of rural change*. Harlow: Longman.
- Winter, M. (1996) *Rural politics*. London: Routledge.

CHAPTER 28

Europe

Kevin Robins

INTRODUCTION

The question of Europe at the present time concerns the possibilities and the implications of European unification and union. A whole array of issues is raised. Some emphasize the economic aspects of the Union, considering the implications of economic integration and the creation of a single market across the continent. For others, the agenda concerns the development of Europe-wide political institutions and considerations of **transnational** democracy. A third issue concerns culture and cultural identity, posing the question of what it means to be a European citizen at the end of the twentieth century. And, of course, these different aspects of the European question are all closely interrelated (the financial and monetary question of the Euro, for example, is clearly a matter of great cultural and political sensitivity).

In this chapter, I want to focus particularly on the cultural and political aspects of Europe. What is at stake now in considering Europe as a cultural and political space? Europe has constantly sought to both confirm and assert its identity, finding 'official' identity resources in both its earliest origins – in what is regarded as its Greek, Roman and Judaeo-Christian heritage – and in the modern achievements of European civilization – its political universalism and democracy and the achievements of scientific and technological culture (Couloubaritsis *et al.*, 1993). At the same time it has struggled to come to terms with

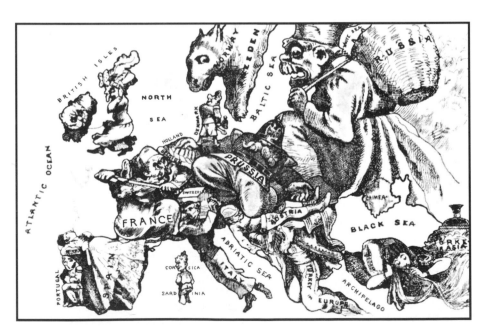

Figure 28.1 Europe of the nation states

some of the more unsavoury aspects of its heritage – its imperial conquests, its involvement in slavery, for example. The identity of Europe is not a given, then. It is a cultural form that European elites have struggled to impose on the diversity and often chaos of their continent. As Timothy Garton Ash puts it, 'like no other continent, Europe is obsessed with its own meaning and direction': 'no continent was externally more ill-defined, internally more diverse, or historically more disorderly. Yet no continent produced more schemes for its own orderly unification' (Garton Ash, 1998: 51, 53). The point I am making, then, is that European cultural identity is, and has been, a construct. 'Europe is less the subject of history,' as Gerard Delanty (1995: 3) observes, 'than its product and what we call Europe is, in fact, a historically fabricated reality of ever-changing forms and dynamics.' The European idea 'expresses our culture's struggle with its contradictions and conflicts' (ibid.: 1).

Here I am concerned with the precise nature of this struggle in the present period. In trying to understand contemporary developments, many have argued that what we are seeeing in Europe now is to do with the decline of the **nation–state**. In the continent that was the historical birthplace of the 'imagined community' (Anderson, 1983) of the nation, it is said, we are now seeing its demise. As to what is happening more positively, there are contrasting accounts. For some, what we are now seeing is the creation of a new pan-European culture and identity – Europe as a cultural melting pot. Others have a quite contrasting idea, emphasizing the growing significance of regionalism and small nationalism in the continent, and thinking of the new Europe as a cultural mosaic (Newhouse, 1997). What is actually happening is far from clear. In the following discussion, I even want to put forward the possibility that the end of the nation–state is far from being a foregone conclusion. Following the arguments of Alan Milward (1992), we might consider whether the development of the European Union has not paradoxically worked to regenerate the nation–state. We might also reflect, I would suggest, on whether the new successor identities – pan-Europeanism, Euro-regionalism – are not themselves simply variants of the national model: we might regard them, that is to say, as 'imagined communities' scaled-up, in the first case, and scaled-down, in the latter.

In exploring these issues, I shall give them a particular focus by concentrating on culture and identity with respect to the mass media, particularly television (see also Chapter 33). For, in the European context, media are absolutely central to the construction of both cultural communities and political public spheres (media corporations are also powerful and significant economic actors). Audiovisual spaces provide an excellent insight into the re-shaping of cultural and political spaces. One possibility is that we are now seeing the transition from national broadcasting systems to a new post-national order, one that may exist in terms of both transnational and local or regional media spaces. This is clearly what appears to be the case. But I shall suggest, in the context of my observations in the previous paragraph, that the transformation is not so clear-cut. What strike me are the difficulties in moving beyond the old framework, the seeming incapacity to transcend the imagined community of the national broadcasting model.

In general, what I am arguing is that, whatever the discourse and rhetoric concerning fundamental cultural transformation, there are in fact significant cultural continuities in contemporary Europe. My narrative is concerned with the constancy of the national framework, or, more specifically, with its perpetuation and perhaps reconstitution in the new European context. What I would argue is that a more creative and meaningful consideration of what Europe could be requires far more radical thinking about both culture and identity. We can no longer just pose the issue in terms of **re-imagining** community. What we have to come to terms with are the challenges of coexistence and encounter in an increasingly **multicultural** Europe. In so far as Europe always has a cultural construct, it is a question of how we might now make it a more **ecumenical** construct.

EUROPEAN CULTURE AND IMAGINED COMMUNITY

Let us first consider how, in the European audiovisual space, the established national cultural communities are being challenged from without (through the free flow of communications across frontiers) and from within (through the reassertion of regionalist aspirations). And let us then consider what kind of challenge this actually amounts to.

First, we have seen new regulations permitting the free flow of media products across Community frontiers, with the aim of creating a pan-European audiovisual market that will be competitive in global terms. From this perspective, as Philip Schlesinger (1994: 30) argues, 'the national level of media production and distribution is seen as an obstacle to be transcended in the interests of forging Europeanness'. Already some 20 years ago, Thomas Guback identified an economic logic struggling to express itself through the European project. The creation of an economically integrated Europe, he argued, 'favours the enlargement of firms to international stature, with concomitant trends toward standardisation, at the expense of small enterprises and a great deal of variety'. And if this is the case, Guback went on,

> then it is obvious that the major emphasis is not upon *preserving* a variety of cultural heritages, but rather upon drawing up a new one which will be in tune with supranational economic considerations. In that case, we had better forget about the past and concentrate upon seeing the creation . . . or fabrication of a new economic European consumer whose needs will be catered to – if not formed – by international companies probably operating with American management and advertising techniques.
>
> *(1974: 10–11)*

This logic is still at work. There is still the belief that it will be possible to fabricate the new model European consumer who will consume new model European programmes.

If it is predominantly an economic logic that drives this project, it is also the case that it carries with it a certain cultural vision, through the expectation that the single market in broadcasting will help to promote the re-imagination of community and identity in Europe. As the free circulation of programmes throughout the European Union reinforces Europe's production and transmission capacity, it is argued, so it will come to promote the ideals of the 'Europe of culture' and the 'citizens' Europe'. For the vision of 'television without frontiers' to become a reality, there must be congruence between the economic space of the Union and its cultural space. 'Programmes intended, from the beginning for all of Europe', the European Commission believes, 'could count on an audience and resources that would never be available at a national level; they would help to strengthen the feeling of belonging to a Community of countries at once different and deeply united' (Commission of the European Communities, 1986: 9). Pan-European television is expected to improve mutual knowledge among the peoples of Europe and to increase their consciousness of the values and the destiny they have in common. Guback (1974: 10) presciently described this project in terms of the 'transform[ation of] a mosaic into a melting pot'. It is about the creation of a synthetic identity that will transcend national particularisms and divisions; about fabricating the continental identity of an enlarged European Community.

If this represents one form of challenge to the established national broadcasting interests, a second comes from within the national cultures themselves. Against the principle of the melting pot, the proponents of European regionalism are fighting to sustain the image of the continent as a cultural mosaic. This

is by no means a novel aspiration. The regionalist case for cultural pluralism and diversity has long been on the European agenda, with the advocates of cultural decentralization preferring the 'personality' of regions to the 'artificial contrivance' of the nation–state (Gilbert, 1951). Within the changing geographical context of the European Union, however, the investment in regionalism is developing a new *élan*. 'A new idea is coming to light', one interested party observes, 'that regionalism can be an antidote to the internationalization of programming; that it will prove indispensable in compensating for the standardization and loss of identity of the big national [and international] networks' (Trelluyer, 1990: 11). There is the belief that the antiquated Europe of national states can now be superseded by a more streamlined 'Europe of the regions'. And there is the further conviction that the complexity of this European cultural mosaic is actually in accord with the cultural logic of **globalization**, which expresses itself in terms of the new global–local nexus.

This argument on behalf of 'decentralization in the global era' puts value on the diversity and difference of identities in Europe, seeking to sustain the variety of cultural and linguistic heritages. Broadcasting has been seen as a major resource in the pursuit of this objective, and since the 1980s we have seen a growing interest in promoting media industries and activities within the regions and small nations of Europe. In most cases, principles of local cultural interest have been mobilized against the market interests of transnational economic forces. In lobbying for support from the European Union, the argument has been put that 'in the particular case of regional TV programming in the European vernacular languages, the criteria should not be based on audience ratings and percentages of the language-speaking population, nor on strict, economic cost-effectiveness' (Garitaonandía, 1993: 291). Since the late 1980s, a certain level of support has been elicited from the European Union, particularly through its MEDIA programme, which provides loans and support for small producers across the continent. Within the Union there has been increasing sensitivity towards cultural difference and commitment to the preservation of cultural identities in Europe. This too is presented as an alternative to the old order of nation–state culture in the European space.

Now, both of these developments – pan-Europeanism and European regionalization, the melting pot and the cultural mosaic – offer interesting possibilities. We should be attentive to whatever potential is inherent in them. But we should also, by the same token, have no illusions about them. To what extent, we should ask, do they really provide an alternative to the old order? To what extent, that is to say, does either present a challenge to the nation–state model (which is what each claims to be transcending)? What I would argue, in response to these questions, is that neither makes a really significant development. What is remarkable in the reorganization of European media is, in fact, the continuity between old and new orders. There is a sense in which we can see contemporary transformations simply in terms of the perpetuation or the revitalization of the national imaginary. Perhaps it is a question of change in order to avoid change?

The supranational project of pan-Europeanism is that which is most evidently seeking to overcome the limitation of national identifications. But to what extent can its ideal of Europe and European Community be said to constitute a next, and more cosmopolitan, stage in the continent's cultural and political life? I doubt that this is happening, and would suggest, as an alternative interpretation, that the project to construct a European Community is, in fact, about trying to re-create the conditions of the national community at a higher order – about the construction of a kind of European nation–state. As Chris Shore (1996: 474) puts it, 'the challenge is to find a European alternative to the axiomatic and **hegemonic** grip that the nation–state continues to hold over the minds of the peoples of Europe'. What is remarkable, he continues, is that

> the new Europe is being constructed on precisely the same symbolic terrain as the old nation–states themselves. Flags, anthems, passports, trophies, maps and coins all serve as icons for evoking the presence of the emergent state, only instead of 'national sovereignty', it is the legitimacy of the EC institutions that is being emphasised and endorsed.
>
> *(ibid.: 481)*

'Europe' is invoked as a new basis for integrating and unifying contradictory and conflicting forces.

The expectation is that the relation between cultural mutuality and political community might be re-negotiated at this higher level. What we see, then, is a kind of transfer or displacement of nationalisms to bring into existence a new and enlarged community, with the same objective and aspirations as the national community of achieving correspondence between state, people and **territory**. In the particular context of broadcasting, this enlarged version of the nation–state model seems to apply too. What the European Union is struggling to create is, in fact, an expanded version of the national broadcasting model, one that seeks to maintain, at a higher level, the congruence between economic, political and cultural spaces of broadcasting. We are told that a European audience will come to enjoy the comforts of imagined community and

Figure 28.2 Euro travel cheques and European currency. Credit: Telegraph Colour Library

solidarity through some kind of supranational identification. Just as in the era of national public service broadcasting, cultural integrity and coherence were conserved, in the face of both regionalist and internationalist pressures, through the national compromise, now, it is argued, a similar integrity may be sustained through a European compromise.

What of the new, or renewed, spirit of Euro-regionalism? To what extent can this be said to be contributing to a meaningful transformation in cultural identification? In one aspect, I think that neo-regionalism does constitute an important contestation of the established nation–state culture. It represents a significant challenge to the dominating force of national governments that have subordinated and marginalized minority cultures within their territory; it constitutes a legitimate reaction against distant government over which citizens have had little or no control. In this respect, it may be seen as an expression of the revitalization of **civil society**, an assertion of more relevant collective identities against the bureaucratic and technocratic culture of the nation–states and now the Brussels super-state. Noting the search for new forms of democratic participation, Julia Kristeva (1992) suggests that such kinds of particularistic attachment may involve 'attempts to close that gap between government and

271

Europe

the man in the street, between politics and the hands-on exercise of responsibility'. To the extent that they do enhance civil and public culture in Europe, these developments are to be welcomed.

But, let us be clear that contemporary regionalism is no longer simply about the reassertion of marginal voices in Europe. Now it is also the wealthier regions that wave their banners against the central state, observes Lothar Baier (1991/92: 86), 'fighting to be free of the obligations to share the costs of modernisation in other regions from which they profit'. 'They combine their regionalist egotism,' Baier continues, 'which once exposed is quite ugly, with thinly veiled attempts at cultural and historical ornamentation.' Peter Sloterdijk (1995: 50, 53) describes a new kind of region-consciousness in Europe now, that of 'regions competing aggressively to become part of the great capitalist experiment' – he calls it 'the slick progressive regionalism of the new community enterprise'. We should also be clear about the proximity of this regionalist assertion to national identification and attachment. The proliferation now of a whole array of ethnic and regionalist sentiments clearly reflects the persistent appeal of nationalist sentiments in a Europe that is being re-shaped by the forces of globalization. To belong to the imagined community of a nation remains a powerful way of belonging in contemporary Europe. One might even venture to argue that Europeanization has actually served to encourage the strengthening and proliferation of national identifications. In the ideal of a Europe of the regions are we not seeing a revitalization of national consciousness – a re-nationalization of the continent?

At the end of the twentieth century, then, it is being said that we are seeing the leaking away of sovereignty from the nation–state both upwards to supra-national institutions, and downwards, to sub-national ones. And, in one sense, this is indeed the case. But whether it is bringing about a significant 're-imagination' or 're-mapping' of cultural and political life and identity is more open to question. I would argue that, in Europe at least, such a cultural transformation still only exists as a potential development, and that what exists in reality is the persistence, and even the reanimation, of the national imaginary. This is apparent, not just in the proliferation of neo-nationalisms, but in the seeming inability to conceive of Europe as anything other than a nation–state writ large. In the particular context of culture and media, too, this national focus is apparent, with both supra-national and sub-national developments in broadcasting, for example, still developing according to the national model. The mass media, and broadcasting in particular in the twentieth century, have been indissociable from national visibility and awareness. It has been one of the fundamental institu-

tions through which the social collectivity has constituted and known itself as 'national'. Even as we now try to conceive of alternative possibilities, in terms of pan-European or Euro-regional media, it seems as if we can only do so in terms of the national model. The question we should be asking ourselves is whether, in the new European context, we can think of media culture in any other way.

Summary

- The increasing economic integration of Europe carries with it a cultural vision promoting the re-imagination of European identity. The mass media play a crucial role in this process of re-imagination.

- Often this European vision is created through the same symbols which help to sustain national identity, such as flags, anthems, passports and currency.

- Reactions to the European vision are helping to revitalize national consciousness. The process of 'Europeanization' is far from straightforward.

THE PROBLEM WITH 'IMAGINED COMMUNITY'

Let us now consider what is fundamentally at issue from a more analytical perspective. There is a certain crisis of the nation–state in Europe, involving a loss of authority and a weakening in popular engagement. And in this context, we are now seeing the emergence of other possibilities for cultural allegiance and attachment. What seems to be the case, however, is that these alternatives sustain the essential features of national identification. On the one hand, there is Europe, conceived according to the model of an enlarged nation–state – though this identification is based on what must be considered a rather abstract condition of cultural unity. And, on the other, there is a turning to the small territories of European regional and stateless nations, which promise more vital and 'meaningful' kinds of involvement. Zygmunt Bauman (1996: 86) describes this in terms of the development of a 'nationalism mark two'.

> It is now the much maligned 'natural communities of origin', *necessarily smaller than the nation–state*, once described by modernising propaganda as parochial backwaters, prejudice-ridden, oppressive and stultifying, which are looked to hopefully as the trusty executors of that streamlining, de-randomising, meaning-saturating of human choices which the nation–state abominably failed to bring forth.
>
> *(ibid.: 84)*

What is most evident in these developments, then, and far more significant than the apparent choice of cultural identities that is on offer, is the sustained appeal of the national form. We are intent, it seems, on reconstituting the national way of belonging on a new and more vigorous basis. Cultural debates have become excessively focused on the meaning and importance of 'imagined communities'. Now this is deeply problematical in times that are throwing up more complex forms of cultural experience, and consequently requiring more open and inventive kinds of response. We should be concerned about the way in which the simple idea of 'imagined community' has come to prevail in contemporary cultural debates. What was once, indeed, a fertile and productive concept has now, I believe, come to inhibit our further understanding of collective cultural experience. We should not let ourselves become comfortable with the idea of 'imagined community'. It draws us into the contemplation of our need for, and entitlement to, shared community, and, by the same token, away from the more difficult question of our cultural responsibilities beyond 'our' cultural world.

Imagined community is not just about belonging, but is associated with a particular *kind* of belonging. It has been centred on the ideal of social integration, committed to the achievement of a sense of coherence and enclosure within a social group. Such a kind of community is characterized by a unitary and bounded culture and identity. It has been through the history of European nationalism that this kind of 'identitarian' thinking has been most fully articulated and realized. The construction of nation–states involved the elimination of complexity, and the extrusion or marginalization of elements that compromised the 'clarity' of national attachment. Monolithic and inward-looking, the unitary state has seemed to be the realization of a desire for purity and integrity of identity. Now we must suspect that the same drive will motivate the new generation of 'mark two' nationalisms. And we should note how this kind of cultural attachment is also being sustained and perpetuated in the imagination of European community. The discourse of Euro-culture is significant: it is that of cohesion, integration, union, security. What is invoked is the possibility of a new European order defined by a clear sense of its own coherence and integrity. European culture is imagined in terms of an idealized wholeness and plenitude, and European identity in terms of boundedness and containment.

This desire for clarity and definition is about the construction of a symbolic geography that will separate the insiders, those who belong to 'our' community, from the outsiders, the others. What is at stake is clearly something more than just territorial integrity:

it is more like the psychic coherence and continuity of the community. Imagined in this sense, the community is always – eternally and inherently – fated to anxiety. Its desired integrity must always be conserved and sustained against the forces of disintegration and dissolution at work in the world. What is emphasized is what is held in common, at the expense of diversity and difference within the community. Such a kind of identification supposes, as Denis-Constant Martin (1992: 587) argues, 'the elimination, the repression, even temporarily, of what can divide'. Difference is experienced, and feared, because it is associated with fragmentation. Hence the prevalence in contemporary European discourse of imagery concerned with the fortification of identity.

Of course, we shall not dispense with the kind of belonging that has been associated with the imagined community of nationalism. As Julia Kristeva emphasizes, the national form will remain a historical reality for the foreseeable future. The point is 'not to reject the idea of the nation in a gesture of wilful universalism but to modulate its less repressive aspects' (Kristeva, 1993: 7). Let us also be clear that there are quite rational reasons for committing oneself to national or neo-national entities. Regions and small nations may now find themselves well placed to benefit from the emerging global economy, and they may also present themselves as more relevant in terms of political aspirations, seeming to provide greater possibilities for political representation and agency. As Tom Nairn (1995) makes clear, nations and nationalisms are far from being outmoded or anachronistic cultural forms. Taking the same attitude as Kristeva, he argues that what is significant is the particular nature of the national form; what is crucial is to promote the general evolution from **ethnic** to civil–political forms of nationhood. We must be pragmatic with respect to the national form of identification, then, but we must also seek to make it more open and accommodating – though how this latter objective might be achieved is far from evident.

But, at the same time, do we not also have to start working against the national form of attachment? For the national community tends to suffer from the illusion that it is self-contained and self-sufficient – and national passions and emotions are quickly aroused in defence against what is perceived as threatening to the imagined integrity of the nation. Now what is thereby denied or disavowed is the reality that particular cultures are constituted in and through their relation to other cultures and identities. Identities 'suppose the other in order to exist and to develop', argues Denis-Constant Martin (1992: 583, 585), 'the affirmation of identity only comes through the exchange and "recuperation" of foreign elements.' And we must be clear that cultural interdependence is not just a reality, but that it is also a value and an obligation. The fundamental issue is not about the right of imagined communities to exist, but about how the communities that assert this right will co-exist. Any kind of meaningful co-existence must depend on the valuation of cultural receptiveness and reciprocity, in the awareness that it is only through their 'valency' that cultures revitalize themselves. And what must further be recognized, in the context of ever-increasing interaction between cultures, is that a community can no longer simply follow the self-interest of its own members: its obligations must now extend beyond itself to 'foreign' citizens, both beyond and within its frontiers.

How, we must ask, might the significance of imagined community be made more relative or provisional? How might the emotional force of the national kind of belonging be dissipated? I cannot answer these immensely difficult questions – for one thing, I do not underestimate the pleasures of imaginary belonging. Instead, I will suggest that, in order to begin to make answers more possible, at least, we need to extend and transform our **discourses** on cultural identity, questioning the now taken for granted idea of 'imagined community', and moving beyond the kind of identity-thinking associated with nationalism and nation-building (see Chapter 21). Can we think of our cultural situation in ways that offer more scope and possibility than does the self-enclosed vision of imagined community? Can we tell ourselves different stories about our cultural past?

Perhaps the developments associated with globalization might now provide us with the basis to think about European history and culture in new ways? This is what James Anderson (1996: 144) is suggesting when he takes up the idea of the contemporary European order as a 'new medieval territoriality', one in which 'space is now regaining some of the fluidity of the medieval era'. What is admitted, within such a perspective, is the possibility of regarding the present in terms of a significant and radical discontinuity with the era of nationalism, and thereby, perhaps, a certain productive disordering can be introduced into our cultural mapping. This subversion of familiar narratives and naturalized categories is also intended in David Rieff's historical revisionism, which revalues the **multiculturalism** of the Ottoman and Hapsburg empires:

As the world flounders in tribalism, those great multicultural empires, the Ottomans and the Hapsburgs, look better and better. At the beginning of the twentieth century, it was nationalism, even nationalism based on ethnicity and confession, that looked modern and these great, inefficient behemoths not only seemed destined for extinction – they were – but seemed to richly deserve their fate. Who

at the time would have sided with the Austrian emperor Franz Josef against the nationalists like Kossuth or Masaryk, or with Sultan Abdul Hamid against the Balkan insurgents who confronted him? But the question must be posed again, as events in Sarajevo close the century that events in Sarajevo opened. And, in retrospect, it may seem that for all its ferocities it was the Ottoman vision and not the nationalist one that holds up better. We shall be polyglot in the next century or we shall kill each other off.

(Rieff, 1993: 14–15)

Slavenka Drakulić (1996: 164) makes the same good point with particular reference to a part of Europe that emerged out of one of those old empires. She refers to the attitudes, and the recent actions, of people living in Istria, a place where 'nationality and identity don't necessarily overlap'. Istrianism, she argues, is a challenge and a confrontation to those who are presently inciting neo-nationalist fervour in that part of the world. Drakulić asks,

[How] can these authorities understand the meaning of Istrianism – the enlarging concept of identity, as opposed to the reducing concept of nationality? To Istrians, identity is broader and deeper than nationality, and they cannot choose a single 'pure' nationality as their identity. Living in the border region, they understand better than anyone else that we all have mixed blood to a greater or lesser extent. They also have suffered from nationalism, and in its worst form – ethnic cleansing – enough to have grown tired of it. Paradoxically, for the first time in their history, at the first elections of the newly independent republic of Croatia, the Istrians felt free to reject the concept of one nation; they felt that the time had come to express what they really consider themselves to be.

(Drakulić, 1996)

Such approaches help to put the national way of belonging into a more relative perspective, allowing us to think of nationalism as something other than the culmination of European cultural and political history.

On this basis, we might have the possibility, at least, of developing a more complex approach to cultural identities. We may find ways to move beyond the singularity of perspective that has characterized the national imaginary. Julia Kristeva (1992) opens up an interesting way of thinking about these things, drawing on the psychological concept of the transitional object.

Imagine the national as a sort of transitional object, an identity aid that provides us with security and at the same time acts as a relay towards others. It would be a question of thinking out and organising a series of bonds reaching out to others without entirely eliminating the archaic one, in this case that of nationhood.

(Kristeva, 1992)

Kristeva puts forward this transitional conception as an alternative to the romantic and integrating conception of the nation that recent events have shaken back to life. The transitional nation provides the security and stability necessary to sustain openness to others beyond its confines. What is necessary, Kristeva is arguing, is that we should work towards the insertion of the national entities, inherited from the past, into higher political and economic wholes. I will also briefly refer to some observations concerning identity made by the French psychoanalyst J.-B. Pontalis, for these, too, provide an interesting bridge between individual and collective identity. Pontalis works with the concept of migration and what he calls the 'migratory capacity' (*capacité migratrice*). He thinks of psychoanalytical experience in terms of the productivity of migration: 'From one language – and one dialect – to another, from one culture to another, from one way of knowing to another – with all the risks that such a *transfer* entails' (Pontalis, 1990: 88). Pontalis puts a value on the movement between positions, and on the production of multiple identifications. In his discourse, 'migration' is a metaphor, drawn from collective culture. Perhaps we can turn it back to where it came from, in order to shift the discussion of cultural identity – which easily becomes a static concept – towards the consideration of cultural exchange and cultural experience?

Finally, we should come back to the new European space, and say something about the significance of broadcasting, particularly, in all of this. Television has been integral to the institution of national community. Historically, it has been mobilized to represent the nation to itself, giving visibility – and thereby substance – to the culture that is held in common. What national audiences get and want, argues Thomas Elsaesser (1994), is 'the familiar – familiar sights, familiar faces, familiar voices . . . television that respects and knows who they are, where they are, and what time it is':

> Locality, language, the conjugation of the day, the seasons and the generations, the respect for lived time, the television community of familiar faces and familiar spaces: this would be one starting point for trying to study how notions of national identity, of belonging and being addressed, intersect on and through television, for it would imply an instinctive respect for the viewer, not just of his or her intelligence, interests and need to be stimulated and entertained, but of his or her existence as a physical as well as a social being.
>
> *(Elsaesser, 1994)*

The pleasures of this kind of investment and involvement in televisual culture are evident. It isn't difficult to understand the desire to replicate such experience in the new regional and national cultures that are now asserting their presence in Europe.

But, in the new European context, is this desire not also very problematical? Should we not be disturbed about the way in which broadcasting works to reinforce what Max Dorra (1996: 32) calls the 'group illusion', 'ferociously eliminating all difference'? Television, in particular, has evolved as a medium that supports cultural integrity and resists cultural promiscuity. It is not a 'travelling' medium – the **hybridization** that we now increasingly associate with other cultural forms is virtually absent from broadcasting culture. It is not a medium of cultural relay and encounter. This ought to be a matter of great concern, and we should now be considering whether broadcasting culture can be re-instituted to accommodate cultural diversity and difference. Rather than sustaining the comforts of (imagined) cultural unity, can television now come to terms with the reality of cultural disunity? This raises the question of whether the challenge to identity can be made as satisfying or desirable as the confirmation has been. Steven Vertovec (1996: 393) describes the exemplary endeavours of Radio Multikulti in Berlin, which is actively seeking to promote openness to, and knowledge about, other European cultures, thereby hoping to 'change the aural space of Berlin'. The station expects that through this approach, 'the city will become more exciting than the banal "side-by-side" of everyday life (*Nebeneinander des Alltags*) makes it seem' (ibid.: 392). In a rapidly changing Europe, isn't this vision of cultural mixture a worthy aspiration? But let us note that this new cultural space is not a regional or national space. It is an *urban* space – for cities are the real spaces of cultural encounter and interaction.

Summary

- The new European identity is being constructed around issues of cohesion, integration, union and security. As such, it is coherent and bounded, as were the old national identities.

- In an increasingly integrated world, imagined communities need to move beyond this exclusive position to embrace difference and diversity in order to sustain openness to others from beyond the boundary.

- This openness can be furthered through cultural exchange and encounter. However, the most appropriate spaces for such interaction may not be the nation or even Europe, but the increasingly diverse urban spaces of Europe's major cities. Thus the new Europe might not be constructed at the continental scale, but at a much more localized urban scale.

Figure 28.3   Credit: Angela Martin

CONCLUSION

The future of European cultural space remains uncertain – Europe is still in the processs of becoming. If we are seeing the promotion of an 'official' agenda from Brussels, it is the case that there are different variants of what it is to be a European citizen. One logic pushes towards homogenization and the melting pot ideal of pan-Europeanism, while another espouses diversity and difference, celebrating the cultural mosaic of a 'Europe of the regions'. At the same time that the European project has been put in motion, we also see the continuing vitality of nation–states and national cultures in the continent. Then we have to take into the account the force of the numerous non-territorial cultures in the European space – the cultures of global migrants in the metropolitan cities – which are having a transformative influence on what it now means to be European. There are possibilities of cultural conservatism and closure – as in the 'fortress Europe' mentality. But there are also possibilities that Europeans can re-invent their identities in new, open and ecumenical ways. Will we hang on to the old imagination of community? Or can we now take advantage of the new conditions of cultural encounter and mixture to revitalize the meaning of Europe?

Further reading

A magisterial survey and analysis of the history of Europe is:
* Davies, N. (1996) *Europe: a history*. Oxford: Oxford University Press.

A useful introductory text highlighting the constructed nature of European culture and identity is:
* Delanty, G. (1995) *Inventing Europe: idea, identity, reality*. Basingstoke: Macmillan.

An excellent selection of articles concerning the past, present and future of European integration is:
* Gowan, P. and Anderson, P. (eds) (1997) *The question of Europe*. London: Verso.

An analysis of tensions in European identity with particular reference to media and communications is:
* Morley, D. and Robins, K. (1995) *Spaces of identity*. London: Routledge.

A coherent collection of articles exploring the tension between unity and diversity in European culture is:
* Wintle, M. (ed.) (1996) *Culture and identity in Europe*. Aldershot: Avebury.

Colonialism and postcolonialism

Richard Phillips

INTRODUCTION: MAPPING EMPIRE

The world map shown in Figure 29.1 appeared in the opening pages of an official *Atlas of Canada*, published by the Dominion of Canada's Department of the Interior in 1915. It is typical of maps that were produced and consumed around much of the world in the first half of this century. People often see maps as straightforward representations of space – facts about geography. But the map makers have made all sorts of choices – what to include; how to depict it; what to leave out.

The first choice the map makers have made, perhaps their most powerful device, is one of colour. To get a sense of the power of colour, you have to look at an original map, perhaps one that is hanging on the wall or unfolded on the table, not a black and white reproduction. In the Canadian world map, originally 40cm by 60cm of full colour, colour is used to represent colonial power, as the key makes clear. Each empire is a different colour. Above all, this is a map of colonial empires, most of them European. The map is cleanly printed. Its even blocks of colour cover regions, nations, even whole continents uniformly. They depict a world that is tidily, uniformly colonized.

The colour scheme on the original map is not arbitrary. The map makers have chosen a colour scheme that reinforces the colonial theme of the map. The colour red seems to occupy much of the land surface of the earth. The key shows that Britain and British colonies are coloured bright red. Other colonial powers appear in more delicate or demure colours: the French Empire, for example, is a gentle

mauve. Each colour has a symbolic and graphic function within the map. Red symbolizes authority (think of 'red tape'), aggression (a red sports car), confrontation (like a 'red rag' to a bull) and England (England's flag, the St George's cross, is white and red); it is also the colour of blood. Graphically, too, it is an aggressive colour, appearing larger mile-for-mile than its more lightly shaded neighbours, also pushing out and spilling into their territory. Looking at this map – imaginatively 'losing oneself' in it – it is easy to imagine a time, not too far into the future, when the colour red will be everywhere. So it is not surprising that the map makers chose the colour red for themselves. Other map makers, such as Germans, often did the same in their own maps.

Notice the words. They are in English. Canada is not and never has been a nation just of anglophones. When the map was made, many Canadians spoke French, as well as a host of languages, some native North American, others imported from Eastern and Northern Europe, Asia and elsewhere. But the map is in English, the dominant, official language. The world is named with the placenames chosen by English-speaking peoples; everywhere is what the English say it is. Thus the map asserts an English way of seeing the world.

Neither is composition merely factual, an innocent artefact of scientific cartography (see Harley, 1992). The map depicts a world with centre and margins; it turns some places into centres and others into margins. Anywhere could be at the centre – Fiji, France, the Falkland Islands. But Europe, and specifically England, is at the centre. The centrality of England is

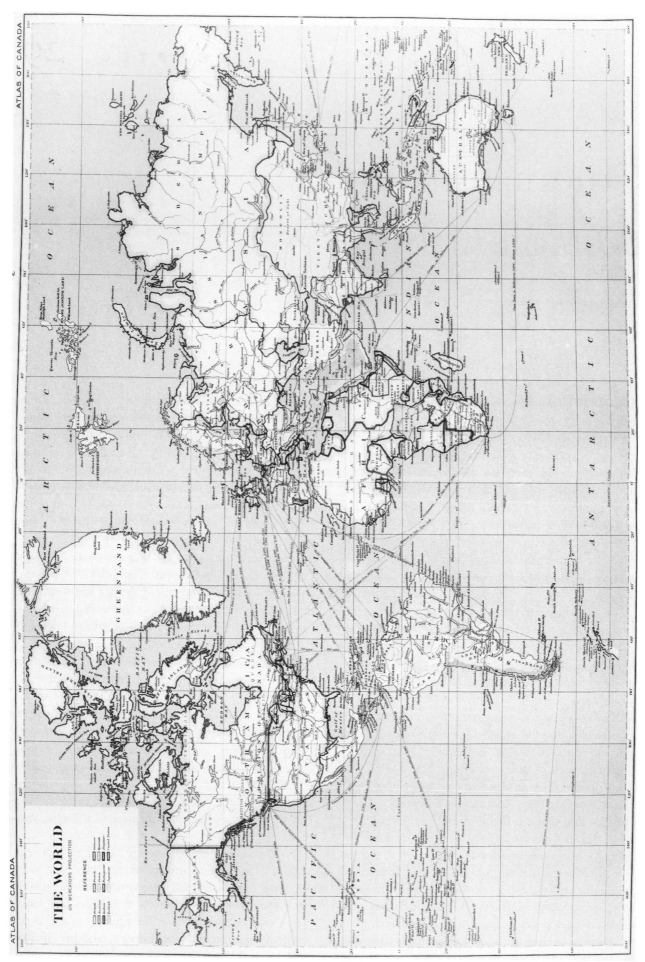

Figure 29.1 This map represents colonial power, and is a vehicle of that power. World map, 1915. Source: *Atlas of Canada*, Ottawa: Government of Canada

also formally marked through lines of *longitude*, which as most people know centre on Greenwich, London. Before 1881 when England got other nations to accept their own *prime meridian*, and thereby established an international convention respected by many (but not all) nations, most map makers drew lines of longitude that were centred upon their own capitals or other cities. Americans, for example, generally used Washington or Philadelphia. Persuading others to respect the Greenwich meridian, the Europeans persuaded them to see London and Britain/Europe as the centre of the world. In case any map reader should miss the point, shipping lines were included, drawing the eye from all corners of the world into Liverpool, Bristol, Southampton and the Thames Estuary, towards a metropolitan centre. The shipping lines also suggest the importance of England as a centre of trade and commerce; every other nation seems to be measured and located according to its links with England.

The projection of the map also adds to the importance of England and other northern countries. Here, the map makers have chosen to use the *Mercator projection*, which was developed in the Netherlands – a leading European imperial power – in the sixteenth century. The Mercator projection distorts space, making areas further from the equator appear bigger (by land area) than they really are. This makes western European nations appear disproportionately large. England, part of a small island in the North Atlantic, is exaggerated in geographical importance, as are British dominions such as Canada, New Zealand and Australia. The projection also serves as a reminder of England's competitor in Eastern Europe and Asia, the Russian Empire, which looks very imposing on the map. Conversely, it makes India appear much smaller, perhaps more easily ruled by the English, than it might otherwise appear.

Summary

- The 1915 World Map represents a world dominated by colonial powers.

- It also functions as an **ideological** vehicle of that colonial power.

EXPERIENCES OF EMPIRE

European colonial empires reached their peak around 1914, before the First World War broke out. Among these empires, the British was the greatest, with the French in a relatively poor if absolutely large second place. Table 29.1 shows the breakdown of British,

Table 29.1 The Colonial Empires at the outbreak of war in 1914

	area in sq. km	pop. in 1 000
1. In British possession		
A. Mediterranean	10	517
B. Asia	5 199	324 114
C. Africa	9 392	50 824
D. America	10 407	10 082
E. Australia and South Seas	8 267	2 508
F. Other	18	599
total:	**33 293**	**388 644**
2. In French possession		
A. Asia	803	14 871
B. Africa	9 499	30 514
C. Other	116	807
total:	**10 418**	**46 192**
3. In Dutch possession	2 036	38 248
4. In Russian possession	16 153	22 605
5. In Japanese possession	288	19 200
6. In German possession	2 954	13 784
7. In American possession	388	10 299
8. In Belgian possession	2 365	10 000
9. In Portuguese possession	2 244	9 146
10. In Italian possession	1 641	1 850
11. In Spanish possession	441	640

Source: Veit Valentin: *Kolonialgeschichte der Neuzeit*, Tübingen, 1915

French and other possessions, by population and area. The British controlled huge areas of land in America (mainly Canada) and Australia, although it was its African and especially its Asian colonies (mainly India) that qualified Britain as as the world's largest imperial power. While British Canada and Australia accommodated just over 12 million increasingly self-governing people, the combined populations of British Africa and India numbered something closer to 400 million, and these colonial subjects were subjected to something closer to absolute colonial rule.

Tables such as this group land and people into columns and aggregate statistics, categorizing them under labels such as 'empires', 'imperial' and 'colonial' – terms we have already found ourselves using. There is, at the outset, a need to clarify this vocabulary. *Imperialism* refers, very broadly, to an 'unequal territorial relationship, usually between states, based on domination and subordination' (Graham Smith in Johnston *et al.* 1994: 274). *Colonialism* is defined more narrowly as 'The establishment and maintenance of rule, for an extended period of time, by a sovereign power over a subordinate and alien people' (Michael Watts in Johnston *et al.* 1994: 75), while *colonization*, more specific still, involves 'the physical settlement of people (i.e. settlers) from the imperial centre to the colonial periphery' (Michael Watts in Johnston *et al.* 1994: 75).

Definitions and labels, like statistics and maps, make the world seem tidier, more ordered and more generic than experience tells us it is. Within the areas painted red, or listed in columns of British statistics, or labelled under headings such as 'colony', there remain a great variety of different experiences and perspectives. These differences – between colonizers, colonies and colonized peoples – must be remembered, because they are often greater than the similarities. As historian Ronald Hyam puts it in *Britain's Imperial Century* (1993: 1) 'When you think about it, there was no such thing as a greater Britain – India, perhaps apart. There was only a ragbag of territorial bits and pieces, some remaindered remnants, some pre-empted luxury items, some cheap samples.' Colonialism, he goes on to explain, is messier than our maps of it. Colonialism takes many different forms, and is experienced by different people in different ways.

The experiences of colonists have been many and varied, as a selection of British colonists illustrates. David Livingstone, a Scotsman, spent much of his life as a missionary and explorer in Africa (see National Portrait Gallery, 1996). Like many colonists, he was driven by faith in Christianity, Civilization and Commerce. He campaigned practically against slavery and for Christianity, in both cases by exploring the continent, thereby opening Africa to trade, development and (a British idea of) progress. As an explorer, Livingstone is remembered for the epic journey to the great waterfall that the British were to name after their Queen Victoria. Most colonists were less famous and less idealistic than Livingsone. Daisy Phillips, for example, was one of many who emigrated from England in search of a new home and a better livelihood in the colonies. Along with her husband, Jack, she embarked upon the long journey from England to the interior of British Columbia, where she did her best to set up home. Frontier life proved difficult and lonely, although Daisy was to remember it fondly. The couple returned to Europe when war broke out, and Jack soon died on the battlefield. Daisy's letters home, reprinted in *Letters From Windermere* (Harris and Phillips, 1984), tell a story of literate, middle-class colonial experience (see also Moodie, 1986). Other emigrants, whose colonial experiences were generally less well documented and considerably less comfortable than Daisy's, include the convicts transported to Australia in the late eighteenth and early nineteenth

centuries. Most were Irish or British, six out of seven were male, and many were transported for petty offences. All found themselves in a brutal world of violence, hard labour and alienation, which Robert Hughes vividly describes in *The Fatal Shore* (1988). When they had served their sentences, few could afford the cost of a passage home, so they stayed as free colonists. Other British colonists had very different experiences, which were also shaped (like Livingstone's) partly by their faith and ideals, and (like Daisy Phillips's) by their class and gender, as well as by other aspects of their identities including their race and sexuality (see McClintock, 1995; Hyam, 1990).

Meanwhile, many other Britons experienced colonialism from an altogether different angle – they stayed at home. There they imagined the colonial empire from a distance, hearing or reading about it. Books such as *Robinson Crusoe* (*see* Fig. 29.2), with exotic settings and lively story lines, narrated and mapped colonial encounters between Europeans and others (Phillips, 1997). Popular geographical narratives such as *Robinson Crusoe* presented an acceptable and exciting face of colonialism to their British, French, German and other readers. They popularized empire, persuading many to support their governments' wider imperial projects and inspiring others to seek adventures of their own, many in actual and would-be colonies. So *imaginative geographies* of empire did not just represent the empire; they helped construct it (see Said, 1993, and Chapter 22 of this volume, in which Felix Driver examines imaginative geographies and their significance in the real world). European armchair and shop-floor imperialists did not just dream about empire; they also consumed it and produced goods for it. Europeans consumed colonial products such as tea from India, sugar from the Caribbean and furs from Canada. They helped to produce the manufactured goods such as railway engines, textiles and guns that Britain shipped to its colonial markets. They posted letters to friends and relatives in (what to them were) far-flung corners of the world. When they received replies, those who noticed stamps often noticed the head of their own monarch. In their personal communications, as in their material life (as producers and consumers) and in their dreams, Europeans participated in a system that was both global and imperial.

The experiences of colonized peoples were equally varied; being colonized meant different things to different people. Chief Sechele, for example, was the only African known to have been converted by the famous British missionary, David Livingstone. Sechele followed his spiritual mentor's advice and agreed to change his polygamous ways; he anulled all but one of his marriages, and in 1848 he was baptized in front of

ROBINSON CRUSOE

IN WORDS OF

ONE SYLLABLE.

BY

MARY GODOLPHIN.

WITH COLOURED ILLUSTRATIONS.

LONDON:
GEORGE ROUTLEDGE AND SONS,
THE BROADWAY, LUDGATE.
NEW YORK: 416, BROOME STREET.
1868.

Figure 29.2　Imaginative geographies accommodated colonial encounters, such as that between Crusoe and Friday, which inspired and legitimized real colonial acts. Source: Frontispiece and title page of *Robinson Crusoe in Words of One Syllable* by Mary Godolphin (1868)

hundreds of weeping spectators. Soon, however, he missed the three wives he had sent away, and resumed sexual relations with at least one of them. Louis Riel was also the leader of a people colonized and fundamentally changed by British imperialism, although of a very different sort. He emerged as leader of the Canadian Métis nation, a **hybrid** people born through centuries of cultural and sexual contact between Amerindians and French-Canadian *voyageurs*. Riel led two uprisings against the British North American authorities, whose plans to settle western Canada threatened the Métis people's way of life. Riel was executed and his people were displaced, as colonists swarmed west. Unlike the Métis, some colonized peoples were entirely wiped out by colonists. This was the fate of many Australian Aboriginal peoples who were displaced by penal colonies. Truganini, the last Tasmanian, saw her countrymen and women broken by English colonists (*see* Fig 29.3). The English hunted down some of the inhabitants and transported the others to an island reserve (Flinders Island, 40 miles to the north) that, for most, was to be their grave. Others experienced colonization in very different ways, which like those of the colonists depended partly upon – and in turn reshaped – their class, race, gender and

sexuality. Some colonized people grew rich from the colonial encounter, while others lost their language, their livelihoods, even their lives. Like the colonists, many colonized peoples consumed stories and maps of the wider world, and participated in an increasingly global economic and cultural system. But unlike the European colonists they were not the principal architects of this world. Their colonial encounters were from positions of relative weakness.

Colonialism transformed the world, simultaneously producing a 'New World' and modernizing the 'Old World'. Changes in the non-European world are perhaps most evident. Some effects of colonialism were superficial – statues of Queen Victoria, for example. Others colonial imprints were more fundamental, and can never be reversed. Enormous population movements, both forced and free, scattered Europeans and Africans, particularly in the Americas. Some 20 million people left Britain and Ireland between 1815 and 1914, in a **diaspora** motivated variously by hunger (particularly during the Irish famine), displacement (notably in the Scottish clearances) and the desire for a better (freer and/or wealthier) way of life (by middle-class emigrants). Europeans and Euro-Americans

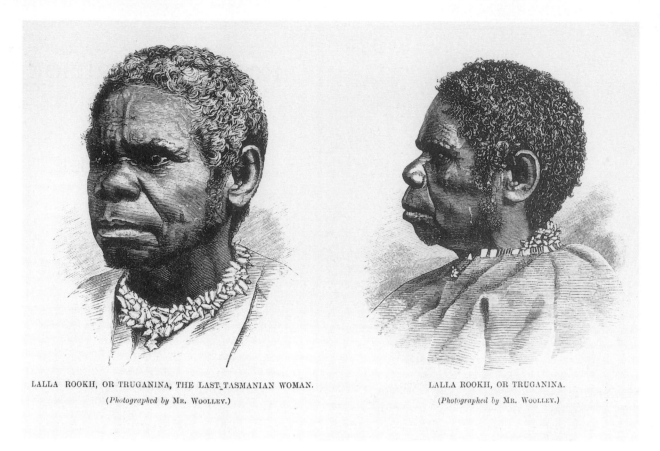

LALLA ROOKH, OR TRUGANINA, THE LAST TASMANIAN WOMAN.
(*Photographed by* MR. WOOLLEY.)

LALLA ROOKH, OR TRUGANINA.
(*Photographed by* MR. WOOLLEY.)

Figure 29.3   Truganini. the last Tasmanian. Source: *The Lost Tasmanian Race* (1884) by James Bonwick (FRGS)

also engineered an African diaspora, in which 12 million Africans were sold to slavery and shipped overseas, mainly to the Americas (see Chapter 30 on diaspora and contemporary migrations; see also Allen and Massey, 1995). Vast tracts of the world were partly or wholly resettled, as aboriginal peoples were deliberately or accidentally wiped out and replaced with free and/or forced immigrants. Movements of capital also transformed and integrated far-flung territories. Railways and shipping lines, which rapidly encircled the Victorian globe, were important both as capital investment, and as infrastructure through which capital and goods, people and information were moved. The result was an increasingly global society and economy, a global geography of development and underdevelopment (see Chapters 7–9 for introductions to geographies of development).

While colonialism transformed the non-European world, it also reshaped Europe. Europeans, the principal architects of the modern **world system**, constructed a modern world in which they were at the centre (see Chapter 17). Europe grew rich, and European lifestyles were enhanced by the non-European products, labour and markets that Europe controlled. Non-European economies supplied the raw materials and provided the markets that enabled Britain and other European countries to industrialize. Buoyed by prosperity, many Europeans grew arrogant, confident of their economic,

political, religious and racial superiority. European confidence, expressed for example in its maps, redefined its place in the world and mapped out its future.

Summary

- European colonization was experienced in many different ways by many different peoples, but when Europeans encountered the peoples of Asia, the Americas, Oceania and Africa, they generally had the upper hand.

- European colonialism established and formalized unequal territorial relationships between peoples, ensuring that modern maps and modern geographies were and are colonial maps and colonial geographies.

'POSTCOLONIAL' GEOGRAPHIES

Now compare the 1915 map (Fig 29.1) with a more up-to-date map of the world (Fig 29.4). What differences do you see? And what similarities? What evidence of colonialism do you see? You may need to look closely.

Perhaps the most striking difference between the two maps is the colour scheme (again the colour originals make this clearer!). The domination of a few European colonial empires, symbolized in the handful

WORLD POLITICAL

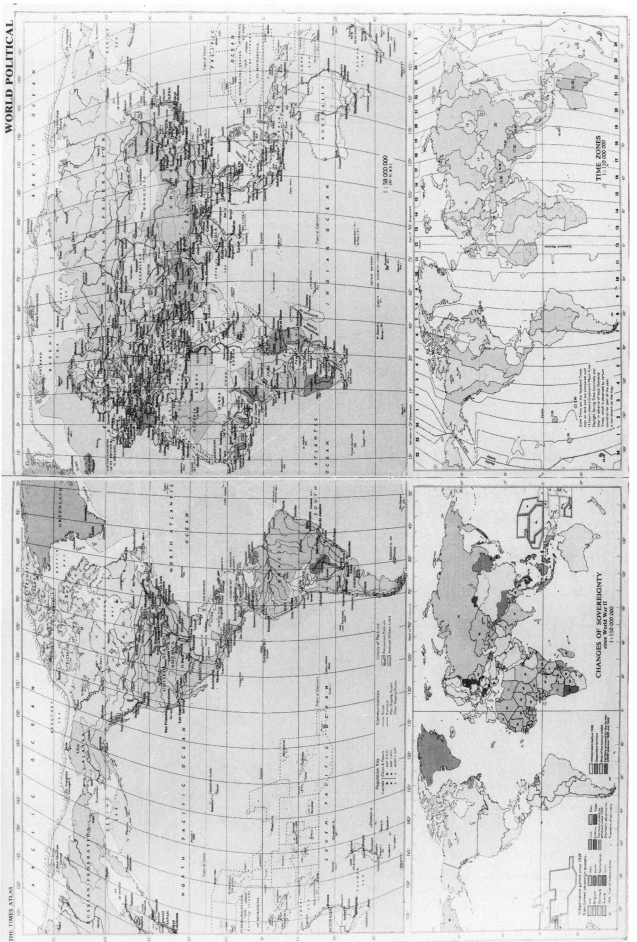

Figure 29.4   Colonialism is less obvious in the 1915 world map, but look closely and you will see legacies and traces of old empires, and more subtle hints of their successors. Source: World map, 1997

of strong colours we have discussed, has gone. In particular, the colour red has receded. Some patches of red (now a softer pink) remain, but you have to look hard to find them. People often say that the sun has now set on the British Empire (a response to the old boast that it never would). In July 1997 the British handed over the control of their last major colony, Hong Kong. *The Guardian* newspaper (*see* Fig 29.5) described the handover of Hong Kong as the end of an era of European empire building that began five centuries ago, 'the eclipse of an empire than lasted more than 400 years'. As European colonial empires were formally dismantled, a chapter of colonial history ended, and according to some observers a ***postcolonial*** chapter began. Postcolonial sometimes refers to that which is after or against colonialism – an ongoing colonial hangover, in which former colonies are plagued by relics and legacies of the defunct colonial order, as well as by some new forms of colonialism.

If you look closely at the new world map you will see relics, traces and new forms of colonialism. First the relics: not all of the old colonies have been handed back. The British, for example, retain dependencies around the world, mainly small islands ranging from Gibraltar (on the southern tip of Spain) to the Falkland Islands (off the coast of Argentina) and Montserrat (in the Caribbean). Britain also claims part of Antarctica. European powers have also held on to their 'internal' colonies. Notice, for example, how Britanny blends seamlessly into France, how Wales blends with its neighbour to the east into a single nation. To many people in Britanny and Wales, French and English colonialisms respectively are very much alive. Internal colonialism can also be seen outside Europe. For example, some people see Tibet and Hong Kong as colonies of China, and the North West Territories as a colony of Canada (Morris, 1990, 1992).

Figure 29.5 An empire closes down but is Hong Kong decolonized? Source: *The Guardian*, 1/7/97. Credit: Associated Press

Now, looking for example at the configurations of states and national borders, you will see traces of European colonialism. Many present-day nations are legacies of colonialism. Canada, for example, is a nation born of British **hegemony** over territory settled by English and French emigrants, and the borders of present-day Canada are a fossil of Empire – albeit one which Quebec separatists would like to dismantle. Other borders reflect the colonial inheritance. The border between India and Pakistan was drawn by a British civil servant, Cyril Radcliffe, a stranger to British India who was given just 36 days to complete his momentous task. Armed only with a pile of outdated maps and some crude census returns, Radcliffe produced the 'Wiggly Line' that initiated one of the world's greatest population movements (15 million Muslims and Hindus crossed the new borders) and set the stage for half a century of animosity between the two nations (Khilnani, 1997). In Africa, as in India, many borders and states are essentially colonial creations transformed into independent states. Their boundaries, shapes and sizes are part of the colonial inheritance (as Griffiths shows in *The Atlas of African Affairs,* 1993; see also Corbridge, 1993).

However closely you look at the 1997 map (Fig 29.4), though, you will not learn very much about new forms of colonialism. For, as overt enthusiasm for the imperial project has receded, map makers and others have tended to downplay and disguise continuing and new forms of colonialism. Stong, confident, aggressive colours are succeded by an apparently arbitrary pattern of equal and different national colours. Still, even though colonial powers have lowered their flags over most former colonies, ending formal colonial rule, **decolonization** has not meant an end to 'unequal territorial relationships'; in other words, it has not meant an end to imperialism (as defined above). As some old imperial powers have fallen, others have risen. A transformed imperial order was revealed in 1956, when Britain and France finally retreated from Egypt, which they had invaded to regain control of the Suez Canal. They retreated because the Soviets had told them to and because the Americans chose not to intervene. Those superpowers continued where their western European imperial ancestors left off. They did not generally set up formal colonial governments, nor did they tend to found new overseas colonial settlements, both of which their precedessors frequently did, so their activities cannot always be labelled colonial. But the external influence and ventures of the USA and the USSR (until its break-up in 1990) can be described as imperial, and they are sometimes said to be *neo-colonial*. Allegedly neo-colonial American ventures involve some of the same tactics that were used by former colonial powers such as the British, for many of the same objectives (including preferential access to resources, labour and markets). These tactics include occasional military intervention to protect investment and trade. The Gulf War, for example, a war in which the USA led an international force that restored territorial boundaries (between Iraq and Kuwait), was partly a war for oil. The invading forces restored boundaries that had been created by former colonial powers in their own interests, and perpetuated largely by the USA in its own interest. In the neo-colonial order, American corporations have access to global markets (Coca Cola and McDonald's, for example) and global labour, which often comes cheap (*see* Fig. 29.6). Though the USA is not overtly

Figure 29.6 No post-colonial utopia: American factory in Honduras producing shorts for the US market. Credit: © Jenny Matthews/Network

imperialist in the manner that Britain once was, it insists upon and enjoys many of the same advantages – those of being the world's greatest power.

Summary

- Empires have shaped modern world maps, dictated and policed borders and configurations of states, controlled systems of production and consumption, and interfered with patterns of culture and settlement.

- Geographies of colonialism and imperialism are not necessarily historical geographies, since the demise of western European colonial empires has been matched by the rise of other, mainly western and northern, imperial and/or neo-colonial powers.

CONCLUSION

Geographies of colonialism and postcolonialism begin in the past, but due to the inertia of old empires and the emergence of new ones, they continue into the present. Colonial and postcolonial perspectives draw together many traditional sub-fields of the discipline – including Third World and development geographies, and geographies of trade, economy, migration, language and culture – in an approach that is synthetic, historical and critical.

Further reading

- Godlewska, A. and Smith, N. (1994) *Geography and empire*. London: Blackwell/IBG.
A series of essays on the relationships between geographical scholarship and empire. Read the editors' introduction and a selection of the others, including Crush's essay on 'Post-colonialism, De-colonisation and Geography'.

- Johnston, R.J., Gregory, D. and Smith, D.M. (eds) (1994) *The dictionary of human geography*. 3rd edition, Oxford: Blackwell.
Useful entries on colonialism, exploration, imperialism, neo-colonialism and postcolonialism, with suggestions for further reading.

- Livingstone, D. (1992) *The geographical tradition*. Oxford: Blackwell.
This substantial contribution to the history and philosophy of geographical knowledge is a must for any undergraduate geographer. Livingstone pays particular attention to relationships between colonialism and the production and use of geographical knowledge.

- Morris, J. (1968–78) *Pax Britannica: trilogy*. London: Faber. *Pax Britannica: the climax of an empire* (Vol 1, 1968); *Heaven's command: an imperial progress* (Vol 2, 1973); *Farewell the trumpets: an imperial retreat* (Vol 3, 1978).
Though superficially more descriptive than critical, Morris's lively and evocative narrative is both entertaining and thought-provoking.

- Phillips, R.S. (1997) *Mapping men and empire: a geography of empire*. London: Routledge.
European colonial encounters were narrated and rehearsed in lively adventure stories, set in faraway lands and seas. This book shows how adventure fiction and non-fiction – forms of popular geographical literature – served the interests of European colonial expansion. Chapter 2, on the geography of *Robinson Crusoe*, provides a readable overview.

- Said, E. (1993) *Culture and imperialism*. London: Vintage.
Real and imagined geographies are at the heart of Said's exploration of relationships between western culture and imperialism. The brief introductory section on 'Empire, Geography, and Culture' (pp. 1–15) is particularly useful. This book develops ideas about relationships between geographical knowledge (in novels, as well as more formal and scholarly works) and colonial power, which Said first introduced in his influential if sometimes difficult *Orientalism* (1978), an accessible summary of which is provided by Massey and Jess (1995: 92–8).

CHAPTER 30

Migrations and diasporas

Claire Dwyer

INTRODUCTION

In November 1997 British East African Asians gathered at the Swaminarayan temple and in Westminster Abbey (*see* Fig. 30.1) to commemorate the twenty-fifth anniversary of their migration to Britain. Many of those present had come to Britain in 1972 as refugees, expelled by Ugandan president Idi Amin. As British passport holders they sought a new home in Britain and this was reluctantly granted to them by British authorities, despite the 1968 Immigration Act which had recently restricted right of entry to Britain to those who also had a parent or grandparent born in Britain. For these Ugandan Asian migrants, and their children, the commemorations were an opportunity to reflect on their own and their families' histories, to meet friends and to remember stories. These family stories are woven around connections and migrations between many different places.

East African Asians migrating to Britain from Uganda, as well as those who came from Kenya and Tanzania, are 'twice migrants' (Bhachu, 1985). They are part of a migration movement between India and East Africa which began during the colonial period, in the late nineteenth century, when indentured labourers from India (as well as from China) were contracted to build railways in East Africa. Once indentured labour ended in 1922 some workers stayed on, establishing a commercial trade and developing a distinctive East African Indian culture sustained by ongoing migration from India. When Kenya, Uganda and Tanzania gained independence in the 1960s East African Asians, who had occupied an ambiguous place in the colonial social hierarchy between European colonizers and African colonized, were offered British citizenship. It was this citizenship which enabled a subsequent migration to Britain. Tracing their own identities in

Figure 30.1 British East African Asians join MPs at Westminster Abbey in November 1997, to commemorate the 25th anniversary of their migration to Britain. Credit: Solo Syndication Ltd

relation to these different migrations, British East African Asians stress connections to multiple places or *homes*. Avtar Brah, a writer, recalls being questioned by a university lecturer: ' "Do you see yourself as African or Indian?" . . . At first this question struck me as somewhat absurd. Could he not see that I was *both*?' (Brah 1996: 2). These connections to different places are also sustained through links with family members who have migrated to other places. While some family members moved to Britain in the 1970s others migrated to Canada, the United States or to India and other parts of South Asia. Thus British East African Asians are linked, through migration, across a number of different national borders.

This account of the histories of British East African Asians is one migration story which opens up some of the questions which this chapter explores – about place, identity, belonging and *home*. These may not be the kinds of questions that you readily associate with the study of geographies of migration. Nevertheless, the approach that I want to take in this chapter is to suggest some different ways in which studying the experiences of migration can lead to different ways of thinking about geography.

In particular, I want to challenge the idea of the bounded **nation–state** as a fixed place of 'home' and as the site for the production of a national culture. Rather than thinking about migrations as being between one bounded place and another I want to think about geographies of flows and geographies of connections between places. These geographies produce webs of connections which cut across existing national boundaries or borders. The people who are part of these webs of connections might be said to have '**transnational**' identities. Such transnational identities suggest links to multiple places and an identity which cannot be confined to one form of national identification or belonging. Through an exploration of these 'transnational' or 'diasporic' identities I hope to provide another perspective on the global–local geographies suggested in Chapter 3. While there are many different ways in which the flows and connections of global–local geographies might be investigated, the focus in this chapter is specifically on the movement of people within global patterns of migration. I begin the chapter by outlining the focus of the study of migration within geography, leading on to look specifically at the significance of migration in relation to processes of *globalization*. Through this focus upon globalization I suggest alternative ways of thinking about place and culture by drawing on the experiences of the transnational identities of the peoples of *diaspora* populations.

GEOGRAPHY AND MIGRATION

Most first-year geography students will be familiar with the study of migration as an integral part of human geography. Population migration might be recognized as a key component of geographical change – whether we consider internal changes within one nation state, such as rural–urban migration, or international migration. Indeed, migration connects to many issues which are often considered central to social science such as the creation of capitalist society and the rise of industrialization, and now the emergence of a global, networked economy (King, 1995: 7).

Traditionally migration has been defined within geography as a *permanent* move in place of residence or 'home' (Johnson and Salt, 1992: 1). This definition has always, perhaps, been subject to debate, in relation to practices such as **transhumance** which involve regular short-term migration. However, contemporary migration patterns increasingly complicate these definitions. Many labour migrations may be 'temporary' migrations but may be established as normal, long-term patterns of employment – for example Caribbean or South American workers employed in the United States, or South African mine workers working in neighbouring states. Such cases increasingly challenge a definition of 'home' as a fixed place of residence, as workers may spend longer 'abroad' than at 'home'. This situation raises some of the questions about place and identity which are explored below.

Studies of migration within geography have focused considerable attention on the motives for migration. These have been explored through three main models; ecological models which have focused on the characteristics of the places of origin and destination; behavioural models, which have looked at the decision-making of individuals; and systems models which have taken a more multi-faceted, often economic approach to understanding migration movements (Johnson and Salt, 1992). All of these models, which might be seen as complementary rather than competing, have fundamentally been aimed at exploring and also sometimes predicting, *why* migration takes place. Less attention has been focused, perhaps, on the experience of migration or its outcome for the migrating population. Predictions of migration flows have played a central part in the geographical study of migration which has used demographic data to understand and model particularly internal migration flows, but also increasingly international population migrations.

For the contemporary geographical study of migration many important new issues are emerging.

Perhaps, most important of these, as I discuss in the next section, is the increasing 'globalization' of migration. Focus on new forms of international migration has included a recognition of the diversity of contemporary migration movements – the extent to which migrant flows are differentiated by skills and gender, for example. The feminization of migration flows (Castles and Miller, 1993) has been of particular interest. Geographers have also studied the role of immigration policies and controls, alongside the perceived increase in *illegal* migration and the extent of migrant trafficking. Finally, the rise in numbers of refugees and asylum seekers has become a particular focus of research.

While the focus of geographical studies of migration remains the measurement and prediction of flows of migrants between places, alternative approaches to migration might include looking at the transformations to our conceptions of place, identity and 'home' which migrations provoke. As peoples and places become increasingly linked through globalization, the traditional focus of migration as a process of movement between two separate and bounded places might be open to scrutiny.

Summary

- Population migration is integral to geographical change and to processes of social change such as industrialization.

- Geographers have focused primarily on explaining why migration patterns occur and predicting future migration flows.

MIGRATIONS AND GLOBALIZATION

It has been argued that we now live in 'an age of migration' (Castles and Miller, 1993) in which global migration movements are accelerating at an unprecedented rate. It has been estimated that in 1992 there were 100 million migrants (ibid.: 4.). This increase in the numbers of migrants is also paralleled by the globalization of migration – more and more countries are becoming incorporated within a global system of population movement. Migrations have been integral to processes of **colonization** (see Chapter 29), industrialization and the development of a capitalist world. Between 1500 and 1800, for example, 2 million Europeans moved between Europe and the 'New World' of North and South America and the Caribbean (Emmer, 1993). These migrations and subsequent settlements accompanied the forced migration of 11 million slaves from Africa to America via the infamous 'triangular

trade' of commodities and peoples between Europe, West Africa and the Americas (Segal, 1995). When slavery was finally abolished, migrations of indentured labourers from India and China to work in European colonies in the Caribbean, Asia and the Americas and in East Africa continued. These migrations can be seen as part of the construction of a global labour market through the globalizing processes of **imperialism** and **colonialism** (King, 1995). These links between countries are evident in the later migration flows between former colonized countries in response to post-war European labour demands within a widening capitalist economy – the recruitment in the 1950s of workers from India, Pakistan and the Caribbean to work in Britain, for example.

In the late twentieth century it is the demands of a global capitalist economy which have accelerated the globalization of migration. Restructuring of the world economy has produced demands for a differentiated labour force, while the impetus of uneven economic development within a globalized system prompts the desire for migration and economic advancement. One of the characteristics of current global migration patterns is the diversity of migrants. Migrants may be both highly educated, skilled and professionals, such as doctors from India or Egypt emigrating to Europe and North America, and casual workers who are unskilled and often work in the informal sector of a progressively deregulated and flexible global economy – such as South Americans coming to work in the USA (Sassen, 1988). With more and more countries incorporated into a global migration system, patterns of migration have become more complex. While flows are predominantly from South to North, and from East to West, some countries which were once countries of out-migration, like Italy, have now become countries of in-migration, particularly for migrants coming from Eastern Europe. Other areas of the world, such as the Gulf States, have also become major new sources of immigration, most notably for workers from South Asia. Another important recent characteristic of migration is the increasing feminization of migration. Women are now a significant part of contemporary migration flows – for example, Filipino women migrating to the Middle East or Thai women to Japan. Finally, an increasingly significant number of contemporary migrants are refugees and asylum seekers – perhaps as many as 20 million of the 100 million migrants estimated for 1992 (Castles and Miller, 1993).

There is an irony in this acceleration of global migration patterns. While the globalization of migration can be seen as evidence of an increasingly integrated world economy, increases in migration flows have been met with an acceleration in measures by richer countries to control immigration (Harris, 1995). For example, the European Union has simultaneously reduced internal borders, through the Schengen Agreement, while increasing its external boundaries to create a 'Fortress Europe'. There are thus contradictions in the globalization of migration. While people may be more closely linked with other places economically and culturally, and be aware of more possibilities for migration, barriers to migration may be stronger than ever before.

Summary

- Migrations have occurred throughout human history and population movement has been integral to processes of colonization, industrialization and world capitalism.

- Migration is increasingly globalized – more countries and more individuals are incorporated into patterns of transnational movement than ever before.

GLOBALIZATION, MIGRATIONS AND CULTURE

Migration movements can be seen as an integral part of globalization – those economic, social, cultural and political processes whereby places across the globe are increasingly interconnected (Giddens, 1990) through the compression of time and space (Harvey, 1989). Migration results in the 'stretching out' of social relations between people across space (Massey and Jess, 1995) as individuals retain links within a community which is spread out across national boundaries. While links are maintained through visits or the exchange of letters, the globalization of telecommunications means that communities can be linked much more immediately. Thus British-Punjabi families in Southall in London exchange videos with family members in India (Gillespie, 1995) while the Internet and e-mail electronic bulletin boards can prove a rapid form of communication for example between Indians residing in the United States and in India (Rai, 1995). Satellite television is also increasingly important in creating shared transnational cultural experiences – such as the British-Jamaican football fans gathered in East London, England to watch Jamaica's world cup qualifying match linked by satellite to watchers in Kingston, Jamaica (Eboda, 1997).

These forms of transnational social relations raise important questions about culture and place. If social relations are stretched out across space, then ideas about culture or cultural identity can no longer be thought of as being bounded in one place – trans-

ported by migrants from one place to another. Instead, these transnational links suggest a more dynamic conceptualization of cultures being formed and transformed across and between national boundaries.

One way of thinking about this might be through the words of Gloria Anzaldúa (1987) a Chicano (Mexican–American) woman who reflects on her own identity in relation to the border between Mexico and the United States, a very important boundary for transnational immigration.

Gloria Anzaldúa
'To Live in the Borderlands Means You . . .

From: *Borderlands/La Frontera: the New Mestiza*, pp. 194–195. San Francisco: Spinsters/Aunt Lute (1987)

To live in the Borderlands means you
    are neither *hispana india negra española*
    ni gabacha, eres mestiza, mulata, half-breed
    caught in the crossfire between camps
    while carrying all five races on your back
    not knowing which side to turn to, run from;

To live in the Borderlands means knowing
    that the *india* in you, betrayed for 500 years,
    is no longer speaking to you,
    that *mexicanas* call you *rajetas*,
    that denying the Anglo inside you
    is as bad as having denied the Indian or Black;

Cuando vives en la frontera
    people walk through you, the wind steals your voice,
    you're a *burra*, *buey*, scapegoat,
    forerunner of a new race,
    half and half – both woman and man, neither –
    a new gender;

To live in the Borderlands means to
    put *chile* in the borscht,
    eat whole wheat *tortillas*,
    speak Tex-Mex with a Brooklyn accent;
    be stopped by *la migra* at the border checkpoints;

Living in the Borderlands means you fight hard to
    resist the gold elixer beckoning from the bottle,
    the pull of the gun barrel,
    the rope crushing the hollow of your throat;

In the Borderlands
    you are the battleground
    where enemies are kin to each other;
    you are at home, a stranger,
    the border disputes have been settled
    the volley of shots have shattered the truce
    you are wounded, lost in action
    dead, fighting back;

To live in the Borderlands means
    the mill with the razor white teeth wants to shred off
    your olive-red skin, crush out the kernel, your heart
    pound you pinch you roll you out
    smelling like white bread but dead;

To survive the Borderlands
    you must live *sin fronteras*
    be a crossroads.

gabacha a Chicano term for a white woman
rajetas literally, 'split,' that is, having betrayed your word
burra donkey
buey oxen
sin fronteras without borders

Anzaldúa reflects on a Chicano identity which is 'in-between' – neither Mexican or American – reflected in her mixing of languages between Chicana, Spanish and English. It is also an identity which is not rooted in one place but draws upon an evocation of several different geographies at the same time.

A concept which is helpful in understanding these 'stretched out' or 'extroverted' geographies of flows and connections between people, culture and places is the idea of *diaspora*. Diaspora refers to the dispersal or scattering of people from their original 'home'. Initially, it was used to describe the dispersal of Jews after the Babylonian exile but it has now been given a much wider currency to describe the dispersals of populations associated with many different forced and unforced migrations. The term diaspora is an attempt to encompass the different and complex belongings of peoples who may be dispersed across geographical boundaries and may have connections to several different places they call 'home' (Clifford, 1994). While the term 'diaspora' suggests a fixed point of origin from which these peoples initially dispersed, the evocation of 'diaspora' is also a challenge to the idea of identities as rooted in fixed places of origin and instead an attempt to explore the ways in which diaspora cultures are created through the fusion and mixing of different cultural elements – as Anzaldúa's poem suggests. This idea of fused or **syncretic** cultures is emphasized by Stuart Hall who argues that the peoples of the new diasporas which result from **post-colonial** migrations 'must learn to inhabit at least two identities, to speak two cultural languages, to translate and negotiate between them' (1992: 310).

In the next section I outline two different examples of how we might begin to think about these 'diaspora cultures'. Through these examples I hope to illustrate how the concept of diaspora and diaspora cultures might suggest different ways of thinking about geography.

Summary

- Migrant communities have 'stretched out' social relations and networks which cross-cut national boundaries and borders.

- Diaspora describes dispersed communities who share multiple belongings to different places or 'homes' in different national spaces.

DIASPORA CULTURES

The black Atlantic

As I outlined above, one of the most important migrations in history was the 'triangular trade' of slaves which dispersed black people from Africa across the Americas. This forced migration formed the basis for what Paul Gilroy (1994) describes as a black Atlantic diaspora. This is a diaspora culture which links African-Americans, Caribbeans, Black British people and peoples in Africa. In his exploration of the cultures of the black Atlantic diaspora, Gilroy uses metaphors of travel – recalling the ships which first took slaves to America as well as those which brought Caribbean migrants to Europe in the post-war period. He argues that the cultures of the black Atlantic are characterized not by a return to African 'roots' but by the interconnection of many interlinked 'routes' between different places.

This is emphasized through his exploration of the connections between different kinds of musical forms. He argues that the music of the black Atlantic diaspora – blues, reggae, jazz, soul, rap – have all been produced through particular fusions of influences in different places. Thus on the plantations African music was transformed particularly through a fusion with other kinds of music such as European religious music. When slaves later migrated to the cities new musical forms such as rhythm and blues and jazz were created, and in each case these were created through new interconnections. Thus New Orleans jazz was a different music from other forms of jazz such as Afro-Cuban jazz. Gilroy traces these connections to music produced in Britain by post-war migrants such as British 'northern soul' or the British adaptations of Jamaican reggae, while noting reggae itself was born of fusions between Jamaican folk music and American rhythm and blues. Similarly, the production of contemporary black music, such as rap, had its origins in 1970s in the fusion of the music of the Jamaican sound systems with the 'talk over' of New York Bronx Djs (Hall, 1995).

Gilroy's argument is that all these musical forms are the results of fusions of different cultural traditions within the black Atlantic diaspora. While all of the musical forms retain some distinctive elements of what might be deemed 'African', they have been transformed within different geographical and national contexts. Through these connections Gilroy unsettles a notion of tradition or fixed origins. The musical forms are not diluted forms of 'traditional' African music but are new syncretic forms produced within a diasporic culture through cultural flows or interconnecting routes. And indeed these routes are not simply one-way – American and British black music is also played and bought in Africa influencing the ways in which contemporary African music is produced.

Gilroy (1987) also traces these interconnections within black British youth cultures – black styles, music, dance, fashion and language – emphasizing the extent to which black British culture must be understood as an integral part of a black Atlantic diaspora. He argues that black cultures 'draw inspiration from those developed by populations elsewhere. In particular, the culture and politics of black America and the Caribbean have become raw materials for creative processes which redefine what it means to be black, adapting it to distinctively British experiences and meanings' (ibid.: 154). In this way black British youth culture is actively made and re-made within a diasporic community of connection across the black Atlantic. At the same time black youth cultures are created within a distinctive British context and increasingly lead the production of a British youth culture which itself becomes part of a transformative diasporic culture.

Bhangra music

Another example of the distinctive cultures of diaspora is provided by focusing upon the music produced within the British Asian diaspora. Bhangra music is a fusion of Punjabi folk music with hip-hop, soul and house. Bhangra, a rural folk music, was originally played by Punjabi migrants to British cities at social events like weddings. In the early 1970s bhangra musicians began to experiment with synthesizers and drum machines to produce a more popular form of dance music particularly aimed at young people (Baumann, 1990). Since the 1980s bhangra artists, such as Apache Indian and Sagoo, have drawn upon musical idioms from the black diaspora, such as reggae, techno, house and soul to produce a distinctive syncretic form of music. While bhangra is an integral part of a British South Asian youth culture it is also being adapted across the South Asian diaspora within specific local contexts such as Toronto, Vancouver or New York as well as in Bombay and Delhi (Gopinath, 1995).

Apache Indian, a British-Asian artist who grew up in Handsworth in Birmingham, performs a fusion of bhangra and reggae which has been dubbed 'bhangra-muffin' (Back, 1995), his lyrics interspersing Jamaican patois with Punjabi phrases and English. Apache Indian's music narrates a diasporic identity which crosses boundaries and evokes connections. This is evident in his name, which references the Jamaican raggamuffin artist Wild Apache, while also playing on the double meanings of the word 'Indian' (Gillespie, 1995: 5). These double meanings are also evoked in the title of his album 'No Reservations' (Gopinath, 1995) (see Fig. 30.2). On the album cover Apache Indian is depicted against the backdrop of the Indian flag combined with the Rastafari colours of red, gold, and green. The back of the cover represents a fusion of different places and histories – maps of Jamaica and the Punjab, old family photos and pictures of Apache Indian with Afro-Caribbean reggae artists. As Gopinath (1995: 311) argues: 'the imagined geography of Apache Indian's past takes as its points of reference multiple sites within both Afro-Caribbean and Asian diasporas'.

Through these multiple sites of identification Apache Indian destabilizes fixed ideas about national boundaries or national cultures. Performing in India, where he commands great popularity, he introduced himself to the audience at a concert in Delhi as an Indian declaring: 'I come to you as an Indian who loves his country and his people, all over the world' (cited in Gopinath, 1995: 313). Through this definition Apache Indian, who is British-born and visited India only as a small child, suggests an idea of 'Indianness' which is not confined to national boundaries but is open to being re-worked and re-fashioned within the diaspora. At the same time he also destabilizes and challenges notions of 'Britishness'. He defines his music as a distinctively British product which emerged from the musical influences he experienced growing up within a multi-ethnic neighbourhood in Birmingham: 'It's a combination of reggae from the streets, Indian music and the language at home, pop because of the country I grew up in. It's a very British music' (cited in Gopinath, 1995: 313).

Apache Indian's music, and his performance of it, can thus be read as an example of how diaspora cultures transform and challenge narratives of national belonging. By asserting that his music is *distinctively British*, Apache Indian resists any suggestion that he is returning to Indian *roots* or origins. Instead, a diasporic identity challenges ideas about fixed origins. Apache Indian's performances in India can be read as transforming Indianness or Indian identity from outside the geographical boundaries of the Indian nation–state from the space of diaspora which cuts

Figure 30.2 Apache Indians 'No Reservations'. Credit: Island Records Ltd

across any prescribed national or cultural boundaries. Simultaneously his assertion of a distinctively British youth culture of fusion and cultural interconnections both asserts membership within British national belonging and also challenges the nature of British culture (Gopinath, 1995: 312).

In both the diaspora cultures of the black Atlantic and the diaspora cultures of South Asia what can be emphasized are cultural forms which are about fusions, interconnections and syncretism. As Gilroy (1994) emphasizes, diaspora cultures are not about a return to a lost 'homeland', a point of origin or to fixed *roots*. Instead disapora cultures are characterized by *routes* which make connections which cut across existing geographical or cultural boundaries. As the example of Apache Indian makes clear, disapora cultures are also about challenging ideas about place, culture and national belonging. The spaces of diaspora cut across existing geographical boundaries and also confront the idea of fixed national cultures within national bounded spaces.

Summary

- Diaspora cultures are characterized by interconnections, fusions and links which cut across or transform geographical boundaries.

- Diaspora cultures challenge the notion of fixed 'roots' or origins and emphasize instead routes linking different places.

- It is in music, art and other cultural forms that diaspora cultures have been most readily recognized.

DIASPORA SPACES

The previous section emphasized that geographies of diaspora are characterized by networks, flows and connections which link multiple locations. These networks are also characterized by specific global–local connections (see Chapter 3). As Apache Indian asserts, it was within the specific local–global nexus of Handsworth that the seeds of his own music emerged from the fusion of different musical influences. In this final section I want to look at how we might understand local–global connections within which particular places might be evoked or imagined as diasporic spaces.

Banglatown

'Banglatown' is an area of East London which might be identified as a diasporic space. Banglatown refers to the area around Brick Lane in the Spitalfields area of the London Borough of Tower Hamlets. Spitalfields is characterized by a history of in-migration and settlement from the French Huguenots in the eighteenth century, Jews and Irish in the nineteenth century and early twentieth century to Bengalis since the 1970s and the area is emblematic of the multicultural history of the city of London

BANGLATOWN
The New Millennium Experience

বাংলাটাউন
নতুন শতাব্দির অভিজ্ঞতা

(Merriman, 1993). Spitalfields, which remains one of the poorest wards in London, is approximately 80 per cent Bengali and the economy of the area is characterized by small family businesses, particularly textiles. 'Banglatown' is a local name used to describe the area around Brick Lane with its collection of Bengali restaurants, music and sari shops. This name creates a space for a Bengali diaspora where individuals assert their identities as simultaneously British and Bengali (Eade 1997), as one local restaurateur remarks: 'Why not call it Banglatown? If anyone comes here they don't see England. They see India, Bangladesh and Pakistan in Europe' (cited by Waugh, 1997). These diasporic connections were re-worked in 1980 when the Bangladeshi president, Ziaur Rahman visited Spitalfields and was presented with a street sign for Brick Lane, with which he promised to name a street in Dhaka (cited in Jacobs, 1996: 96).

In 1997 'Banglatown' was used as a 'concept' (the term used by the development company, Cityside Regeneration Limited) to drive redevelopment in the Spitalfields area (*see* Fig. 30.3). As regeneration of the Spitalfields market takes place, particularly through the introduction of city investment (see Jacobs, 1996), Tower Hamlets Council is backing an initiative to develop Brick Lane and the surrounding area in partnership with private investment. While this development encompasses shops, businesses, office spaces and housing, it is linked together through the imaginative geography of 'Banglatown' – symbolized by the construction of an arch as a gateway into Brick Lane which joins together Indian, Islamic and classical

Figure 30.3 Banglatown, Brick Lane. Credit: Tower Hamlets Council

Figure 30.4 The building at the corner of Brick Lane and Fournier St in London's East End was built by Huguenot refugees in 1743 as a church. It later became a synagogue and is now used as a mosque by the Bengali community in Whitechapel. It stands today as a symbol of centuries of immigration to Britain. Credit: © David Hoffman. Source: The Commission for Racial Equality, 1996: *Roots of the Future*

European architectural styles. Projects planned for the area include a 'Moorish Market Bazaar', a 'rich-mix centre' which will celebrate ethnic and cultural diversity through a range of educational projects and a South Asian music and dance academy.

The use of the concept of Banglatown has been supported by many local Bengalis who see it as an opportunity to bring much needed investment to the area (Jacobs, 1996: 100). They recognize the need to attract city investment into the Tower Hamlets area and therefore have been keen to embrace the concept of 'Banglatown' as a marketing tool. At the same time, there are also fears that the marketing of 'Banglatown' may be a means by which the distinctive 'Bengaliness' of Banglatown is **essentialized** or commodified as 'exotic' or 'typical'. Thus the space of Banglatown would no longer be a dynamic space of diasporic culture but instead a place in which **commodified** notions of 'Bengaliness' were simply represented to others – for example, through dress or food.

It is perhaps too early to say what the outcome of the investment in Banglatown will be. However, you might want to consider this example of diaspora space as well as others that you might be able to think about. (For example, you could also consider Notting Hill Carnival as a diasporic space – Alleyne-Dettmers, 1997; Cohen, 1982; Hall, 1995). In what ways might you consider the development of the space of Banglatown, or the taking over of the streets of Notting Hill by the Carnival as an example of a challenge and transformation to fixed or stable notions of 'British culture'?

Summary

- Diaspora cultures can be explored within specific spaces within which global-local connections can be recognized.

- Such diasporic spaces are characterized by the coming together or fusion of different cultural ideas or connection.

- Diaspora spaces can also be created or 'invented' through the commodification of cultural difference as 'exotic'.

CONCLUSION: POSSIBILITIES OF DIASPORA?

In this chapter I have suggested that the globalization of migration has produced diasporas. Diasporas are defined as peoples dispersed across geographical boundaries but linked through connections to particular places which are imagined as multiple 'homes'. I have emphasized the ways in which thinking about diasporas challenges many geographical notions. The evocation of multiple 'homes' unsettles fixed geographical and national boundaries and the association of culture with particular places. Instead diasporic cultures are celebrated for their fusion of differences both across different borders but also coming together within particular places to create distinctive syncretic cultures. Thus the geographies of diaspora are about the flows and connections between particular places and of networks which link up global and local spaces. While diaspora cultures are characteristic of particular groups I have also suggested that diaspora cultures may be transformative of all national cultures since they challenge the very notion of fixed national cultures within geographical boundaries.

Further reading

- King, R. (1995) Migrations, globalisation and place. In Massey, D. and Jess, P. *A place in the world?* Open University Press: Milton Keynes, 5–44.
This article develops the ideas of globalization and migration presented in the first section in greater detail.

- Gillespie, M. (1995) *Television, ethnicity and cultural change.* London: Routledge.
This is an ethnography of young British Punjabis living in Southall. Worth dipping into as an example of the 'everyday' ways in which the transnational linkages of a diasporic community are maintained. It focuses particularly on the role of television.

- Hall, S. (1995) New cultures for old. In Massey, D. and Jess, P. (eds) *A place in the world?* Open University Press: Milton Keynes, 175–214.
Stuart Hall, a sociologist, is one of the foremost writers associated with ideas about diaspora and cultural transformation. This article, part of a student reader, gives further examples, such as cricket, of the notion of diaspora cultures.

Travel and tourism

Luke Desforges

INTRODUCTION

It is not difficult to justify the inclusion of studies of travel and tourism as part of Human Geography. To start with, touring distant places has long provided an impetus to academic Geography, and doubtless many people feel drawn to the subject because of the opportunities it offers for 'exploration' through field trips and overseas research. Moreover, travel and its accompanying industries are caught up in the formation of human geographies across the world, being enmeshed with the environmental, social, economic, cultural and political fabric of the lives and landscapes of literally millions of people. One way of presenting the importance of tourism to Human Geography is through figures concerning its economic role in the contemporary world (*see* Fig. 31.1). Mills (1989) claims that world-wide tourism is growing at about 5–6 per cent per annum, and that globally it will be the largest source of employment by the year 2000 (Urry, 1990). Recently deindustrialization in the First World has highlighted to geographers the economic importance of the service sector, a sector in which tourism, heritage and leisure play a considerable role.

In both senses Geography's travel stories play an important part in our understandings of the contemporary world. Within the discipline, geographical knowledge concerning the **globalization** and expansion of tourism has often been oriented towards the processes of planning and policy. Some geographers would see their role as one of providing information about the development of tourism for agencies such as **non-governmental organizations**, pressure groups, and the local and national state. A range of geographers have set themselves the task of dealing with the social and environmental consequences of travel, where our critical analysis is used in the resolution of tourism's problems.

However, although this practical response to the problems of tourism is perhaps the most popular in the discipline, it is not the only possibility. This chapter moves onto slightly different ground. Rather than providing insights into the processes of tourism planning, it sets off on a slightly different tack. Tourism Geography's practical bent has often led it to be divorced from other insights that have been developed elsewhere in Human Geography at a more theoretical level. Such insights may at first glance appear to have little to say about tourism and travel. Abstract arguments about the nature of place, about mobility, displacement and spatial interconnections almost seem a distraction from very real environmental and social concerns about tourism. But in this chapter I argue that these insights encourage us to think through some of the criticisms that have been directed at tourism. Theoretical insights from Human Geography offer a way of rethinking the geography of tourism and travel.

The chapter, then, starts by outlining some of the criticisms of tourism and travel that circulate in both academic texts and popular culture (such as newspaper articles). It then looks at some of the theoretical insights offered by Human Geography. The final section looks at the uses of these insights to Tourism Geography, and some recent case studies of tourism out in 'the field'.

CRITICISMS OF TRAVEL AND TOURISM

In contemporary commentaries on tourism, common criticisms are circulated across a number of different arenas, from academic writing to non-governmental

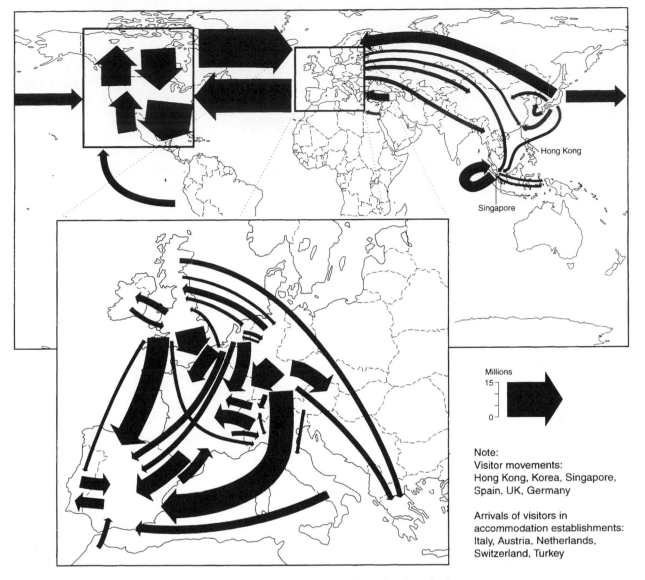

Millions
15

0

Note:
Visitor movements:
Hong Kong, Korea, Singapore,
Spain, UK, Germany

Arrivals of visitors in
accommodation establishments:
Italy, Austria, Netherlands,
Switzerland, Turkey

Figure 31.1 Major international tourist movements, 1995. Source: World Tourism Organization

organizations to broadsheet journalism. To give an example, concern is voiced about the growth of long-haul travel to countries such as Nepal, which in common with many developing countries has used tourism in the quest to fulfil its development goals. In 1970 a total of 1500 tourists visited the country, but in 1993 journalist David Nicholson-Lord reported that more than 300 000 travellers would spend £40 million, amounting to one-sixth of Nepal's foreign earnings (*The Independent on Sunday*, 3 October 1993).

Nicholson-Lord's report offers a number of important criticisms of tourism in Nepal. First, he writes about the impact of tourism on the Nepalese environment, describing the 22 tonnes of non-biodegradable, non-burnable rubbish which is dumped on trekking trails every year. The

Sagarmatha National Park around Mount Everest is described as 'the world's highest trash pit', and trekkers who are unsure of their route are advised to follow the 'toilet paper trails'. Even more problematic is the deforestation of the Nepalese forest. The demand for trekking lodges with hot showers and cooked food means that the average trekker is estimated to consume five to ten times more wood than a Nepali, according to the Nepal Tourist Watch Centre (Netwac). The result is that an area the size of a county such as Hampshire is cleared each year. Tourism Geographers articulate similar concerns for the environment. David Zurick, for example, writes about how

[the] adverse effects of tourist overpopulation in the Khumbu region . . . – trailside litter, forest cutting for fuel-

wood and construction, trail erosion – are now commonly observed throughout heavily visited parts of the Nepal Himalaya . . . Trail litter, whilst unsightly to tourists and perhaps offensive to local people, may not be important enough to curtail tourist activity, but other environmental disturbances such as increased tree cutting for tourist lodges and other forms of land degradation are serious concerns.

(Zurick, 1992: 622)

These two authors also identify a number of social problems caused by tourism to a 'Nepali way of life'. In particular there is a concern about the consequences of increased money flowing through the country and the **commodification** of everyday life. David Nicholson-Lord, for example, expresses a fear for traditional Nepalese family structures as a 'get rich quick' mentality sweeps the country. According to Nicholson-Lord, family members are drawn away from the land to earn a living in teahouses and lodges. Children leave school to become touts or guides, and husbands go into portering jobs. A representative of Netwac is quoted as saying '[t]wenty years ago Nepali people didn't know about asking for money. Now they see a white face, they think: "The money is coming." But they are neglecting the future for quick gain today.' Netwac argue that tourists and locals engage in an 'uncomprehending culture clash'. Some Nepalis are offended by tourists in scanty clothing who smoke or take drugs, because they show a lack of understanding and a disrespect for local culture. On the other hand, Netwac's representative thinks that local culture is fast disappearing:

[t]hey see the tourists, they watch satellite television, they think everyone in the West must be wealthy and sophisticated . . . They don't know about the three million unemployed in Britain. In villages on trekking routes, you see young Nepali men wearing jeans and dark glasses. They cannot read or write – but they think they look very smart.

David Zurick adds to these criticisms. For example, he argues that although traditional Nepalese handicrafts can still be seen in the villages, 'tourism may safeguard the artefacts of a culture but destroy the spirit that created them' (1992: 618). Similarly, the wealth of the tourist economy means that despite international finance for the reconstruction of local religious sanctuaries, 'religious life in Khumba appears less attractive than in former times'. Zurick does, however, see some grounds for optimism about the ability of Nepalis to 'preserve' their local culture and environment. For example, in the Annapurna region of Nepal, Zurick argues that the need to conserve woodland resources has been supported by conservation strategies, such as bans on fuelwood cutting and the development of nursery plantations. These policies are legitimated through sacred decrees based on the belief that spirits guard the mountains and the forests (1992: 622; for a more extensive account of tourism in Nepal, see Stevens, 1993; Adams, 1992).

Summary

- Important critiques of tourism are circulated through a network of geography academics, journalists and non-governmental organizations.

- Tourism is thought to have an adverse impact on the nature and culture of places throughout the world.

- As far as Nepal is concerned, tourism's impact on nature is in terms of rubbish and deforestation.

- Tourism's impact on Nepali culture is to commodify cultural life, meaning that the exchange of cultural traditions and artefacts for money becomes part of a 'westernized' local economy.

RETHINKING CRITIQUES OF TOURISM

The argument that tourism impacts adversely on local environments and cultures would appear to be transferable to many places around the globe where a tourist economy is developing, whether it is the mass tourist destinations of the Spanish Costa or the exploration of long-haul exotica such as Bali. The spread of tourism to seemingly 'unspoilt' places raises strong reactions from those interested in the politics of travel. At times the battlefields are clearly drawn between 'on the one hand, the tourists, tourist agencies, traffic industries and ancillary services, to say nothing of governments anxious to raise foreign currencies; and all those who care about preserving natural beauty on the other' (Mishan, 1969, quoted in Urry, 1990: 41). The seemingly endless spread of tourism is presented as 'a competitive scramble to uncover all places of once quiet repose, of wonder, of beauty and historic interest to the money flushed multitude, [which] is in effect literally and irrevocably destroying them' (Mishan, 1969, quoted in Urry, 1990: 42).

As far as local cultures are concerned, Davydd Greenwood argues that:

Treating cultures as a natural resource or a commodity over which tourists have rights is simply perverse, it is a violation of the people's cultural rights . . . what must be remembered is that in its very essence [culture] is something people believe in *implicitly*. By making it part of the tourism package it is turned into an explicit and paid performance and no longer can be believed in the way it was before. Thus, the commoditisation of culture in effect robs people of the very meaning by which they organise their lives.

(1989: 179)

These authors take a strong line on tourism. Metaphors of tourism as a form of **colonialism** are found (see for example, Urry, 1995: 190), and indeed Mishan argues for a ban on international air travel. When faced with the very real problems of tourism found in places such as Nepal, there is a temptation to base criticisms of tourism in a desire to return to a pre-tourist culture, a culture and environment which have been untouched by the presence of 'Western' money and bodies. Not all criticisms of tourism use this argument, and indeed many geographers and policy-makers are concerned with the 'proper management' of tourism and its consequences rather than its suppression. But theoretical insights from Human Geography and elsewhere mean that these kinds of criticisms of the impact of global tourism on local cultures and environments may need to be rethought.

In particular, rethinking tourism means addressing the idea that places are either 'authentic' or 'inauthentic'. I've put these two words in inverted commas to signify that their original meaning has come to be questioned by many who deal with tourism, and it is through thinking about these terms that critiques of tourism have been reworked. 'Authentic' places are often associated with 'the primitive, the folk, the peasant and the working class' (Frow, 1991: 129, see also Culler, 1988) whose culture is seen as sticking to its roots and traditions in an unself-conscious way. In particular, authentic places are thought to contain cultural objects and practices that have not been produced for sale, but for other members of the local community. Such places are represented as having a shared culture which is untouched by outside forces. They are captured in phrases such as 'the real India' or 'unspoilt France'. Conversely, 'inauthentic' places are characterized by a sense of self-conscious design for others, for example, making souvenirs for sale to tourists, rather than for the use of the local community. 'Inauthentic' places have been 'spoilt' by their contact with people from 'outside'.

What is being presented here is a relationship between culture and place that is criticized by Human Geographers in this book and elsewhere. The idea that places are either 'authentic' or 'inauthentic' implies that to remain unspoilt a culture must be 'bounded off' from neighbouring influences, particularly globalizing Western cultures. 'Authentic' cultures are those that emerge 'organically' from a place without interference that 'corrupts' their original form. The differences between places, then, are seen

as created internally. In Chapter 3, Philip Crang uses the metaphor of a 'mosaic' to discuss this way of thinking about places and their differences.

As Crang argues, the idea that spatial differences are the result of boundaries around places can be questioned factually, and this is as true of the spaces of tourism as any other. To give an example: the area in which I live and work, in and around Aberystwyth, is renowned as a bastion of Welsh cultural difference. According to the 1991 census, around 70 per cent of the people living in Ceredigion speak Welsh, the constituency has a Welsh nationalist Plaid Cymru Member of Parliament, and the town of Aberystwyth is often presented as a regional capital for Welsh culture. The heart of Welsh-speaking Wales is associated with the small villages and towns of Ceredigion, which have fostered Welsh-speaking communities for generations.

This image of Welsh culture is used to market Ceredigion as a tourist destination (*see* Fig. 31.2, Fig. 31.3, Fig. 31.4). The tourist office brochure talks of a warm Welsh welcome, of local hospitality and friendliness, and of a landscape that is unspoilt and natural. Visiting Ceredigion provides the opportunity to escape the stresses and strains of modern life in the city, a chance to spend a week away from it all. And yet the character of Ceredigion is very much the result of links to elsewhere. The church up the road from my house has gravestones showing that Ceredigion men in the nineteenth century earned their wages by sailing to and from the United States, Australia and South America, and indeed many of them died abroad. The improved pasturelands and quaint cottages typical of Ceredigion were built on the back of these connections (Jenkins, 1982). Likewise, slightly further south in Pembrokeshire, the shipping of roadstone to England and the rest of Europe built communities associated with quarrying, ports and boats. More recently, in the early to mid-twentieth century, when the railways used to run directly from London to the west coast, profits were made by sending family members to open dairies, which supplied the capital with milk. Cows in milk were sent down to London, and those which needed to calve again were sent back (Colyer, 1976; Knowles, 1997: 71–87).

In addition to west Wales's connections elsewhere, the rest of the world has long come to west Wales. Travellers have visited Ceredigion for centuries, bringing their own 'impacts' and 'influences' recorded in travel books from Giraldus Cambrensis' eleventh-century *A Journey Through Wales* to George Borrow's nineteenth-century *Wild Wales* to Jan Morris's twentieth-century *The Matter of Wales*. Aberystwyth as a town, with its pier and promenade,

Discover the difference
Gwelwch y gwahaniaeth

A warm Welsh welcome - or as we say 'Croeso' - awaits you in Ceredigion. Welcome to a land of spectacularly different landscapes, people, culture and language.

Welcome to a different world, a far cry from everyday stress and noisy, polluted cities. Welcome to fresh mountain air, sea breezes and a chance to discover the difference of Ceredigion's magnificent coast and country.

Figure 31.2 Discover the difference.
Source: Wales Tourist Board

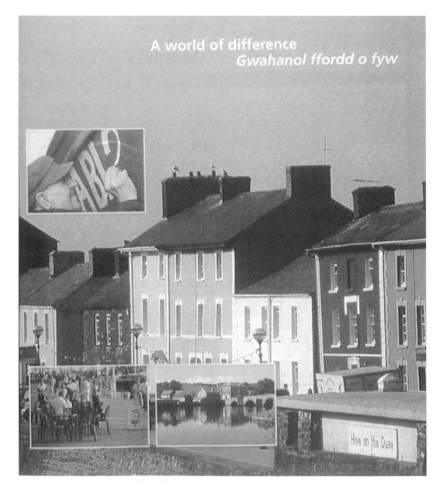

Figure 31.3 A world of difference.
Source: Wales Tourist Board

Figure 31.4 Towns and villages.
Source: Wales Tourist Board

was built on the back of the railway from Birmingham and its countless holidaymakers. Tourism as a social formation is older than is often realized (see Adler, 1989).

The point of these details is that whilst west Wales's unique culture and landscape are seemingly the result of isolation, and resistance to forces of homogenization, in fact the very characteristics of the place which are used to attract tourists are a result of Ceredigion's connections to the rest of the world. Tourist places, then, like all other places, have never been 'authentic' because they have never been bounded places shut off from the rest of the world. Their character and difference have been formed through a whole series of connections.

In fact, as MacCannell notes (1989: 104), criticisms of tourism that draw on notions of spoilt authenticity are as old as tourism itself. Divisions between the 'real' traveller who encounters authentic places, and mere tourists who only get to see the spoilt remains are not only popular figures quoted by Judith Chalmers and the Lonely Planet alike, but also by John Ruskin who in 1865 complained that 'Going by railroad I do not consider as travelling at all; it is

merely being "sent" to a place, and very little different from being a parcel' (quoted in Boorstin, 1992: 87). While this may seem a bit odd today, when travelling by steam train is considered a quintessential symbol of 'real travel', it illustrates the point that a yearning for 'untouched' places has long been a feature of commentary on tourism.

To round off this section it is clear that in critiquing tourism, we cannot refer back to a 'golden age' of authenticity which existed before tourists 'invaded'. Tourism and travel are merely a new form of interconnection between places. However, this does not mean denying the problems caused by tourism. If we take the view that places are formed through their interconnections (Massey, 1994), then the character of places clearly changes as their relationships with the global become mediated through tourism. To critique tourism, we need to think about what happens when places become connected to the global tourist industry. Who benefits from these new(ish) interconnections? Who is able to use them to earn their livelihood? And who is excluded? How are tourism's 'goods' and 'bads' distributed?

Summary

- Critiques of tourism have often revolved around notions of 'authentic' places and the ways in which tourism 'spoils' them.

- The idea of a place as 'authentic' relies on a conceptualization of places as bounded. Such notions have been criticized in Human Geography.

- Tourist places have always been interconnected to other places; for example, a tourist destination such as west Wales has long been connected to London, Birmingham, Europe, the former British empire and beyond.

- Criticisms of tourism as a harbinger of inauthenticity are as old as tourism itself. Critiques of tourism that refer back to a pre-tourist 'golden age' are difficult to sustain.

TOURIST PLACES AND TRAVEL INTERCONNECTIONS

In this final section I look at some recent conceptual and empirical work on tourism that tries to tease out what it means for places to be interconnected via the global tourist industry. To do this we need to think about the ways in which these interconnections are entangled in the formation of tourist places and the livelihoods and landscapes of those who live and work in them. It would seem obvious from the figures presented in the introduction to this chapter that tourism is often associated with a profound shift in the intensity of relationships with the global. But it is worth identifying the ways in which the nature of tourism produces very specific sorts of interconnections that empower and disempower different social groups.

Tourism produces a unique set of interconnections between places because of its relationship between producers and consumers. The transfer of material goods often forms the nature of interconnections between places (think of west Wales's export of roadstone and cows). Consumers of those goods are influential in the formation of far distant communities and their livelihoods. But they will probably never meet or even know about the producers of those goods. In tourist places, by contrast, tourist consumers and the producers of tourist experiences often meet face to face (they are spatially and temporally co-present).

In addition, the product that is sold at tourist sites is very often intangible. The tourist product is an experience that cannot physically be taken home. John Urry (1990) argues that on the whole tourists consume places visually (although see Veijola and Jokinen, 1994), an experience which he says is structured by 'the tourist gaze'. If tourism is based on the desire to experience new, different and extraordinary places, then we are constantly informed about what constitutes an extraordinary experience by films, television, books, friends and videos. The tourist gaze is socially structured. Urry goes on to argue that when we travel we look for things which symbolize this extraordinariness. As Jonathon Culler puts it

> the tourist is interested in everything as a sign of itself . . . All over the world the unsung armies of semioticians [**semiotics** is the study of symbolic meaning], the tourists, are fanning out in search of signs of Frenchness, typical Italian behaviour, exemplary Oriental scenes, typical American thruways, traditional English pubs.
>
> *(Culler, 1988: 155)*

Earning a livelihood in tourist places means producing symbolic meaning for tourists. Lash and Urry (1994) describe tourism as part of 'an economy of signs'. They see the recent development of tourism as part of a shift to a **post-industrial** sign economy in which places earn their income through the flow of information, words, images and texts. In other words, it is through the circulation of culture as a product that tourist destinations are able to build up new relationships and interconnections with the global economy.

Those who are empowered by tourist connections, then, are those that are able to produce the necessary symbolic and cultural meanings. Recent empirical work suggests that far from being forced to perform these meanings simply as a commodity, the production and circulation of cultural meanings for tourists become entangled in the lives of those working in the industry in many complicated ways. To give an example, Cynthia Cone writes about two women in Mexico who make Mayan craft objects such as rugs and pottery for the local tourist market. Cone argues that the skills and cultural resources used by the two women have allowed them independence in more ways than one. First, they have become economically independent, second, they have had some freedom to develop new craft objects which have drawn on Mayan traditions at the same time as innovating new aesthetic forms, and third, they have been able to feel themselves to be independent people (for example, both have resisted the demand that they take up the roles of housewife and mother). As Cone puts it '[t]hey have constructed their lives, their identities, their life stories, with a sense of adventure and increasing self-confidence in their abilities to rise to new occasions and encounters' (1995: 325). For further examples of the ways in which tourism becomes entangled with identities and livelihoods, see Crang (1997, 1994), Bouquet (1987) and Hochschild (1983).

> 'In their pioneering efforts, Manuela and Pasquala have developed relationships with outsiders that have paved their way to relative economic independence and to their role as innovators in transforming local craft objects into items suitable in the world market ... Both have functioned as artisans, as teachers, and for a time, as intermediaries to merchants. Both have experienced ethnicity as a 'contested terrain'; but because of the particular constraints and opportunities of their lives, they have resolved the contest differently. Manuela presents herself as wholly Mayan. She is pleased with her accomplishments and expresses contentment with her life ...
>
> To the extent that Pasquala has rejected her Mayan heritage, she has increased her freedom to operate as an entrepreneur and to aspire to a more middle class life. Pasquala perceives herself in dramatic terms as a survivor. She has lived by her wits, accommodating herself to a series of foreigners with whom she has intensely identified. 'Thank God for strangers!' she exclaims.
>
> *(Cone, 1995: 325)*

At the same time, however, there are those who are excluded from the 'economy of signs' inaugurated by tourism. Recently the UK pressure group Tourism Concern has run a campaign highlighting the situation of those who are displaced by tourism. For example, in Tanzania the Serengeti Game Park has been set up at the expense of previous inhabitants. For many conservationists, the Serengeti has come to symbolize 'authentic nature', or a version of nature in which humans do not play any part. The German zoologist Professor Bernhard Grzimek has been quoted as arguing that 'A national park must remain a primordial wilderness to be effective. No men, not even native ones, should live within its borders' (*The Independent*, 15 September 1997). Here Grzimek is making a familiar argument about the need to return to an 'authentic' past in the face of 'modern' threats.

The important thing to look at in this case is who has the power to make a living (in the widest sense used by Cone) out of the space of land encompassed by the Serengeti. Before the National Park was set up, the land was used to make a living by local people through their trading networks. Once the National Parks came, tourists were allowed in but local people were excluded because they didn't fit into the visitor's imagination of a notional pure wilderness. Consequently those who were transporting tourists in and providing accommodation were making money out of the Serengeti. The new interconnections which tourism introduced acted to disempower local people. Such acts of 'eco-colonialism' have been criticized by the World-Wide Fund for Nature, who argue that the group's work in such areas is not only

to conserve nature, but to attempt to make the global economy work for local people.

Summary

- Tourism provides a set of interconnections between the global and the local which are different from many other interconnections.

- This means that tourism places are structured differently. They are part of an 'economy of signs'.

- The tourist 'economy of signs' can empower those who are able to use their resources to produce cultural products.

- The tourist 'economy of signs' can disempower and displace those who do not fit into the symbols expected of tourist places.

CONCLUSION

In summary, this chapter has made the case that tourism geography cannot rely on the argument that tourists 'spoil' places in order to critique tourism and travel. Insights from Human Geography and elsewhere suggest that notions of authenticity rely on the idea of 'bounded places'. In fact, tourism is merely one more way in which places are connected to the wider world. That said, tourism interconnections do structure places in unique ways, because of tourism's position as part of an 'economy of signs'. In order to discover the ways in which tourist interconnections empower, disempower and displace people, we need to turn to empirical work which looks at the ways in which tourism entangles itself into the lives and landscapes of those living and working in tourist destinations.

Further reading

- Urry, J. (1990) *The tourist gaze: leisure and travel in contemporary society*. London: Sage.
A good introduction to both theories of tourism and empirical work on recent changes in the tourism economy.

- Britton, S. (1991) Tourism, capital and place: towards a critical geography of tourism. *Environment and Planning D: Society and Space* **9**: 451–78.
Although slightly difficult in places, this paper provides a powerful argument that geographers should consider the role of tourism in shaping both the local and the global.

- Crang, P. (1997) Performing the tourist product. In Rojek, C. and Urry, J. (eds) *Touring cultures: transformations of travel and theory*. London: Routledge.
This chapter provides a useful account of research into what it is like working in an 'economy of signs' and the ways in which culture is produced for the tourism market.

- May, J. (1996) In search of authenticity off and on the beaten track. *Environment and Planning D: Society and Space* **14**:709–36.
Although my chapter does not deal directly with tourist consumption of places, there is a growing literature about research on the topic. Jon May's article is a good and accessible place to start.

- Boorstin, D. (1992) From traveller to tourist: the lost arts of travel. In *The image: a guide to pseudo-events in America*. New York: Vintage Books, Random House (second edition).
The original rant against tourism. An example of what this chapter is arguing against.

Commodities

Michael Watts

> A commodity appears at first glance a self-sufficient, trivial thing. Its analysis shows that it is a bewildering thing, full of metaphysical subtleties and theological capers.
>
> *(Karl Marx*, Capital, *1867)*

A COMMODITY IS A BEWILDERING THING: THE CAPITALIST COSMOS AND THE WORLD OF COMMODITIES

With its price tag, said the great German critic Walter Benjamin, the commodity enters the market. In the capitalist societies, that is to say the market economies, which we inhabit this appears perfectly obvious. The *Oxford Dictionary* defines a commodity as something *useful* that can be turned to *commercial advantage* [significantly, its Middle English origins invoke profit, property and income]; it is an article of trade or commerce, a thing that is expedient or convenient. A commodity, in other words, is self-evident, ubiquitous and everyday; it is something that we take for granted.

Commodities surround us and we inhabit them as much as they inhabit us. They are everywhere, and in part define who and what we are. It is as if our entire cosmos, the way we experience and understand our realities and lived existence in the world, is mediated through the base realities of sale and purchase. Indeed, virtually *everything* in modern society *is* a commodity: books, babies (is not adoption now a form of negotiated purchase?), debt, sperm, ideas (intellectual property), pollution, a visit to a National Park, and human organs are all commodities. An Italian tourist company recently offered the experience of war – a two-week tour of ethnic cleansing in Yugoslavia – as a commodity for sale. Even things

that do not exist can become items for purchase and sale. For example, I can buy a 'future' on a basket of major European currencies which reflects the average price (the exchange rate) of those national monies at some distant point in time.

As someone once said: in America virtually everything is for sale . . . which means virtually everything is a commodity. This may be of little comfort to you. But one way of thinking about contemporary capitalist societies like the United States, in which virtually everything is a commodity (i.e. for sale), is that it is a *commodity economy*. It is, in other words, a system of commodities producing commodities. So why examine commodities if they are so trivial and ubiquitous? And why might they be of interest to geographers?

Well, one issue is that commodity-producing societies are quite recent inventions historically speaking, and many parts of the world, while they may produce for the market, are not commodity societies in the same way as our own. Socialist societies (and perhaps parts of China and Cuba today), stood in a quite different relationship to the commodity than so-called advanced capitalist states. Low-income countries, or the Third World so-called, are 'less developed' precisely because they are not mature commodity-producing economies. In the peasant village in which I lived in northern Nigeria in the 1970s, much of what was produced by family farmers (i.e. peasants) did not pass through the market at all. It was directly consumed or entered into complex circuits of gift giving and non-market exchange. The rural household as a unit of production was not a commodity producer; it was not fully *commoditized*.

So the commodity form as a way of organizing social life has little historical depth; that is to say it appeared in the West within the last two hundred years. It is derivatively part of what Max Weber called the 'spirit of modern **capitalism**', but it remains an unfinished project if viewed globally. Over large parts of the earth's surface the process of *commodification* – of ever greater realms of social and economic life being mediated through the market as a commodity – is far from complete. Perhaps there are parts of our existence, even in the heart of **modernity** (see Chapters 16 and 17), that never will take a commodity form. I after all do not purchase my wife's labour power or affection; neither do I buy the ability to play soccer with my young son. But in a commodity economy in which the logic of the market rules, the prospect of converting social intimacy into a commodity is always present. Indeed, it is happening before our eyes.

Another peculiarity of a commodity economy is that some items are traded as commodities but are not intentionally produced as commodities. Cars and shoes are produced to be sold on the market. But labour, or more properly labour power, is also sold – I sell something of myself to my employer, the University of California – and yet it (which is to say me as a person) was not conceived with the intention of being sold. Since I am not a slave, I was not in any meaningful sense produced, like a manufactured good or a McDonald's hamburger. This curious aspect of labour as a commodity under capitalism is as much the case for land or Nature. These sorts of curiosities are what Karl Polanyi in his book *The Great Transformation* (1947) called *'fictitious commodities'*. Polanyi was of the opinion that market societies which do not regulate the processes by which these fictitious commodities becomes commodities will assuredly tear themselves apart. The unregulated, free market, commodity society would eat into the very fabric which sustains it by destroying Nature and by tearing asunder the most basic social relationships. We need look no further than the booming trade in human organs. In her book *Contested Commodities*, Margaret Radin shows how the fact that a poor Indian woman sells her kidney and other organs out of material desperation 'threatens the personhood of everyone' (1996: 125).

And not least, there is the tricky matter of price, which after all is the *meaning* of the commodity in the marketplace, how it is fixed, and what stems from this price fixing. For example, the only pair of running shoes that a poor inner city kid wants is Air Nike which costs slightly more than the Ethiopian GNP per capita and perhaps more than his mother's weekly income. Or the fact that a great work of art, Van Gogh's *Wheat Field,* is purchased for the astonishing sum of $57 million for investment purposes.

The problem of the determination of prices and their relations to *value* lay at the heart of nineteenth-

century classical **political economy** but it is an enormously complex problem that really has not gone away or in any sense been solved. The 'metaphysical subtleties' that Karl Marx refers to are very much about the misunderstandings which arise from the way we think about prices (doesn't it have something to do with supply and demand?) and what we might call the sociology of commodities. But if there is more to commodities than their physical properties and their prices which are derived from costs of production or supply and demand curves, then there is a suggestion that commodities are not what they seem. Commodities have strange, perhaps 'metaphysical' effects. For example, the fact that a beautiful Carravagio painting is a commodity – and correlatively, that it is private property and only within the means of the extravagantly rich – fundamentally shapes my experience of the work, and of my ability to enjoy its magnificent beauty in some unalloyed way. Its commodity status has tainted and coloured my appreciation of it.

A commodity, then, appears to be a trivial thing – here's a car for sale, it has these fine qualities – but it is in fact bewildering, even theological. The commodity, said Walter Benjamin, has a phantom-like objectivity, and it leads its own life after it leaves the hands of its maker. What on earth might this mean?

Summary

- A commodity is something useful which enters the market.

- Commodification refers to the process by which more and more of the material, cultural, political, biological and spiritual world is made rendered as something for sale.

- A commodity producing economy is one in which the logic of commodification is dominant.

A COMMODITY BIOGRAPHY: THE CHICKEN AND U.S. CAPITALISM

A century ago you'd eat steak and lobster when you couldn't afford chicken. Today it can cost less than the potatoes you serve it with. What happened in the years between was an extraordinary marriage of technology and the market.

(John Steele Gordon)

Once in a while I will bring into my undergraduate class a freshly dressed chicken – oven-ready in poultry parlance – and ask students to identify this cold and clammy creature which I've tossed upon the lectern.

After five minutes of 'it's a chicken', 'it's fryer', 'it's a dead bird', 'it's a virtual Kentucky Fried Chicken', I solemnly pronounce that it is none of the above: it is in fact a bundle of social relations.

So let's examine the humble chicken. According to the latest Agricultural Census, 7 billion chickens were sold in the United States in 1994 (roughly 30 per person). In 1991, chicken consumption per capita exceeded beef, for the first time, in a country which has something of an obsession with red meat. The fact that each American man, woman and child currently consumes roughly 1.5 pounds of chicken each week reflects a complex vectoring of social forces in post-war America. First, a change in taste driven by a heightened sensitivity to health matters and especially the heart-related illnesses associated with red meat consumption. Second, the fantastically low cost of chicken meat which has in real terms *fallen* since the 1930s (a century ago Americans would eat steak and lobster when they could not afford chicken). And not least the growing extent to which chicken is consumed in a panoply of forms (Chicken McNuggets, say) which did not exist twenty years ago and which are now delivered to us by the massive fast food industry – a fact which itself points to the reality that Americans eat more and more food outside of the home (food consumption 'away from home' is, by dollar value, 40 per cent of the *average* household food budget).

The vast majority of chickens sold and consumed are broilers (young chickens) which, it turns out, are rather extraordinary creatures. In the 1880s there were only 100 million chickens. In spite of the rise of commercial hatcheries early in the century, the industry remained a sideline business run by farmer's wives until the 1920s. Since the first commercial sales (by a Mrs Wilmer Steele in 1923 in Delmarva who sold 357 in one batch at prices five times higher than today), the industry has been transformed by the feed companies, which began to promote integration and the careful genetic control and reproduction of bird flocks, and by the impact of big science, often with government backing. The result is what was called in the 1940s the search for the 'perfect broiler'. Avian science has now facilitated the mind-boggling rates at which the birds add weight (almost five pounds in as many weeks!). The average live bird weight has almost *doubled* in the last 50 years; over the same period the labour input in broiler production has fallen by 80 per cent! The broiler is the product of a truly massive R&D campaign; disease control and regulation of physiological development have fully industrialized the broiler to the point where it is really a cyborg: part nature, part machine (think *Terminator*). Our understanding of chicken nutrition

now exceeds that of any other animal, *including* humans! Applied poultry science and industrial production methods have also been the key to the egg industry. A state of the art hen house holds 100 000 birds in miniscule cages stretching the length of two football fields; it resembles a late twentieth-century high-tech torture chamber (*see* Fig. 32.1). The birds are fed by robot in carefully controlled amounts every two hours around the clock. In order to reduce stress, anxiety and aggression (which increases markedly with confinement), the birds wear red contact lenses which for reasons that are not clear reduces feed consumption and increases egg production. It's pretty weird.

Broilers are overwhelmingly produced by family farmers in the USA but this turns out to be a deceptive statistic. They are grown by farmers under contract to enormous **transnational corporations** – referred to as 'integrators' in the chicken business – who provide the chicks and feed. The growers (who are not organized into unions and who have almost no bargaining power) must borrow heavily in order to build the broiler houses and the infrastructure necessary to meet contractual requirements. Growers are not independent farmers at all. They are little more than underpaid workers – what we might call 'propertied labourers' – of the corporate producers who also dominate the processing industry. Work in the poultry processing industry, in which the broilers are slaughtered and dressed and packaged into literally hundreds of different products, is some of the most underpaid and dangerous in the country (in this morning's *New York Times*, February 9, 1998, p.A12, a US

Figure 32.1   Battery farm chickens

Government report cites almost two-thirds of all poultry processing plants as in violation of overtime payment procedures!). Immigrant labour – Vietnamese, Laotian, Hispanic – now represents a substantial proportion of workers in the industry. The largest ten companies account for almost two-thirds of broiler production in the USA. Tyson Foods, Inc., the largest broiler producer, accounts for 124 million pounds of chicken meat per week, and it controls 21 per cent of the US market with sales of over $5 billion (two-thirds of which go to the fast food industry). According to Don Tyson, the CEO of Tyson's, his aim is to 'control the center of the plate for the American people'. Pretty scary.

The heart of the US chicken industries is in the ex-slave-holding and cotton-growing South. Until the Second World War the chicken industry was located primarily in the Delmarva peninsula in the mid-Atlantic states (near Washington, DC). During the 1940s and 1950s the industry moved south and with it emerged the large integrated broiler complexes – what geographers call flexible or **post-Fordist** capitalist organization. The largest producing region is Arkansas – the home state of President Clinton – and the chicken industry has been heavily involved in presidential political finance and lobbying, including a recent case in which the Secretary of Agriculture was compelled to resign. The lowly chicken reaches deep into the White House.

The USA is the largest producer and exporter of broilers with a sizable market share in Hong Kong, Russia and Japan. Facing intense competition from Brazil, China and Thailand, the chicken industry is now global driven by the lure of the massive Chinese market and by the newly emerging and unprotected markets of Eastern Europe and the post-Soviet states. Actually, the world chicken market is highly segmented: Americans prefer breast meat, while US exporters take advantage of foreign preference for leg quarters, feet and wings to fulfil the large demand from Asia. The chicken is a thoroughly global creature – in its own way not unlike the global car or global finance.

You start with a trivial thing – the chicken as a commodity for sale – and you end up with a history of post-war American capitalism.

THE COMMODITY

One way to think about the commodity is derived from Karl Marx who begins his massive treatise on capitalism (Volume 1 of *Capital*) with a seemingly bizarre and arcane examination of the commodity, with what he calls the 'minutiae' of bourgeois society.

The commodity he says is the 'economic cell form' of capitalism. It is as if he is saying that in the same way that the DNA sequence holds the secret to life, so the commodity is the economic DNA, and hence the secret of modern capitalism. But the commodity itself is a queer thing because while it has physical qualities and uses, and is the product of physical processes which are perceptible to the senses, its *social* qualities – what Marx calls the social or value form – are obscured and hidden. 'Use value' is self-evident (this is a chair which I can use as a seat and has many fine attributes for the comfort of my ageing body) but value form – the social construction of the commodity – is not. Indeed this value relation – the ways in which commodities are constituted, now and in the past, by social relations between people – is not perceptible to the senses. Sometimes, says Marx, the social properties things acquire under particular circumstances are seen as inherent in their natural forms (that is, in the obvious physical properties of the commodity). The commodity is not what it appears. There is, then, a hidden life to commodities and understanding something of this secret life might reveal profound insights into the entire edifice – the society, the culture, the political economy – of commodity-producing systems.

Let's return to the chicken as a commodity. It has two powers. First, it can satisfy some human want (my need for a chicken curry). That is what Adam Smith called a *use value*. Second, it has the ability to command other commodities in exchange. This power of exchangeability Marx called *exchange value* or value form. Use values coincide with the *natural form* of the commodity – its chicken-ness – whereas the value form expresses its *social form*. Use values express the qualitative incommensurability of commodities – the uses of chicken can never be commensurable with the uses of a car – whereas exchange value expresses quantitative commensurability (I can exchange 20 000 chickens for one car). But commodity exchange in turn requires a universal equivalent to facilitate this quantitative commensurability: which is to say, money. The commensurability of commodities is expressed phenomenally through money, that is in the form of a price (a chicken is $1.00 per pound and turkey is $2.00 per pound which means that chicken exchanges for turkey at the rate of two chicken for one turkey). *Commodity circulation* refers to the process by which a commodity is exchanged for money which in turn permits the purchase of another different commodity.

But the basis of comparing commodities through price is only its phenomenal form. The real basis is *value*. But what is value? For Marx it turns on what he calls abstract labour – any labour whatsoever viewed as a process of consumption of human energy – so that value is expended labour. Capitalism, however, is a reality not an abstraction and rests upon commodities producing commodities in quite specific sorts of ways. More specifically, a capitalist starts with money, purchases labour power and the **means of production**, and produces commodities which are sold for money. This process generates more money for the capitalist than he began with (i.e. profit). Put differently, money and commodities circulate as expressions of the expansion of capital (not just the circulation of commodities) which rests upon the existence of a commodity-producing system and the emergence of a universal equivalent (money) (see Part II).

Exchange value is, to summarize, bound up with the particular form of social labour as it exists in a commodity-producing economy (capitalism). This particular form of social labour is wage labour – that is to say there are a class of individuals (workers) who sell their labour power to another class who organize production (capitalists). It is this aspect of labour – as a commodity, labour power, sold to someone else at a price called the wage – which is unique to a commodity-producing economy. A commodity can thus be seen as being composed of three aspects or relations: variable capital (that paid to labour as wages), constant capital (that covering fixed capital costs such as machines) and surplus value (profits). Marx tried to explain where this surplus came from and how it emerged from the disparity between the value that the workers embody in the commodities they produce and the value they require for their own reproduction.

Whether Marx's theory of value or account of the origins of profit is right or wrong is of less relevance than the fact that the commodity allows us to analyse the forms which arise on the basis of a well-developed commodity economy. Capitalism is unique because it rests upon commodities which are fictitious. Labour, capital and money are all commodities (the wage, rate of interest and exchange rates determines their respective market value) but are not produced as commodities. The commodity is the way into the problem of value (its origins and its forms) and it establishes a sort of toolkit with which we can understand something of the distinctiveness of living in capitalist societies. Property, money, value in its various forms, class interests, the circulation of commodities, they are all implied in the '**dialectical** union' of use and exchange value which the commodity contains.

Summary

- A commodity is a unity of exchange value and use value.
- Exchange value expresses the social, and hidden, form of the commodity.
- Commodity exchange requires a universal equivalent (money).

THE COMMODITY CIRCUIT

As tangible, physical things – as the embodiment of particular uses and values – commodities have lives, or *biographies*. They are made, born or fabricated; they are fashioned and differentiated in a variety of ways; they are sold, retailed, advertised and ultimately consumed or 'realized' (and perhaps even recycled!). The life of the commodity typically involves movement through space and time, during which it adds values and meanings of various forms. Commodities are therefore pre-eminently geographical objects.

Let's return again to the US chicken. It is possible to construct a diagrammatic 'biography' of the broiler from production to consumption which depicts many of the actors involved in the commodity's complex movements and valuations. This is a *commodity circuit* or a *commodity chain* (in French it is referred to as a *filière*) (Friedland *et al.*, 1981). Figure 32.2 depicts the US broiler *filière* (Boyd and Watts, 1996). At the centre of this figure is the broiler complex and the large transnational integrator – Tysons Foods, Perdue Farms, ConAgra, and so on – but there are obviously a multiplicity of other actors: the public extension systems, the R&D sector, the fast food chains, the exporters, the retailers, the service providers, the state and local government. The starting point of the commodity circuit might be the breeding units but this itself is a collaborative effort which has involved a half century of genetics research to produce breeding flocks. What is bred is a peculiar creature which has been 'industrialized' to maximize productivity (but with the danger of massive disease problems which itself generates one large part of the commodity circuit devoted to chicken 'health care'). The terminus is the consumption of chicken in a bewildering array of forms: as a complement to other foods in a TacoBell burrito, as an 'organic free range chicken' bought ready to cook in a yuppy store, or as 'mass' chicken parts destined for institutions like schools or hospitals.

Commodity circuits can depict different types of commodity chains and contrasting commodity dynamics. Figures 32.3 and 32.4 depict global commodity circuits for cocaine and for apparel and automobile sectors. In the former the peasant grower of coca leaf in Colombia is linked through a series of agents (processors, wholesalers, transporters) to the

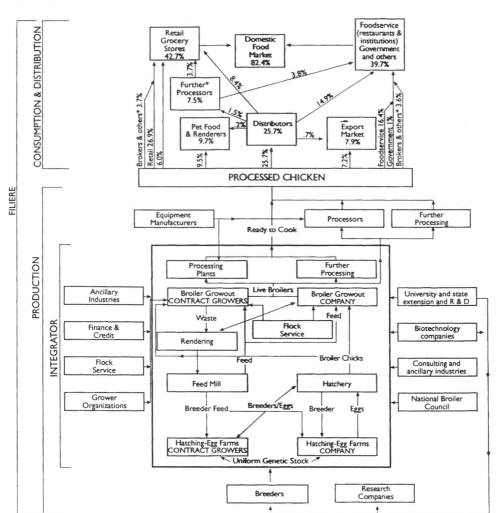

Figure 32.2 The broiler filière, c. 1999. Source: Boyd and Watts, 1996, p. 205

street dealer in say Detroit. It is an *illegal* commodity chain which links the Third World as producer to the First World as consumer. This has historically been the case for many Third World drug commodity circuits (tea, coffee, sugar) which are, however, usually typically legal and dominated by rather different agents and actors (agribusiness companies rather than the Medellin Cartel). Figure 32.4 reveals different dynamics within two contrasting commodity circuits (Gereffi, 1995). The buyer circuit for which the apparel industry is the prime case is dominated by the *retailers*. The high end fashion retailers (Armani, Donna Karan) typically produce apparel in sweat shops in the core countries or in newly industrialized countries like Hong Kong. Low-end supermarket retailers (WalMart) have producers in particularly poor and low wage countries (Sri Lanka, Philippines). Producers and retailers are often held together by complex subcontracting arrangements. In producer-driven chains conversely – the automobile is an exemplar – transnational companies (TNCs), as integrated industrial enterprises, play the central role. Toyota or Ford have inte-

grated production complexes embracing literally thousands of parent, subsidiary and subcontractor firms dotted around the world. The producer-driven commodity chain produces the 'world car' in which component parts are produced in multiple locations though the final commodity – say, the Ford Fiesta – is assembled at one single site.

My earlier discussion of the US chicken industry highlights a number of key geographical aspects of commodity circuits. First, that different actors and agents in the circuit are linked together in complex market *and* non-market relations. At particular locations within the commodity circuit there are especially dense sets of social and institutional relations: contracts between growers and integrators, co-operative relations between companies and government, and so on. Second, throughout the course of the circuit the commodity itself is differentiated in enormously complicated ways into a panoply of new products and processes. According to the industry there are now literally hundreds of 'chicken products'. Third, the process of moving through the com-

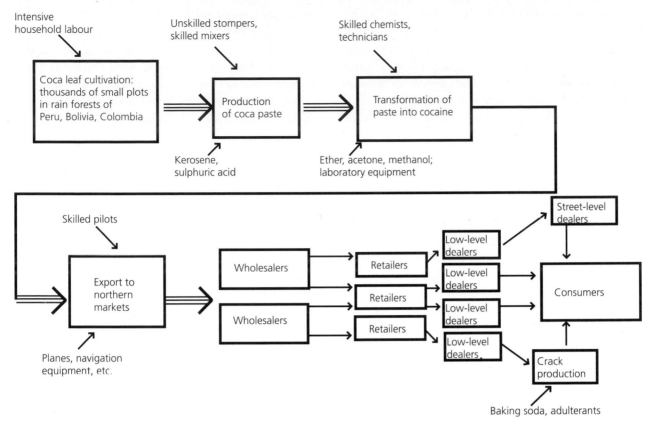

Figure 32.3 Cocaine commodity chain. Source: Morales, 1989, p. 104

Producer-driven commodity chains

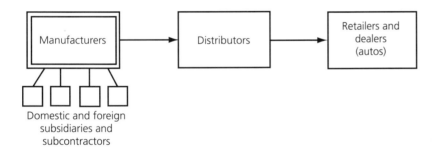

Buyer-driven commodity chains

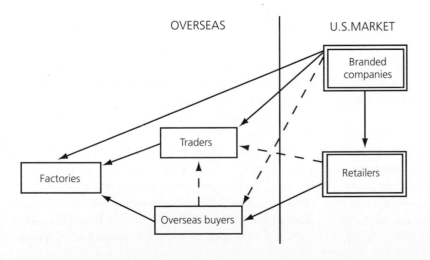

Figure 32.4 Producer-driven commodity chain. Source: G. Gereffi 1995: p.114

modity chain is simultaneously a process of adding value. Growers command a low proportion of the final product price which raises the question of who captures the value-added (in cocaine it is clearly the wholesalers and not street level dealers or peasant growers). And finally, within each commodity chain there are particular *nodes* – they can be seen as 'sinks' of especially intense activity – in which the commodity acquires particular meanings and attributes (these are values but not necessarily economic values). In the broiler *filière* for example these qualities may have to do with freshness, or the acquisition of a brand name (the 'Rocky Road' chicken), or the attachment of 'quality'. Quality is typically about status (think of the cachet of Michael Jordan Air Nikes), and the **semiotics** of the commodity (this is a *real* Stilton cheese produced *traditionally* in Leicestershire). A commodity circuit can, then, display both the space–time attributes of the commodity – many of which are now global of course – and also of what Marx called the 'social' (and partially hidden) qualities of commodities: their value, their meanings and their fetish qualities.

COMMODITIZATION/COMMODIFICATION

[T]he idea of a self-adjusting market implied a stark utopia. Such an institution could not exist for any length of time without annihilating the human and natural substance of society.

(Karl Polanyi, The Great Transformation, 1947)

The process by which everything becomes a commodity – and therefore everything comes to acquire a price and a monetary form (*commoditization/commodification*) – is not complete, even in our own societies where transactions still occur outside of the marketplace. But the reality of capitalism is that ever more of social life is mediated through and by the market. Karl Polanyi referred to this process as the *embedding* of social relations in the economy. On the one hand, he said more of social life is embedded in the logic of a commodity economy – industries are given permits to pollute which can be bought and sold, Nature is patented by private companies – and, on the other, the market itself, if left to its own devices, becomes *disembedded* from social institutions.

Karl Polanyi was concerned to show that societies dominated by the self-adjusting market, in which individuals relentlessly pursued their own interests as Adam Smith suggested, would be no society at all. Rather, every person is pitted against each other in a state of quasi-war. Smith recognized the costs of unbridled accumulation, and saw civil society as the

necessary saviour of a market system whose powers he so admired. If the genesis of market-regulated societies carries the prospect of disembedded markets and economically embedded social relations, they remain tendencies rather than inevitabilities. The case of the chicken industry revealed of course that in a highly competitive and market-driven broiler industry, markets and commodities are indeed socially embedded. Economics is, as Polanyi put it, *an instituted process*. This institutedness takes the forms of social alliances, networks and studied forms of trust between actors in the commodity system (see also Chapter 10). Firms build up relations of trust and cooperation between one another; the relationship between grower and integrator is contractual, and not a pure market relationship. The vertically integrated, patrimonial Korean conglomerates (*chaebol*) such as Daewoo and the Taiwanese flexible, contractually linked family firms are the socially embedded forms of market, commodity-producing behaviour associated with the so-called Asian miracle. Commodities are always fashioned in institutional and cultural ways.

In the far-flung corners of the globe, there are societies which are not commoditized at all, or at least the pursuit of 'commercial advantage' represents a minor part of their social existence. Some contemporary Indian communities in the Amazon for example, or hunter–gatherer communities in Zaire, produce almost nothing for the market and buy little in the way of consumer goods. Some back-to-the-land communes in California also aspire to self-sufficiency. However, even these non-market (or non-capitalist) societies are typically commodified in some way. When I worked among isolated pastoral nomads in West Africa in the 1970s – small mobile families who depended entirely on livestock for their survival – these seemingly traditional communities did view cattle and other animals as property, and indeed would sell limited numbers of animals, particularly during the dry season when lactation rates of their animals had fallen due to the deterioration of pasture, in order to buy grain, and tea and sugar.

Statistically speaking, one of the largest classes of people in the world is the peasantry, and they are defined specifically by their *partial commoditization*. Peasants own the means of production – they directly work their land with their own family labour which means that they do sell their labour power as a commodity. But peasants *are* involved with the market to some degree, selling part of what they grow (often export crops such as cotton or tea) to acquire money to buy clothes, pay taxes and cover school fees for their children. This partial commoditization can have some unusual consequences. Henry Bernstein (1978) pointed out that in such circumstances, a peasant family may produce commodities in order to gain cash but should the price of this commodity fall (the price of bananas on the world market falls by 40 per cent), the peasant may be forced to either produce more of a commodity whose price is falling or work harder just to meet the irreducible family needs. For a family with a small plot of land this may mean working longer and harder and exploiting the soil in order to, as it were, stay in the same place. Bernstein referred to this conundrum of price squeezes (commodity prices falling) and partial commoditization (household enterprises with irreducible consumption goals) as '*the simple reproduction squeeze*'.

Polanyi's 'great transformation' was the process by which economy and society were separated during the course of the rise of the self-adjusting market: the emergence of the market economy as a *totality* in contradistinction to the patchwork market economy prior to the Industrial Revolution. The reach of the market – which is to say the extent to which everything has become a commodity – has deepened since the 1750s. What Marx called the 'vulgar commodity rabble' now encompasses much of what we do and feel. Human life has become dependent upon the market and commoditization inevitably charges inward into the refuges of social life. Jürgen Habermas (1993) calls this the 'colonization of the lifeworld'.

But this process is always resisted, especially in societies undergoing a radical and rapid transition to a market economy. The **moral economy** – the embeddedness of economic relations – in pre-capitalist or partially capitalist societies (whether peasant communities in contemporary Borneo or in late eighteenth-century France) breaks down slowly and unevenly in face of the onslaught of commodity production (Thompson, 1991). In my peasant community in Nigeria, for example, land was rarely sold in large part because of the cultural and spiritual meanings attached to the land and the anti-social character of land sale (this remains an issue for example in Native American Indian communities in the United States). Poor families would steadfastly deny that they sold their labour because of the shame surrounding the fact of being perceived to be not self-sufficient. These are elements of a larger moral economy in which there is some effort to guarantee subsistence rights and also strong sentiments for a just price (for bread, for example). The process by which this moral economy has been undercut by the commodity economy – by growing commoditization – has often generated social conflict and strife. Some have argued that the great peasant rebellions of this century – Tonkin in the 1930s, Mexico earlier in the century – were efforts to defend the moral economy of the peasantry against the onslaught of the world market and the irrepressible logic of the commodity economy (Scott, 1976; Thompson, 1991). Of course the moral economy is still not entirely dead; something of this remains in English villages or the crofting communities of the Hebrides. The efforts by the state to provide services (i.e. unemployment relief) outside of the market and through moneyless exchange can also be seen as an effort to *decommoditize* some realms of social life (Offe and Heinze, 1992).

COMMODITY FETISHISM AND THE COMMODITY SPECTACLE

In 1993, in a media stunt for her animal rights book, Rebecca Hall offered four men $2500 each to live like battery hens for a week – in other words barefoot in a small wire cage with a sloping floor, with 24-hour light, automated food delivery and a cacophonous noise (*see* Fig. 32.5). They lasted 16 hours. Two years later one of the major US broiler firms hired a nationally known chef to use 'fresh' oven-ready chicken to play bowls outside of the US Congress, to demonstrate the fact that its competitors were in fact selling purportedly fresh chickens that were frozen as hard as tita-

Figure 32.5 Four men attempt to live like chickens in Rebecca Hall's cage. Credit: Martin Argles/Guardian

nium. Or take a look at the advertisement in Figure 32.6 for an 89 Male broiler: top 'livability', superior growth, excellent efficiency. Chicken as machine.

These chicken tales are each speaking to quite different aspects of the commodity under capitalism. A fetish is a material object invested with magical powers. Marx invoked *commodity fetishism* to describe the ways in which commodities have a phantom objectivity. The social character of their making is presented in a 'perverted' form. By this he meant a number of complex things. First, that the social character of a commodity is somehow seen as a natural attribute intrinsic to the thing itself. Second, that the commodities appear as an independent and uncontrolled reality apart from the producers who fashioned them. And third, in confusing relations between people and between things, events and processes are represented as timeless or without history, they are **naturalized**. Another way to think about this is that commodity production – the unfathomable swirl of commodity life – produces particular forms of **alienation** and **reification**. Let me elaborate.

We come to accept the creature we buy as an oven-ready chicken as a natural product which stems from its use-value (as food). In fact, it is a sort of machine, something created by science to be an input (of particular proportion, colour, efficiency, and so on) into industrial manufacture; something which, far from being natural, is a social artefact containing many and complex forms of value. Moreover, the social value of the chicken appears, as it is under capitalism, in terms of relations between things (growth, efficiency, white meat). In our society virtually all of our existence appears as a thing – it is reified – but these reified things interact with us to give the impression that the social really *is* natural. The chicken futures market is 'up' this week – as though the market has a life of its own – which is confirmed to me because my shares in Tysons Foods carry increased dividends. This confusion or obfuscation of relations between people – the huge number of people involved in the community filière – with relations between things is central to the alienation rooted in a world in which everything is for sale, and everything is a thing.

In his book *Society of the Spectacle* (1977) Guy Debord argues that in a world of total commodification, life presents itself an as immense accumulation of *spectacles*. The spectacle, says Debord is when the commodity has reached the total occupation of social life and appears as a set of relations mediated by images (*see* Fig. 32.7 and Chapter 12). The great

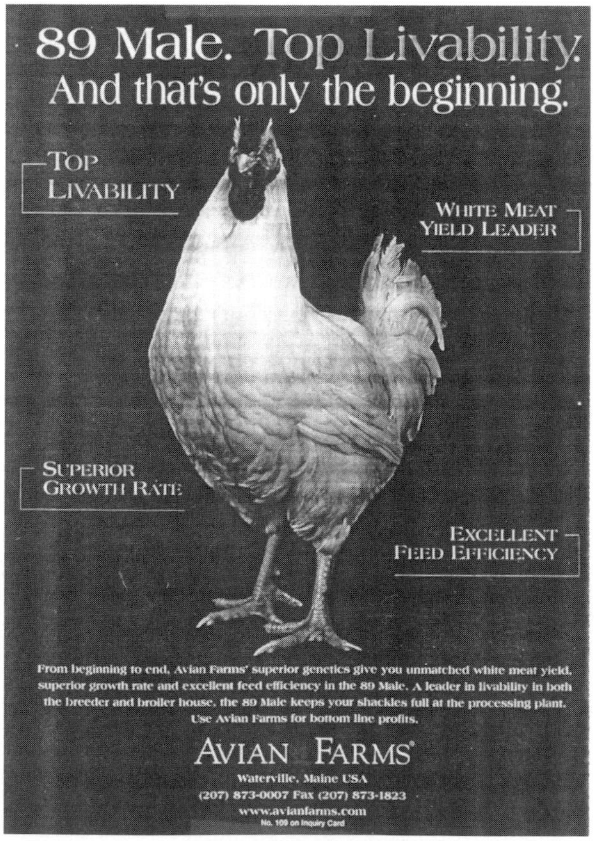

7

Figure 32.6 The Industrial Chicken advert. Source: *Broiler Industry,* August 1997

Figure 32.7 Society as spectacle. Source: John Berger, 1972: *Ways of Seeing,* p. 129, Penguin. Credit: © Sven Blomberg

world exhibitions and arcades of the nineteenth-century were forerunners of the spectacle, celebrating the world as a commodity. But in the contemporary epoch in which the representation of the commodity is so inextricably wrapped up with the thing itself, the commodity form appears as spectacle, or as a spectacular event, whether four men trying to be chicken or a chef playing bowls with a frozen broiler.

Once they leave the confines of their makers, commodities take on a life of their own.

Summary

- A commodity circuit is a methodological device to reveal a commodity as an organized space-time system.

- Commodification always involves complex processes of economic and social embedding and disembedding.

- Commodity fetishism refers to the ways in which alienation and reification operate in commodity-producing societies.

COMMODITIES AND THE IMMENSE COSMOS OF CAPITALIST ACCUMULATION

We began with the commodity as a trivial thing and have ended with a world of commodities that 'actually conceals, instead of disclosing, the social character of private labour, and the social relations between the individual producers' (Marx, *Capital*, Vol. 1, pp. 75–6). But this hidden history of the commodity allows us to expose something unimaginably vast, namely the dynamics and history of capitalism itself. The proliferation of commodities as exchange values presupposes a universal equivalent, that is, money as an expression of value (see Chapter 11). But this itself has its own preconditions – property is implied by money-mediated exchange and correlatively the entire superstructure which sustains property relations – and poses the knotty problem of the relations between price and value. In exploring the question of value we have seen that it turns in large part on the peculiarities of labour and money being themselves commodities. Whether Marx was right that labour is a special commodity which has the ability to produce greater value than it has, the relations between commodity, price and value do nonetheless lead inexorably to the centrality of class relations between labour and capital as a fundamental aspect of capitalism. This itself poses the question of how capital and wage labour come to be and how they are reproduced under conditions of contradictory interest (profit versus wage). Wow. We may as reasonable men and women differ in our accounts of how capitalism as a class system secures the conditions of its own reproduction, and at what cost, but the commodity as its 'cellular form' is surely one of the keys to unlocking the secrets of what Max Weber (1958) called the capitalist cosmos.

Further reading

- Appadurai, A. (ed.) (1987) *The social life of things*. Cambridge: Cambridge University Press.
- Buck-Morss, S. (1990) *The dialectics of seeing*. Cambridge, MA: MIT Press.
- Comaroff, J. and J. (1990) *Ethnography and the historical imagination*. Boulder, CO: Westview Press.
- Harvey, D. (1982) *The limits to capital*. Oxford: Blackwell.
- Mintz, S. (1985) *Sweetness and power*. New York: Viking Books.
- Ollman, B. (1972) *Alienation*. Cambridge: Cambridge University Press.
- Pred, A. and Watts, M. (1992) *Reworking modernity*. New Brunswick: Rutgers University Press.
- Taussig, M. (1980) *The devil and commodity fetishism in Latin America*. Durham, NC: University of North Carolina Press.
- Weiss, B. (1997) *The making and unmaking of the Haya lived world*. London: Duke University Press.

CHAPTER 33

The media

James Kneale

INTRODUCTION

The mass media seem rather trivial compared to other geographical issues like the exploitation of the South by **neo-colonialism**. In fact, what have they got to do with Geography at all? In this chapter I want to argue that the media do have a place in our studies and that they do have geographies which can help us to understand wider social and cultural issues – including the example above.

The media are significant because they are so commonplace. The UK General Household Survey reveals that in 1977 97 per cent of the adults interviewed had watched television in the four weeks before the survey. From 1987 onwards the figure has been 99 per cent. Again, in 1977 54 per cent of the adults interviewed had read a book in the four weeks before the survey; by 1993 this figure was 65 per cent (OPCS, 1977, 1987, 1993). Stephen Heath asks 'Can anyone in our societies be outside television, beyond its compulsions?' and the same could be said of other mass media like print (1990: 283).

This significance is not restricted to Western(-ized) societies, either; the media play a vital role in **globalization** (see Chapter 3). They help to diffuse products and ideas across space, and act as global **commodities** in their own right. They transmit US soaps to distant countries and bring the wider world into Western homes. The television in the living room is one place 'where the global meets the local' (Morley, 1991) and from Marshall McLuhan's 'global village' (1964) to David Harvey's 'time-space compression' (1989) the media are credited with shrinking the world through faster communications.

The global reach of the media is extensive (see Silverstone, 1994: 88–92 for a review) but we must not assume that they have infiltrated every corner of the globe. As Doreen Massey points out (1994), local experiences of globalization are very different, and the media are no exception. The homogenizing power of Western media should not be exaggerated; India's 'Bollywood' film industry and public access cable shows in the USA are 'local' challenges to corporate film and television. And there is also a convincing argument for the media as preservers of national and international identity – the BBC as the voice of Britain, or the way that European broadcasting may constitute a pan-European identity (see Chapter 28). Even such a cursory consideration shows the media to be caught up in a number of interesting geographical issues.

We should also consider the arguments put forward by Jacquie Burgess in a key piece of research in media geography (1990). She argues that we need to get to grips with the media because they are an important source of environmental knowledge (see Chapter 15); because advertisers use **representations** of nature to sell commodities; and because visual media depict landscapes to provide realism or convey a particular feeling. Burgess concludes that 'media texts are, in fact, saturated with geographical messages and meanings' (1990: 141), and this observation underlies one strand of geographical research into the media. Many media texts have a setting – newspaper stories are tied to a location, TV programmes must have some kind of background – which passes on geographical information to its consumers. The most obvious examples are holiday programmes which effectively 'sell' places and which make us 'armchair travellers'. Uncovering the **ideology** of mediated representations has been one task geographers have become interested in (see for example Burgess and Gold, 1985).

There is another way we can look at media,

though. Burgess notes that media texts have to be produced and consumed; they are part of a social process of meaning creation. Where do the meanings of these texts come from, and where do they end up? There is another kind of social geography of the media, which sees that they connect up different spaces, flowing through space to diffuse ideas to different groups of people in different places. The final question which could be asked here is: do the media actually play a part in making the spaces they flow through?

To examine these ideas, I'll be looking at a historical example first. Once some of the possibilities of this approach have been drawn out, the chapter will concern itself with explorations of the spaces of the contemporary media.

GEOGRAPHIES OF THE BOOK

Recent histories of the book since the printing revolution of the mid-fifteenth century have indicated that print media can tell us a great deal about the geography of knowledge as well as wider social change.

One aspect of these histories which I find fascinating is the fact that until the sixteenth and seventeenth centuries, most people who could read did so out loud, and often to an audience. The significance of this is that the shift to private and silent reading was also a change in the way that people related to one another; public reading could form communities of readers and listeners (*see* Fig 33.1). In particular, 'reading aloud was a way of structuring family life' (Chartier, 1989: 152) since it served to reproduce the authority of parent and patriarch. Private reading contributed to the changing nature of the family, though this was a complicated process; silent reading allowed individuals to cut themselves off from one another while reading. This even created new spaces within the home, like the private library and the nineteenth-century study, which were places for seclusion where men in particular could retreat from the outside world.

Beyond these kinds of changes to domestic places and social interaction, histories of the book can tell us something about wider geographies of knowledge. For example, the European book market after 1500 illustrates how particular areas became well supplied with books and information, while others remained outside of this circulation of media (Martin, 1994). To broaden the range of their stock, booksellers developed collective networks of exchange with other sellers, sometimes bartering books for other books. Regional and then national networks began to

Figure 33.1 Engraving by Marillier of reading in public in eighteenth century France. Source: A. Manguel 1996: *A History of Reading*. London: HarperCollins. Credit: © Bibliotheque Nationale/Paris/Archives Ceuil

appear, like the one in northern France organised by Parisian printers (ibid.: 249).

In Germany the rise of a national network followed the Reformation; the immense appetite for books and tracts for or against Martin Luther meant that popular titles outstripped scholarly ones, and booksellers' networks had to extend into rural areas through peddlers (ibid.: 253–4). Because of this 'a cleavage opened between regions won over to Lutheranism (or Calvinism) and southern lands that had remained faithful to the Roman Church and that would be inundated with a mass literature of their own during the Catholic revival' (ibid.: 254).

Later mappings of the circulation of books in society show a similar picture. Robert Darnton (1979) shows that the eighteenth century distribution of Diderot's *Encyclopédie* indicates that most purchasers would have been well-educated members of the urban upper and middle classes (*see* Fig. 33.2). Nigel Thrift suggests that this meant that knowledge about the world was strongly localized (both spatially and socially), forming a map of places with 'stocks' of information and spaces with relatively little (1996).

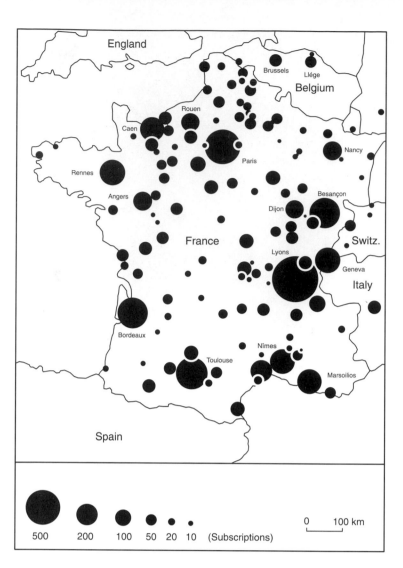

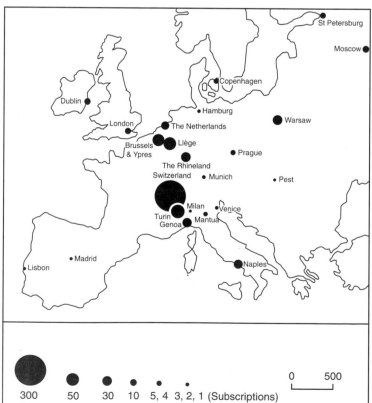

Figure 33.2 Diffusion of Diderot's Encyclopédie in and outside France, 1777–82. Source: N. Thrift 1996: *Spatial Formations*. London: Sage, 112–13.

And since the *Encyclopédie* contained a treasury of **Enlightenment** thinking, this revolution in thought was also highly unevenly distributed.

The Enlightenment itself was built on books – rediscoveries of classic texts and the production and circulation of new ones. The men and women whose writing was part of this great shift – not only in how people thought about the world but how they related to it – recommended books to one another, wrote them for one another, exchanged them and passed on information about booksellers and publishers. The movements of these books across Europe (and outside it) is part of a wider history of social change, but these changes themselves partly depended on the accumulation of knowledge by a particular class of people.

Summary

- Changes in print culture played their part in the 'privatization' of the household.

- European networks of bookselling were geographically uneven and socially unequal, and show us geographies of print-based knowledge.

- The geography of the book can help us to understand the nature of the social changes associated with the Enlightenment.

CIRCUITS, FLOWS AND SPACES

How exactly do the media link places together? Perhaps the best way of thinking about this is through Richard Johnson's 'circuit of culture' (1986: 283). Figure 33.3 shows how Johnson illustrates the way that the meaning of a media text changes as it moves through society. Let's take the British BBC soap opera *EastEnders* as an example. This programme does contain geographical meanings as a text: the programme begins and ends with a map of the East End of London, and is situated in Walford, a mythical part of this area; it is also restricted to particular public and private spaces within this setting, chiefly Albert Square, the Queen Vic, and so on. But this text is only one of the 'moments' in the circuit of this particular programme.

EastEnders is produced by a whole host of people – scriptwriters and continuity people, directors and sound technicians, and many others. The finished product is the text, the transmitted programme. The audience for *EastEnders* – which is exported around the world – consume the text, making their own meanings from it. These meanings are then taken away from the TV set, into the lived culture; viewers talk about it on the bus or at work, for example. The

circuit then begins again, as producers draw upon this culture to see how popular the programme is, commissioning market research to gather feedback from viewers: is a character working out? Is a storyline unpopular?

Obviously, we can use the circuit to track transformations in the geographical meanings of media texts, as Burgess suggests. But it also shows how the media can join spaces together. *EastEnders* flows through scriptwriters' studies at home, BBC offices, Pinewood Studios and onto the audience's screen; in living rooms, kitchens, bedrooms and pubs the programme is consumed; and in countless other spaces it is discussed, criticized, and picked over.

Seen in this way, media texts are just like other commodities: they flow through social spaces, sometimes altering them, sometimes being transformed themselves. Writing about the history of the book, Roger Chartier notes 'their peregrinations ... about the social world' (1994: x). It is these wanderings that I want to concentrate on now, rather than the geographical meanings of texts. Two geographies of the media are particularly interesting: the household, because it tells us something about the way that the media affect particular 'bounded' spaces; and second, some of the networks which connect these spaces together and flow across their boundaries.

Summary

- Media flow around the 'circuit of culture', from production to consumption and into wider social life.

- They therefore connect different places and people together as they flow through them.

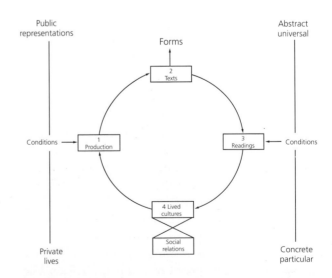

Figure 33.3 The circuit of culture. Source: R. Johnson 1986: The story so far. In D. Punter (ed.) *Introduction to Contemporary Cultural Studies*. Harlow: Longman.

- Media shape spaces and places, but spaces and places shape media.

DOMESTIC MEDIA

Having already noted the globalizing effects of the media, it might seem rather parochial to concentrate upon the home – and the Western home at that. However, most of the research which examines social spaces of the media is based upon the households of the 'First World'. Much of this work is of interest to the geographer who wishes to develop an understanding of the inter-relationships between the media and social space.

There is a very important reciprocal relationship between domesticity and the media in these household spaces. Take television, for example:

Television is a domestic medium. It is watched at home. Ignored at home. Discussed at home. Watched in private and with members of family or friends. But it is part of our domestic culture in other ways too, providing in its programming and its schedules models and structures of domestic life, or at least of certain versions of domestic life.

(Silverstone, 1994: 24)

This domestic life is already structured by power relationships, often those of the family.

When media consumption takes place in the family ... it takes place in a complex social setting in which different patterns of cohesion and dispersal, authority and submission, freedom and constraint, are expressed in the various sub-systems of conjugal, parental or sibling relationships and in the relationships that the family has between itself and the outside world.

(ibid.: 33)

Domesticity depends especially upon particular forms of gender relations and upon a gendered division of labour; 'Men dominate the television at home because home is where they relax, where they are looked after' (ibid.: 39). Many women, on the other hand, have to fit their media consumption around a schedule of domestic housework: watching TV with the children not because they want to watch it but because they might be needed, or keeping one eye on a daytime magazine programme while preparing a meal.

We can look at the relationships between TV and domesticity in a number of ways. First, TV, like radio before it, produces a routine for living because it is scheduled on an hourly, daily, and weekly basis. In particular these media are associated with women's

domestic work: 'the schedules of radio and television were not arbitrary, but were designed in accordance with certain structures associated with housework itself' (Seiter et al., 1989: 229). TV schedules still fix our days. Though the family meal is an increasingly rare occurrence, household members may still gather to watch the news, evening soaps or sitcoms. These meetings are like the public reading of the early modern period; they reproduce the structures of the family and domestic life around the 'electronic hearth' (see Fig 33.4). Finally, the content of some programmes reinforces domesticity; soaps and sitcoms usually rely on family life for their material, even when the families represented are far from happy ones.

There are a number of objections that could be made to these arguments. First, I've been using the words 'family', 'household' and 'domesticity' fairly loosely; the overlap between them is not as neat as it might be. Silverstone notes, for example, that only one-third of UK households consist of families in the sense of two adults plus children; other households are occupied by adults sharing, by lone parent families, or by one adult living alone (1994: 33). As a result we shouldn't assume that the consumption of domestic media is dominated by nuclear families; there are different social relationships within different kinds of households.

Second, haven't technologies like the VCR weakened the power of television to structure domestic time, by allowing us to watch what we want when we want? Well, to some extent this might be true, but research has shown that the VCR is not an innocent technology. Like the remote control (Morley, 1986), it tends to be controlled by men, so that women have less freedom to use it to alter their schedule (Gray, 1992). And to a large extent, TV programming has ignored the ubiquity of videos and continues to follow the division of the day into 'work' (daytime programmes, aimed primarily at women and, at certain times, children) and 'leisure' (evening programmes, which include 'family' TV as well as news, sport, and documentaries, which tend to be more popular with men).

Finally, it should also be pointed out that the media can also threaten domesticity. There have been a number of recent 'moral panics' concerning television in the UK; the satirical *Brass Eye* was condemned for wasting Parliamentary time over the made-up drug 'cake', and *Teletubbies* has been castigated for 'dumbing-down' toddlers' telly. The media are also used to launch panics over the state of the family: *Child's Play III* was charged with inspiring the Jamie Bulger murder, but the killers' families were also criticized for allowing them to watch the video. What lies behind many of these fears is the way the media

Figure 33.4 Who is in control? Credit: Angela Martin

connect the household with the wider world, or how it brings the public into the private. The reaction to this threatening material is usually to demand that it be kept away from susceptible members of the audience, usually children, through a number of regulatory mechanisms: the nine o'clock watershed, video certificates, the editing of films shown on television, child locks on VCRs, and so on.

The media can also highlight and antagonize gender relations within the home. A good example comes from Janice Radway's *Reading the Romance* (1984), a study of women readers of romantic fiction. For many of them, the pleasure of reading was accompanied by guilt. Unlike watching TV, reading demands absorption in the book, and therefore cannot usually be done in conjunction with domestic work. Radway's interviewees therefore valued their reading as 'times for oneself' but worried that it meant neglecting their 'duties' – a feeling often strengthened by the hostility of their partners towards reading. The privacy of reading is, in fact, often an ambiguous pleasure:

> Something in the relationship between a reader and a book is recognized as wise and fruitful, but it is also seen as disdainfully exclusive and excluding, perhaps because the image of an individual curled up in a corner, seemingly oblivious of the grumblings of the world, suggests impenetrable privacy and a selfish eye and singular secretive action
>
> (Manguel, 1996: 21)

A less guilty group of women can be found in Henry Jenkins' study of US science fiction television fandom (1992). Discussing their first encounters with the TV

series *Beauty and the Beast*, members of the programme's Boston area fan club clearly identified its importance in terms of 'neglecting' their domestic work:

> I was sitting there and I didn't get any work done! My chin was on the ground! I couldn't believe what I was seeing.
>
> (ibid.: 58)

> I sat there and I sat there and I'm like this and my kids come in and I remember them going past me vaguely but I didn't let it distract me at all.
>
> (ibid.: 58–60)

In these and other ways the relationship between the media and domesticity can be seen to be rather complex.

Summary

- Domestic media consumption involves very specific social geographies.

- Domestic media fit into and reproduce the inequalities of gender, age and class which make up 'the family' and 'the household'.

- The media can also threaten the stability of the household either through their interruption of domestic schedules or their transmission of seemingly 'threatening' material.

AUDIENCES AS SOCIAL NETWORKS

As Silverstone reminds us, households are 'leaky' social spaces in the sense that they aren't enclosed and self-sufficient. Members must leave the home to work, socialize, and so on. It is at this point that media are carried into the lived culture of Johnson's circuit.

Media texts flow away from the original site of consumption and are re-used elsewhere, in social networks outside the household. One of the best examples of this comes from Dorothy Hobson's interviews with women soap viewers, where she points out that 'talking about soap operas forms part of the everyday work culture of both men and women. It is fitted in around their working time or in lunch breaks', becoming part of their experience of work spaces (1989: 150). As Silverstone points out, 'the life of the soap opera pervades life in the home, life over the garden wall, on the street, in the pub, in the canteen and in the factory' (1994: 74).

In fact these social relationships can actually lead people to watch television when they might not have otherwise bothered. Hobson interviewed one woman who began watching *EastEnders* because her colleagues discussed it so much at work:

> When a storyline is so strong that it is a main topic of conversation it is reason enough to get someone watching it so as not to be left out of the conversation which takes place at work
>
> *(1989: 161)*

These kinds of conversations about media take on a new significance in fan culture, where relationships between fans become organised to the extent that they operate as important social networks in their own right. The extent of these networks is much further than a chat over the garden wall; fans receive letters and fan produce from other countries around the world. And these networks are formed primarily because of the media, as opposed to the media fitting into networks established at work or elsewhere. Jenkins discusses one aspect of this:

> The exchange of videotapes has become a central ritual of fandom, one of the practices helping to bind it together as a distinctive community . . . No sooner do two fans meet at a convention than one begins to offer access to prized tapes and many friendships emerge from these attempts to share media resources
>
> *(1992: 71).*

A similar network underpins the riot grrrl subculture; young women produce fanzines and distribute them to one another, becoming friends in the process (Leonard, 1998). The fanzine (*see* Fig 33.5) allows a sense of community to develop: 'whilst [they] may reach people in other countries, their content and scale of production give the impression of conversing with a close group of friends' (108).

Both riot grrrls and SF fans also use the Internet (see

Chapter 34); but both media SF fans (mostly women; literary SF fans are mostly men) and riot grrrls express suspicion of this forum because of its masculine character. Jenkins usefully compares *Star Trek* fandom and *Twin Peaks* Internet discussion groups to show that the two communities seek different things from these programmes, talk about them in different ways, and form different kinds of social relationships with them precisely because the former group is made up mainly of women while the latter are mostly men (ibid.: 77–9; 109–15). Here **gender** is crucial to an understanding of these dispersed geographies.

One of the challenges facing geographers interested in the media is evaluating the role that they play in constituting social networks, and the ways that these networks cross space and link places.

Summary

• Media flow out from households into places of work and leisure.

• Talking about the media forms an important part of social relationships outside the home.

• If these networks are more formalized, as is the case for fan culture, they can acquire a global reach.

Figure 33.5   Riot Grrl zines. Source: T. Skelton and G. Valentine (eds) 1998: *Cool Places: Geographies of Youth Cultures*. London: Routledge. Credit: © Rob Strachan

CONCLUSION

The media saturate the social spaces of many people around the world with meanings. These meanings are of interest to geographers because they often depict spaces or landscapes, and can be interpreted as ideological representations. However, they also create and are created by other social geographies. In particular they help to construct domestic or family spaces, where they reproduce (and sometimes transform) gender relations as well as ideas of the 'public' and the 'private'.

These spaces are not isolated from the outside world, however; by talking about media texts, consumers make them part of their lives outside the home. Many social relationships involve discussions of the media; some are initially built upon a shared interest in a programme, while others draw upon the media as one of many resources for producing social spaces. In some cases, like fan communities, this goes beyond talking about media at work, and the geographies created by video exchanges, writing to fanzines, or going to conventions span international distances.

I've discussed these media geographies and the interpretation of the geographical meanings of texts as if they are separate concerns. However, to fully understand the meanings of texts, we need to know how their meanings are transformed as they circulate through space and around the circuit of culture – and we don't know enough about this yet. There's still a lot to do – but it's a very interesting and challenging project.

Further reading

• The 'Landscapes as Television and Popular Culture' special number of *Landscape Research* (1987) **12**(3).
Includes a number of useful introductions to the meanings of media texts, particularly Burgess J, 'Landscapes in the living-room: television and landscape research', pp1–7, and Higson A, 'The Landscapes of Television', pp 8–13.

• Burgess, J. (1990) The production and consumption of environmental meanings in the mass media: a research agenda for the 1990s. *Transactions of the Institute of British Geographers* **15**: 139–61.
A strong argument for studying the media, and a clear introduction to these issues, though it does concentrate upon environmental meanings.

• Morley, D. (1986) *Family television: cultural power and domestic leisure*. London: Comedia.
One of the studies which inspired 'ethnographies' of the domestic audience; extremely interesting material on the contested nature of household viewing.

• Morley, D. (1991) Where the global meets the local: notes from the sitting room. *Screen* **32**, 1–15.
One of the best short articles written about the media and local-global relations; worth reading to get a feel for work in media studies which discusses geographical issues.

• Silverstone, R. (1994) *Television and Everyday Life*. London: Routledge.
Quite taxing in places, but a sustained investigation which has a great deal to say about domesticity and the role of the media in everyday social space.

CHAPTER 34

Cyberspace and cyberculture

Ken Hillis

INTRODUCTION

Imagine you need a reference for a geography paper due at nine the following morning. The deadline looming, you head for the computer lab, and through a web-based browser search through on-line abstracts and articles. You navigate between different sites, some located on computer servers halfway around the world. After finding the perfect citation, you enter an Internet Relay Chat (IRC) chat room and talk with friends you've made there.

For some, this scenario is a fantasy of the wealthy. For others, however, it is already a daily practice that is contributing to a reformulation of how individuals and communities think about the real world. Computerized data bases, web surfing, entering a chat room, and even sending an e-mail raise the issue of an imaginary or metaphorical space called **cyberspace**. Cyber means immaterial, virtual, nowhere, not really there. The term space reflects an implicit human understanding that, partially because we have bodies, we need to be somewhere. We organize human relationships by using space, which seems a fundamental component, if not principle, of reality itself. Merging cyber and space suggests something novel, yet it also allows new and unfamiliar technologies and social practices to seem less threatening and unknown.

New information technologies communicate all manner of information, in digital form, across space, including aspects of individual identity. They also introduce possibilities for new forms of human expression and are part of a larger cultural dynamic by which images, signs and façades – representations which are part of reality – come to stand in for *all* of reality, space included. Simulation seems as good as being there. Deep-seated, taken for granted assumptions about what is real are unsettled by this confusion

between images and what they represent. When web surfing students, for example, travel through cyberspace to foreign places without leaving their homes, they acquire a form of cultural knowledge partially, if not wholly, disengaged from the political, material, and social realities which exist in those places. They also experience a sense of inclusion in a wider public sphere made up of information, along with a feeling – supported by the idea of an interactive network – that they can control this world even though data transmission is open to transcription and review by corporate and/or State agencies.

Cyberculture, the move toward all things cyber and virtual, refers to the various cultural practices that occur in cyberspace. Cyberculture is an ongoing process; its possibilities are still being explored, and the implications only beginning to be mapped. This chapter discusses the relationship between cyberspace and cyberculture and looks at how implicit promises for cyberculture help fuel rapid acceptance of new information technologies and social practices. These technologies, along with how industry and advertising suggest we use them, influence how people think about time and space, history and culture; they contribute to a reorganization of the ways we live, talk to one another, and think about ourselves – when alone, in private, and in public. The pervasive hype surrounding new technologies, however, can be so persuasive that many individuals buy into the hype without an adequate consideration of what is at stake.

THE PROMISES OF CYBERCULTURE

Cyberspace and cyberculture confirm each other's existence. All cultures and cultural activity require a space for interaction to take place among their members, and cyberspace is a metaphor for the non-geographic place in which digital interactions happen between people, and between people and machines. It's a kind of electronic nowhere – the wires that link people sitting at home alone chatting to people perhaps living nearby or across the Atlantic. It's the virtual reality you enter when you put on the VR helmet at the arcade. It's the space you set up when you telephone a friend. In short, it's where you are when you deal with electronic communication technologies.

Cyberspace is also where cyberculture is produced. Now, culture is both an idea *and* a material practice. For Raymond Williams (1983), culture is 'one of the two or three most complicated words in the English language'. Williams isolates three modern usages: first, 'a general process of intellectual, spiritual and aesthetic development'; second, 'a particular way of life, of either a people, a period or a group'; and third, 'the works and practices of intellectual and especially artistic activity'. One may also speak of material culture – the objects a culture produces. Culture includes how Patsy Stone dresses on *Absolutely Fabulous*, and the kinds of food we eat. It's attending church, synagogue or mosque. Watching a movie or an opera is part of culture, as are the ideas in which we believe. Culture is all of these things, and more. Broadly defining the idea of culture permits examining all sorts of human activities and analysing and critiquing them. It's useful to bear in mind that the cultures we live in influence the ways we interpret or read the world. If we spend more and more time in cyberculture, jacked into the quasi-private, quasi-public Internet and its world of images, what kinds of influences might this have, for example, on how we treat homeless people living their lives in public spaces?

The word cyberspace was coined by William Gibson in 1984 in his science fiction bestseller *Neuromancer*. Gibson defines cyberspace as a 'consensual hallucination' (ibid.: 51) – an important idea I'll return to below. Douglas Rushkoff (1994) writes that in cyberspace 'the limitations of time, distance and the body are meaningless' (ibid.: 3); it's clear he's implicitly speaking about a utopia. He argues that cyberspace and the technologies that take us there will transform our lives in wonderful ways. The idea of utopia comes from Thomas More's book *Utopia*, written in 1516, and it means 'no place'. It's an imaginary place of ideal perfection, and the idea of a perfect place assumed by cyberspace's proponents has a long history.

Ralph Schroeder, a sociologist who writes critically about cyberculture, notes that

the artificial worlds created by information and communication networks . . . may allow for new forms of cultural

expression . . . [C]yberculture promotes an alternative understanding of the role of technology that aims to subvert military and corporate uses and seeks instead to use technology to explore novel forms of consciousness suited to a new age in which science merges with art and technology and becomes a tool for the political imagination.

(Schroeder, 1996: 7)

From Schroeder's text, we can deduce what people who like cyberspace – cyberpunks, hackers, futurists, and certain academics – see as five utopic promises for cybercultural practices, which will:

1. provide new forms of culture;

2. revitalize politics;

3. allow for subversive or alternative forms and uses of technology;

4. enable humans to explore new forms of consciousness; and

5. be part of the heralding-in of a new age.

What lies behind the utopian idea that new technology itself will take us to the promised land? As James Carey and John Quirk (1970) have argued, in the United States, utopian attitudes towards technology can be traced to the country's beginnings. From the start, technology was seen as a universal saviour. People believed it possible to use machines for good purposes while avoiding the horrors of the European Industrial Revolution – people thrown off their land, abusive child labour, 20-hour work days, and polluted air, water and food. But when Americans saw the same conditions emerging in cities such as Baltimore and Philadelphia, belief that machines could be connected to a natural world of peace had to be reconsidered. For people like Thomas Jefferson, Nature increasingly came to be seen as a remedy for industrialization's ills, and it was projected onto the 'pure place' to the west of the Appalachian mountains. But when settlers pushed westward, when all of Nature had been turned into real estate, there was no more frontier, although the idea of Nature as purifying remained. Imaginary and metaphysical spaces were created as substitutes, including cyberspace.

One must always separate the utopic notions of perfect technology from the ways in which technologies actually get used. Cyberculture, by promoting an ideal picture of technology, places technology on the same high pedestal where people like Thomas Jefferson once placed Nature. As a consequence, cyberculture only criticizes *how* technology is used; technology itself is never critiqued. In cyberculture, the machine itself, if we could somehow merge with it, would purify us. People like Jefferson, looking aghast at the Industrial Revolution, argued that the USA would be different; that Nature could absorb and humanize the machine. In cyberculture, it is technology that replaces Nature, absorbs it, and takes on qualities earlier ascribed to Nature: open space, harmony, communion, peace, becoming whole. In many ways, the implicit promise of cyberculture is individual renewal or rebirth. To look at what that might mean, I'm going to address how cyberculture's promises are put into practice in a real place.

Summary

- Cyberspace, a geographic metaphor for the virtual space created by electronic communication technologies, and cyberculture, new cultural practices made possible by these technologies, are seen by proponents as heralding in a new age.

- This utopian thinking is part of a long tradition of seeing technology as a means to perfect humans, and does not fully consider how technology may influence social relations.

CYBER-REALITY

Consider how Figure 34.1 might be an example of cyberculture. Is there anything about the image that seems odd, considering that what's pictured looks like a city street from almost any urban area in the United States? Consider the generic signage – Books and Coffee – and the absence of automobiles. Figure 34.1 is a photograph of CityWalk, the make-believe urban downtown which is really on the Universal Studio Lot in Universal City in suburban Los Angeles. You *must* arrive by car, park for $10.00, and enter CityWalk through the parking structure. CityWalk is a private retail environment made to seem like a public street.

I noted earlier that cyber means immaterial, something that doesn't really materially exist. If this is a city scene, then as a cybercity, it doesn't really exist. Of course CityWalk is materially real — people really are walking down what seems to be a main or high street – but real downtowns have no private entrance or admission fee disguised as a parking charge. CityWalk is in Los Angeles, the place famous for making images of the American dream, and partially because of Los Angeles' vast scale, cyberculture and the communication technologies upon which it relies are popular there.

In certain ways, CityWalk is a commercialized utopia. There is no crime, no homeless people. It's spotlessly clean and pure – just like Jefferson's view of

Figure 34.1 CityWalk, Universal Studios, Los Angeles. Credit: C. Brannstrom, 1993

Nature. CityWalk isn't contaminated by the problems that plagued industrialized Philadelphia in 1797 or Los Angeles in 1997. Universal Studios understands how much people – white, middle-class people in particular – yearn for a perfect world they could enter as if by magic. So Universal, for a price, offers an effortless utopia in which, for an hour or two, you can be at the centre of your own little utopia, a star in your own Hollywood movie.

Figure 34.2 can help us understand why some people, particularly those who grew up in the 1960s and 1970s, might want to go to CityWalk. It shows a giant poster on the side of a building organized around political concerns. A hippie woman wears bell bottoms. Signs read 'NO MORE WAR', and 'KEEP EARTH FRESH'. A peace sign is imprinted over a pizza, and an astronaut walks on one. The sign is a façade for a pizza restaurant. But the images are also nostalgic – they suggest a collapse of time between the late 1960s and today. A lot of people shopping at CityWalk are babyboomers, now middle-aged, who like the comfort of a place like CityWalk, where they can eat a pizza and be reminded of when they were young and committed to changing the world. The sign speaks to several pivotal events:

Figure 34.2 Giant sign/building, CityWalk, Los Angeles. Credit: C. Brannstrom, 1993

- The Environment. The first Earth Day was in 1971.

- The Vietnam War. The women and peace sign refer to the protests against the war. There were over 21,000 demonstrations and bombings in the US in 1969 alone.

- The Moon Landing. The first landing occurred in 1969 – during the height of protests against the Vietnam War – and it created tremendous excitement. The astronaut with his foot in the pizza refers to what Neil Armstrong, the first human to set foot on the moon, said at the time: 'That's one small step for man, one giant leap for mankind'. It seemed like the promise of technology was true; technology triumphed while politics in Vietnam failed. The landing promoted confidence in the future, and it was the dream come true of John Kennedy, who had announced the moon race as part of the Cold War with the Soviet Union. Kennedy was assassinated in 1963, so in a way the moon landing was a culmination of American idealism. Both the landing and the peace sign share a belief that the future will be one of peace on earth – a Camelot come to pass.

Recalling cyberculture promise number 2, the revitalization of politics, the pizza façade at CityWalk suggests how oppositional politics can get taken up by the dominant culture and slightly altered for its own purposes. The restaurant is trading on the images that people would like to see in an undisguised effort to get them to buy a pizza. Is this about the revitalization of politics?

Assuming the price of admission, you can visit CityWalk, obey the rules, shop the shops, and look at signs the size of buildings designed to make you recall past events. This requires an idealized present only possible in a private, controlled space. It might even be asked whether signs for Steele's Motel refer to anything more real than the one for Hollywood Freezway (*see* Fig. 34.3). Steele's Motel is a nostalgic appeal to a time that is past. Remember Rushkoff's comment that in cyberspace time doesn't matter?

William Gibson's definition of cyberspace as a 'consensual hallucination' seems applicable to CityWalk. From this example, one can see why cyber is an important modification to the word culture. Although the images depict real places, and real signs that may or may not be duplicates of ones from the past, the idea of real becomes difficult to pin down. Particularly so when it's possible to take images of the past that are about important political events and use them to sell pizza, and for people to consent to the hallucination that they're downtown when they're really in Universal's real-life version of Hollywood TV. If cyberculture is indeed the culture of the immaterial, then it becomes difficult to see how it will revitalize politics in the real world. Moreover, CityWalk trades on the same cultural understandings and utopian forgetting about the nature of corporate control that underpin many of the promises of cyberculture.

Wired, the popular San Francisco-based magazine specializing in all things digital, is the brainchild of editor Kevin Kelly, who saw a need for a magazine that deals with aspects of cyberculture. The 1993 premiere issue identifies Marshall McLuhan as *Wired*'s patron saint. Its opening pages quote from McLuhan's *The medium is the message* (1967):

> The medium, or process, of our time – electric technology – is reshaping and restructuring patterns of social interdependence and every aspect of our personal life. It is forcing us to reconsider and re-evaluate practically every thought, every action, and every institution formerly taken for granted. Everything is changing, you, your family, your education, your neighborhood, your job, your government, your relation 'to the others'. And they're changing dramatically.

Wired adds that these social changes prophesied by McLuhan are so profound that they're akin to the dis-

(a)

(b)

Figure 34.3 Advertising Nostalgia (a) Hollywood freezway (b) Steele's Motel, CityWalk, Los Angeles. Credit: C. Brannstrom, 1993

covery of fire. *Wired* continued its promotion of cyberspace and cyberculture, and its editorial commentary in Issue number 3, by Mitchell Kapor, informs readers that

> Life in cyberspace is more egalitarian than elitist, more decentralized than hierarchical . . . it serves individuals and communities, not mass audiences . . . We might think of life in cyberspace as shaping up exactly like Thomas Jefferson would have wanted it: founded on the primacy

of individual liberty and a commitment to pluralism, diversity, and community.

Comparing recent issues of *Wired* to Issues number 1 and number 3 reveals an almost threefold increase in the magazine's page count and a 600 to 700 per cent increase in advertising. Just as CityWalk appropriates an aspect of counterculture to sell pizza, *Wired* takes on cyberculture and uses its popularity to sell more magazines. The pure definition of cyberculture gets transformed by the dominant society cyberculture claims to oppose into something less pure. What are the implications for the five promises of cyberculture listed above? How oppositional or alternative is the pizza sign at CityWalk? How different is the vision of the digital generation that produces *Wired* magazine from the one on view at CityWalk?

It should be noted that new Information Technologies are having a real influence on politics. People are using the Internet to organize themselves politically. Even *Wired* promotes this decentralized opening up of the culture to new forms of expression and conversation. And these new forms of culture do touch on promises of revitalized politics and exploring new forms of consciousness. However, technologies can have a lot of hopes placed onto them that they will make the world a better place. And this is happening at a time when many people have lost faith in the political process, especially in the United States, though this is not confined solely to that country. It would be nice if technology were able to solve all problems. But it needs to be asked if such a wish isn't really an example of utopian thinking, if it's not a 'consensual hallucination' that people share when faced with real-life difficult political choices they'd rather not address.

Cyberculture promises to allow for subversive or alternative forms and uses of technology. Consider the 1991 beating of African-American motorist Rodney King by Los Angeles Police Department officers. King had been stopped for supposedly swerving between lanes. Officers later testified they beat him because they feared he was dangerous, and believed King had control of the situation, not themselves. The famous video of the attack was taped by George Holliday, who lived beside the freeway site where King was beaten. The video and the still images and montages taken from it became part of cyberculture; their world-wide dissemination was made possible by electronically mediated communication.

Holliday's tape pushed the political process in ways that relate to promises number 2 and number 3 on the list. It was impossible for the LAPD to deny the violence, and an unanticipated use of a camcorder exposed police racism on network TV. But look at what corporate imagination makes of the event (*see* Fig. 34.4). Sharp Electronics capitalizes on Holliday's use of an off-the-shelf technology. On one hand, the advert is cynical. 'Drive Safely' at the top of the picture on the right refers to King, and plays on people's fears that they need to carry a camcorder to protect themselves from police brutality. On the other hand, the sign can be read as saying that knowing how to operate these technologies is part of what it takes to survive. A videocamera becomes more than just a tool; it becomes a part of our *identity*. This advert, then, embraces the third promise of cyberculture – enabling humans to explore new forms of consciousness – and Sharp hopes that these forms will be brought to you, in part, by Sharp.

The Sharp advert co-opts a cyberculture event – the aftermath of the King beating – to sell cameras. Sharp aligns itself with cyberculture; by criticizing the police, and by implication the State, Sharp suggests that technology is separate from and better than politics. The advert also implicitly suggests that cyberculture can shift the political process, and that the best way to manage this shift is to move towards greater hands-on engagement with technology.

Summary

- Including material places into how cyberculture is understood and practised allows us to see both its hype and promise.

Figure 34.4 You can't do this with just any camcorder. Source: M. McDougall 1995: Banalities of information. In J. Brook and I.A. Boal (eds) *Resisting the Virtual Life: the Culture and Politics of Information*. San Francisco: City Lights, p.212

- A fantasy environment such as CityWalk – which relies on the same utopian desires that drive cyberculture – can be easily consumed, temporarily satisfying utopian desires and mitigating disillusionment with the real world.

- In a similar fashion, cyberculture's promoters promise that not only will technology succeed where politics have failed, but that technology can protect us from corrupt political processes.

CONCLUSION: RESISTANCE IS ESSENTIAL

I noted that cyberculture criticizes how technology gets used but not technology itself. In so doing, cyberculture reflects a belief in technological determinism, a view that technology on its own causes important cultural changes, in this case for the better. This forgets that people produce technologies for specific reasons. E-mail or Virtual Reality are parts of activities or cultural processes which take place in historical context. In other words, technologies are always in bed with politics, and this means they are always part of the real world, cyberculture and its own utopian politics included. Technological determinism conveniently forgets this in order to suggest a vision that never really comes to pass, but always seems better than what is at hand.

Is cyberculture a new form of culture? Does it herald a new age? In many ways, yes. The passage in the box is from a newsgroup on the net and posted by a woman with the nickname 'outrider'. Her posting is part of cyberculture; she's using an information technology to criticize the paranoid, control-obsessed world she believes will result from an over-reliance on the very technologies she's using to make her critique. Yet the future she describes has many of today's social and spatial problems – issues like fear of being in public that bring the crowds to CityWalk. Outrider describes much of what this chapter has addressed. Her words 'never give in never submit', remind me of the Borg originally depicted on *Star Trek*. The Borg, a species that has merged with technology – an idea cyberculture promotes – travel the galaxy consuming cultures and forcing them to join the technology-driven collective. The main message to cultures the Borg cannibalizes is 'Resistance is futile'. The Borg exemplifies how an alternative idea of cyberculture can be appropriated by Hollywood and made over with a more mainstream message, in a similar way to how Universal merges sanitized images of alternative politics with a utopian image of a sanitized city to make CityWalk more appealing to consumers.

Never give in, never submit. Or just never go out of your house anymore. In twenty years this will be Life: stay home all the time because its too dangerous to go out … get all stimuli, info, human contact, groceries, money, etc. on your computer. All materials will be delivered by heavily armed people in tanks: they must cross the moat filled with piranha, crocodiles, and weird water-borne disease organisms, and also pass the security check that keeps them from getting Swiss-cheesed by the remote control firepower in the gun turrets at the razorwire perimeter, then they have to pass the DNA identity scanner at the last portal … After a pleasant meal of micronuked frozen blah, you can jump onto the Net and read the Daily Horros in the form of movingpicto-news; go to the library and download the original French version of *Madame Bovary* and a decent French dictionary. Read in the comfort of your cozy warm bed, safe behind triple-wall steel constructed building … Sleep. Dream of a more lifelike life . . . remember the olden days when you could walk outside in the Night and go places, when you could drive safely from here to there . . .

'outrider' (Dery, 1993: 563)

Looking at cyberculture non-critically conforms to a tradition that avoids criticizing technology. This contributes to a frustration with politics because if technology is a kind of utopia upon which we can focus our desires, even though the utopia's promise never fully materializes, a series of ever newer technologies works to renew technology's promise and deflect attention from thinking about how politics might be revitalized in the real world. Yet precisely because new communication technologies *do* offer many benefits, it's important to think critically about how they are used. The reality of cyberculture and cyberspace can never be as totally oppositional as that depicted by its proponents. Aspects of cyberculture's promises, however, are available to us if we work with the technology. But the promise of cyberculture and cyberspace starts to recede if, like the Borg, we allow the technology to consume us.

Further reading

- Brook, J. & Boal, I.A. (eds) (1995) *Resisting the virtual life: the culture and politics of information*. San Francisco: City Lights. A provocative set of articles about the meaning of information/overload to contemporary cultures.

- Hillis, K. (1996) A geography of the eye: the technologies of virtual reality. In Shields, R. (ed.) *Cultures of Internet: virtual spaces, real histories, living bodies*. London: Sage, pp. 70–98. A critical history of the history behind the development of virtual reality technology.

- Stephenson, N. (1992) *Snow crash*. New York: Bantam.
Excellent science fiction account of cyberspace and its
implications. Written by a geographer.

- Turkle, S. (1995) *Identity in the age of the Internet*. New York:
Simon and Schuster.
Thorough investigation of the implications of electronic
communication technologies on identity formation.

- Woolley, B. (1992) *Virtual worlds: a journal in hype and
hyperreality*. Oxford: Blackwell.
Easy-to-read account of the hype and science behind virtual
reality.

Postscript: your human geographies

We began this book by talking about how beginning to study Human Geography at the university level involved confronting many unfamiliar ideas and topics. But Human Geography is never just a distant body of knowledge that you have to be introduced to. It is part of your life already, and it is something you can contribute to. It is not just 'ours' on this side of the word processor, but yours too. Those of us lucky enough to teach the subject know that we don't just pass ideas on to our students. The ways they respond to the ideas we present, the arguments they raise, the examples they talk about, all end up making us re-think those ideas. Human Geography is, or certainly should be, a meeting ground for many voices, not a series of monologues. It has certainly been our intention that much of the material in this book should intersect with the everyday lives of those who read it. We believe that Human Geography should be in this realm of the visible, audible, touchable, experience-able features of the lives of those doing the Human Geography. While these features will vary between different people and different places, nevertheless the approaches we have outlined should be relevant across the spectrum of that variation.

So, on the basis of what you have read in this book, we would urge you to be much more sensitive to the human geographies going on around you. Don't be fooled into thinking that the subject is best represented by crusty old abstract notions found in crusty old books written by crusty old men! Instead, start with your own experiences and then work outwards from that. Be aware of the human geographies wrapped up in and represented by the food you eat, the news you read, the movies you watch, the music you listen you, the television you gaze at. Be aware of the places that you live in, or travel to, or see images of. Be aware of the person you are, the company you keep, the society you live in, the nature of your and others' living, working and play spaces. And in your being aware, take note of what or who is being omitted, marginalized or 'othered' in your narratives. Take note of the power of discourse; how stories are told, and how people or places are categorized, are powerful weapons which can lead to advantage/disadvantage, belonging/not belonging and visibility/invisibility. Take care to listen to voices other than your own and those who agree with you. Take care to question whether politics should simply empower pleasure, or whether key issues of economic and social distribution should be addressed in the political realm.

In the end, all introductions to Human Geography are partial, even those as large as this. All come from particular perspectives, times and places. Ours presents an anglophonic Human Geography at the start of the twenty-first century. We hope you are inspired by it. But we do not ask that you agree with all the ideas and interpretations in it. Rather, we encourage you to think through and spell out your own. Human geographies are yours to learn from, but also yours to contribute to. Geography matters, you matter, your geography matters. Having been introduced to Human Geography by this book, why not start to practise Human Geography for yourself?

Glossary

This glossary provides brief definitions of key terms used but not fully explained in the course of this collection's chapters. Terms in the glossary are marked in bold when first used in a particular chapter. Here, in the glossary itself, cross-referencing between entries is facilitated by capitalizing terms that are separate glossary entries in their own right.

alienation: a term with two interrelated meanings, the second being a more specific formulation of the first. Firstly, then, alienation refers to a sense of estrangement or lack of power felt by people living in the MODERN world. In this respect it is often used to describe the experience of modern urban living, in which traditional forms of social cohesion and belonging supposedly break down. Secondly, and drawn from MARXIST social thought, alienation refers more particularly to the separation of labour from the MEANS OF PRODUCTION under CAPITALISM. To explain, since under capitalism it is capitalists who own the resources required for economic production (land, machinery, money etc.) workers have no control over their productive lives: how production is organized, what is produced, what the product is used for, and how they relate to other workers. They are therefore alienated from their work and its products.

autoethnography: the processes by which the Human Geographer chooses to make explicit use of her or his own POSITIONALITY, involvements and experiences as an integral part of ETHNOGRAPHIC research.

biodiversity: describes the variability in nature, including of genes, individuals, species and ecosystems. Biodiversity exists both in particular living things (such as individual rare species) and also in biological processes and dynamic ecological systems.

capital accumulation: the prime goal of CAPITALISM as a MODE OF PRODUCTION. It is the deployment of capital to convert it into new (and more) capital. This process is often designated by the simple formula MCM′ where M is money or liquid capital which is invested in C which is commodity capital (land, labour, machines) with a view to profit, realized as M′ which is more money (or more technically 'expanded liquidity').

capitalism: an economic system in which the production and distribution of goods is organized around the profit motive (*see* CAPITAL ACCUMULATION) and characterized by marked inequalities in the social division of work and wealth between private owners of the materials and tools of production (capital) and those who work for them to make a living (labour) (*see* CLASS).

civil society: a concept with a long and changing history of meanings, civil society has been used in the last decade to emphasize a realm of social life and a range of social institutions that are separate from the NATION-STATE.

class: a collection of people sharing the same economic position within society, and/or sharing the same social status and cultural tastes. The precise ways in which one's economic position – for example as a worker, a capitalist or a member of the land-owning aristocracy – is related to one's social status or cultural tastes has been much debated. However, Human Geographers have studied class and its geographies from all these perspectives: as an economic, social and cultural structuring of society.

Cold War: a period conventionally defined as running from the end of the Second World War until the fall of the Berlin Wall, during which the globe was structured around a binary political geography that opposed U.S. CAPITALISM to Soviet communism. Although never reaching all-out military confrontation, this period did witness intense military, economic, political and ideological rivalry between the superpowers and their allies.

colonialism: the rule of a NATION-STATE or other political power over another, subordinated people and place. This domination is established and maintained through political structures, but may also shape economic and cultural relations. *See also* NEOCOLONIALISM and IMPERIALISM.

colonization: the physical settling of people from a colonial power within that power's subordinated colonies (*see* COLONIALISM). *See also* DECOLONIZATION.

commodification: this term is used in two interrelated ways: (a) as the conversion of any thing, idea or person into a COMMODITY (the term 'commoditization' is often preferred for this sense); and (b) a wider societal process whereby an ever increasing number of things, human relationships, ideas and people are turned into commodities. Both meanings see the process of commodification as symptomatic of the penetration of CAPITALISM into the everyday lives of people and things.

commodity: something that can be bought and sold through the market. A commodity can be an object (a car, for example), but can also be a person (the car production worker who sells their labour for a wage) or an idea (the design or marketing concepts of the car). Those who live in capitalist societies are used to most things being commodities, though there are still taboos (the buying and selling of sexual intercourse or grandmothers, for example) (*see* CAPITALISM and COMMODIFICATION). This should not disguise the fact that the 'commodity state' is a very particular way of framing objects, people and ideas.

commodity fetishism: the process whereby the material origins of commodities are obscured and they are presented 'innocent' of the social and geographical relations of production that produced them.

context of the commodity: the real or imaginary geographical setting in which a commodity is displayed, such as an advertisement or shop-window. The setting is typically designed to attach a particular set of associations onto the commodity. *See also* COMMODIFICATION and COMMODITY.

cultural ethnocide: the permanent disappearance of a particular ethnic group, associated with the eradication of their cultural practices, brought about by the effects of economic and/or political policies. *See also* ETHNICITY.

cultural landscape: traditionally this phrase has meant the impact of cultural groups in fashioning and transforming the natural landscape. More recently it has been suggested that landscape itself is a cultural image, a way of symbolizing, representing and structuring our surroundings.

cyberspace: the forms of space produced and experienced through new computational and communica-

tion technologies. Exemplary are the SPATIALITIES of phenomena such as the Internet, Virtual Reality programmes, hypertext and hypermedia, or the science-fictional accounts that draw on these for inspiration. Human Geographers have approached these spatialities in at least three distinct ways: in terms of the kinds of places users encounter within them; by analysing the kinds of communicational networks they facilitate; and in terms of the geographical location of their infrastructures.

decolonization: this term has two meanings: (a) the ending of formal colonial rule by one power over another (*see* COLONIALISM); (b) the departure of a settler population from a colonized territory (*see* COLONIZATION). In both cases, however, processes of decolonization are in fact likely to be far less of a 'clean break' than these definitions suggest. Legacies from, and new forms of, colonialism are still central to POST-COLONIAL experiences.

de-industrialization: this term usually refers to a decline in manufacturing industry. It can designate either a fall in manufacturing output or, more commonly, a fall in the number and share of employees in manufacturing industry as a result of plant closures, layoffs etc. It is possible for de-industrialization to occur at the same time as manufacturing output and exports rise if automation is associated with jobless growth. Areas which have suffered from de-industrialization are characterized by large numbers of vacant derelict factories. *See also* POST-INDUSTRIAL.

demesne: the parkland attached to a country house or stately home.

deterritorialization: the uncoupling of political and economic processes from particular national spaces, as when banks move offshore or when economic decisions are entrusted to the World Bank or the G7 powers. *See also* TERRITORY.

developmentalism: a set of propositions or policies which demand (or provide for) the transformation of pre-modern societies into MODERN societies. INDUSTRIALIZATION is often considered to play a key role in the process of development. *See also* POST-DEVELOPMENTALISM.

devolution: the process of devolving some political power to more local levels of government, often associated with the formation of local and regional assemblies. This can take place at many scales: at a supra-national scale, for example within the European Union, power can be devolved to individual nations or to regions; and within the NATION STATE power can be devolved to regions and localities.

diaspora: the dispersal or scattering of people from their original home. As a noun it can be used to refer to a dispersed 'people' (hence the Jewish diaspora or the Black diaspora). However, it also refers to the actual processes of dispersal and connection that produce any scattered, but still in some way identifiable, population. In this light it also can be used as an adjective, diasporic, to refer to the senses of home, belonging and cultural identity held by a dispersed population.

discourse, discursive: drawing on the work of the French philosopher Michel Foucault, Human Geographers define discourses as ways of talking about, writing or otherwise representing the world and its geographies (*see also* REPRESENTATION). Discursive approaches to Human Geography emphasize the importance of these ways of representing. They are seen as shaping the realities of the worlds in which we live, rather than just being ways of portraying a reality that exists outside of language and thought. They are also seen as connected to questions of practice, that is what people actually do, rather than being confined to a separate realm of images or ideas. More specifically, Human Geographers have stressed the different ways in which people have discursively constructed the world in different times and places, and examined how it is that particular ways of talking about, conceptualizing and acting on people and places come to be seen as natural and common-sensical in particular contexts.

dialectic, dialectical: a dialectic is a process through which two opposites are generated, interrelated and eventually transcended. This process can be purely intellectual, in so far as it is a procedure of thought. Examples in this vein would include the opposites used as the starting points for all the chapters in Part I of this collection (nature–culture, society–space and so on), which are then worked through and beyond in the course of each chapter. Dialectical processes can also be identified in the wider world, though. An example would be the combination in a COMMODITY of both use and exchange values, in which opposites become interpenetrated in the commodity form.

ecocentrism: a perspective favouring a humble and cautious outlook towards the scope for interfering with the planet, and arguing for a smaller-scale, more communitarian style of living.

ecology: a way of studying living things (plants, animals or people) that emphasizes their complex and dynamic inter-relationships with each other and the environment. As well as their use in Biogeography, ecological theories about competition for resources, invasion and distribution have also been applied in Urban Geography.

ecosystem: an ecosystem comprises: a set of plants, animals and micro-organisms among which energy and matter are exchanged; and the physical environment with which, and within which, they interact. *See also* ECOLOGY.

ecumenical: taking its meaning from the noun ecumene ('the habitable earth'), Human Geographers have used the adjective ecumenical to refer to a sense of self and place identity that is open to the world's differences and seeks to transcend the boundedness of people and places.

embodied: this concept suggests that the self and the body are not separate, but rather that the experiences of any individual are invariably, shaped by the active and reactive entity which is their body - irrespective of whether this is conscious or not. The arguement then runs that the uniqueness of human experience is due, at least in part, the unique nature of individual bodies.

Enlightenment: a philosophical and intellectual movement usually dated to the seventeenth and eighteenth centuries and centred in Europe which advanced the view that the world could be rendered knowable and explained systematically by the application of rational thought (science). Revolutionary in its challenge to the religious beliefs and superstitions that then held sway, it has since been much criticised for projecting rationality as a universal, rather than situating reasoning processes in particular social and material contexts.

environmental audit: the attempt to calculate, and list for comparison, all the likely effects of any proposed action on the surrounding ECOSYSTEMS and populations.

environmental determinism: a school of thought which holds that human activities are controlled by the environment in which they take place. Especially influential within Human Geography at the end of the nineteenth and beginning of the twentieth centuries, when the approach was used in particular to draw a link between climatic conditions and human development. Some authors used this link to argue that climate stamps an indelible mark on the moral and psyiological constitution of different races (*see* RACE). This in turn was used to assert the superiority of western civilization, and hence to justify and support the imperial drive of nineteenth century Europe (*see* COLONIALISM and IMPERIALISM).

environmentalism: a social and political movement aimed at harmonizing the relationship between human endeavour and the presumed limits to interference of planetary life support systems.

environmental risk: real and imagined threats to existing human, social and ecological systems posed by physical events, such as earthquakes; or from the unforeseen consequences of human activity, such as releasing genetically modified organisms into the environment.

epistemology, epistemological: epistemology is the study of knowledge. This study seeks to connect up questions of content (what people know) with structures of belief (why and how they know) and issues of authority (how and why knowledges are valued and justified). Human Geographers are especially interested in epistemological questions in terms of the geographical knowledges held by both academics and lay publics, and their relationship to wider systems of belief and authority.

essentialize, essentialization: to reduce a person, people or place to their supposed essences or essential qualities. The term is often used negatively, to emphasize the simplifications and stereotypings that this can produce.

ethnicity: a criterion of social categorization that distinguishes different groups of people on the basis of inherited cultural differences. Ethnicity is a very complex idea that needs careful consideration. For instance, in popular usage ethnicity often becomes a synonym for RACE, but in fact there is a crucial distinction in so far as race differentiates people on the grounds of physical characteristics, and ethnicity on the grounds of learnt cultural differences. Moreover, whilst everyday understandings of ethnicity often treat it as a quality only possessed by some people and cultures (for instance 'ethnic minorities' and their 'ethnic foods' or 'ethnic fashions') in fact these differential recognitions of ethnicity themselves need explanation. The complexities of the concept are further emphasized by recent debates within and beyond Human Geography over the extent to which new forms of ethnicity are emerging through the cultural mixing associated with processes of GLOBALIZATION (*see also* HYBRID and SYNCRETIC).

ethnocentric: an adjective describing the tendency for people to think about other cultures, societies and places through the assumptions of their own culture, society or place. An example of ethnocentricism is the production of theories about the whole world based on the specific model of Western development (*see also* DEVELOPMENTALISM).

ethnography: the processes which use qualitative methods to provide in-depth explorations and accounts of the lives, interactions and 'textures' associated with particular people and places.

feminism: a series of perspectives, which together draw on theoretical and political accounts of the oppression of women in society to suggest how GENDER relations and human geography are interconnected (*see also* PATRIARCHY).

Fordism: a form of industrial capitalism dominating the economies of the US and Western Europe from the end of World War II to the early 1970s. It was characterized by mass production and mass consumption, where high levels of productivity (often promoted by new assembley line production) sustained high wages, which in turn led to high levels of demand for industrial products. This demand fed into higher production, which supported higher productivity, and the 'virtuous circle' began all over again. It was an economic system which was often underpinned by a political compromise between capiatal and labour, and by subsequent state policies towards wages, taxes and welfare provision which helped to sustain mass production and consumption. *See also* POST-FORDISM.

gender: a criterion of social categorization that distinguishes different groups of people on the basis of their femininity or masculinity. As a concept, gender is usually used in Human Geography in distinction to that of sex (i.e. femaleness and maleness) in order to emphasize the SOCIAL CONSTRUCTION of women's and men's roles, relations and identities. Human Geographers' accounts of the world have always been shaped through understandings of gender (*see* MASCULINISM) but explicit analyses of the geographies of gender and the gendering of geographies are comparatively recent, and associated with the growth of Feminist Geography (*see* FEMINISM).

gentrification: an urban geographical process commonly taken to have two main attributes. The first is the invasion (or replacement) of traditional inner city working CLASS residential areas by middle CLASS inmigrants; and the second is the upgrading, improvement, and renovation of the existing housing, whether done by the new residents or by developers. Commercial gentrification refers more specifically to the replacement of older, traditional, low rent, retail and other uses by new, stylish, fashionable boutiques, cafes, bars and other retail outlets.

geographical imagination: an awareness of the role of space, place and environment in human life. This phrase is sometimes used in the definite singular – the geographical imagination – to refer to the distinctive intellectual concerns and contributions of Geography. It is also used in the plural – geographical imaginations – to emphasize the many different ways in which academics, students and lay publics alike can develop their sensitivities to human geographies.

geopolitics: an approach to the theory and practice of statecraft which considers certain laws of geography (e.g. distance, proximity and location) to play a central part in the formation of international politics. Although the term was originally coined by Swede Rudolf Kjellen in 1899, it was popularized in the early twentieth century by British geographer Halford Mackinder.

global city: a term used by Saskia Sassen in the early 1990s to denote those cities which play a key role in the operation of the capitalist world economy, particularly in terms of financial and business services. The term builds upon the concept of 'world city', put forward by Freidmann and Wolff, to capture the command and control centres of late CAPITALISM, particularly in terms of the concentration of headquarters of multi- and TRANSNATIONAL CORPORATIONS.

globalization: the economic, political, social and cultural processes whereby: (a) places across the globe are increasingly interconnected; (b) social relations and economic transactions increasingly occur at the intercontinental scale (*see* TRANSNATIONAL); and (c) the globe itself comes to be a recognizable geographical entity. As such globalization does not mean everywhere in the world becomes the same. Nor is it an entirely even process; different places are differently connected into the world and view that world from different perspectives. Globalization has been occurring for several hundred years, but in the contemporary world the scale and extent of social, political and economic interpenetration appears to be qualitatively different to international networks in the past.

gravity model: a mathematical model based on a rather crude analogy with Newton's gravitational equation which posited a constant relationship between the gravitational force operating between two masses, their size and the distance between them. Geographers have used this model to predict and account for a range of flows between two or more points, especially those to do with migration and transport (*see also* SPATIAL SCIENCE).

hegemony, hegemonic, hegemon: hegemony is an opaque power relation relying more on leadership through consensus than coersion through force or its threat, so that domination is by the permeation of ideas. For instance, concepts of hegemony have been used to explain how when 'the ruling ideas are the ideas of the ruling class' other classes will willingly accept their inferior position as right and proper (*see* CLASS). Hegemonic is the adjective attached to the institution which possesses hegemony. For instance, under CAPITALISM the bourgeoisie are the hegemonic class. Hegemon is a term used when the concept of hegemony is applied to the competition between NATION-STATES: a hegemon is a hegemonic state. For instance, the USA has been described as the hegemon of the world economy in the mid-twentieth century.

historicity: to recognize the historicity of human geographies is to recognize their historical variability. This involves both an emphasis on historical specificity – on how historical periods differ from each other – and on historical change – on how human geographies are re-made and transformed over time.

humanist: an outlook or system of thought which emphasizes human, rather than divine or supranatural, powers in understanding the world. Associated with the ENLIGHTENMENT, humanism marks human beings off from other animals and living things by virtue of supposedly distinctive capacities for language and reasoning. While underscoring progressive social changes, like the idea of human rights, it is criticised for making universal claims about human nature; privileging the individual over the social relations of human being; and licensing human abuse of the natural world.

Humanistic Geography: a theoretical approach to Human Geography that concentrates on studying the conscious, creative and meaningful activities and experiences of human beings. Coming to prominence in the 1970s, Humanistic Geography was in part a rebuttal of attempts during the 1960s to create a law-based, scientific Human Geography founded on statistical data and analytical techniques (*see* SPATIAL SCIENCE). In contrast it emphasized the subjectivities of those being studied and, indeed, the Human Geographers studying them.

hybrid, hybridity, hybridization: hybrids are the products of the combination of usually distinct things. These terms are often used to describe new plant types but are used in Human Geography to emphasize the equal and positive mixing of cultures rather than negative ideas of cultural assimilation, dilution, pollution or corruption. *See also* SYNCRETIC.

hyper-reality: a phrase most associated with the Italian semiotician (*see* SEMIOTICS) and novelist Umberto Eco, which suggests the development of simulated events and representations which out do the 'real' events they are meant to be depicting. Thus in hyper-reality the REPRESENTATION exceeds the original, being more sensational, more exciting or so forth. *See also* SIMULACRUM.

icon: a visual image, landscape feature or other material form that comes to symbolize or stand for a wider set of meanings or phenomena.

iconography: This term means both: (a) the study of the symbolic meanings of a picture, visual image or landscape; and, less often, (b) the system of visual meaning thereby being studied. Developed in particular within disciplines such as art history, Human Geographers have extended iconography to the analysis of landscape symbolism and meaning. This analysis combines examination of the symbolic elements of a landscape with consideration of the social contexts in which a landscape is produced and viewed (*see also* SEMIOTICS).

ideological: a meaning, idea or thing is ideological in so far as it helps to consititute and maintain relations of domination and subordination between two or more social groups (CLASSES, GENDERS, age groups etc.).

ideology: a meaning or set of meanings which serves to create and/or maintain relationships of domination and subordination, through symbolic forms such as texts, landscapes and spaces.

idyll–ized: the process by which dominant myths about places and spaces come to reflect circumstances of picturesque beauty, tranquillity and harmonious living conditions. The term is often used in relation to rural spaces, where social problems can be hidden by the impression of idyllic life in close-knit communities and close to nature.

imagined geographies: the impressions formed in the mind (and then often acted upon) by various REPRESENTATIONS of places, spaces and the people who inhabit them. Imagined geographies often constitute a dominant impression of people and places, particularly those of which the individual has no direct personal experience.

imperialism: a relationship of political, and/or economic, and/or cultural domination and subordination between geographical areas. This relationship may be based on explicit political rule (*see* COLONIALISM), but it need not be.

industrialization: the process through which societies develop an economy based on the mass and mechanized production of goods. *See also* DE-INDUSTRIALIZATION and POST–INDUSTRIAL.

inter-generational equity: inter-generational equity refers to the concept of equality in access to resources or wealth between one generation of people and another (usually a future generation, or people as yet unborn).

intra-generational equity: intra-generational equity refers to the concept of equality in access to resources or wealth within a single generation (usu-ally between different groups of people today, for example between rich and poor classes in one country, or between rich and poor countries such as between 'First World ' and 'Third World' countries).

linguistic (or cultural) turn: this phrase is used to describe changes in the social sciences and Human Geography over the last 30 years. It refers to the adoption of interpretive (qualitative) approaches to explore the ways in which meanings, values and knowledges are constructed through language and other forms of communication.

marxist: social and economic theories influenced by the legacy of the leading nineteenth century political philosopher, Karl Marx. Highly influential in the framing of critical geography, these theories focus on the organization of capitalist society and the social and environmental injustices which can be traced to it. *See also* CAPITALISM, MODE OF PRODUCTION, ALIENATION and COMMODIFICATION for examples of the influence of Marxist thinking in Human Geography.

masculinist: an adjective describing a form of thought or knowledge which, whilst often claiming to be impartial, comprehends the world in ways that are derived from men's experiences and concerns. Many Feminist Geographers have argued that Human Geography has traditionally been masculinist (*see* FEMINISM).

means of production: the resources which are indispensable for any form of production to occur. Typically this would include land, labour, machines, money capital, knowledge/information. In MARXIST thinking the means of production and labour power together constitute the 'forces of production'.

mental map: a mental map describes our everyday notions about our spatial location. People rarely picture their spatial location to themselves through the images of a formal map. But of course we are all spatially located and aware. A mental map thus relates the elements we see as important and misses places we do not visit. Studies have shown that instead of a bird's eye view mental maps are organized around paths and landmarks that help us find our way in daily life. *See also* PERCEPTION.

metaphor: the use of a word or symbol from one domain of meaning to apply to another. Thus a rose (a botanical term) is translated into the domain of human relationships to symbolize love. Metaphors involve this movement of concepts from their normal realm to a new realm.

mode of production: a term taken from MARXIST social thought referring to the distinctive ways in which production has taken place in different types of society and in different historical epochs. Thus CAPITALISM is seen as one 'mode of production' distinguishable, for example, from feudalism and communism. In Marxist Geography, the mode of production is seen as the fundamental determinant of the kind of society and the kind of human geographies a person has to live with and through.

modern, modernity, modernism: ideas of the modern are most commonly defined through their opposition to the old and the traditional. In this light, the adjective 'modern' is synonymous with 'newness'; 'modernity' refers both to the 'post-traditional' historical epoch within which 'newness' is produced and valued, as well as to the economic, social, political and cultural formations characteristic of that period; and 'modernism' applies more narrowly to artistic, architectural and intellectual movements that centrally explore ideas of 'newness' and develop 'new' aesthetics and ways of thinking to express these. Modernity has been most commonly located in Euro-American societies from the eighteenth-century onwards, and thereby associated with their characteristic combination of capitalist economies (see CAPITALISM), political organization through NATION-STATES, and cultural values of secularity, rationality and progress (see ENLIGHTENMENT). However, increasingly Human Geographers are recognizing that modernity is a global phenomenon that has taken many different forms in different times and places.

moral economy: a term used by the historian Edward Thompson to describe the fact that economic relations have a normative aspect. For instance, the moral economy in the peasant economy would refer to the idea of a subsistence ethic in which the social relations of the poor attempt to provide forms of local security and support to prevent starvation. The concept of the moral economy has connections to Karl Polanyi's idea that the economy is an instituted process embedded in social relations.

multicultural, multiculturalism: multicultural is an adjective used to describe a place, society or person comprised of a number of different cultures. Multiculturalism is a body of thought that values this plurality. As the Human Geographer Audrey Kobayashi has noted, both the multicultural and multiculturalism can be conceived of in a number of ways: as 'demographic', i.e. simply reflecting a diversity of population; as 'symbolic', i.e. as about the presence or absence of the symbols associated with particular cultural groups within wider societal or national culture (for example on the media, or in museums, or in school curricula); as 'structural', in so far as institutions

are established to reflect a multicultural society and pursue multiculturalism; and 'critical', to the extent that multiculturalism itself critiques the assumptions of distinct, separate cultural groups sometimes attached to notions of the multicultural. Within Human Geography, increasing emphasis is being placed on developing this 'critical multiculturalism' through examining the HYBRID or SYNCRETIC cultural forms emerging within multicultural societies.

nation-state: a form of political organization which involves (a) a set of institutions that govern the people within a particular TERRITORY (the state), and (b) which claims allegiance and legitimacy from those governed, and from other states, on the basis that they represent a group of people defined in cultural and political terms as a nation.

naturalization, naturalized: this term has two distinct and different meanings: (a) the way in which social relations, cultural norms or institutions are made to seem 'natural' rather than SOCIAL CONSTRUCTIONS; and (b) the process through which species become established in new environments, sometimes also applied to human life to refer to the formal integration of immigrants in new societies.

neocolonialism: economic and political ties, continuing after formal independence, between metropolitan countries and the South, that work to the benefit of the North. See also COLONIALISM.

neoliberal: pertaining to an economic doctrine that favours free markets, the deregulation of national economies, and the privatization of previously state-owned enterprises (e.g. education, health). A doctrine which, in practice, favours the interests of the powerful (e.g. TRANSNATIONAL CORPORATIONS) against the less powerful (e.g. peasants) within societies.

non-governmental organization (also NGOs): service-providing organizations staffed by non-state professionals, and working in a broad range of development activities.

Others, Otherness: usually typographically capitalized, an Other is that person or entity which is understood as opposite or different to oneself; Otherness is the quality of difference which that Other posseses. A rather abstract conceptual couplet, potentially applicable at scales varying from the individual person to the global political bloc, these terms have been used in Human Geography to emphasize how ideas about human and geographical difference are structured through oppositions of the Self/Same versus the Other/Different. They also stress how the Other is often defined in terms of its relations to that Self – as its negative, everything it is not – rather than in its own terms. For example, a number of studies have examined how dominant ideas about GENDER are based on a logic in which Woman is Other to Man; how ideas about global politics and culture frame the East as Other to the West, and the South as Other to the North; and so on. A crucial problem being debated in Human Geography today is therefore how to recognize difference and 'Otherness' but in ways which do not simply present that difference in terms derived from the Self. For example Feminist Geographers are theorizing how to consider femininity in ways that do not simply see it as the opposite of masculinity; and Cultural and Political Geographers have been considering how to understand the East (the Islamic world and the Orient) in ways that do not simply see it as a cultural and political opposite to the West.

partnerships: an arrangement between a number of agencies and institutions in which objectives are shared and a common agenda is developed in pursuit of a common purpose. The partnership approach encourages collaboration and integration, and the aim is that by blending and pooling their resources the different agencies will be able to produce a capacity for action which is more than simply the sum of their individual parts.

patriarchy: a social system in which men oppress and exploit women. The term was first coined in analyses of households headed by men and organized to the benefit of those 'patriarchs' (for example through an unequal division of domestic work, or through women's marriage vows 'to obey', or through the legality of rape by husband of wife). However, the term is now used in a wider sense to think about how unequal power relations between men and women are established through realms stretching from the social organization of reproduction and childcare, to the organization of paid work, the operations of the state, cultural understandings of GENDER differences, the regulation of human SEXUALITY, and men's violence towards women.

pauperization: the progressive impoverishment of people owing to the impacts of certain development programmes. For example, the displacement of peasant and tribal peoples from their sources of livelihood (e.g. land) to make way for the construction of hydroelectric dams.

perception: the process through which people form mental images of the world. Often assumed to be both one directional (from the world to us) and biological (neurologically controlled), many academic studies have emphasized the role of cultural filters or frames in altering how we form pictures of the world.

performance: the manner of an individual's conduct, presentation and behaviour. It may vary over time and space for any individual according to a mixture of self identity and social pressure - the resulting 'performance' often combines elements of the image they wish to project and the image they feel is expected of them.

planning betterment: the negotiated outcome between a planning authority and a developer to achieve social gain through the attachment of advantageous provisions to a development proposal.

political-economy: the study of how economic activities are socially and politically structured and have social and political consequences. Political-economic approaches in Human Geography have paid particular attention to understanding capitalist economies and their geographical organization and impact (*see* CAPITALISM and also MARXIST GEOGRAPHY). Central to such analyses have been questions concerning the class-based nature of the human geographies of capitalist societies (*see* CLASS).

positionality: the personal experiences, beliefs, identities and motives of the Human Geographer which influence her or his work and the way in which her or his knowledge is situated.

post-colonial: Sometimes hyphenated, sometimes not, this term has two distinct meanings: (a) the post-colonial era, i.e. the historical period following a period of COLONIALISM (*see also* DECOLONIZATION); (b) post-colonial political, cultural and intellectual movements, and their perspectives which are critical of the past and ongoing effects of European and other colonialisms.

post-developmentalism: a radical critique of DEVELOPMENTALISM which demands the self-empowerment of poor or marginalized people in opposition to the powers of the state or capital.

post-Fordism: refers to the forms of production, work, consumption and REGULATION which have emerged out of the crisis of mass, standardized forms of capitalist production (FORDISM) during the 1970s. In terms of production and work, Post-Fordism turns on the importance of flexibility in work and other institutional forms of productive organization. Economic Geographers have analysed how this flexibility has been driven both by versatile and programmable machines, and by forms of 'vertical disintegration' of some firms in some sectors which make greater use of strategic alliances and subcontracting. Accompanying these changes in production are changes in consumer demand (the centrality of quality over standardization), in labour markets, in finance and legal structures, and in the broad social contract which characterized post-war Fordism.

post-industrial: a description applied to the new economic, social and cultural structures emerging in late capitalist societies in the late twentieth century, highlighting in particular, trends away from manufacturing, manual work and the mass production of physical goods (*see* INDUSTRIALIZATION) and towards the tertiary sector, forms of service employment and the production of experiences, images, ideas and relationships.

post-materialist: a philosophy relating the quality of existence, and peaceful relations between people and between people and the planet, to the manageable acquisition of material goals.

post-modern, post-modernity, post-modernism: the British national newspaper, *The Independent*, sarcastically defined post-modernism thus: 'This word has no meaning. Use it as often as possible'. In fact, the main problem for a glossary entry such as this is that ideas of the post-modern have been used so often with so many different meanings! Nonetheless, one can generalize to say that notions of the post-modern (sometimes hyphenated, sometimes not) are used to suggest a move beyond 'modern' society and culture (*see* MODERN, MODERNITY, MODERNISM). More specifically: (a) post-modern is an adjective used to describe social and cultural forms that eschew 'modern' qualities of order, rationality and progress in favour of 'post-modern' qualities of difference, ephemerality, superficiality and pastiche; (b) 'post-modernity' is the contemporary epoch, after a period of 'modernity', in which such post-modern forms supposedly predominate, an epoch characterized both by the loss of an overall sense of social direction and order and by the triumph of the media image over reality (*see* HYPER-REALITY and SIMULACRUM); and (c) 'post-modernism' refers more narrowly to a collection of artistic, architectural and intellectual movements that promote post-modernist values, aesthetics and ways of thinking. If that is not complicated enough, whilst some view all things post-modern as signs of a radically new historical era of post-modernity, others see them more as recent twists to the history of modernity. So perhaps a revised version of *The Independent* definition might be: 'This word has a host of meanings. This makes it interesting, but also means it should be used with care'!

post-productionist (also post-productivist): the era beyond that (productionist/productivist) which is dominated by production. Post-productionist spaces are those where processes of consumption have become very important or dominant.

psychoanalytic: largely associated with the work of Freud, psychoanalytic theory concerns itself with the mental life of individuals rather than with any overt observable behaviour, and argues that the most important elements of such mental lives are the unconscious ones. It posits that the unconscious parts of the mind (the 'id') are in perpetual conflict with both the more rational and conscious elements (the 'ego') and with those parts of the mind concerned with conscience (the 'superego'). Psychological disturbances can then be traced to these conflicts, and can be remedied through psychoanalytical therapy which is able to give an individual insight into their unconscious mental life.

race: a criterion of social categorization that distinguishes different groups of people on the basis of particular secondary physical differences (such as skin colour). Human Geographers have studied questions of race in a number of ways including: (a) the extent, causes and implications of the spatial segregation of different racial groups within cities, regions or nations; (b) the role played by geographical understandings of place and environment in the construction both of ideas of race per se and of ideas about particular races; and (c) the forms of racism and inequality that operate through these geographical patterns, processes and ideas. Increasingly Human Geographers have emphasized how racial categories, whilst having very real consequences for people's lives, cannot simply be assumed as biological realities, having instead to be recognized as SOCIAL CONSTRUCTIONS. *See also* ETHNICITY.

reductionist: a model or theory is said to be reductionist if it attempts to explain a complex idea, process or structure in terms of simpler components or elements within it. In other words the multi-dimensional character of the object under investigation is 'reduced', to one or more of its component parts.

reflexivity: a process through which we are able to reflect on what we know, how we come to know it, and how we interact with others. The key point is that we are able to change aspects of ourselves and the structures which make up society in the light of these reflections.

regulation: the arrangement of an economy into more or less cohesive forms of production, consumption, monetary circulation and SOCIAL REPRODUCTION. Derived initially from a disparate group of French political-economists (*see* POLITICAL-ECONOMY) a 'regulationist approach' therefore attends to the different ways of socially, politically and culturally organizing economic activity, especially capitalist economic activity (*see* CAPITALISM). These different ways of regulating capitalist economies are identified both historically and geographically (*see also* FORDISM and POST-FORDISM).

reification: the act of transforming human properties, relations, and actions into properties, relations and actions which are seen to be independent of human endeavour, and which then come actually to govern human life. The term can also refer to the transformations of human beings into 'thing-like' beings. Reification is therefore a form of ALIENATION.

relativize, relativization: to place equal value on different phenomena. Relativization therefore stands in opposition to proclaiming some ways of life and some knowledges as better than others.

representation: the cultural practices and forms by which human societies interpret and portray the world around them and present themselves to others. In the case of the natural world, for example, these representations range from pre-historic cave paintings of the creatures that figured in the lives of early human groups to the televisual images and scientific models that shape our imaginations today. *See also* DISCOURSE.

semiotics: this term has two interrelated meanings: (a) the study of forms of human communication and the ways they produce and convey meaning; and (b) those forms of human communication and systems of meaning themselves. Whilst including spoken and written language, semioticians deliberately also analyse how other social phenomena – including dress, architecture and the built environment, visual art, social gatherings and events, landscapes – are communication systems whose 'languages' can and should be analysed. Human Geographers have drawn on ideas from semiotics (*see* SIGN, SIGNIFICATION, REFERENT) but parallel work in art history on the language of painting has been more directly influential (*see* ICONOGRAPHY).

sexuality: sexual attitudes, preferences, desires and behaviours. Human Geographers have emphasized how our sexualities are not simply a biological given but complex sociocultural constructs (*see also* SOCIAL CONSTRUCTION). They have examined how, on the one hand, these constructs sexualize our encounters with places and environments in personally and socially significant ways and, on the other hand, how our sexualities themselves are shaped through experiences and understandings of the geographies of the body, the home, the city, the nation, travel etc.

sign, signification, referent: these are ideas relating to the DISCURSIVE construction of geographical worlds. The sign is a concept or word which is signif-

icant in the understanding of everyday meanings and places and people (for example 'rurality'). Signification is the process by which significant meanings are attached to signs (for example, social representations and interpretations of 'rurality'). The referent is the geographical phenomenon which is being signalled (for example, rural localities).

simulacrum: refers to the notion of a copy without an original. If we say at the first point we have an original object, then any image represents it, or replicas are copies of it. We thus have a pattern of original and copy. However, often it is the copies themselves that are copied. The products may begin to diverge from the original. The idea of a simulacrum, or of many simulacra, takes this a step further by emphasizing the presence in contemporary POST-MODERN landscapes of many copies with no original. An example might be the shopping mall 'Parisian cafe' that clearly imitates a Parisian cafe, but for which there is no single original it imitates; or Disneyland's main street, which is again a REPRESENTATION for which there was no original. *See also* HYPER-REALITY.

social construct: a set of specific meanings which become attributed to the characteristics and identities of people and places by common social or cultural usage. Social constructs will often represent a 'loaded' view of the subject, according to the sources from which, and the channels through which, ideas are circulated in society.

social construction: a catch-all term that emphasizes how both human geographies in the world 'out there' and the knowledges Human Geographers have of these geographies are the outcome of social practices and forces. For example, a variety of very different approaches to understanding nature would nonetheless agree that the natural world is socially moulded in material ways (*see* CULTURAL LANDSCAPE) and that knowledges of the natural world – including those of science – are the outcome of intricate social practices which make nature real to us in very particular (and contestable) ways. Whilst it runs the risk of reproducing a crude dichotomy of society and nature, an emphasis on social construction has therefore been widely drawn on in Human Geography in order to emphasize how: (a) the things, situations and ideas that surround us are not innate but the products of social forces and practices that require explanation; and (b) nor are they inevitable, instead being open to the possibility of both critique and change.

social reproduction: the processes through which societies sustain themselves in social and material terms across space and time. In purely material terms, people need to consume various commodities (housing, food, clothing, recreation etc.) in order to sustain (reproduce) themselves and these commodities have to be made available though various forms of exchange (e.g. wholesaling, retailing, sites of consumption like cinemas, concert venues, clubs etc.) But, crucially, social reproduction is necessarily also a social process. It involves the development and transmission of norms and 'rules' of behaviour around the circuits of social reproduction which give direction to, and make sense of, their activities within such circuits. Social reproduction also suggests that the purpose of economic activity is not the generation of particular moments of production, consumption or exchange but the sustenance (the reproduction) of these activities over space and time.

space of flows: first coined by the urban theorist Manuel Castells in distinction to the 'space of places', this self-confessed 'cumbersome expression' emphasizes how the character and dynamics of a bounded place are reliant upon a host of connections and flows that go beyond its boundaries. These include flows of people (through many forms of travel and migration), of capital and money (think of the impacts of the global networks of the international financial system, for example), of ideas, of media imagery and of objects, amongst many others. The notion of the 'space of flows' is therefore a complement and corrective to Human Geographers' long-standing interest in bounded places and territories (*see* TERRITORY), perhaps particularly important in an age of intensified GLOBALIZATION.

spatiality: socially produced space. This term is used by Human Geographers to emphasize how space is socially constructed and experienced, rather than being an innate backdrop to social life (*see also* SOCIAL CONSTRUCTION). As such it is a central concept of contemporary Human Geography. It is sometimes used in the plural, spatialities, in order to stress the many different ways in which space can be constructed and experienced.

spatial science: an approach to Human Geography which became influential in the 1960s by arguing that geographers should be concerned with formulating and testing theories of spatial organization, interaction and distribution. The theories were often expressed in the form of models – of, for instance land use, settlement hierarchy, industrial location and city sizes. If validated, these theories were then accorded the status of universal 'laws'. Through this manouevre, the advocates of spatial science claimed that Human Geography had been shifted from an essentially descriptive enterprise concerned with the study of regional differences to a predictive and explanatory science. Critics claim that in its attempts to formulate universally applicable laws, spatial science ignored the social and economic context within which its spatial variables were located.

structuration: an approach to social theory which stresses the interconnection between knowledgeable human agents and the wider social structures within which they operate. Although used across the social sciences, the notion of structuration is most closely associated with the British sociologist Anthony Giddens, who developed structuration theory in a number of writings in the late 1970s and early 1980s. A key point of such theory is that the wider structural properties of social systems are both the medium and the outcome of the social practices that constitute the systems.

syncretic: an adjective applied to a culture or cultural phenomenon that is composed of elements from different sources, and which combines them in such a way as to create something new and different from those sources. Drawn from anthropological literatures, notions of the syncretic are very similar to those of HYBRIDITY, but are often preferred for their less biological undertones.

technocentrism: an outlook of optimism towards any environmental or other geographical challenge on the basis of the adaptiveness and innovation of human endeavour and technological know-how.

temporality: socially produced time. This term is used by Human Geographers to emphasize how time is socially constructed and experienced, rather than being an innate backdrop to social life (*see also* SOCIAL CONSTRUCTION).

terrains of resistance: the material and/or symbolic ground upon which collective action (e.g. by social movements) takes place. This can involve the economic, political, cultural, and ecological practices of resistance movements as well as the physical places where their resistance occurs.

territory: a more or less bounded area over which an animal, person, social group or institution claims and attempts to enforce control.

transaction costs: the costs involved in engaging in transactions, normally between productive firms. Such costs (which normally do not include the costs of commodities being exchanged or the costs of transport between points of supply and points of demand) may be associated with factors like distance. Distance may serve to increase costs as delays or misinformation may be more common when transactions are taking place between partners located some way away from each other. By contrast, geographical and social proximity may help to reduce transaction costs.

transhumance: the agricultural and migratory practice of periodically moving flocks and herds of animals, along with human populations, from region to region, usually on the basis of climatic seasonality.

transnational: an adjective used to describe human geographical processes that have escaped the bounded confines of the NATION-STATE. These have been identified in the realms of the economy (*see* TRANSNATIONAL CORPORATIONS), in politics (for instance through the political agency of groups in relation to a nation-state they do not reside in; e.g. Kurdish exiles campaigning for Kurdish nationalism) and in culture (for example through the identification of 'transnational communities' that have dispersed from an originary homeland into a number of other countries but which also have strong linkages across this DIASPORA).

transnational corporations: very large companies with offices or plants in several countries; and/or companies that make decisions and accrue profits on a global basis (sometimes called multinational or global corporations).

venture capital: is the provision of finance by an investor to businesses not quoted on stock markets and is in the form of shares. The aim of venture capitalists is make a very high return on their investment by selling their shares at a later date. Venture capital is a highly risky way to invest and, partly because of this, investors typically expect returns of over 30%.

white flight: this term refers to the ethnically specific nature of outmigration from urban areas, particularly in the USA. In many of the major metropolitan areas, large scale ethnic immigration into the inner cities from the 1950s onwards was followed by outmigration by affluent whites. The result has been a growing ethnic differentiation of urban space between poor minority urban areas and white suburbs. *See also* ETHNICITY.

world–system: an integrated international economic system, founded upon mercantile then industrial CAPITALISM, originated in Europe around 1450 and spread to cover most of the world by 1900. World-systems analysis, which examines this system, treats the world as a single economic and social entity, the capitalist world economy.

References

Adams, V. (1992) Tourism and Sherpas, Nepal: reconstruction of reciprocity. *Annals of Tourism Research*, 19, 534–54.

Adams, W. (1990) *Green development: environment and sustainability in the Third World*. London: Routledge.

Adams, W. (1996) *Future nature*. London: Earthscan.

Adler, J. (1989) Origins of Sightseeing. *Annals of Tourism Research*, 16, 7–29.

Agnew, J. (1987a) *Place and Politics*. Boston: Allen and Unwin.

Agnew, J. (1987b) *The United States in the world economy: a regional geography*. Cambridge: Cambridge University Press.

Agnew, J. (1989) The devaluation of place in social science. In Agnew, J. and Duncan, J. (eds) *The power of place*. London: Unwin Hyman.

Agnew, J. (1995) The rhetoric of regionalism: the northern league in Italian politics, 1983–94. *Transactions of the Institute of British Geographers*, 20(2), 156–72.

Agnew, J. (1998) *Geopolitics*. London: Routledge.

Agnew, J. and Corbridge, S. (1989) The new geopolitics: the dynamics of geopolitical disorder. In Johnston, R. and Taylor, P. (eds) *A world in crisis?* (Second edition) Oxford: Blackwell.

Agnew, J. and Corbridge, S. (1995) *Mastering space: hegemony, territory and international economy*. London: Routledge.

Aitken, S. and Zonn, L. (1993) Weir(d) sex: representations of gender-environment relations in Peter Weir's *Picnic at Hanging Rock* and *Gallipoli*. *Environment & Planning D: Society and Space* 11, 191–212.

Albers, P. and James, W. (1988) Travel Photography – a methodological approach. *Annals of Tourism Research*, 15, 134–58.

Allen, J. (1995) Global worlds. In Allen, J. and Massey, D. (eds) *Geographical worlds*. Oxford: Oxford University Press.

Allen, J. and Massey, D. (eds.) (1995) *Geographical worlds*. Oxford: Oxford University Press.

Alleyne-Dettmers, P. (1997) Tribal arts: a case study of global compression in the Notting Hill Carnival. In Eade, J. (ed.) *Living the global city*. London: Routledge.

Amin, S. (1990) *Maldevelopment: anatomy of a global failure*. London: Zed Books.

Amsden, A. (1989) *Asia's next giant: South Korea and late-industrialisation*. New York: OUP.

Anderson, A. (1997) *Culture, media and environmental issues*. London: UCL Press.

Anderson, B. (1983) *Imagined communities: the origins and spread of nationalism*. London: Verso.

Anderson, B. (1991) *Imagined communities: reflections on the origins and spread of nationalism*. London: Verso.

Anderson, J. (1988) Nationalist ideology and territory. In Johnston, R.J., Knight, D.B. and Kaufman, E. (eds) *Nationalism, self-determination and political geography*. London: Croom Helm.

Anderson, J. (1996) The shifting stage of politics: new medieval and postmodern territorialities. *Environment and Planning D: Society and Space*, 14(2), 133–53.

Anderson, K. (1991) *Vancouver's Chinatown: racial discourse in Canada 1875–1980*. Montreal: McGill-Queen's University Press.

Anderson, K. and Gale, F. (eds) (1992) *Inventing places: studies in cultural geography*. London: Longman.

Ang, I. (1985) *Watching Dallas*. London and New York: Methuen.

Ang, I. (1991) *Desperately seeking the audience*. London and New York: Routledge.

Ansprenger, F. (1989) *The dissolution of the colonial empires*. London: Routledge.

Anzaldúa, G. (1987) *Borderlands/La Frontera: the new Mestiza*. San Francisco: Aunt Lute.

Appadurai, A. (1990) Disjuncture and difference in the global cultural economy. *Theory, Culture and Society*, 7, 295–310.

Arblaster, A. (1984) *The rise and fall of Western liberalism*. Oxford: Blackwell.

Arrighi, G. (1990) The three hegemonies of historical capitalism. *Review*, 13, 365–408.

Arrighi, G. (1994) *The long twentieth century*. London: Verso.

Atkinson, R. and Moon, G. (1994) *Urban policy in Britain*. London: Macmillan.

Augé, M. (1998). *A sense for the other*. (Trans. A. Jacobs.) Stanford, CA: Stanford University Press.

Back, L. (1995) X amount of Sat Siri Akal! Apache Indian, reggae music and the cultural intermezzo. *New Formations*, 27, 128–47.

Baier, L. (1991/92) Farewell to regionalism, *Telos* 90 (winter), 82–8.

Baker, N. C. (1984) *The beauty trap*. London: Piatkus.

Baker, S. (1993) *Picturing the beast: animals, identity and representation*. Manchester: Manchester University Press.

Baldwin, E., Longhurst, B., McCracken, S., Ogborn, M. and Smith, G. (1999) *Introducing cultural studies*. Hemel Hempstead: Prentice Hall, Chapter 5, Culture, time and history.

Barbier, E.B., Burgess, J.C. and Folke, C. (1994) *Paradise lost? The ecological economics of biodiversity*. London: Earthscan.

Barnes, C. (1991) *Disabled people in Britain and discrimination*. London: Hurst and Company.

Barnes, M. (1997) *Care, communities and citizens*. Harlow: Longman.

Barnes, T.J. and Duncan, J.S. (eds) (1992) *Writing worlds: discourse, text and metaphor in the representation of landscapes*. London: Routledge.

Barrell, J. (1980) *The dark side of the landscape: the rural poor in English painting 1730–1840*. Cambridge: Cambridge University Press.

Barrell, J. (1990) The public prospect and the private view: the politics of taste in eighteenth-century Britain. In Pugh, S. (ed) *Reading landscape: country – city – capital*. Manchester: Manchester University Press.

Barthes, R. (1972) The great family of man. In *Mythologies*. London: Jonathan Cape.

Barton, H. (1996) The Isles of Harris superquarry: concepts of environment and sustainability. *Environmental Values*, 5, 97–122.

Baudrillard, J. (1988a) *America*. London: Verso.

Baudrillard, J. (1988b) Consumer society. In Poster, M. (ed.) *Selected writings*. Cambridge: Polity Press.

Bauman, Z. (1990) *Thinking sociologically*. Cambridge, MA: Blackwell.

Bauman, Z. (1996) On communitarians and human freedom, *Theory, Culture and Society*, 13(2), 79–90.

Baumann, G. (1990) The re-invention of bhangra. *World Music*, 32(2), 81–95.

Beauregard, B. (1994) *Voices of decline: the postwar fate of US cities*. Oxford: Blackwell.

Beck, U. (1992) From industrial society to the risk society: questions of survival, social structure and ecological enlightenment. *Theory, Culture and Society*, 9, 97–123.

Bell, D. (1995) Pleasure and danger: the paradoxical spaces of sexual citizenship. *Political Geography*, 14(2), 139–54.

Bell, D. (1997) Anti-idyll: rural horror. In Cloke, P. and Little, J. (eds) *Contested countryside cultures*. London: Routledge.

Bell, D. and Valentine, G. (1995) Queer country: rural lesbian and gay lives. *Journal of Rural Studies*, 11, 113–22.

Benjamin, W. (1978) *Reflections*. New York: Harcourt Brace Jovanovitch.

Bennett, B. and Routledge, P. (1997) Tibetan resistance 1950–present. In Powers, R.S., Vogele, W.B., Kruegler, C. and McCarthy, R.M. (eds) *Protest, power, and change*. New York: Garland Publishing.

Berger, J. (1972) *Ways of seeing*. Harmondsworth: Penguin.

Berman, M. (1982) *All that is solid melts into air: the experience of modernity*. London: Verso.

Bernstein, H. (1978) Notes on capital and peasantry. *Review of African Political Economy*, 10, 53–9.

Bernstein, R. (1992). *The new constellation: the ethical – political horizons of modernity/postmodernity*. Cambridge, MA: MIT Press.

Berry, B.J.L. (1970) The geography of the United States in the year 2000, *Transactions of the Institute of British Geographers*, 51, 21–54.

Bhabha, H.K. (1994) *The location of culture*. London: Routledge.

Bhachu, P. (1985) *Twice migrants*. London: Tavistock Publications.

Billig, M. (1995) *Banal nationalism*. London: Sage.

Binnie, J. (1995) Trading places: consumption, sexuality and the production of queer space. In Bell, D. and Valentine, G. (eds) *Mapping desire*. London: Routledge.

Blaut, J. (1993) *The colonizer's model of the world*. New York: Guilford.

Bonwick, J. (1884) *The lost Tasmanian race*. London: Sampson Low, Marston, Searle and Rivington.

Boorstin, D. (1992) From traveller to tourist: The lost arts of travel. In *The image: a guide to pseudo-events in America*. Second edition. New York: Vintage Books, Random House.

Booth, D. (ed.) (1994) *Rethinking social development: theory, research and practice*. Harlow: Longman.

Borrow, G. (1989) *Wild Wales: the people, language and scenery*. Century Hutchinson.

Bouquet, M. (1987) Bed, breakfast and an evening meal: commensality in the nineteenth and twentieth century farm household. In Bouquet, M. and Winter, M. (eds) *Who from their labour's rest? Conflict and practice in rural tourism*. Aldershot: Avebury.

Bourdieu, P. (1984) *Distinction: a social critque of the judgement of taste*. Cambridge, MA: Harvard University Press.

Bourdicu, P. (1990) *The logic of practice*. Stanford, CA: Stanford University Press.

Boyce, J.K. (1992) Of coconuts and kings: the political economy of an export crop. *Development and Change*, 23(4), 1–25

Brah, A. (1996) *Cartographies of diaspora*. London: Routledge.

Brantlinger, P. (1985) Victorians and Africans: the genealogy of the myth of the Dark Continent. *Critical Inquiry*, 12, 166–203.

Brecher, J. and Costello, T. (1994) *Global village or global pillage*. Boston: South End Press.

Brown, K., Turner, R.K., Hameed, H. and Bateman, I. (1997) Environmental carrying capacity and tourism development in the Maldives and Nepal. *Environmental Conservation*, 24, 316–19.

Brundtland, H. (1987) *Our common future*. Oxford: Oxford University Press.

Bunn, D. (1994) 'Our wattled cot': mercantile and domestic space in Thomas Pringle's African landscapes. In Mitchell, W.J.T. (ed.) *Landscape and power*. Chicago, IL: University of Chicago Press.

Burgdorf, M.P. and Burgdorf, R. (1975) A history of unequal treatment: the qualifications of handicapped persons as a suspect class under the equal protection clause. *Santa Clara Lawyer*, 15, 855–910.

Burgess, J. (1985) News from nowhere: the press, the riot and the myth of the inner city. In Burgess, J. and Gold, J. (eds) *Geography, the media and popular culture*. London and Sydney: Croom Helm.

Burgess, J. (1990) The production and consumption of environmental meanings in the mass media. *Transactions of the Institute of British Geographers*, 15, 139–61.

Burgess, J. (1993) Representing nature: conservation and the mass media. In Goldsmith, B. and Warren, A. (eds) *Conservation in progress*. Chichester: John Wiley.

Burgess, J. and Gold, J. (eds) (1985) *Geography, the media and popular culture*. London and Sydney: Croom Helm.

Burgess, J. and Unwin, D. (1984) Exploring the living planet with David Attenborough. *Journal of Geography in Higher Education*, 8(2), 93–113.

Burgess, J., Harrison, C.M. and Filius, P. (1998) Environmental communication and the cultural politics of environmental citizenship. *Environment and Planning*, A, 30.

Butler, R. (1998) Rehabilitating the images of disabled youths. In Skelton, T. and Valentine, G. (eds) *Cool places: geographies of youth culture*. London: Routledge.

Butler, R. and Bowlby, S. (1997) Bodies and spaces: an exploration of disabled people's use of public space. *Environment and Planning D: Society and Space*, 15(4), 411–33.

Butler, T (1997) *Gentrification and the middleclasses*. Aldershot: Ashgate.

Cairncross, F. (1977) *The death of distance*. Cambridge, MA: Harvard Business School Press.

Cambrensis, G. (1978) *The journey through Wales and the description of Wales*. Harmondsworth: Penguin.

Campbell, D. (1992) *Writing security: United States foreign policy and the politics of identity*. Minneapolis: University of Minnesota Press.

Carey, J. and Quirk, J. (1970a) The mythos of the electronic revolution. *American Scholar*, Winter, 219–41.

Carey, J. and Quirk, J. (1970b) The mythos of the electronic revolution. *American Scholar*, Summer, 395–424.

Carney, G. (1998). *Baseball, barns and bluegrass: a geography of American folklife*. Oxford: Rowman and Littlefield.

Carr, E.H. (1961) *What is history?* London: Macmillan.

Carson, R. (1964) *Silent spring*. Boston; Little, Brown.

Castells, M. (1983) *The city and the grassroots*. London: Edward Arnold.

Castells, M. (1989) *The informational city*. Oxford: Blackwell.

Castells, M. (1996) *The rise of the network society. Volume I, The information age: economy, society and culture*. Cambridge, MA and Oxford: Blackwell.

Castells, M. (1997) *End of millennium: Volume III, The information age: economy, society and culture*. Oxford: Blackwell.

Castles, S. and Miller, M. (1993) *The age of migration*. Basingstoke: Macmillan.

Castree, N. (1995) The nature of produced nature. *Antipode*, 27(1), 12–48.

Cell, J. (1982) *The highest stage of white supremacy*. Cambridge: Cambridge University Press.

Champion, A.G. (1989) Counterurbanisation in Europe, *The Geographical Journal*, 155, 52–59.

Chaplin, M. (1998) Authenticity and otherness: the Japanese theme park. *Architectural Design*, 131, 76–9.

Charlesworth, A. (1994) Contesting places of memory: the case of Auschwitz. *Environment and Planning D: Society and Space*, 12, 579–93.

Chartier, R. (1989) The practical impact of writing. In Chartier, R. (ed.) *Passions of the Renaissance*. Cambridge, MA: Belknap Press.

Chartier, R. (1994) *The order of books: readers, authors, and libraries in Europe between the fourteenth and fifteenth centuries*. Cambridge: Polity.

Chatterjee, P. (1986) *Nationalist thought and the colonial world*. London: Zed Books.

Chatterjee, P. and Finger, M. (1994) *The earth brokers: power, politics and world development*. Routledge: London.

Chennault, C. (1948) Why we must help China now. *Reader's Digest,* April, 121–2.

Chertow, M. and Esty, D. (eds) (1997) *Thinking ecologically: the new guardians of environmental policy*. New Haven, CT: Yale University Press.

Cheshire, P (1995) A new phase of urban development in Western Europe: The evidence for the 1980s, *Urban Studies*, 32, 7, 1045–64.

Choi, S.R., Park, D. and Tschoegl, A.E. (1996) Banks and the world's major banking centres 1990. *Weltwirtschaftliches Archiv*, 132, 774–93.

Churchill, W. (1993) *Struggle for the land*. Monroe, ME: Common Courage Press.

Clark, W.A.V. (1998a) *The California cauldron: immigration and the fortunes of local communities*. New York: Guilford Press.

Clark, W.A.V. (1998b) Mass migration and local outcomes: is international migration to the United States creating a new urban underclass. *Urban Studies*, 35, 3, 371–84.

Clastres, P. (1998) *Chronicle of the Guayaki Indians*. (Trans. by P. Auster.) London: Faber and Faber.

Clifford, J. (1994) Diasporas. *Cultural Anthropology,* 9(3), 302–38.

Cloke, P. (1993) The countryside as commodity: new spaces for rural leisure. In Glyptis, S. (ed.) *Leisure and the environment*. London: Belhaven.

Cloke, P. (1997) Poor country: marginalization, poverty and rurality. In Cloke, P. and Little, J. (eds) *Contested countryside cultures*. London: Routledge.

Cloke, P. and Little, J. (eds) (1997) *Contested countryside cultures*. London: Routledge.

Cloke, P. and Milbourne, P. (1992) Deprivation and lifestyles in rural Wales II: rurality and the cultural dimension. *Journal of Rural Studies*, 8, 360–74.

Cloke, P. and Perkins, H. (1998) 'Cracking the canyon with the Awesome Foursome': representations of adventure tourism in New Zealand. *Environment and Planning D: Society and Space*, 16, 185–218.

Cloke, P., Philo, C. and Sadler, D. (1991) *Approaching human geography*. London: Paul Chapman Publishing.

Cloke, P., Crang, P., Goodwin, M., Painter, J. and Philo, C. (2000) *Practising human geography*. London: Sage.

Cohen, A. (1982) A polyethnic London carnival as a contested cultural performance. *Ethnic and Racial Studies* 5, 23–41.

Cohen, S. (1980) *Folk devils and moral panics: the creations of the Mods and Rockers*. Second edition. Oxford: Martin Robertson.

Colyer, R. (1976) *The Welsh cattle drovers: agriculture and the Welsh cattle trade before and during the nineteenth century*. Cardiff: University of Wales Press.

Commission of the European Communities (1986) Television and the audiovisual sector: towards a European policy. *European File*, 14(86), August–September.

Cone, C. (1995) Crafting Selves: the lives of two Mayan women. *Annals of Tourism Research,* 22(2), 314–27.

Conversi, D. (1990) Language or race? The choice of core values in the development of Catalan and Basque nationalisms. *Ethnic and Racial Studies*, 13(1), 50–70.

Coombes, A.E. (1994) National unity and racial and ethnic identities: the Franco–British Exhibition of 1908. In *Reinventing Africa*. New Haven, CT: Yale University Press.

Corbett, J. (1994) A proud label: exploring the relationship between disability politics and gay pride. *Disability and Society,* 9(3), 343–57.

Corbridge, S. (1993) Colonialism, post-colonialism and the political geography of the Third World. In Taylor, P. (ed.) *Political geography of the twentieth century*. London: Bellhaven Press.

Corrigan, P. (1997) *The sociology of consumption: an introduction*. Thousand Oaks: Sage.

Cosgrove, D. (1985) Prospect, perspective and the evolution of the landscape idea. *Transactions of the Institute of British Geographers*, 10, 45–62.

Cosgrove, D. (1994) Contested global visions: one-world, whole-earth and the Apollo space photographs. *Annals of the Association of American Geographers*, 84, 270–94.

Cosgrove, D. and Daniels, S. (eds) (1988) *The iconography of landscape: essays on the symbolic, design and use of past environments*. Cambridge: Cambridge University Press.

Cotgrove, S. and Duff, A. (1980) Environmentalism, middle class radicalism and politics. *Sociology Review*, 28, 335–51.

Couloubaritsis, L., De Leeuw, M., Noel, E., and Sterckx, E. (1993) *The origins of European identity*. Brussels: European Interuniversity Press.

Cowen, T. (1998) *In praise of commercial culture*. Cambridge, MA: Harvard University Press.

Crang, M. (1998) *Cultural geography*. London: Routledge.

Crang, P. (1994) It's showtime: on the workplace geographies of display in southeast England. *Environment and Planning D: Society and Space*, 12, 675–704.

Crang, P. (1997) Performing the tourist product. In Rojek, C. and Urry, J. (eds) *Touring cultures: transformations of travel and theory*. London: Routledge.

Cresswell, T. (1994) Putting women in their place: the

carnival at Greenham Common. *Antipode,* 26(1), 35–58.

Cronon, W, (1995) The trouble with wilderness: or getting back to the wrong nature. In Cronon, W. (ed.) *Uncommon ground: towards reinventing nature.* New York: W.W. Norton.

Culler, J. (1988) *Framing the sign: criticism and it institutions.* Oxford: Basil Blackwell.

Cunningham, H. (1992) *Children of the poor: representations of childhood since the seventeenth century.* Oxford: Polity Press.

Dalby, S. (1988) Geopolitical discourse: the Soviet Union as Other. *Alternatives,* XIII.

Dalby, S. (1990) American security discourse: the persistence of geopolitics. *Political Geography Quarterly,* 9(2), 171–88.

Dalby, S. (1994) Gender and geopolitics: reading security discourse in the new world order. *Environment and Planning D: Society and Space,* 12(5), 525–46.

Daly, H.E. (1990) Toward some operational principles of sustainable development. *Ecological Economics,* 2, 1–6.

Daniels, S. (1993) *Fields of vision: landscape imagery and national identity in England and the United States.* Cambridge: Polity Press.

Daniels, S. and Seymour, S. (1990) Landscape design and the idea of improvement 1730–1900. In Dodgson, R.A. and Butlin, R.A. (eds) *An historical geography of England and Wales.* London: Academic Press.

Daniels, S. and Cosgrove, D. (1993) Spectacle and text: landscape metaphors in cultural geography. In Duncan, J.S. and Ley, D. (eds), *Place/culture/representation.* London: Routledge.

Dann, G. (1996) The people of tourist brochures. In Selwyn, T. (ed.) *The tourist image: myths and myth making in tourism.* London: Wiley.

Darnton, R. (1979) *The business of the Enlightenment: a publishing history of the encyclopaedia.* Cambridge, MA: Harvard University Press.

Davidson, B. (1992) *The black man's burden.* New York: Times Books.

Davis, M. (1990) *City of quartz: excavating the future in Los Angeles.* London: Verso.

Debord, G. (1977) *Society of the spectacle.* Detroit: Black and Red Books.

Delanty, G. (1995) *Inventing Europe: idea, identity, reality.* Basingstoke: Macmillan.

Department for International Development (1997) *Eliminating world poverty: a challenge for the 21st century.* London: UK Government Stationery Office.

Der Derian, J. (1992) *Anti-diplomacy: spies, terror, speed and war.* Oxford: Blackwell.

Dery, M. (1993) Flame wars. In Dery, M. (ed.) *Flame wars: the discourse of cyberculture.* Durham, NC: Duke University Press.

Deskins, D.R. (1996) Economic restructuring, job opportunities and black social dislocation in Detroit. In O'Loughlin, J. and Friedrichs, J. (eds) *Social polarization in post-industrial metropolises.* Berlin and New York: de Gruyter.

Dicken, P. (1998) *Global shift.* Third edition. London: Paul Chapman.

Dicken, P. and Thrift, N. (1992) The organization of production and the production of organization: why business enterprises matter in the study of geographical industrialization. *Transactions of the Institute of British Geographers* NS 17(3), 279–91.

Dodd, L. (1992) Heritage and the 'Big House': whitewash for rural history. *Irish Reporter,* 6, 9–11.

Dodd, L. (1993) Interview by Nuala Johnson, Strokestown Park House, September 2.

Doel, M. (1994) Deconstruction on the move: from libidinal economy to liminal materialism. *Environment and Planning A,* 26, 1041–59.

Domosh, M. (1991) Towards a feminist historiography of geography. *Transactions of the Institute of British Geographers,* NS 16, 94–104.

Dorra, M. (1996) La traversée des apparences, *Le Monde Diplomatique,* June, 32.

Douglas, M. and Isherwood, B. (1979) *The world of goods.* New York: Basic Books.

Downs, A. (1972) Up and down with ecology: the issue-attention cycle. *The Public Interest,* 28, 38–50.

Drakulic, S. (1996) *Café Europa.* London: Abacus.

Driver, F. (1992) Geography's empire: histories of geographical knowledge. *Environment and Planning D: Society and Space,* 10, 23–40.

Driver, F. and Gilbert, D. (1998) Heart of empire? Landscape, space and performance in imperial London. *Environment and Planning D: Society and Space,* 16, 11–28.

Dryzek, J. (1997) *The politics of the earth.* Oxford: Oxford University Press.

Dunbar, G. (1974) Geographical personality. *Geoscience and Man,* V, 25–33.

Duncan, J. (1990) *The city as text: the politics of landscape interpretation in the Kandyan kingdom.* Cambridge: Cambridge University Press.

Duncan, J. (1995) Landscape geography 1993–94. *Progress in Human Geography,* 19, 414–22.

Duncan, S. and Goodwin, M. (1988) *The local state and uneven development.* Cambridge: Polity Press.

Dyer, R. (1997) *White.* London: Routledge.

Dymski, G.A. and Veitch, J.M. (1996) Financial transformation and the metropolis: booms, busts and banking in Los Angeles. *Environment and Planning A* 28, 1233–60.

Eade, J. (1997) Identity, nation and religion: educated

young Bangladeshi Muslims in London's East End. In Eade, J. (ed.) *Living the global city*. London: Routledge.

Eboda, M. (1997) Rum do as Reggae Boyz blow hot. *The Observer*, 18 November 1997, 12.

Eco, U. (1985) How culture conditions the colours we see. In Blonsky, M. (ed.) *On signs*. Oxford: Blackwell.

Ecologist, The (1972) *A Blueprint for Survival*. Harmondsworth: Penguin.

Edwards, E. (1996) Postcards: greetings from another world. In Selwyn, T. (ed.) *The tourist image: myths and myth making in tourism*. London: Wiley.

Ehrlich, P.R. (1972) *The population bomb*. London: Ballantine.

Eksteins, M. (1990) *Rites of spring, the Great War and the birth of the modern age*. New York: Bantum Books.

Elsaesser, T. (1994) European television and national identity: or 'what's there to touch when the dust has settled'. Paper presented to the European Film and Television Studies Conference, London, July.

Emmer, P. (1993) Intercontinental migration as a world historical process. *European Review,* 1(1), 67–74.

English Nature (1993) Position statement on sustainable development. November.

Enloe, C. (1989) *Bananas, beaches and bases: making feminist sense of international relations*. Berkeley, CA: University of California Press.

Entrikin, J.N. (1991) *The betweeness of place: towards a geography of modernity*. London: Macmillan.

Entrikin, J.N. (1994) Place and region. *Progress in Human Geography,* 18(2), 227–33.

Escobar, A. (1992) Culture, economics, and politics in Latin American social movements theory and research. In Escobar, A. and Alvarez, S.E. (eds) *The making of social movements in Latin America*. Boulder, CO: Westview Press.

Escobar, A. (1995) *Encountering development: the making and unmaking of the Third World*. Princeton: Princeton University Press.

Falah, G. (1989) Israeli 'Judaization' policy in Galilee and its impact on local Arab urbanization. *Political Geography*, 8(3), 229–54.

Fisher, A. (1996) Deutsche Bank in Asia facing a rough ride in the East. *The Financial Times,* 13 November, 25.

Fisher, A. (1997) A reluctant departure. *The Financial Times*, 13 May, 21.

Fisher, A. (1998) Deutsche Bank warns of lower profits. *The Financial Times*, 29 January, 36.

Florida, R. and Smith, D.F. (1993) Venture capital formation, investment and regional industrialisation. *Annals of the Association of American Geographers*, 83, 434–51.

Foucault, M. (1980) *Power/knowledge*. London: Harvester Wheatsheaf.

Fox, J. (1996) How does civil society thicken? The political construction of social capital in rural Mexico. *World Development*, 24, 1089–1103.

Frank, A.G. (1967) *Capitalism and underdevelopment in Latin America*. London: Monthly Review Press.

Frank, T. and Weiland, M. (1997) *Commodify your dissent: salvos from the baffler*. New York: W.W. Norton and Company.

Friedland, W., Barton, A. and Thomas, R. (1981) *Manufacturing green gold*. Cambridge: Cambridge University Press.

Friedmann, J. and Wolff, G. (1982) World city formation: an agenda for research and action. *International Journal of Urban and Regional Research*, **6**, 3, 309–44.

Frow, J. (1991) Tourism and the semiotics of nostalgia. *October*, 57, 123–51.

Gabriel, Y. and Lang, T. (1996) *The unmanageable consumer: contemporary consumption and its fragmentations*. Thousand Oaks: Sage.

Gadgil, M. and Guha, R. (1995): *Ecology and equity*. London: Routledge.

Garitaonandía, G. (1993) Regional television in Europe. *European Journal of Communication*, 9(3), 277–94.

Garreau, J. (1991) *Edge city: life on the new frontier*. New York: Doubleday.

Garton Ash, T. (1998) Europe's endangered liberal order. *Foreign Affairs*, 77(2), 51–65.

Gedicks, A. (1993): *The new resource wars*. Boston: South End Press.

Gereffi, Gary (1995) Global production systems and Third World development. In Stallings, B. (ed.) *Global change, regional response,* Cambridge: Cambridge University Press.

Gertler, M. (1997) The invention of regional culture. In Lee, R. and Wills, J. (eds) *Geographies of economies*. London: Arnold.

Gibson, W. (1984) *Neuromancer*. New York: Ace.

Giddens, A. (1990) *The consequences of modernity*. Cambridge: Polity Press.

Giddens, A. (1991) *Modernity and self identity: self and society in the late modern age*. Cambridge: Polity Press.

Gilbert, E.W. (1951) Geography and regionalism. In Taylor, G. (ed.) *Geography in the twentieth century*. London: Methuen.

Gilbert, E.W. (1960) The idea of the region. *Geography,* 45(3), 157–75.

Gilbert, E.W. (1972) British regional novelists and Geography. In *British pioneers in geography*. Newton Abbott: David and Charles.

Gilderbloom, J.I. and Rosentraub, M.S. (1990) Creating the accessible city: proposals for providing housing and transportation for low income, elderly and disabled people. *American Journal of Economics and Sociology,* 49(3), 271–82.

Gillespie, M. (1995) *Television, ethnicity and cultural change*. London: Routledge.

Gilroy, P. (1987) *There ain't no black in the Union Jack*. London: Routledge.

Gilroy, P. (1993a) *The black Atlantic: modernity and double consciousness*. London: Verso.

Gilroy, P. (1993b) *Small acts*. London and New York: Serpent's Tail.

Ginsburg, F. (1991) Indigenous media: Faustian contract or global village? *Cultural Anthropology*, 94–114.

Ginsburg, F. (1993) Aboriginal media and the Australian imaginary. *Public Culture*, 5, 557–78.

Gladwell, M. (1997) The coolhunt. *The New Yorker*, March 17: 78–88.

Glassner, M.I. (1993) *Political geography*. Chichester: John Wiley.

Godlewska, A. and Smith, N. (eds) (1994) *Geography and empire*. London: Blackwell/IBG.

Godolphin, M. (1868) *Robinson Crusoe in words of one syllable*. London: George Routledge.

Goffman, E. (1963) *Stigma*. Englewood Cliffs, NJ: Prentice Hall.

Goodwin, M. (1991) Replacing a surplus population: the policies of London Docklands Development Corporation. In Allen, J. and Hamnett, C. (eds) *Housing and Labour Markets*. London: Unwin and Hyman.

Goodwin, M. and Painter, J. (1996) Local governance, Fordism and the changing geographies of regulation. *Transactions of the Institute of British Geographers*, 21(4), 635–49.

Gopinath, G. (1995) 'Bombay, UK, Yuba City': Bhangra music and the engendering of diaspora. *Diaspora*, 4(3), 303–22.

Goss, J. (1993) The magic of the mall: form and function in the retail built environment. *Annals of the Association of American Geographers*, 83(1), 18–47.

Goss, J. (1999) Once upon a time in the commodity world: an unofficial guide to Mall of America. *Annals of the Association of American Geographers*.

Gowan, P. and Anderson, P. (eds) (1997) *The question of Europe*. London: Verso.

Gray, A. (1992) *Video playtime: the gendering of a leisure technology*. London: Routledge.

Green, D. (1991) *Faces of Latin America*. London: Latin America Bureau.

Green, E., Hebron, S. and Woodward, W. (1990) *Women's leisure, what leisure?* Basingstoke: Macmillan.

Greenwood, D. (1989) Culture by the pound: an anthropological perspective on tourism as cultural commoditization. In Smith, V. (ed.) *Hosts and guests*. (Second edition) Philadelphia: University of Pennsylvania Press.

Gregory, D. (1982) *Regional transformation and industrial revolution: a geography of the Yorkshire woollen industry*. London: Macmillan.

Gregory, D. (1994) *Geographical imaginations*. Oxford: Blackwell.

Gregory, D. (1995) Imaginative geographies. *Progress in Human Geography*, 9, 447–85.

Griffiths, I. (ed.) (1993) *The atlas of African affairs*. London: Routledge.

Griffiths, J. (1997) F1 probe says threat to relocate racing was groundless. *The Financial Times*, 15 December, 1.

Gruffudd, P. (1994) 'Back to the land': historiography, rurality and the nation in inter-war Wales'. *Transactions of the Institute of British Geographers*, 19(1), 61–77.

Guback, T. (1974) Cultural identity and film in the European Economic Community. *Cinema Journal*, 13(1), 2–17.

Guha, R. (1989) The Problem. *Seminar*, March, 12–15.

Habermas, J. (1993) *The structural transformation of the public sphere*. Cambridge, MA: MIT Press.

Halfacree, K. (1993) Locality and social representation: space, discourse and alternative definitions of the rural. *Journal of Rural Studies*, 9, 23–38.

Halfacree, K. (1996) Out of place in the country: travellers and the 'rural idyll'. *Antipode*, 28 (1), 42–72.

Hall, S. (1991) Old and new identities, old and new ethnicities. In King, A.D. (ed.) *Culture, globalization and the world system*. London: Macmillan.

Hall, S. (1992) The question of cultural identity. In Hall, S., Held, D. and McGrew, T. (eds) *Modernity and its futures*. Oxford: Polity.

Hall, S. (1992) The West and the rest. In Hall, S. and Gieben, B. (eds) *Formations of modernity*. Oxford: Polity.

Hall, S. (1995) New cultures for old. In Massey, D. and Jess, P. (eds) *A place in the world?* Milton Keynes: Open University Press.

Hamnett, C. and Cross, D. (1998) Social polarisation and inequality in London: earnings evidence, 1979–95. *Environment and Planning C, Government and Policy*, 16, 659–680.

Hardin, G. (1968) The tragedy of the commons. *Science*, 162, 1243–8.

Harley, J.B. (1992) Deconstructing the map. In Barnes, T. and Duncan, J. (eds) *Writing worlds: discourse, text and metaphor in the representation of landscape*. London: Routledge.

Harris, N. (1995) *The new untouchables: immigration and the new world order*. Harmondsworth: Penguin.

Harris, R.C. and Phillips, E. (eds) (1984) *Letters from Windermere 1912–1914*. Vancouver: UBC Press.

Harrison, C.M., Burgess, J. and Filius, P. (1996) Rationalising environmental responsibilities: a comparison of lay publics in the UK and the Netherlands. *Global Environmental Change*, 6(3), 215–34.

Harriss, J., Hunter, J. and Lewis, C. (eds) (1995) *The new institutional economics and Third World development*. London: Routledge.

Harvey, D. (1969) *Explanation in geography*. London: Edward Arnold.

Harvey, D. (1985) Paris, 1850–1870. In Harvey, D. *Consciousness and the urban experience*. Oxford: Blackwell.

Harvey, D. (1989) *The condition of postmodernity: an enquiry into the origins of cultural change*. Oxford: Blackwell.

Harvey, N. (1995) Rebellion in Chiapas: rural reforms and popular struggle. *Third World Quarterly*, 16(1), 39–72.

Hay, C. (1995) Structure and agency. In Marsh, D. and Stoker, G. (eds) *Theory and methods in political science*. Basingstoke: Macmillan.

Heath, S. (1990) *Representing television*. In Mellencamp, P. (ed.) *Logics of television*. Bloomington: Indiana University Press.

Hechter, M. (1975) *Internal colonialism: the Celtic fringe in British national development, 1536–1966*. London: RKP.

Henry, N. and Pinch, S. (1997) *A regional formula for success? The innovative region of motor sport valley*. Edgbaston: University of Birmingham.

Henry, N., Pinch, S. and Russell, S. (1996) In pole position? Untraded interdependencies, new industrial spaces and the British Motor Sport Industry. *Area* 28.1, 25–36.

Herrington, J. (1984) *The outer city*. London: Harper and Row.

Hobsbawm, E. (1983) Introduction: inventing traditions. In Hobsbawm, E. and Ranger, T. (eds) *The invention of tradition*. Cambridge: Cambridge University Press.

Hobson, D. (1989) Soap operas at work. In Seiter, *et al.* (eds) *Remote control*. London and New York: Routledge.

Hochschild, A. (1983) *The managed heart: commercialisation of human feeling*. Berkeley, CA: University of California Press.

hooks, b. (1991) *Yearning: race gender and cultural politics*. London: Turnaround.

Hooson, I.D. (ed.) (1994) *Geography and national identity*. Oxford: Blackwell.

Hudson, B. (1977) The new geography and the new imperialism, 1870–1918. *Antipode,* 9(2), 12–19.

Huggins, J., Huggins, R. and Jacobs, J.M. (1995) Kooramindanjie: place and the postcolonial. *History Workshop Journal*, 39, 164–81.

Hughes, R. (1988) *The fatal shore*. New York: Vintage.

Hyam, R. (1990) *Empire and sexuality*. Manchester: Manchester University Press.

Hyam, R. (1993) *Britain's imperial century*. Basingstoke: Macmillan.

Ignatieff, M. (1993) *Blood and belonging: journeys into the New Nationalism*. London: BBC Books/Chatto & Windus.

Inglehart, R. (1977) *The silent revolution*. Princeton, NJ: Princeton University Press.

Inkeles, A. and Smith, D.H. (1974) *Becoming modern*. Cambridge, MA: Harvard University Press.

Irwin, A. (1995) *Citizen science: a study of people, expertise and sustainable development*. London: Routledge.

IUCN (1980) *The world conservation strategy, international union for the conservation of nature and natural resources*. Gland: World Wildlife Fund and United Nations Environment Programme.

Jackson, P. (1989) *Maps of meaning*. London: Routledge.

Jackson, P. and Penrose, J. (eds) (1993) *Constructions of race, place and nation*. London: UCL Press.

Jacobs, J.M. (1988) Politics and the cultural landscape: the case of Aboriginal land rights. *Australian Geographical Studies*, 26, 249–63.

Jacobs, J.M. (1996) Authentically yours: de-touring the map. In Jacobs, J.M. *Edge of empire: postcolonialism and the city*. London: Routledge.

Jacobs, J.M. (1996) *Edge of empire: postcolonialism and the city*. London: Routledge.

Jacobs, M. (1993) *Sense and sustainability: land use planning and environmentally sustainable development*. London: Council for the Protection of Rural England.

Jacobs, M. (1995) Sustainable development, capital substitution and economic humility: a response to Beckerman. *Environmental Values*, 4, 57–68.

Jenkins, H. (1992) *Textual poachers: television fans and participatory culture*. London and New York: Routledge.

Jenkins, J.G. (1982) *Maritime heritage: the ships and seamen of southern Ceredigion*. Llandysul: Gomer Press.

Johnson, J. and Salt, J. (1992) *Population migration*. Walton-on-Thames: Thomas Nelson.

Johnson, N.C. (1996) Where geography and history meet: heritage tourism and the big house in Ireland. *Annals of the Association of American Geographers*, 86, 551–66.

Johnson, R. (1986) The story so far: and further transformations? In Punter, D. (ed.) *Introduction to contemporary cultural studies*. London and New York: Longman.

Johnston, R.J., Gregory, D. and Smith, D.M. (eds) (1994) *The dictionary of human geography*. Third edition. Oxford: Blackwell.

Jones, J.-P. and Moss, P. (1995). Democracy, identity, space. *Environment and Planning D: Society and Space,* 13, 253–7.

Karp, I. and Lavine, S. (1991) (eds) *Exhibiting cultures: the poetics and politics of museum display*. Washington DC: Smithsonian Institute.

Keat, R. and Urry, J. (1975) *Social theory as science*. London: Routledge & Paul.

Kern, S (1983) *The culture of time and space, 1880–1918*. Cambridge, MA: Harvard University Press.

Khilnani, S. (1997) *The idea of India*. London: Hamish Hamilton.

King, R. (1995) Migrations, globalisation and place. In Massey, D. and Jess, P. *A place in the world?* Oxford: Oxford University Press.

Kinsman, P. (1995). Landscape, race and national identity: the photography of Ingrid Pollard. *Area*, 27, 300–10.

Knowles, A. (1997) *Calvinists incorporated: Welsh immigrants on Ohio's industrial frontier*. Chicago: University of Chicago Press.

Knox, P. (1991) The restless urban landscape: economic and sociocultural change and the transformation of Metropolitan Washington DC. *Annals of the Association of American Geographers*, 81(2), 181–209.

Kristeva, J. (1992) Le temps de la dépression. *Le Monde des Débats*, October.

Kristeva, J. (1993) *Nations without nationalism*. New York: Columbia University Press.

Lacey, C. and Longman, D. (1993) The press and public access to the environmental debate. *Sociological Review*, 41, 207–43.

Lamm, S.E., Reed, D.A. and Scullin, W.H. (1996) *Real-time geographic visualization of World Wide Web traffic*. http://www.pablo.cs.uiuc.edu/Projects/Mosaic/ww3/.

Landau, J.M. (ed.) (1984) *Atatürk and the modernization of Turkey*. Boulder, CO: Westview Press.

Lash, S. and Urry, J. (1994) *Economies of signs and space*. London: Sage.

Lee, R. (1998) Shelter from the storm? Mutual knowledge and geographies of regard (or legendary economic geographies). Paper presented to the RGS-IBG Annual Conference. University of Surrey, Guildford, 6 January.

Lee, R. (1999) *Access to the gods? Social relations and geographies of material life*. Routledge: London.

Leftwich, A. (1993) Governance, democracy and development in the Third World. *Third World Quarterly*, 14(3), 605–24.

Leftwich, A. (1995) Governance, the state and the politics of development. *Development and Change*, 25, 363–86.

Lehmann, D. (1997) An opportunity lost: Escobar's deconstruction of development. *Journal of Development Studies*, 33(4), 568–78.

Lélé, S.M. (1991) Sustainable development: a critical review. *World Development*, 19, 607–21.

Leonard, M. (1998) Paper planes: travelling the new grrrl geographies. In Skelton, T. and Valentine, G. (eds) *Cool places: geographies of youth cultures*. London and New York: Routledge.

Levy, R. (1995) Finding a place in the world economy. Party strategy and party vote: the regionalization of SNP and Plaid Cymru support, 1979–92. *Political Geography*, 14(3), 295–308.

Lewis, W.A. (1955) *The theory of economic growth*. London: George Allen and Unwin.

Ley, D. (1974) The black inner city as frontier outpost: images and behaviour of a Philadelphia neighbourhood. *Association of American Geographers, Monograph Series* 7, Washington DC.

Ley, D. (1977) Social geography and the taken-for-granted world. *Transactions of the Institute of British Geographers* NS 2, 498–512.

Ley, D. (1996) *The new middle class and the remaking of the central city*. Oxford: Oxford University Press.

Ley, D. and Cybriwsky, R. (1974) Urban graffiti as territorial markers. *Annals of the Association of American Geographers*, 64, 491–505.

Ley, D. and Samuels, H. (eds) (1978) *Humanistic geography: prospects and problems*. London: Croom Helm.

Leyshon, A. (1996) Dissolving difference? Money, disembedding and the creation of global financial space. In Daniels, P. and Lever, W.F. (eds) *The global economy in transition*. London: Longman.

Leyshon, A. and Thrift, N.J. (1997) *Money/space*. London: Routledge.

Liebes, T. and Katz, E. (1990) *The export of meaning: cross-cultural readings of Dallas*. Oxford: Oxford University Press.

Lilley III, W. and De Franco, L.J. (1997) No guarantees for F1's 'Sport Valley'. *The Financial Times*, 31 December.

Lipsitz, G. (1994) Kalfou Dangere. In *Dangerous crossroads: popular music, postmodernism and the poetics of place*. London: Verso.

Little, J., Peake, L. and Richardson, P. (1988) *Women in cities*. London: Macmillan.

Livingstone, D.N. (1992) *The geographical tradition*. Oxford: Basil Blackwell.

Livingstone, D.N. and Harrison, R.T. (1981) Hunting the snark: perspectives on geographical investigation. *Geografiska Annaler*, 63B, 69–72.

Logan, J. and Molotch, H. (1987) *Urban fortunes*. Beverley, CA: University of California Press.

Longhurst, R. (1995) The body and geography. *Gender, Place and Culture*, 2(1), 97–105.

Lowe, P. and Rudig, R. (1986) Political ecology and the social sciences: the state of the art. *British Journal of Sociology*, 16, 513–50.

Lowenthal, D. (1991) British national identity and the English landscape. *Rural History*, 2, 205–30.

Lowenthal, D. (1994) Identity, heritage and history. In Gillis, J.R. (ed.) *Commemorations: the politics of national identity*. Princeton, NJ: Princeton University Press.

Lowenthal, D (1996) *The heritage crusade and the spoils of history*. London: Viking.

Lowenthal, D. (1997) *Geographical Journal*, 163, 355.

Luke, T. (1997) At the end of nature: cyborgs, 'humachines' and environments in postmodernity. *Environment and Planning A*, 29, 1367–80.

Lutz, C and Collins, J. (1993) *Reading National Geographic*. Chicago, IL: University of Chicago Press.

MacCannell, D. (1989) *The tourist: a new theory of the leisure classes*. (Second edition). New York: Schocken.

MacCannell, D. (1992) *Empty meeting grounds: the tourist papers*. Routledge: London.

MacKenzie, J. (1995) *Orientalism: history, theory and the arts*. Manchester: Manchester University Press.

Maffesoli, M. (1996) *The time of the tribes: the decline of individualism in mass society*. Trans. D. Smith. Thousand Oaks: Sage.

Manguel, A. (1996) *A history of reading*. London: HarperCollins.

Martin, D.-C. (1992) Le choix d'identité. *Revue Française de Science Politique*, 42(4) 582–93.

Martin, H.-J. (1994) *The history and power of writing*. Chicago: University of Chicago Press.

Martin, R. and Minns, R. (1995) Undermining the financial basis of regions: the spatial structure and implications of the UK pension fund system. *Regional Studies*, 29, 125–44.

Martinez-Allier, J. (1990) Ecology and the poor: A neglected dimension of Latin American history. *Journal of Latin American Studies*, 23, 621–39.

Martinussen, J. (1997) *Society, state and the market: a guide to competing theories of development*. London: Zed Books.

Marx, K. (1976 [1867]). *Capital*. Volume 1. Harmondsworth: Penguin Books.

Marx, K. and Engels, F. (1967 [1848]) *The communist manifesto*. Harmondsworth: Penguin Books.

Maslow, H. (1970) *Motivation and personality*. New York: Harper and Row.

Mason, C. and Harrison, R. (1998) Financing entrepreneurship: venture capital and regional development. In Martin, R.L. (ed.) *Money and the space economy*. Chichester: John Wiley.

Massey, D. (1993) Power-geometry and a progressive sense of place. In Bird, J., Curtis, B., Putnam, T., Robertson, G. and Tickner, L. (eds) *Mapping the futures: local cultures, global change*. London: Routledge.

Massey, D. (1994 [1991]) A global sense of place. In *Space, place and gender*. Oxford: Polity Press.

Massey, D. (1995) The conceptualization of place. In Massey, D. and Jess, P. (eds) *A place in the world?* Oxford: Oxford University Press.

Massey, D. (1995) *Spatial divisions of labour: social structures and the geography of production*. Second edition. Basingstoke and London: Macmillan.

Massey, D. and Jess, P. (1995) Places and cultures in an uneven world. In Massey, D. and Jess, P. (eds) *A place in the world?* Oxford: Oxford University Press.

Massey, D. and Jess, P. (eds) (1995) *A place in the world?* Oxford: Oxford University Press.

Massey, D.S. and Denton, N.A. (1993) *American apartheid*, Cambridge, MA: Harvard.

Matless, D. (1992) An occasion for geography: landscape representation and Foucault's corpus. *Environment and Planning D: Society and Space*, 10, 41–56.

McClintock, A. (1995) *Imperial leather: race, gender and sexuality in the colonial contest*. New York: Routledge.

McCormick, J. (1988) America's third world. *Newsweek*, 8 August, 20–24.

McCormick, J. (1991) *British politics and the environment*. London: Earthscan.

McCormick, J.S. (1992) *The global environmental movement: reclaiming paradise*. London: Belhaven.

McDowell, L. (1997) *Capital culture*. Oxford: Blackwell.

McKibben, B. (1990) *The end of nature*. Oxford: Oxford University Press.

McLuhan, M. (1964) *Understanding media*. London: Routledge and Kegan Paul.

McNay, L. (1994) *Foucault: a critical introduction*. Cambridge: Polity Press.

Mearsheimer, J. (1990) Why we will soon miss the Cold War. *The Atlantic* 266(2), 35–50.

Meinig, D. (ed.) (1979) *The interpretation of ordinary landscapes*. Oxford and New York: Oxford University Press.

Merriman, N. (ed.) (1993) *The peopling of London: fifteen thousand years of settlement from overseas*. London: Museum of London.

Meyrowitz, J. (1985) *No sense of place: the impact of electronic media on social behaviour*. Oxford: Oxford University Press.

Milbourne, P. (ed.) (1997) *Revealing rural others*. London: Pinter.

Miller, D. (1992) The young and the restless in Trinidad. A case study of the local and the global in mass consumption. In Silverstone, R. and Hirsch, E. (eds) *Consuming technologies*, London: Routledge.

Miller, D. (1994) *Modernity: an ethnographic approach*. Oxford: Berg.

Miller, D. (1995) Consumption as the vanguard of history. In Miller, D. (ed.) *Acknowledging consumption: a review of new studies*. London: Routledge.

Miller, D. (1997) *Capitalism: an ethnographic approach*. Oxford: Berg.

Mills, S. (1989) Tourism and Leisure – setting the scene. *Tourism Today*, 6, 18–21.

Milward, A. (1992) *The European rescue of the nation state*. London: Routledge.

Mishan, E. (1969) *The costs of economic growth*. Harmondsworth: Penguin Books.

Mitchell, W.J.T. (1994a) Introduction. In Mitchell, W.J.T. (ed.) *Landscape and power*. Chicago and London: University of Chicago Press.

Mitchell, W.J.T. (1994b) Imperial Landscape. In Mitchell, W.J.T. (ed.) *Landscape and power*. Chicago and London: University of Chicago Press.

Mohanty, C. (1991) Cartographies of struggle, Third World women and the politics of feminism. In Mohanty,C., Parker, A. and Russo, A. (eds) *Cartographies of struggle, Third World women and the politics of feminism*. London: Routledge.

Mollenkopf, J. and Castells, M. (1991) *Dual city? Restructuring New York*. Russell Sage Foundation.

Moodie, S. (1986) *Roughing it in the Bush*. London: Virago

Moody, R. (ed.) (1988) *The indigenous voice*. (2 vols.) London: Zed Books.

Moore, R. (1992) Marketing alterity. *Visual Anthropology Review*, 8(2), 10–26.

Morales, E. (1989) *Cocaine*. Tucson: University of Arizona Press.

Morley, D. (1986) *Family television: cultural power and domestic leisure*. London: Comedia.

Morley, D. (1991) Where the global meets the local: notes from the sitting room. *Screen*, 32, 1–15.

Morley, D. (1992) Where the global meets the local: notes from the sitting room. In Morley, D. *Television audiences and cultural studies*. Routledge.

Morley, D. and Robins, K. (1995) *Spaces of identity: global media, electronic landscapes and cultural boundaries*. London: Routledge.

Mormont, M. (1990) Who is rural? Or how to be rural: towards a sociology of the rural. In Marsden, T., Lowe, P. and Whatmore, S. (eds) *Rural restructuring*. London: David Fulton.

Morris, J. (1986) *The matter of Wales: epic views of a small country*. Harmondsworth: Penguin Books.

Morris, J. (1990) *Hong Kong: epilogue to an empire*. London: Penguin Books.

Morris, J. (1991) *Pride against prejudice*. London: The Women's Press.

Morris, J. (1992) *O! Canada*. London: Hale.

Moser, C. (1993) *Gender planning and development*. Routledge, London.

Murdoch, J. and Pratt, A. (1993) Rural studies: mod-ernism, postmodernism and the 'post-rural'. *Journal of Rural Studies*, 9, 411–28.

Murdoch, J. and Pratt, A. (1997) From the power of topography to the topography of power: a discourse in strange ruralities. In Cloke, P. and Little, J. (eds) *Contested countryside cultures*. London: Routledge.

Murgatroyd, L. and Neuburger, H. (1997) A household satellite account for the UK. *Economic Trends*, 527, October, 63–71.

Nairn, T. (1977) *The break-up of Britain*. London: New Left Books.

Nairn, T. (1995) Breakwaters of 2000: from ethnic to civic nationalism. *New Left Review*, 214, 91–103.

Nandy, A. (1984) Culture, state and rediscovery of Indian politics. *Economic and Political Weekly*, 19(49), 2078–83.

National Portrait Gallery (1996) *David Livingstone and the victorian encounter with Africa*. London: National Portrait Gallery.

Newby, H. (1988) *The countryside in question*. London: Hutchinson.

Newhouse, J. (1997) Europe's rising regionalism. *Foreign Affairs*, 76(1), 67–84.

Newman, D. (1989) Civilian and military presence as strategies of territorial control: the Arab-Israeli conflict. *Political Geography*, 8(3), 215–28.

Nijman, J. (1994) Nicholas Spykman. In O'Loughlin, J. (ed.) *Dictionary of geopolitics*. Westport, CT: Greenwood Press.

Nijman, J. (1996) Ethnicity, class and the economic internationalization of Miami. In O'Loughlin, J. and Friedrichs, J (eds) *Social polarization in post-industrial metropolises*. Berlin and New York: de Gruyter.

Nora, P. (1989) Between memory and history: *Les Lieux de Mémoire*. *Representations*, 26, 7–25.

Norris, C. (1992) *Uncritical theory: postmodernism, intellectuals and the Gulf War*. Amherst, MA: University of Massachusetts Press.

Nye, D. (1991) The emergence of photographic discourse: images and consumption. In Nye, D. and Pedersen, C. (eds) *Consumption and American culture*. Amsterdam: VU University Press.

Oakley, A. (1981) *Subject women*. London: Martin Robertson.

O'Brien, R. (1992) *Global financial integration: the end of geography?* London: Pinter/ RIIA.

Offe, K. and Heinze, R. (1992) *Beyond employment*. London: Polity Press.

Office of Population Censuses and Surveys, Social Survey Division (1993–97) *General household survey*. London: HMSO.

Ogborn, M. (1998) *Spaces of modernity: London's geographies 1680–1780*. New York: Guilford Press.

Oppenheim, C. (1990) *Poverty: the facts*. London: Child Poverty Action Group.

O'Riordan, T. (1976) *Environmentalism*. London: Pion.

O'Riordan, T. (1988) The politics of sustainability. In: Turner, R.K. (ed.) *Sustainable environmental management*. London: Belhaven Press.

O'Riordan, T. (ed.) (1999) *Environmental science for environmental management*. Second edition. Harlow: Longman.

O'Riordan, T. and Jordan, A, (1998) Kyoto in Perspective. *ECOS*, 18, 314, 38–42.

O'Riordan, T. and Voisey, H. (eds) (1998) *The transition to sustainability*. London: Earthscan.

Osborne, P. (1996) Modernity. In Payne, M. (ed.) *A dictionary of cultural and critical theory*. Oxford: Blackwell.

Ó Tuathail, G. (1996) *Critical geopolitics*. Minneapolis: Minnesota University Press.

Ó Tuathail, G. and Agnew, J. (1992) Geopolitics and discourse: practical geopolitical reasoning in American foreign policy. *Political Geography*, 11(2), 190–204.

Overton, M. (1994) Historical geography. In Johnston, R.J., Gregory, D. and Smith, D.M. (eds) *The dictionary of human geography*, third edition. Oxford: Blackwell.

Owens, S. (1997) Negotiated environments: needs, demands, and values in the age of sustainability. *Environment and Planning A*, 29, 571–80.

Painter, J. and Philo, C. (1995) Spaces of citizenship: an introduction. *Political Geography*, 14(2), 107–20.

Park, R. (1926) The urban community as a spatial pattern and a moral order. In Burgess, E.W. (ed.) *The urban community*. Chicago, IL: University of Chicago Press.

Parker, J. and Smith, C. (1940) *Modern Turkey*. London: George Routledge & Sons.

Peach, C. (ed.) (1975) *Urban social segregation*. London: Longman.

Peach, C. (1996) Does Britain have ghettoes? *Transactions of the Institute of British Geographers*, 21(1), 216–35.

Pearce, D. *et al.* (1988) *Blueprint for a green economy*. London: Earthscan.

Pearce, D. *et al.* (1993) *Blueprint three: measuring sustainable development*. Earthscan: London.

Peet, R. (1989) World capitalism and the destruction of regional cultures. In Johnston, R.J. and Taylor, P. (eds) *The world in crisis?* Second edition, Oxford: Blackwell.

Peet, R. and Thrift, N. (1989) Political economy and human geography. In Peet, R. and Thrift, N. (eds) *New models in geography: volume 1. The political-economy perspective*. London: Unwin Hyman.

Peet, R. and Watts, M. (eds) (1996) *Liberation ecologies: environment, development and social movements*. London: Routledge.

Pepper, D. (1984) *The roots of modern environmentalism*. London: Routledge.

Pepper, D. (1996) *Modern environmentalism: an introduction*. London: Routledge.

Perlmutter, T. (1993) Distress signals: a Canadian story – an international lesson. In Dowmunt, T. (ed.) *Channels of resistance: global television and local empowerment*. London: BFI/Channel 4.

Phillips, R.S. (1997) *Mapping men and empire: a geography of empire*. London: Routledge.

Philo, C. (1992). Neglected rural geographies: a review. *Journal of Rural Studies*, 8, 193–207.

Philo, C. (1995) Animals, geography, and the city: notes on inclusions and exclusions. *Environment and Planning D: Society and Space* 13(6), 644–81.

Philo, C. (1997). Of other rurals. In Cloke, P. and Little, J. (eds) *Contested countryside cultures: otherness, marginality and rurality*. London: Routledge.

Philo, G. (1993) From Buerk to Band Aid: the media and the 1984 Ethiopian famine. In Eldridge, J. (ed.) *Getting the message: news, truth and power*. London: Routledge.

Pietz, W. (1988) The 'post-colonialism' of Cold War discourse. *Social Text*, 19/20(Fall), 55–75.

Pile, S. (1993) Human agency and human geography revisited: a critique of 'new models' of the self. *Transactions of the Institute of British Geographers*, NS 18, 122–39.

Pile, S. (1996) *The body and the city: psychoanalysis, space and subjectivity*. London: Routledge.

Pile, S. and Thrift, N. (1995) *Mapping the subject: geographies of cultural transformation*. London: Routledge.

Pinckney, D.H. (1958) *Napoleon III and the rebuilding of Paris*. Princeton, NJ: Princeton University Press.

Platteau, J-P. (1994) Behind the market stage where real societies exist: Parts I and II. *Journal of Development Studies*, 30, 533–77 and 753–817.

Pocock, D. (1981) Place and the novelist. *Transactions of the Institute of British Geographers*, NS 6, 337–47.

Polanyi, K, (1947) *The great transformation*. Boston: Beacon Books.

Pollard, J.S. (1996) Banking at the margins: a geography of financial exclusion in Los Angeles. *Environment and Planning A*, 28, 1209–32.

Pollock, G. (1988) Modernity and the spaces of femininity. In Pollock, G. *Vision and difference: femininity, feminism and the histories of art*. London: Routledge.

Pontalis, J.-B. (1990) *La force d'attraction*. Paris: Seuil.

Pratt, G. and Hanson, S. (1994) Geography and the construction of difference. *Gender, Place and Culture*. 1, 5–29.

Pratt, J., Leyshon, A. and Thrift, N.J. (1996) Financial exclusion in the 1990s II: geographies of financial inclusion and exclusion. *Working Papers on Producer Services*, 38.

Pratt, M.L. (1986) Scratches in the face of the country; or, what Mr Barrow saw in the lands of the Bushmen. In Gates, H.L. Jr (ed.) *'Race', writing and difference*. Chicago, IL: Chicago University Press.

Pred, A. (1990) *Lost words and lost worlds: modernity and the language of everyday life in late nineteenth-century Stockholm*. Cambridge: Cambridge University Press.

Probyn, E. (1993) *Sexing the self: gendered positions in cultural studies*. London: Routledge.

Raban, J (1990) *Hunting Mr Heartbreak*. London: Pan Books.

Radcliffe, S. and Westwood, S. (eds) (1993) *Viva: women and popular protest in Latin America*. London: Routledge.

Radin, M. (1996) *Contested commodities*. Cambridge, MA: Harvard University Press.

Radway, J. (1984) *Reading the romance: women, patriarchy, and popular literature*. Chapel Hill, NC: University of North Carolina Press.

Rai, A. (1995) India on-line: electronic bulletin boards and the construction of a diasporic Hindu identity. *Diaspora*, 4, 31–58.

Rawcliffe, P. (1998) *Swimming with the tide: environmental groups in transition*. Manchester: Manchester University Press.

Redclift, M. (1984) *Development and the environmental crisis: red or green alternatives?* London: Methuen.

Redclift, M. (1987) *Sustainable development: exploring the contradictions*. London: Methuen.

Reed, H.C. (1981) *The pre-eminence of international financial centers*. New York: Praeger.

Relph, E. (1976) *Place and placelessness*. London: Pion.

Rhodes, R. (1996) The new governance: governing without government. *Political Studies*, XLIV, 652–67.

Rieff, D. (1993) Notes on the Ottoman legacy written in a time of war. *Salmagundi*, 100, 3–15.

Riley, R. (1994) Speculations on the new American landscapes. In Foote, K., Hugill, P., Mathewson, K. and Smith J (eds) *Re-Reading cultural geography*. Austin: University of Texas Press.

Robins, K. (1996) Interrupting identities: Turkey/Europe. In Hall, S. and du Gay, P. (eds) *Questions of cultural identity*. London: Sage.

Robinson, N. (ed.) (1993) *Agenda 21: Earth's action plan*. New York: Ocean Publications.

Rose, G. (1993) *Feminism and geography*. Cambridge: Polity Press.

Rose, G. (1995) Place and identity: a sense of place. In Massey, D. and Jess, P. (eds) *A place in the world?* Oxford: Oxford University Press.

Rose, G. (1997) Looking at landscape: the uneasy pleasures of power. In Barnes, T. and Gregory, D. (eds) *Reading human geography*. London: Arnold.

Ross, J. (1995) *Rebellion from the roots*. Monroe, ME: Common Courage Press.

Rostow, W. (1960) *The stages of economic growth: a non-communist manifesto*. Cambridge: Cambridge University Press.

Rowlands, J. (1997) *Questioning empowerment: working with women in Honduras*. Oxford: Oxfam.

Rushkoff, D. (1994) *Cyberia: life in the trenches of hyperspace*. New York: HarperSanFrancisco.

Ryan, J. (1997) *Picturing empire: photography and the visualization of the British Empire*. London: Reaktion Books.

Sack, R. (1992) *Place, modernity and the consumer's world: a relational framework for geographical analysis*. Baltimore: Johns Hopkins University Press.

Sack, R. (1997) *Homo Geographicus*. Baltimore: Johns Hopkins University Press.

Said, E. (1995 [1978]) *Orientalism*. London: Penguin Books.

Samuel, R. (1994) *Theatres of memory. Volume 1: past and present in contemporary culture*. London: Verso.

Sassen, S. (1988) *The mobility of capital and labour*. Cambridge: Cambridge University Press.

Sassen, S. (1990) Finance and business services in New York City: international linkages and domestic effects. *International Social Science Journal*, 42, 287–306.

Sassen, S. (1991) *The global city: New York, London, Tokyo*. Princeton, NJ: Princeton University.

Scannell, P. (1988) Radio times: the temporal arrangements of broadcasting in the modern world. In Drummond, P. and Paterson, R. (eds) *Television and its audience: international research perspectives*. London: BFI, 15–31.

Schama, S. (1987) The Enlightenment in the Netherlands. In Porter, R. and Teich, M. (eds), *The Enlightenment in national context*. Cambridge: Cambridge University Press.

Schick, I.C. and Tonak, E.A. (eds) (1987) *Turkey in transition: new perspectives*. New York: Oxford University Press.

Schirmer, J. (1994) The claiming of space and the body politic within national–security states. In Boyarin, J. (ed.) *Remapping memory: the politics of timespace*. Minneapolis: University of Minnesota Press.

Schlesinger, P. (1994) Europe's contradictory communicative space. *Daedalus*, 123(2), 25–52.

Schor, J.B. (1998) *The overspent American: upscaling, downshifting and the new consumer*. New York: Basic Books.

Schroeder, R. (1996) *Possible worlds: the social dynamic of virtual reality technology*. Boulder, CO: Westview Press.

Schumacher, E.F. (1973) *Small is beautiful: a study of economics as if people mattered*. London: Blond and Briggs.

Scott, J. (1976) *The moral economy of the peasantry*. New Haven, CN: Yale University Press.

Scott, R.A. (1969) *The making of blind men: a study of adult socialization*. London: Transaction Books.

Scottish Natural Heritage (1993) *Sustainable development and the natural heritage: the SNH approach*. Edinburgh: Scottish Natural Heritage.

Seager, J. (1994) *Earth follies*. London: Routledge.

Segal, R. (1995) *The black diaspora*. New York: Farrar, Straus and Giroux.

Seiter, E., Borchers, H., Kreutzner, G. and Warth, E.-M. (1989) 'Don't treat us like we're so stupid and naïve': towards an ethnography of soap opera viewers. In Seiter, E. *et al.* (eds) *Remote control*. London and New York: Routledge.

Sen, G. and Grown, C. (1987) *Development crises and alternative visions. Third World women's perspectives*. New York: Monthly Review Press.

Seymour, S., Daniels, S. and Watkins, C. (1994) Estate and empire: Sir George Cornewall's management of Moccas, Herefordshire and La Taste, Grenada, 1771–1819. Working Paper 28, University of Nottingham, Department of Geography.

Shakespeare, T. (1994) Cultural representations of disabled people: dustbins for disavowal? *Disability and Society*, 9(3), 283–99.

Shapiro, M. (1989) Representing world politics: the sport/war intertext. In Der Derian, J. and Shapiro, M.J. (eds) *International/intertextual relations: postmodern readings of world politics*. Lexington MA: Lexington Books.

Sharp, J. (1993) Publishing American identity: popular geopolitics, myth and the *Reader's Digest. Political Geography*, 12(6), 491–503.

Sharp, J. (1996) Hegemony, popular culture and geopolitics: the *Reader's Digest* and the construction of danger. *Political Geography*, 15(6/7), 557–70.

Sherman, D.J. and Rogoff, I. (1994) *Museum culture: histories, discourses and spectacles*. Minneapolis: University of Minnesota Press.

Shilling, C. (1993) *The body and social theory*. London: Sage.

Shore, C. (1996) Transcending the nation-state? The European Commission and the (re)-discovery of Europe. *Journal of Historical Sociology*, 9(4), 473–96.

Shortridge, J.R. (1991) The concept of the place-defining novel in American popular culture. *The Professional Geographer*, 43, 280–91.

Shurmer-Smith, P. and Hannam, K. (1994) *Worlds of desire, realms of power: a cultural geography*. London: Edward Arnold.

Sibley, D. (1995) *Geographies of exclusion: society and difference in the West*. London: Routledge.

Silverstone, R. (1994) *Television and everyday life*. London: Routledge.

Simmons, I. (1996, first edition 1989). *Changing the face of the earth. Culture, environment, history*. Oxford: Basil Blackwell.

Simon, D. (1997) Development reconsidered: new directions in development thinking. *Geografiska Annaler*, 79B(4), 183–201.

Sloterdijk, P. (1995) World markets and secluded spots: on the position of the European regions in the world-experiment of capital. In Büchler, P. and Papastergiadis, N. (eds) *Random access: on crisis and its metaphors*. London: Rivers Oram Press.

Smith, A.D. (1991) *National identity*. Harmondsworth: Penguin Books.

Smith, B. (1985) *European vision and the South Pacific*. Second edition. New Haven and London: Yale University Press.

Smith, M.P. (ed.) (1995) *Marginal spaces*. New Brunswick, NJ: Transaction Publishers.

Smith, N. (1990, first edition 1984) *Uneven development*. Oxford: Basil Blackwell.

Smith, N. (1994) Geography, empire and social theory. *Progress in Human Geography*, 18(4), 491–500.

Smith, N (1996) *The new urban frontier: gentrification and the revanchist city*. London: Routledge.

Smith, N. and Godlewska, A. (1994) Introduction: critical histories of Geography. In Godlewska, A. and Smith, N. (eds) *Geography and empire*. Oxford: Blackwell.

Smith, S.J. (1989) *The politics of 'race' and residence*. Cambridge: Polity Press.

Smith, S.J. (1993) Residential segregation and the politics of racialisation. In Cross, M. and Keith, M. (eds) *Racism, the city and the state*. London: Routledge.

Smith, S.J. (1994) Citizenship. In Johnston, R., Gregory, D. and Smith, D.M. (eds) *The dictionary of human geography*, third edition. Oxford: Blackwell.

Smith, S.J. (1997) Beyond geography's visible worlds: a cultural politics of music. *Progress in Human Geography* 21, 502–29.

Smith, S.J. and Mallinson, S. (1996) The problem with social housing: discretion, accountability and the welfare ideal. *Policy and Politics*, 24, 339–58.

Smith, W.D. (1984) The function of commercial centres in the modernisation of European capitalism: Amsterdam as an information exchange in the seventeenth century. *Journal of Economic History*, 44, 985–1005.

Soja, E (1992) Inside exopolis: scenes from Orange County. In Sorkin, M. (ed.) *Variations on a theme park*. New York: Noonday.

Soja, E.W. (1996) *Thirdspace*. Oxford: Blackwell.

Sorkin, M. (ed.) (1992) *Variations on a theme park: the new American city and the end of public space*. New York: The Noonday Press.

Stallybras, J. (1996) *Gargantua: vision and mass culture*. London: Verso.

Stevens, J.E. (1988) *Hoover Dam: an American adventure*, Norman, OK: University of Oklahoma Press.

Stevens, S. (1993) Tourism, change and continuity in the Mount Everest region. *Geographical Review*, 83, 410–27.

Stoddart, D. (1986) Geography, exploration and discovery. In *On geography and its history*. Oxford: Blackwell.

Stoddart, D. (1987) To claim the high ground: geography for the end of the century. *Transactions of the Institute of British Geographers*, NS 12, 327–36.

Stoker, G. (1996) Governance as theory: five propositions. Mimeo (available from the author at the Department of Government, University of Strathclyde).

Storper, M. and Walker, R. (1989) *The capitalist imperative*. New York and Oxford: Basil Blackwell.

Synnott, A. (1993) *The body social: symbolism, self and society*. London: Routledge.

Taylor, G. (1949) *Urban Geography*. London: Methuen.

Taylor, H. (1951) No watchdog for America. *Reader's Digest*, Feb, 85–7.

Taylor, P. (1989) The error of developmentalism in human geography. In Gregory, D. and Walford, R. (eds) *Horizons in human geography*. Basingstoke: Macmillan.

Taylor, P.J. (1996) *The way the modern world works: world hegemony to world impasse*. Chichester: John Wiley.

Tendler, J. (1997) *Good government in the tropics*. Baltimore: Johns Hopkins University Press.

Thede, N. and Ambrosi, A. (eds) (1991) *Video the changing world*. Montreal and New York: Black Rose Books.

thee data base (1996) The K Foundation: why we burnt a million pounds. *thee data base* [online], 7, Available from http://members.xoom.com/databass/KFound.htm

Thomas, H. (1997) *The slave trade*. London: Macmillan.

Thomas, W., Sauer, C., Bates, M. and Mumford, L. (1956) *Man's role in changing the face of the earth*. Chicago, IL: University of Chicago Press.

Thompson, E. (1991) *Customs in common*. London: Penguin Books.

Thrift, N. (1983) On the determination of social action in space and time. *Environment and Planning D: Society and Space*, 1, 23–57.

Thrift, N. (1994) On the social and cultural determinants of international financial centres. In Corbridge, S., Thrift, N.J. and Martin, R. (eds) *Money, power and space*. Oxford: Blackwell.

Thrift, N. (1996, orig. 1985) 'Flies and Germs: a geography of knowledge', in *Spatial Formations*. London: Sage.

Thrift, N. (1996) *Spatial formations*. London: Sage.

Tickell, A. and Peck, J. (1996) The return of the Manchester men: men's words and men's deeds in the remaking of the local state. *Transactions of the Institute of British Geographers*, 21(4), 595–616.

Tiffen, M., Mortimore, M.J. and Gichugi, F. (1994) *More people, less erosion: environmental recovery in Kenya*. Chichester: John Wiley.

Toye, J. (1993) *Dilemmas of development: the counterrevolution in development theory and policy*. Second edition. Oxford: Blackwell.

Trelluyer, M. (1990) La télévision regionale en Europe. *Dossiers de l'Audiovisuel*, 33, 10–55.

Tuan, Yi-Fu. (1977) *Space and place: the perspective of experience*. Minneapolis: University of Minnesota Press.

Turner, T. (1991) The social dynamics of video media in an indigenous society: the cultural meaning and personal politics of video-making in Kayapo communities. *Visual Anthropology Review*, 7(2), 68–76.

UN Centre for Human Settlements (1996) *An urbanizing world: global report on human settlements 1996*. Oxford: Oxford University Press.

United Nations Development Programme (1997) *Human development report 1997*. Oxford: OUP-UNDP.

UPIAS (1976) *Fundamental principles of disability*. London: Union of the Physically Impaired Against Segregation.

Urry, J. (1990) *The tourist gaze: leisure and travel in contemporary society*. London: Sage.

Urry, J. (1995) *Consuming places*. London: Routledge.

Valentine, G. (1993a) Desperately seeking Susan: a geography of lesbian friendships. *Area*, 25(2), 109–16.

Valentine, G. (1993b) Negotiating and managing multiple sexual identities: lesbian time-space strategies. *Transactions of the Institute of British Geographers*, 18, 237–48.

Valentine, G. (1996) Children should be seen and not heard: the production and transgression of adult's public space. *Urban Geography* 17(3), 205–20.

Veijola, S. and Jokinen, E. (1994) The body in tourism. *Theory, Culture and Society*, 11, 125–51.

Vertovec, S. (1996) Berlin Multikulti: Germany, 'foreigners' and 'world-openness'. *New Community*, 22(3), 381–99.

Vidal, J. (1997) The long march home. *The Guardian Weekend*, April 26, 14–20.

Wade, R. (1990) *Covering the market: economic theory and the role of government in East Asian industrialisation*. Princeton: Princeton University Press.

Wadham-Smith, N. (1996) Geography re-invented. *British Studies Now*, 7, 3–8.

Wallerstein, I. (1979) *The capitalist world economy*. Cambridge: Cambridge University Press.

Wallerstein, I. (1984) *The Politics of the World-Economy*. Cambridge, UK: Cambridge University Press.

Waquant, L. (1993) Urban outcasts: stigma and division in the Black American ghetto and the French urban periphery. *International Journal of Urban and Regional Research*, 17(3), 366–83.

Warf, B. (1988) Regional transformation, everyday life, and Pacific Northwest lumber production. *Annals of the Association of American Geographers*, 78, 326–46.

Wark, M. (1994) *Virtual geography*. Bloomington and Indianapolis: Indiana University Press.

Waugh, P. (1997) 'Banglatown'. *The Evening Standard*, 23 January, 15.

Weber, M. (1958) *The Protestant ethic and the spirit of capitalism*. New York: Scribners.

Wells, H.G. (1902) *Anticipations of the reaction of mechanical and scientific progress upon human life and thought*. London: Chapman and Hall.

Wells, H.G. (1920) *The outline of history*. London: Cassell.

Which? (1989) No Entry. October, 498–501.

Williams, A. and Shaw, G. (eds) (1988) *Tourism and economic development*. London: Belhaven.

Williams, C.H. and Smith, A.D. (1983) The national construction of social space. *Progress in Human Geography*, 7(4), 502–18.

Williams, R. (1983) *Keywords*. New York: Oxford University Press.

Wilson, A. (1992). *The culture of nature*. London: Routledge.

Wilson, E (1991) *The Sphinx in the city*. Berkeley, CA: University of California Press.

Wilson, W.J. (1987) *The truly disadvantaged, the inner city, the underclass and public policy*. Chicago: University of Chicago Press.

Wilson, W.J (1996) *When work disappears: the world of the new urban poor*. New York: Alfred A. Knopf.

Winchester, H. and White, P. (1988) The location of marginalised groups in the inner city. *Environment and Planning D: Society and Space*, 6, 37–54.

Wolch, J. and Emel, J. (eds) (1998) *Animal geographies*. London: Verso.

Women and Geography Study Group (1997) *Feminist geographies: explorations in diversity and difference*. Harlow: Longman.

Worcester, R.M. (1993) Public and elite attitudes to environmental issues. *International Journal of Public Opinion Research*, 5, 315–34.

World Bank (1992) *World development report, 1992*. Oxford: Oxford University Press/World Bank.

World Bank (1994) *Adjustment in Africa: reforms, results and the road ahead*. Oxford: OUP-World Bank.

Wright, P. (1985) *On living in an old country*. London: Verso.

Wright, T. (1995) Tranquility city: self organisation, protest and collective gains within a Chicago homeless encampment. In Smith, M.P. (ed.) *Marginal spaces*. New Brunswick, NJ: Transaction Publishers.

Wrigley, N. (1998) Leveraged restructuring and the economic landscape: the LBO wave in US food retailing. In Martin, R.L. (ed.) *Money and the space economy*. Chichester: John Wiley.

Wynne, B. (1993) Public uptake of science: a case for institutional reflexivity. *Public Understanding of Science*, 2, 321–37.

Yapa, L. (1996) What causes poverty? A postmodern view. *Annals of the Association of American Geographers*, 86, 707–28.

Young, I.M. (1990a) The ideal of community and the politics of difference. In Nicholson, L. (ed.) *Feminism/postmodernism*. London: Routledge.

Young, I.M. (1990b) Throwing like a girl: a phenomenology of feminine body comportment, motility and spatiality. In *Throwing like a girl and other essays in feminist philosophy and social theory*. Bloomington: University of Indiana Press.

Young, J. (1993) *The texture of memory: holocaust memorials and meaning*. London: Yale University Press.

Zerubavel, Y. (1995) *Recovered roots: collective memory and the making of Israeli national tradition*. Chicago, IL: Chicago University Press.

Zukin, S. (1982) *Loft living: culture and capital in urban change*. Baltimore: Johns Hopkins University Press. (London: Radius, 1988).

Zurick, D. (1992) Adventure travel and sustainable tourism in the peripheral economy of Nepal. *Annals of the Association of American Geographers*, 82(4), 608–28.

Index